T0392947

Biodegradable Polymers
Concepts and Applications

Editors

Margarita del Rosario Salazar Sánchez
Departamento de Ciencias Agroindustriales
Facultad de Ingenierías y Tecnológicas
Universidad Popular del Cesar
Cesar, Colombia

José Fernando Solanilla Duque
Departamento de Agroindustria
Facultad de Ciencias Agrarias
Universidad del Cauca
Popayán, Colombia

Aidé Sáenz Galindo
Facultad de Ciencias Química
Universidad Autónoma de Coahuila
Saltillo, México

Raúl Rodríguez Herrera
Universidad Autónoma de Coahuila
Saltillo, México

CRC Press is an imprint of the
Taylor & Francis Group, an **informa** business
A SCIENCE PUBLISHERS BOOK

Cover illustration reproduced by kind courtesy of Margarita del Rosario Salazar Sánchez.

First edition published 2023
by CRC Press
6000 Broken Sound Parkway NW, Suite 300, Boca Raton, FL 33487-2742

and by CRC Press
4 Park Square, Milton Park, Abingdon, Oxon, OX14 4RN

© 2023 Margarita del Rosario Salazar Sánchez, José Fernando Solanilla Duque, Aidé Sáenz Galindo and Raúl Rodríguez Herrera

CRC Press is an imprint of Taylor & Francis Group, LLC

Reasonable efforts have been made to publish reliable data and information, but the author and publisher cannot assume responsibility for the validity of all materials or the consequences of their use. The authors and publishers have attempted to trace the copyright holders of all material reproduced in this publication and apologize to copyright holders if permission to publish in this form has not been obtained. If any copyright material has not been acknowledged please write and let us know so we may rectify in any future reprint.

Except as permitted under U.S. Copyright Law, no part of this book may be reprinted, reproduced, transmitted, or utilized in any form by any electronic, mechanical, or other means, now known or hereafter invented, including photocopying, microfilming, and recording, or in any information storage or retrieval system, without written permission from the publishers.

For permission to photocopy or use material electronically from this work, access www.copyright.com or contact the Copyright Clearance Center, Inc. (CCC), 222 Rosewood Drive, Danvers, MA 01923, 978-750-8400. For works that are not available on CCC please contact mpkbookspermissions@tandf.co.uk

Trademark notice: Product or corporate names may be trademarks or registered trademarks and are used only for identification and explanation without intent to infringe.

Library of Congress Cataloging-in-Publication Data (applied for)

ISBN: 978-1-032-13714-8 (hbk)
ISBN: 978-1-032-13715-5 (pbk)
ISBN: 978-1-003-23053-3 (ebk)

DOI: 10.1201/9781003230533

Typeset in Times New Roman
by Radiant Productions

Preface

As a consequence of the negative impact of synthetic polymers on the environment, not only for the present generation but also for future generations, the need has arisen to develop biodegradable polymers that not only meet most of the quality standards for the manufacture of products but also have a degradation period much shorter than that of synthetic polymers. This book appears as a collaborative project between different institutions of higher education and research in order to bring together the different topics related to the production, application and final destination of biodegradable polymers, and their impact on modern life and on future generations.

The book of biodegradable polymer science reveals the basic concepts of biodegradable polymer science, describing the techniques, standards and analysis to be performed to characterize biodegradable polymeric materials, highlighting that it is important to further develop and/or innovate processes considering the environment. Pointing to this, this book focuses on the engineering of biopolymers in food processing and explores the processing technology associated with biopolymer applications. The difference between a normal book on biodegradable polymers and the present book is that in our version, there are some very interesting chapters that are framed on the rheological and structural analysis, in addition to the applications in different areas of knowledge such as: health, agriculture and technology.

The present book will be useful for the biodegradable polymer's industry because it covers from films to composites, as well as in academia and research including framing processes of analysis and technology which could be used to improve biodegradable polymer production, where the reader will find different tools to become familiar with biopolymers analysis and implementation of these analyses. The contribution of this book would complement the themes developed by a reference book on biodegradable polymer bases because this book refers to treatment topics among which are Compounding and Additives, Powdering, Chemical Treatment, Surface Treatment, Adhesive Compositions, Biopolymer Engineering in Food Processing and Biodegradation, other product developments from these biopolymers are also covered. All applications are shown from a sustainability and sustainability approach, it is important to highlight that biodegradability has a great burden when it involves substituting, modifying and/or designing existing processes in harmful and polluting processes. The book concludes with a reflection on the development of biodegradable polymers in different areas of knowledge and trends.

Contents

Introduction

Zainul Akmar Zakaria,[1] *Siti Hajjar Che Man,*[1] *Rocio Castillo-Godina*[2] and *Raul Rodriguez-Herrera*[2]

Biopolymers are produced by living organisms and should be differentiated for its term synthetic biodegradable polymers. The monomer units of most biopolymers usually consist of recurring molecules of either nuclei acid of nucleotides (deoxyribonucleic acid, DNA and ribonucleic acid, RNA), amino acid proteins (collagen, gelatin, gluten) or saccharides derived from sugars such as cellulose, chitosan, chitin (Samrot et al., 2020). Increasing depletion of petroleum-based resources to produce various types of polymers as well as its associated non-biodegradability features, has become the major driving force in the continuous development in the area of biopolymers production, processing and applications. Biopolymer offers an interesting and viable solution for this problem based on its biodegradability features. Notably, some weaknesses such as high cost, limited rate of production and poor mechanical properties have been identified as focus for further investigation in this area. Biopolymers are comprehensively segregated into two principle groups namely biodegradable and non-biodegradable (Rai et al., 2021). They can also be classified as elastomers, thermosets and thermoplastics according to their physical and chemical changes when stressed in different thermal conditions (George et al., 2020). Other researchers have classified biopolymers according to its various forms of either composites, blends or laminates. Some examples for commercially-produced biopolymers include polyesters (polyhydroxybutyrate, PHB) and (polybutylene succinate, PBS), bio-polyolefins, Bio-PE (polyethylene), bio-polyamides, Bio-PA (homopolyamides - Bio-PA 6, Bio-PA 11) and biopolyurethanes, Bio-PUR.

Biopolymers have been established and gained significant interest in various applications such as biomedical, nanomaterials, food industry as well as water treatment. The ability of this polymer to be tailored with other materials to suit certain properties made them more attractive and feasible. In biomedical applications (Moohan et al., 2020), biopolymers such as starch, cellulose, chitosan, polylactic acid (PLA) (Ghalia and Dahman, 2017) and Poly (Glycolide Acid) (PGA) have been extensively used for tissue engineering, pharmaceutical carriers and medical devices. These polymers are known for their advantageous features that include cytocompatibility and the ability to degrade in the body without releasing harmful substances. Tyler et al. (2019) reviewed the applications of PLA and its copolymers as nanoparticle drug carriers, such as liposomes, polymeric nanoparticles, dendrimers and micelles. The ability of PLA to be moldable, allowing its applications to take on numerous shapes, i.e., scaffolds, sutures, micelle, etc., to suit such applications. Other than PLA, biopolymers such as gelatin and hydroxyapatite (HA) has been long used as artificial scaffolds, due to their compatibility to the natural bone tissue (Hajinasab et al., 2018; Szcześ et al., 2017). Despite

[1] School of Chemical and Energy Engineering, Faculty of Engineering, Universiti Teknologi Malaysia, 81310 UTM Johor Bahru, Johor, Malaysia.

[2] Autonomous University of Coahuila, School of Chemistry, Blvd. V. Carranza and J. Cardenas s/n, Col. Republica Ote. Saltillo Coahuila 25280. México.

* Corresponding author: raul.rodriguez@uadec.edu.mx

their advantages, there are several limitations that restrict their use for these particular applications due to the lack of mechanical strength. To date numerous studies have been reported to improve the issue with the property's instability of this biopolymers (Hamad et al., 2015; Reddy et al., 2015).

Recently, biopolymers are preferred to be used in food packaging coating to replace the traditional non-degradable materials such as polypropylene, polyethylene, etc. The demand for higher shelf life as well as better packaging quality have led to the increased interest in this area. Biopolymers such as starch, polyhydroxyalkanoates (PHA), polyhydroxybutyrates (PHB), PLA and so on, are among the commercially viable materials in food packaging, which can be processed via conventional equipment. Barrier properties is an extremely important characteristic especially for biobased food packaging materials. Biopolymers are known as hydrophilic materials with inadequate moisture resistance. Thus, various improvements have been reported to enhance the barrier properties as well as the mechanical properties. The development of coatings from natural polymers for food packaging have received significant interest to overcome the issue. Various polysaccharides and protein-based compounds have been investigated for food packaging coating which shown good grease, good affinity to the packaging substrate as well as a positive effect on mechanical properties (Nechita et al., 2020). Chemical and physical crosslinking such as grafting and coating have been reported to improve the compatibility of the coating and substrate (Reddy et al., 2015). The incorporation of nanofillers into biopolymers matrices not only to enhance the mechanical and barrier properties but also to impart other attractive properties such as antimicrobial agent and biosensor (Othman, 2014; Qamar et al., 2020).

Preparation of nanomaterials derived from cellulose (CNs) have recently gained significant interest, due to their excellent inherent and physical properties. In addition, CNs may serve as an inexpensive, renewable and biodegradable in comparison to its counterpart carbon nanotubes (CNTs). It was reported that CNs exhibited a greater axial elastic modulus than Kevlar and its mechanical properties are within the range of other reinforcement materials. The presence of reactive surface of –OH side groups in the structure of CNs enables the grafting of chemical species to achieve surface functionalization which allows the tailoring of particle surface chemistry to facilitate self-assembly (Moon et al., 2011). To date, the potential of CNs have been recognized in various fields, i.e., paper and packaging, automotive, medical, construction, personal care, textile industries as well as waste water treatment (Carpenter et al., 2015; Mohan et al., 2020; Tayeb et al., 2018).

In other applications, biopolymers have been incorporated in the fabrication of lithium-ion battery separators due to the excellent properties particularly low cost, high thermal stability, excellent mechanical properties, non-toxic, light-weight and excellent wettability to the electrolyte (Xu et al., 2014; Zhang et al., 2019). Whereas in biosensor application, biopolymer is a great material to be used as it can act as good immobilization matrix for entrapping biorecognition units such as enzymes, whole cells and others and provide good adhesion of the composite to electrode for carrying out sensing measurements.

Next the most important facts of the book chapters will be mentioned. In Chapter 3 on the science of biodegradable polymers, it is emphasized that synthetic polymers in general meet many needs of the human population, which is why they are materials of mass consumption, however, characteristics such as: high resistance to corrosion, water and bacterial decomposition makes them difficult to eliminate, and consequently, a serious environmental problem that lasts for years (Valero-Valdivieso et al., 2013). On the other hand, there has been a demand for products that replace petroleum-derived plastics with alternatives based on renewable resources and above all biodegradable. This focuses an approach of multiple scientific disciplines and the so-called science of biodegradable polymers arises. Biodegradable polymers are those that have the ability to decompose in the presence of enzymes produced by microorganisms such as: bacteria, fungi and algae. It is necessary to differentiate between degradable, which refers to decomposition due to chemical or physical changes, while biodegradable is due to degradation by biological mechanisms (Niaounakis, 2013; Karamanlioglu et al., 2017).

The latest trends in polymer production include the use of renewable sources (Chapter 4) instead of traditional polymers based on fossil sources, responsible for CO_2 emissions into the atmosphere and which also generate non-biodegradable waste, taking years for its decomposition, which therefore represents an environmental problem (Peplow, 2016; Llevot et al., 2016). Biodegradable polymers can be produced by biological systems (microorganisms, plants and animals) or they can be synthesized from biological raw materials (for example, corn, sugar, starch, etc.), also known as bio-based (Gómez and Yory, 2018). In general, the production of these biopolymers includes synthetic polymers obtained from renewable resources, such as: polylactic acid (PLA); biopolymers produced by microorganisms such as polyhydroxyalkanoate (PHA) and natural biopolymers such as starch or proteins (Rudin and Choi, 2013).

Biodegradable polymers can be classified as follows:

- Polymers extracted or removed directly from biomass: polysaccharides such as starch and cellulose. Proteins like casein, keratin and collagen.
- Polymers produced by classical chemical synthesis using biological monomers from renewable sources.
- Polymers produced by microorganisms, native or genetically modified producing bacteria.

In Chapter 5, the authors focus on plastics technology, specifically those of biological origin: PHAs and polylactic acid (PLA). PHAs are also called "double green polymers." PLA is a natural monomer produced by fermentative ways from elements rich in sugars, cellulose and starch. Bioplastics have physicochemical and thermoplastic properties equal to those of polymers made from oil, but once deposited under favorable conditions, they biodegrade (Díaz del Castillo, 2012). Polymerization is the process in which small molecules of a single unit (monomers) or of a few units (oligomers) are chemically united to create bigger molecules, in which the atoms are strongly linked by a covalent bond (Díaz del Castillo, 2012). Currently, enzymatic polymerization represents a great approach to functionalizing polymers and biopolymers and preventing the generation of waste through the use of catalytic processes with high selectivity, as well as preventing or limiting the use of dangerous organic reagents (Peponi et al., 2015).

The technological methods of manufacturing plastics are determined by their rheological properties and will depend on whether the material in question is thermoplastic or thermoset. Some of these methods are: injection molding, blow molding, rotational molding, blow processing methods, calendaring, casting, coatings, extrusion, film techniques, foam forming, lamination and low-pressure molding, filling techniques, plasticizers and other additives, an example of the latter such as antioxidants and colorants (Billmeyer, 2020).

Polymers obtained from both oil and renewable natural resources need to be analyzed using different methodologies, which are discussed in Chapter 6, within these techniques are: Analytical methods in which the process used in polymers is not different from the techniques used in low molecular weight organic compounds. The physical analysis of polymers consists of different techniques: mass spectrometry and gas chromatography, infrared spectroscopy, in which the emission and absorption spectra of the molecules are determined, X-ray diffraction analysis, nuclear magnetic resonance spectroscopy and electrical spin, thermal analysis, which includes experimental methods of calorimetry, differential thermal analysis, microscopy and physical tests that include mechanical properties, fatigue tests, thermal properties, optical properties, electrical properties and chemical properties (Billmeyer, 2020).

The structure of polymers is covered in Chapter 7, which depends on the shape of the chains. Based on the structure, polymers can be classified as linear, branched or cross-linked (López-Carrasquero, 2004). Linear polymers are those in which the monomeric units are linked side by side in a single direction. Under certain conditions or with certain types of monomers, polymers with another type of architecture can be obtained which are characterized by having branches that are generated from the main chain, this characteristic has significant effects on many physical properties of the polymer, for example, in decreasing crystallinity. Branched polymers cannot easily fit into a crystal lattice as linear polymers do. On the other hand, branched polymers are much less soluble than their linear counterparts and cross-linked polymers are insoluble materials. Crosslinking can occur during the polymerization process or later through various chemical reactions. Crosslinking is used to impart good elastic properties in some elastomers, as well as to provide rigidity and dimensional stability to some materials called thermoplastics.

Biopolymers have macromolecules (proteins, carbohydrates, etc.), that generate dispersions that have certain rheological behaviors (Zambrano-Herrera, 2020). Rheology studies the fundamental and practical knowledge of the deformation or flow of matter (Hernández, 2014). Based on the above, Chapter 8 discusses the rheological properties of biopolymers. This branch of physics allows the characterization of flow patterns during processing, which facilitates handling and operating conditions, in addition to the properties of the final product (Zambrano-Herrera, 2020).

One of the most common forms of bioplastics is in the form of thin sheets obtained from blowing films. In this procedure, a thin-walled tube is extruded and then expanded by increasing the internal pressure of the tube (Mendoza and Velilla, 2011). In Chapter 9, the authors report the research on thin films and mixtures to obtain biodegradable polymers. In this regard, starch is widely used, given its availability and low cost. Starch is used as a thermoplastic in the production of biodegradable plastics (Funke, 1998). During its production, it must be mixed with a plasticizer and undergo a de-structuring procedure to be processed by injection, blow molding and extrusion (Thuwall, 2006).

In Chapter 10, the authors deal with composites which are composite materials that are characterized by exceeding the properties of the materials if they were used individually and by possessing specific properties and characteristics. These materials are made up of two phases; a continuous one called matrix and another dispersed called reinforcement. The reinforcement provides the mechanical properties to the composite material and the matrix provides thermal and environmental resistance. The matrix and the reinforcement are separated by the interface. Composite materials are classified according to their structural components: fibrous, laminated and particulate (Lubin, 2013).

In the gradual interest of finding biodegradable materials to replace EPS expanded polystyrene foam, different studies have focused on developing starch foams, which have been industrially obtained by extrusion, although these biodegradable foams have properties similar to polystyrene materials. Some studies have been carried out to improve its mechanical properties by adding cellulose (Motloung et al., 2019), chitosan, sugar cane fibers (Debiagi et al., 2011), Yucca flour, corn fiber (Sumardiono et al., 2021), bioplastic from banana peel (Castillo et al., 2015) and other biodegradable polymers. This topic on biodegradable foams is addressed in the Chapter 11. Many of the alternatives seek to take advantage of agricultural residues to obtain a sustainable and biodegradable product in order to reduce the consumption of non-renewable sources in the coming years.

The variation of different external factors, especially temperature, cause changes in the shape of polymers during their processing. Thus, heat-shrinkable polymers result from different heat stimuli and therefore changes in their shape in Chapter 12 describes the advantages and disadvantages of this kind of polymers. Currently, some studies focus on replacing synthetic materials with flexible films from biodegradable sources (Montilla et al., 2016; Montilla and Joaquí, 2016), because of the multiple applications that heat-shrinkable biodegradable polymers have. These heat-shrinkable polymers have application in many fields: packaging (Khankrua et al., 2019), medical devices, drug administration, intravenous needles (Xiao et al., 2019), electronic devices, digital storage media (Cui et al., 2017), etc.

Some of the fundamental characteristics of polymers to be used in medicine, is that they must be biocompatible, that its mechanical properties are suitable for use, that they are excretable and non-toxic for the recipient and that they can be sterilized (Labeaga, 2018). Based on these antecedents in Chapter 14 the health applications of biodegradable polymers are discussed. The most common application of biopolymers in surgeries is for development of absorbable sutures, some examples are polyglycolic and polylactic acids and polycaprolactone, although the copolymers between them have been studied more. Another application is the release of drugs, to improve their effect in the body to be administered, for example, some polyesters such as PLA, PGLA or PCL. In addition, PLA is applied in regenerative medicine, the function of these polymers is to totally or partially replace damaged tissues in the body such as bones, tendons, cartilage or heart valves (Tim et al., 1999). PLA is also used commercially as a bone fixative.

The agriculture applications of biodegradable polymers are discussed in Chapter 15. Some of the uses are for plastic films or soil mulch, since young plants must be covered to avoid frost and conserve moisture, raise soil temperature, reduce weeds and thus improve the plant growth rate. In this sense, biodegradable polymers have high biodegradability (Chiellini et al., 2003), whose characteristics allow them to leave the films in the soil at the end of the growing season and thus be biodegraded (Mazollier and Taullet, 2003). Another use of biodegradable polymers is a controlled release system of pesticides, nutrients, fertilizers or pheromones to repel insects, so the active agent can be dissolved, dispersed or encapsulated by the polymeric matrix or coating.

Chapter 16 deals with biodegradation. Natural sources of biodegradable polymers are mainly polysaccharides such as starch or cellulose, others sources can be proteins. These natural sources can be chemically modified to improve their mechanical properties or their rate of degradation. An example is the biodegradation of proteins, which is achieved by protease enzymes in an amine degradation reaction. Polysaccharides such as chitin, one of the most abundant, is degraded by chitinases, while, chitosan processed from chitin is degraded by enzymes such as chitosanases or lysozymes. Another example is starch, a polysaccharide from plant sources, whose biodegradation is achieved by enzymes via hydrolysis at the acetal bond (Yukuta et al., 1990). The α-1,4 bond is attacked by amylases, while the α-1,6 bond by glucosidases. Cellulose biodegradation occurs through enzymatic oxidation with peroxidase secreted by fungi or bacteria.

The tendencies on biodegradable polymers are discussed in the Chapter 17. Research on biodegradable biomaterials is being focused on the synthesis of polymers with certain properties for specific biomedical applications (Song et al., 2018) such as: temporary prostheses, 3D porous scaffolds for tissue engineering and for drug delivery. Biodegradable polymers can incorporate microbial technology for food packaging, thus extending its preservation and shelf life. Other trends focus on the use of bio-based polymers composed from renewable resources formulated through a bacterial fermentation process (Babu et al., 2013). As for the plastics industry, it is intended that they be more biodegradable but at the same time, maintain their strength and durability, thus focusing on plastics made from synthetic-based polymers such as polylactic acid and polycaprolactone, adding starch nanoparticles (Abioye et al., 2019), being in this way, apt for a wide variety of applications.

There is a need to evaluate different aspects of a polymeric material, especially those compounds with potential toxic risk. For this reason, in Chapter 18 the authors review and discuss this important aspect of biodegradable polymers. The products subject to evaluation in a polymeric material are: the base polymer, e.g., polyethylene, polycarbonate, residual monomers, when present above the relevant threshold, oligomers of known concern, e.g., styrene trimers and dimers, all additives, residual catalyst, intentionally added compounds, mercury, hexavalent chromium, halogenated organic compounds, phthalates, blowing agents or coloring agents when present in any concentration. These evaluations can be carried out by techniques such as GC/MS chromatography, infrared spectroscopy, crystallinity index (Gamba et al., 2017).

Bioremediation is a process that can mitigate the environmental effects of synthetic plastics in abundance. The raw materials used for the production of biodegradable polymers are mainly from agricultural sources: such as starch, cellulose and flax fibers. In this way, natural materials can be incorporated into synthetic plastic matrices acting as their biodegradable components, in Chapter 19, the current technology on bioremediation is revised. On the other hand, the raw material from a microbial source, which produces biopolymers through microbial fermentation, the products are naturally degradable and substitutes for synthetic plastics that are environmentally friendly. These materials also have the same resistance characteristics of a synthetic plastic (Ashish and

Priyanka, 2012). Polyhydroxy Alkanoates (PHA's) are produced by bacteria, and Polylactic Acid (PLA), is also produced by microbial fermentation.

References

Abioye, A. A., Fasanmi, O. O., Rotimi, D. O., Abioye, O. P., Obuekwe, C. C., Afolalu, S. A. and Okokpujie, I. P. (2019). Review of the development of biodegradable plastic from synthetic polymers and selected synthesized nanoparticle starches. In *Journal of Physics: Conference Series*, 1378(4): 042064.

Ashish, C. and Priyanka, C. (2012). Bioremediation of biopolymers. *Journal of Bioremediation and Biodegradation*, 3: e127.

Avendaño-Romero, G. C., López-Malo, A. and Palou, E. (2013). Propiedades del alginato y aplicaciones en alimentos. *Temas selectos de Ingenierías de Alimentos*, 7(1): 87–96.

Babu, R. P., O'connor, K. and Seeram, R. (2013). Current progress on bio-based polymers and their future trends. *Progress in Biomaterials*, 2(1): 1–16.

Billmeyer, E. W. (2020). *Ciencia de los polímeros*. Reverté. 610 p.

Carpenter, A. W., de Lannoy, C. F. and Wiesner, M. R. (2015). Cellulose nanomaterials in water treatment technologies. *Environmental Science & Technology*, 49(9): 5277–5287.

Castillo, R., Escobar, E., Fernández, D., Gutiérrez, R., Morcillo, J., Núñez, N. and Peñaloza, S. (2015). Bioplástico a base de la cáscara de plátano. *Revista de Iniciación Científica,* 1: 1–4.

Chiellini, E., Chiellini, F., Cinelli, P. and Ilieva, V. (2003). Bio-based polymeric materials for agriculture applications. pp. 185–220. *In*: Chiellini, E. and Solaro, R. (eds.). Biodegradable Polymers and Plastics. Kluwer Academic/ Plenum Publishers: New York, USA.

Cui, J., Adams, J. G. M. and Zhu, Y. (2017). Pop-up assembly of 3D structures actuated by heat shrinkable polymers. *Smart Mater. Struct.* 26(12): DOI: 10.1088/1361-665X/aa9552.

Debiagi, F., Mali, S., Grossmann, M. V. E. and Yamashita, F. (2011). Biodegradable foams based on starch, polyvinyl alcohol, chitosan and sugarcane fibers obtained by extrusion. *Brazilian Archives of Biology and Technology*, 54(5): 1043–1052.

Funke, U., Bergthaller, W. and Lindhauer, M. G. (1998). Processing and characterization of biodegradable products based on starch. *Polymer Degradation and Stability*, 59: 293–296.

Gamba, A. M., Fonseca, J. S., Méndez, D. A., Viloria, A. C., Fajardo, D., Moreno, N. C. and Rojas, I. C. (2017). Assessment of different plasticizer–polyhydroxyalkanoate mixtures to obtain biodegradable polymeric films. *Chemical Engineering Transactions*, 57: 1363–1368.

George, A., Sanjay, M. R., Srisuk, R. Parameswaranpillai, J. and Siengchin, S. (2020). A comprehensive review on chemical properties and applications of biopolymers and their composites. *International Journal of Biological Macromolecules*, 154: 329–338.

Ghalia, M. A. and Dahman, Y. (2017). Biodegradable poly (lactic acid)-based scaffolds: synthesis and biomedical applications. *Journal of Polymer Research*, 24(5): 74.

Gómez-Ayala, S. L. and Yory-Sanabria, F. L. (2018). Aprovechamiento de recursos renovables en la obtención de nuevos materiales. *Ingenierías USBMed,* 9(1): 69–74.

Hajinasab, A., Saber-Samandari, S., Ahmadi, S. and Alamara, K. (2018). Preparation and characterization of a biocompatible magnetic scaffold for biomedical engineering. *Materials Chemistry and Physics*, 204: 378–387.

Hamad, K., Kaseem, M., Yang, H. W., Deri, F. and Ko, Y. G. (2015). Properties and medical applications of polylactic acid: A review. *Express Polymer Letters*, 9(5).

Hernández, L. M. J. (2014). Caracterización reológica de hidrogeles de MCC-NaCMC + almidón. Tixotropía y sinergismo. Ph. D. Dissertation Universitat of Valencia. Proquest, Miami EE.UU. 232p.

Jobling, S. (2004). Improving starch for food and industrial applications. *Current Opinion in Plant Biology*. 7(2): 210–218.

Karamanlioglu, M., Preziosi R. and Robson, G. D. (2017). Abiotic and biotic environmental degradation of the bioplastic polymer poly (lactic acid): a review. *Polym. Degrad. and Stab.*, 137: 122–130.

Khankrua, R. et al. (2019). Development of PLA/EVA reactive blends for heat-shrinkable film. *Polymers*, 11(12): 1925. doi.org/10.3390/polym11121925.

Labeaga Viteri, A. (2018). Polímeros biodegradables. Importancia y potenciales aplicaciones. Master Thesis, Universidad Nacional de Educación a Distancia (España). Facultad de Ciencias. Departamento de Química Inorgánica e Ingeniería Química. 48p.

Llevot, A., Dannecker, P. K., von Czapiewski, M., Over, L. C., Söyler, Z. and Meier, M. A. R. (2016). Renewability is not enough: recent advances in the sustainable synthesis of biomass-derived monomers and polymers. *Chem. Eur. J.,* 22(33): 11510–11521.

López Carrasquero, F. (2004). Fundamentos de polímeros. *Escuela Venezolana para la enseñanza de la Química*. Universidad de Los Andes, *Mérida*, Ven. 49–51pp.

Lu, D. R., Xiao, C. M. and Xu, S. J. (2009). Starch-based completely biodegradable polymer materials. *Express Polymer Letter*. 3(6): 366–375.

Lubin, G. (2013). *Handbook of Composites*. Springer Science & Business Media.

Mazollier, C. and Taullet, A. (2003). Paillages et ficelles biodégradables : une alternative pour le maraîchage bio. *Alter. Agric*., 59: 10–13.

Mendoza-Quiroga, R. and Velilla-Díaz, W. (2011). Metodología para la caracterización termo-mecánica de películas plásticas biodegradables. *Prospect*., 9(1): 46–51.

Montilla, C. and Joaquí, D. (2016). Evaluation of flexible films heat shrinkage obtained from starch, plasticizer and polylactic acid. *Vitae*, 23: S614.

Montilla, C., Joaquí, D., Delgado, K. and Villada, H. (2016). Efecto de la relación de estiramiento en el termoencogimiento de películas flexibles de almidón termoplástico. *Agronomía Colombiana*, 34(1Supl): S161–S163.

Moohan, J., Stewart, S. A., Espinosa, E., Rosal, A., Rodríguez, A., Larrañeta, E., Donnelly, R. F. and Domínguez-Robles, J. (2020). Cellulose nanofibers and other biopolymers for biomedical applications. A review. *Applied Sciences*, 10(1): 65; doi.org/10.3390/app10010065.

Moon, R. J., Martini, A., Nairn, J., Simonsen, J. and Youngblood, J. (2011). Cellulose nanomaterials review: structure, properties and nanocomposites. *Chemical Society Reviews*, 40(7): 3941–3994.

Motloung, M. P., Ojijo, V., Bandyopadhyay, J. and Ray, S. S. (2019). Cellulose nanostructure-based biodegradable nanocomposite foams: A brief overview on the recent advancements and perspectives. *Polymers*, 11(8): 1270, doi.org/10.3390/polym11081270.

Nechita, P. (2020). Review on polysaccharides used in coatings for food packaging papers. *Coatings*, 10(6): 566. doi.org/10.3390/coatings10060566.

Niaounakis, M. (2013). Biopolymers: reuse, recycling and disposal. pp. 77–94. *In*: Andrew, W. (ed.). Oxford: Elsevier Inc.

Othman, S. H. (2014). Bio-nanocomposite materials for food packaging applications: types of biopolymer and nano-sized filler. *Agriculture and Agricultural Science Procedia,* 2: 296–303.

Peplow, M. (2016). The plastics revolution: how chemists are pushing polymers to new limits. *Nature*, 536(7616): 266–268.

Peponi, L., Barrera-Rivera, K., Navarro-Baena, I., Alpizar-Negrete, A. G., Marcos-Fernandez, A., Kenny, J. M., López, D. and Martinez-Richa, A. (2015). Polimerización enzimática para la síntesis de biopolímeros. *Revista de Plásticos Modernos*, 110(703): 1–6.

Qamar, S. A., Asgher, M., Bilal, M. and Iqbal, H. M. (2020). Bio-based active food packaging materials: Sustainable alternative to conventional petrochemical-based packaging materials. *Food Research International*, 137: 109625. doi.org/10.1016/j.foodres.2020.109625.

Rai, P, Mehrotra, S., Priya, S., Gnansounou, E. and Sharma, S. K. (2021). Recent advances in the sustainable design and applications of biodegradable polymers. *Bioresource Technology.* 325: 124739. doi.org/10.1016/j.biortech.2021.124739.

Reddy, N., Reddy, R. and Jiang, Q. (2015). Crosslinking biopolymers for biomedical applications. *Trends in Biotechnology*, 33(6): 362–369.

Rudin, A. and Choi, P. (2013). The Elements of Polymer Science and Engineering. Third edition, Oxford: Elsevier Inc., pp. 521–535.

Samrot, A. V., Sean, T. C., Kudaiyappan, T., Bisyarah, U., Mirarmandi, A., Fardjeva, E., Abubakar, A., Ali, H. H., Angalene, J. L. A. and Kumar, S. S. (2020). Production, characterization and application of nanocarriers made of polysaccharides, proteins, bio-polyesters and other biopolymers: A review. *International Journal of Biological Macromolecules,* 165: 3088–3105.

Scarfato, P., Di Maio, L. and Incarnato, L. (2015). Recent advances and migration issues in biodegradable polymers from renewable sources for food packaging. *Journal of Applied Polymer Science,* 132(48): doi: 10.1002/APP.42597.

Song, R., Murphy, M., Li, C., Ting, K., Soo, C. and Zheng, Z. (2018). Current development of biodegradable polymeric materials for biomedical applications. *Drug Design, Development and Therapy*, 12: 3117–3145.

Sumardiono, S., Pudjihastuti, I., Amalia, R. and Yudanto, Y. A. (2021). Characteristics of biodegradable foam (Bio-foam) made from cassava flour and corn fiber. *Mater. Sci. Eng.,* 1053: 1–7. doi:10.1088/1757-899X/1053/1/012082.

Szcześ, A., Hołysz, L. and Chibowski, E. (2017). Synthesis of hydroxyapatite for biomedical applications. *Advances in Colloid and Interface Science*, 249: 321–330.

Tayeb, A. H., Amini, E., Ghasemi, S. and Tajvidi, M. (2018). Cellulose nanomaterials—Binding properties and applications: A review. *Molecules*, 23(10): 2684.

Thuwall, M., Boldizar, A. and Righdahl, M. (2006). Extrusion processing of high amylose potato starch materials. *Carbohydrate Polymers*, 65(4): 441–446.

Tim, S., Stock, U., Hrkach, J., Shinoka, T., Lien, J., Moses, M., Stamp, A., Taylor, G., Moran, A., Landis, W., Langer, R., Vacanti, J. and Mayer, J. (1999). Tissue engineering of autologous aorta using a new biodegradable polymer. *Tissue Engineered Aortic Autografts*. 68: 2298–2305.

Tyler, B., Gullotti, D., Mangraviti, A., Utsuki, T. and Brem, H. (2016). Polylactic acid (PLA) controlled delivery carriers for biomedical applications. *Advanced Drug Delivery Reviews*, 107: 163–175.

Valero-Valdivieso, M. F., Ortegón, Y. and Uscategui, Y. (2013). Biopolímeros: avances y perspectivas. *DYNA*, 80(181): 171–180.

Xiao, G., Kim, J., Cai, X. and Cui, T. (2019). Shrink-induced highly sensitive dopamine sensor based on self-assembly graphene on microelectrode. 20th Int. Conf. Solid-State Sensors, Actuators Microsystems Eurosensors. EUROSENSORS XXXIII 1120–1123.

Xu, Q. et al. (2014). Polydopamine-coated cellulose microfibrillated membrane as high-performance lithium-ion battery separator'. *RSC Advances*. The Royal Society of Chemistry, 4(16): 7845–7850.

Yukuta, T., Akira, I. and Masatoshi, K. (1990). Developments of biodegradable plastics containing polycaprolactone and/or starch. Polym. *Mater. Sci. Eng.*, 63: 742–749.

Zambrano-Herrera, W. (2020). Reología de polímeros. *Revista Agrollania de Ciencia y Tecnología*, 19: 47–53.

Zhang, T. W. et al. (2019). Recent advances on biopolymer fiber-based membranes for lithium-ion battery separators. *Composites Communications*, 14: 7–14.

Chapter 1
Biodegradation of Polymers

Sarai Agustín Salazar,[1] *Sabu Abdulhameed*[2] and
Margarita del Rosario Salazar Sánchez[3,*]

1. Introduction

The modern world is facing a major problem of environmental pollution, which arises due to several anthropogenic activities. Nowadays, most of the human activities involve the use of plastics. Polymeric materials do not decompose easily, and this causes the accumulation of plastics in landfill and oceans (Bahl et al., 2020, 2021; Eskander and Saleh, 2017). For several years, the search for reducing the environmental impact caused by human activities have been conducted to the development of sustainable methodologies and eco-friendly products, especially to reduce plastic pollution.

The development of green energy-saving materials and sustainable plastic waste management are currently key research topics from the perspective of sustainability, in response to the rapidly growing burden of energy consumption and environmental pollution (Eskander and Saleh, 2017; Matjašič et al., 2021; Mi et al., 2020; Polman et al., 2021; Quecholac-Piña et al., 2020; Urbanek et al., 2021; Walker and McKay, 2021). A particular category of environmentally important plastics is those used in agricultural applications (greenhouses, mulching, small tunnels, silage, nets, pipes, etc.), and food packgaging (Briassoulis et al., 2020; Europe, 2021). Incorrect agriculture practices and the growing industrial sector contribute to the emission of various contaminants into the environment and the accumulation of plastic waste in the environment. Many of the pollutants that enter the environment undergo a degradation process through the action of microorganism's presents in the terrestrial or aquatic system, which is commonly known as biodegradation. Since several environmental microorganisms are able to quickly respond to available nutrients, they play a significant role in degrading organic substances in the ecosystem, such as cellulose or lignin (Ahmed et al., 2018; Eskander and Saleh, 2017; Polman et al., 2021; Quecholac-Piña et al., 2020; Urbanek et al., 2021).

Biodegradation is considered nature's way of recycling of wastes. Generally, biodegradation is defined as the breakdown of organic material by microorganisms or biological media. Through this natural process, toxic pollutants are converted into less toxic or non-toxic compounds, aerobically or anaerobically. Microorganisms are considered as the natural forces of biodegradation, and they alter the structure of the chemical compounds either through metabolic or enzymatic action (Eskander and Saleh, 2017). Microorganisms can interact with the pollutants both physically and chemically.

[1] Departamento de Ingeniería Química y Metalurgia, Universidad de Sonora, Hermosillo, México.
[2] Department of Biotechnology and Microbiology, School of Life Sciences, Kannur University, Kannur, Kerala. India.
[3] Departamento de Ingenierías, Universidad Popular del Cesar, Cesar, Colombia.
* Corresponding author: mdelrosariosalazar@unicesar.edu.co

Many of the microorganisms found in the soil can degrade environmental pollutants. The biodegradation process occurs naturally due to several reasons. Sometimes a limitation of the essential nutrient for the microorganisms in the soil environment may occur. For example, there may be a decrease in the amount of carbon needed for the microorganism. In such situations, if a new carbon source is available in the form of organic pollutant, the organism is capable of utilizing the carbon from such sources. Then the proportion of such microorganisms capable of using the organic pollutants may go on, increasing to a dominant level. In some cases, the pollutant may show resemblance to the substrate of the enzymatic system of the microorganism which will induce the degradation process (Alshehrei, 2017).

Bacteria and actinomycetes are the main microorganisms involved in the biodegradation processes (Diez, 2010). In addition to microorganisms, there are studies that involve organisms such as fungi (Diez, 2010), algae (Chia et al., 2020), worms (Peng et al., 2020), snails (Song et al., 2020) and even insects (*Ulomoides dermestoides*) (Salazar-Sánchez et al., 2019) (Fig. 1.1). Complete degradation results in the formation of carbon dioxide and water. Partial degradation results in the formation of intermediate products which will later get released into the environment (Bagi, 2018; Urbanek et al., 2021). Pollutants make their entry into the environment as solids, liquids or gases, and depending on the environmental conditions, they get transported into different compartments.

The biodegradation mechanism as well as its efficiency depends on the involved organism and some abiotic or chemical factors, such as media (in soil or in aquatic system), UV irradiation, physical stress, temperature, pH, oxygen (aerobia or anaerobia) and the pollutant nature. Biodegradation efficiency could also be a result of the sum of two or more factors (Bagi, 2018; Diez, 2010; Polman et al., 2021; Quecholac-Piña et al., 2020; Urbanek et al., 2021; Wilkes and Aristilde, 2017).

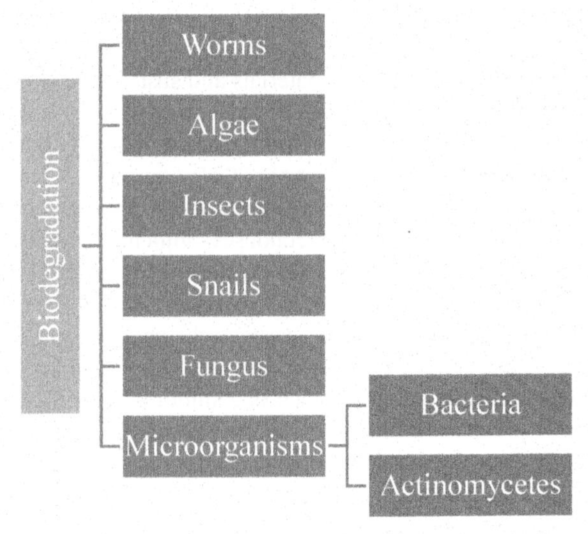

Figure 1.1. Biological organisms studied for their biodegradation activity.

2. Plastic pollutants

Among pollutants, plastics have a great impact in the environment. Plastics are organic compounds of high molecular mass, made of long chain polymeric molecules. It is structured by a main organic chain that could be linked through specific molecular groups. Often, other organic and inorganic compounds are also added (additives, plasticizers, fillers, etc.). Plastics used in great quantity are polyethylene, polystyrene, polyvinyl chloride, polypropylene and polyethylene terephthalate (Bahl et al., 2021; Yang et al., 2018).

Plastic pollution poses a serious threat to human and animal's health. Chemical ingredients of more than 50% of plastics are hazardous. These harmful chemicals leeched from plastic wastes or in the form of small or micro plastic debris impact ecologically soils (declining crop production), water bodies (deteriorate fishing industry and damage aquatic environment) and air (emission of air pollutants such as volatile organic compounds, soil, polycyclic aromatic hydrocarbons, polychlorinated dibenzofurans, dioxins, etc., due to plastics combustion) (Yang et al., 2018).

The production and use of plastic brought considerable advantages in the technological development of packaging, construction, automotive, electronics, agriculture, etc. At the same time, the over-production of plastics resulted in serious environmental consequences in the last years (Eskander and Saleh, 2017; Serrano-Ruiz et al., 2021; Urbanek et al., 2021; Zhong et al., 2020). Plastic packaging is one of the largest contributors to the environmental impact. Reducing production of new single-use plastics will thus reduce plastic pollution and curb gas emissions (Walker and McKay, 2021).

Adopting recyclable alternative packaging materials or reducing consumption of single-use plastic items will also help reduce these unintended environmental impacts of plastic pollution. Unfortunately, waste management alone will not be sufficient to reduce the growing global plastic footprint. Then, the reduction and substitution at source should come first to complement reuse and recycling (Alshehrei, 2017; Walker and McKay, 2021; Zhong et al., 2020).

3. Biodegradable and compostable plastics

The use of biodegradable plastics (bioplastics) as an alternative to conventional non-degradable polymeric materials, such as polyethylene (LDPE, HDPE), polypropylene (PP), among others, is increasing largely. Indeed, even non-biodegradable plastics, when placed in the soil become brittle due to the action of light, heat, water and oxygen, and start to break down into smaller pieces. These plastic fragments have an impact on the environment, with effects of their size poorly known, but pointing to higher impact of microplastics on plants than on other soil organisms. Biodegradable plastics are designed to degrade under environmental conditions or in biological waste treatment facilities. The main degradation process in biodegradable plastics results in the fragmentation of material via microbial enzymatic activities and bond cleavage (Folino et al., 2020; Serrano-Ruiz et al., 2021; Urbanek et al., 2021; Zhong et al., 2020).

There are very few studies about degradation on conventional plastics, due to their nature, however, Yang et al. (2018) reported an interesting effect of mealworms those can degraded ingested PS in their gut up to 50% within the relatively short time (15 hours). Figure 1.2 illustrates PS biodegradation is likely gut microbe dependent, but that despite some rare examples, traditional polymers (PE, PP, PS, PVC, etc.) are non-biodegradable, and therefore it is necessary to develop new polymeric materials to replace traditional plastics with degradable ones.

Microorganisms like bacteria, fungi, algae, especially blue green algae are equally involved in the biodegradation process. Fungi convert the chemical pollutants into nontoxic compounds by introducing small structural changes. The bio-transformed chemical compound is then subjected to

Figure 1.2. Styrofoam-eating mealworms from various sources (A) Beijing, China, (B) Harbin, China; and (C) Compton, California, USA (Yang et al., 2018).

bacterial degradation (Diez, 2010). Some microorganisms which may work for one type of pollutant may not work for the other. Both bacteria and cyanobacteria are actively involved in the degradation of petroleum hydrocarbons in the marine environments (Urbanek et al., 2021).

The introduction of biodegradable plastics, especially in food packaging, includes biodegradable or compostable polymers, such as poly (lactic acid) (PLA), poly(ε-caprolactone) (PCL), poly (butylene succinate) (PBS) and Poly (Butylene Succinate-co-butylene Adipate) (PBSA) and biomaterial-based (starch, lignocellulose) plastics. The main advantage of these materials is due to their susceptibility to attack by microorganisms, which are ubiquitous in the environment (Agustin-Salazar et al., 2018; Angelini et al., 2016; Polman et al., 2021; Urbanek et al., 2021).

The capacity for biodegradation of plastics has been found in many genera of bacteria (e.g., *Brevibacillus*, *Bacillus*, *Streptomyces*, *Clostridium*, *Pseudomonas*, *Rhodococcus*, *Comamonas*) and fungi (e.g., *Cladosporium*, *Debaryomyces*, *Eupenicillium*, *Fusarium*, Mucor, *Penicillium*, *Pullularia*, *Cryptococcus*, *Rhizopus*, *Aspergillus*) as a natural capability (Urbanek et al., 2021).

The biodegradation by microorganisms is related to extracellular and intracellular secretion of enzymes. Assimilation and uptake of plastic fragments depends on metabolic pathways that occur in the cells, whereas plastic particles are broken down outside the cells. Many enzymes have been demonstrated to be involved in the biodegradation process (lipases, esterases, proteases, dehydrogenases and hydrolases). Microorganisms can degrade different types of plastic under suitable conditions, but very often due to the hardness of these polymers and their non-solubility in water, achieving this biodegradation is still an issue. Most of the microorganisms have an incredible flexibility and the ability to adapt to unfavorable conditions for survival. Thus, it is assumed that the increasing amount of plastic waste leaking may provide a new substrate for microorganisms and force them to adapt to new available nutrients (Urbanek et al., 2021; Wilkes and Aristilde, 2017).

In polymers' biodegradation, very often hydrolyzable bonds between the polymer monomers are required, and the released monomers are used as energy and carbon sources by soil microorganisms to grow. This process involves three main steps: (1) microbial colonization of the polymer surface, mainly bacteria and fungi, (2) depolymerization by microorganism extracellular enzymes and, (3) microorganism consumption of the hydrolysis products (Serrano-Ruiz et al., 2021). However, to predict which bioplastic is the most accessible for microorganisms is quite difficult, and depends on the specific abilities of the isolated strain (Matjašič et al., 2021; Urbanek et al., 2021; Wilkes and Aristilde, 2017).

Table 1.1 enlists some biodegradation methodologies carried out to assess biodegradation in polymers. It is notable that not only microorganisms have been tested in polymers biodegradation, but also worms, insects, algae and snails.

The biodegradation on bio-composite materials have been also investigated (Iwańczuk et al., 2015). The behavior of two biodegradable polymers, poly (lactic acid) and poly(3-hydroxybutyrate-co-3-hydroxyvalerate) and polyethylene and their composites with flax fibers in anaerobic digestion was tried (37°C for 72 days). In that study, composites underwent biodegradation faster than the matrix polymer; however, the extent of degradation is lower if the fibers are more resistant to biodegradation than the polymer. Microorganisms caused changes in the internal structure of polymers, with a reduction in tensile properties when the degradation time increases.

However, composting biodegradable plastics also represent a half solution due the gas emission during degradation and many oxo-biodegradable or biodegradable formulations, comprising up to 25% of petroleum-based plastics, degrade into microplastics and few jurisdictions have industrial composting equipment to properly handle them (Walker and McKay, 2021). Recent advances and issues in making bioplastics are elaborated comprehensively, including the addition of natural additives, such as lignocellulosic materials to enhance the biodegradability of polymers and at the same time reinforcing the final products. Lignocellusic biomass has the possibility to function as a plasticizer, stabilizer or bio-compatibilizer in bioplastics and to enhance its miscibility with

Table 1.1. Biodegradation studies on polymers.

Plastic	Biodegradation conditions	Source
PLA	Bacteria: *Arthrobacter sulfonivorans, Serratia plymuthica* and two fungi: *Clitocybe* sp., *Laccaria laccata*, at 26°C in soil.	(Janczak et al., 2020)
	Bacteria: *Fusarium moniliforme, Penicillium Roquefort, Amycolatopsis* sp., *Bacillus brevis, Rhizopus delemer.*	(Bahl et al., 2021)
PBS	Fungal isolates: Geomyces, Fusarium and Sclerotinia	(Urbanek et al., 2021)
	Bacteria: *Pseudomonas chlororaphis, Amycolatopsis* sp., *Micrcobispora rosea, Excellospora japonica, E. viridilutea, Schlegelella thermodepolymerans, Caenibacterium thermophilum,* at different degradation conditions.	(Bahl et al., 2020)
PCL	Fungus: *Alternaria alternata* ST01, 72.09 and 56.49% within 15 days under shaking and stationary conditions, respectively.	(Abdel-Motaal et al., 2014)
	Fungal isolates: Geomyces, Fusarium and Sclerotinia.	(Urbanek et al., 2021)
	Bacteria: *Penicillium oxalicum*, 10 days at 30°C	(Li et al., 2012)
	Fungal strain: *Filobasidium uniguttalatum*, 62% (at pH = 5 and 30°C) in 15 days. Bacterial strains: *Bacillus megaterium, Alcaligenes aquatilis* and *Shewanella haliotis* (59, 56 and 53% weight loss, respectively). At 30°C in 15 days.	(Nchedo Ariole and George-West, 2020).
PET	Bacteria: *Arthrobacter sulfonivorans, Serratia plymuthica* and two fungi: *Clitocybe* sp., *Laccaria laccata*, at 26°C in soil.	(Janczak et al., 2020).
EPS	Snails: *Achatina fulica* at 25°C and air humidity of 60%.	(Song et al., 2020).
	Worms: *Larvae of Zophobas atratus* at 25°C.	(Peng et al., 2020).
LDPE	Gram-negative bacteria microbulbifer hydrolyticus at 37°C.	(Li et al., 2020).
	Worms: *Larvae of Zophobas atratus* at 25°C.	(Peng et al., 2020).
PVC	Bacterial strains: *Pseudomonas citronellolis* and *Bacillus flexus*, under aerobic conditions, at 30°C for 45 days.	(Giacomucci et al., 2020).
PSLP	Cassava starch-based bioplastics reinforced with crude kaolin or heat treated kaolinitic: bacterial soil at T = 25°C, P = 1 bar.	(Méité et al., 2021).
	Thermoplastic starch: aspergillus bacteria, bacillus fungus, and amylase enzyme.	(Polman et al., 2021).
	Cellulose-acetate: cellulases enzymes, and *pseudomonas* bacteria.	(Polman et al., 2021).
	Lignin based plastics: biodegradation under standard composting conditions.	(Polman et al., 2021).
PHAs:	Biodegradation in marine, soil and sludge media at different conditions.	(Meereboer et al., 2020).
	Bacteria: *Pseudomonas lemoignei, Streptomyces* sp., *Alcaligenes faecalis, Pseudomonas stutzeri, Fusarium solani.*	(Bahl et al., 2021).

PLA: Poly (Lactic Acid), PBS: Poly (Butylene Succinate), PCL: poly(ε-caprolactone), PET: Poly (Ethylene Tereftalate), EPS: expanded polystyrene, LDPE: low density polyethylene, PVC: polyvinyl chloride, PSLP: polysaccharide and lignocellulosic plastics, PHA: polyhydroxy alcanoates.

biopolymers considerably by functionalization of hydroxyl groups (Agustin-Salazar et al., 2018; Álvarez-Chávez et al., 2017; Sánchez-Acosta et al., 2019; Yang et al., 2019).

Pseudomonas sp. is one of the most used bacteria a bioremediating microorganism that is capable of degrading various types of plastics (Bahl et al., 2020; Giacomucci et al., 2019, 2020; Li et al., 2020; Polman et al., 2021). Li et al. (2020) have particularly used these bacteria in the biodegradation of polyphenylene sulfide plastic beads in a short reaction time of 10 days.

As oceans are cold environments that are predominant over the Earth, there are studies analyzing microbial adaptations to oceans, such as structural adjustment of enzymes, maintenance of membrane fluidity and expression of cold shock proteins. The use of highly specific groups of microorganisms such as extremophiles, which include cold-adapted microorganisms, may have both

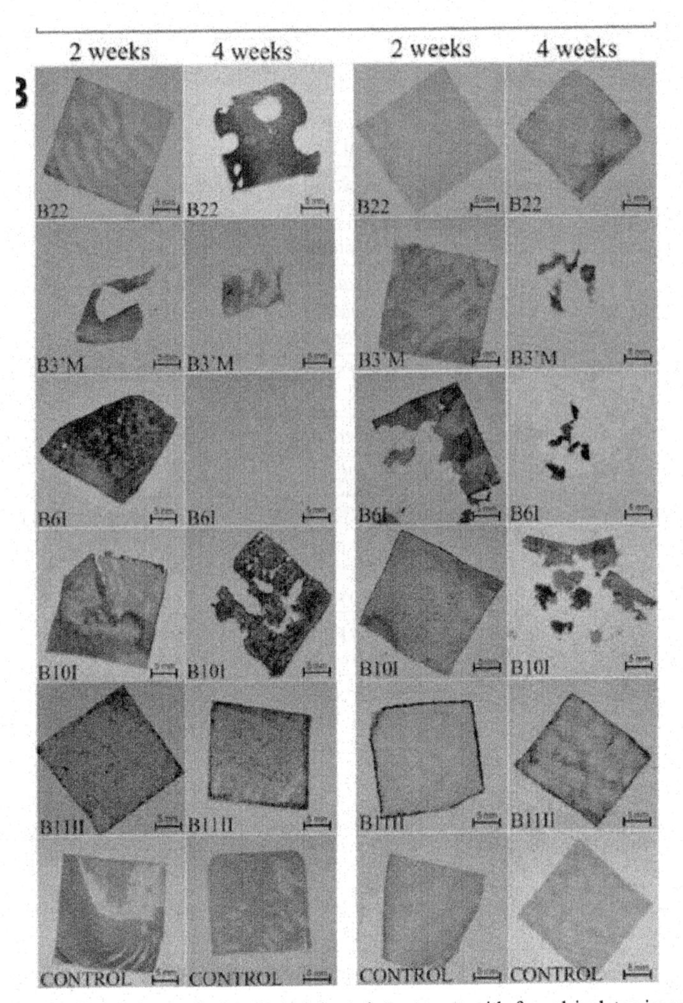

Figure 1.3. PBSA and PCL films recovered after 2- and 4-week treatment with fungal isolates in comparison to control films. (A) Growth of fungi on MM medium with PBSA and PCL films as the only carbon source at room temperature; (B) surface of recovered films with visible damage. Fungal isolates marked as B22, B6I and B10I were described as belonging to Geomyces, B3'M to Fusarium and B11IV to Sclerotinia (Urbanek et al., 2021).

ecological and economic effects. During processes with the application of such microorganisms or cold-active enzymes produced by them, lower temperatures may be used, which reduces energy consumption (Urbanek et al., 2021).

Urbanek et al. (2021) have tested the biodegradability rate of several polymers by three fungal strains (*Sclerotinia* sp., *Fusarium* sp. and *Geomyces* sp.), at three different temperature values of 4, 20 and 28°C. Figure 1.3 shows the biodegradation effects on PBSA and PCL films. After 4 weeks of incubation, for *Sclerotinia* sp. and *Fusarium* sp. Strains, the optimal temperature for the biodegradation process was 20°C, where the biodegradability rate reached 49.68% for PBSA. At 4°C, all fungal strains growth. Figure 1.4 shows the biodegradation effects on PBSA and PCL after 4 weeks of incubation at 100x and 5000x magnifications (Urbanek et al., 2021).

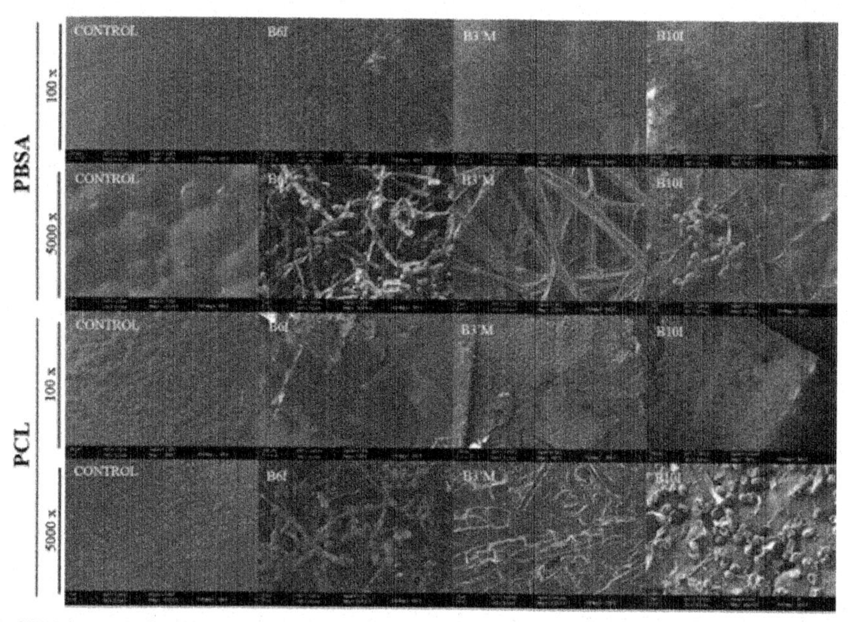

Figure 1.4. SEM images of PBSA and PCL films recovered from plate cultures after 4 weeks of incubation at room temperature in comparison to control. Magnification of 100x and 5000x. Fungal isolates marked as B6I and B10I were described as belonging to Geomyces, B3'M to Fusarium.

4. Role of microorganisms in the biodegradation process

Microorganisms like bacteria and fungi are involved in the biodegradation of both natural and synthetic plastics. Polymers like plastics serve as potential substrates for heterotrophic microorganisms. The first step in the biodegradation of plastics is the conversion of polymers into its monomers so that they can pass through the cell membrane and get biodegraded within the microbial cells. Microbial enzymes play an important role in the depolymerization of polymers (Alshehrei, 2017).

Microorganisms have the ability to assimilate carbon in natural or synthetic organic substances as a source of nutrients and energy for growth and reproduction. This ability is due to the microorganism having co-existed with a variety of organic compounds for billions of years (Chinaglia et al., 2018). Degradation of contaminants by microorganisms limits its spread into different areas thereby reducing the extent of pollution (Eskander and Saleh, 2017). Polluted sites contain a wide variety of contaminants instead of a single compound. Microbial communities consisting of phylogenetically and physiologically different microorganisms are involved in the degradation of such contaminants. In such cases the degradation takes place by the coordinated activities of the microorganisms (Vogt and Richnow, 2013). Continuous exposure of microorganisms to pollutants may sometimes help them to evolve certain mechanisms for the biodegradation of the pollutant. For example, some soil microorganisms have the capacity to degrade pesticides when they are continuously applied into the soil (Eskander and Saleh, 2017).

Microorganisms are able to carry out a wide variety of enzymatic activities. They mediate the biodegradation process through the production of enzymes, vesicles and storage bodies, chelating agents and cell surface agents (Alshehrei, 2017; Matjašič et al., 2021; Urbanek et al., 2021). The presence of contaminants either induces or depresses the enzymatic system of microorganisms which mainly depends up on the microbial community and the structural and functional groups on the chemical compounds (Singh et al., 2014). Enzymes involved in the biodegradation process can be intracellular, membrane bound or extracellular. For intracellular and membrane bound enzymes to work, the pollutants must be in contact with the cell, while for the extracellular enzymes to work, there is no need for the pollutant to enter the cell since the action of these enzymes takes place outside the cell. Microorganisms like fungi produce extracellular enzymes such as laccases and peroxidases which are involved in the biodegradation of pollutants (Matjašič et al., 2021; Tanjung et al., 2018; Urbanek et al., 2021). In certain situations, an inhibition may occur to the enzyme activity or to the multiplication process of microorganisms involved in the degradation process. This is due to the competition between microorganisms for the nutrient sources or the antagonistic interaction between microorganisms or by the predation of microorganisms by bacteriophage and protozoa (Tahri et al., 2013).

Morphological and physiological adaptations may occur in microorganisms in response to environmental changes. In some cases, to increase the contact area with the pollutants, microorganisms develop multi-dimensional structure. Mycelia of fungi which enables the efficient mobilization of chemicals in the soil is a typical example. Pollutants become bioavailable to the bacteria through these mycelia mediated transport. Surface active molecules are sometimes released by the microorganisms to increase the bioavailability of the pollutant. These biosurfactants helps to reduce the interfacial tensions of the pollutant, thereby increasing its solubility and mobility. Biosurfactants also help to increase the level of hydrophobicity on the cell surface and the binding with hydrophobic substrates. Microorganisms may also exhibit some behavioural adaptations like chemotaxis and swimming modes. The movement of bacteria towards the contaminant through chemotaxis increases the bioavailability of the pollutant and the biodegradation efficiency (Matjašič et al., 2021; Ren et al., 2018; Serrano-Ruiz et al., 2021; Urbanek et al., 2021).

5. Factors affecting the biodegradation process in natural environments

The efficiency and pace of biodegradation in the soil or in the aquatic system are determined by a number of factors. First, the characteristics of both the pollutant and the microorganism affects the biodegradation process. Apart from that, various environmental factors also play an important role in the biodegradation of pollutants (Bagi, 2018; Polman et al., 2021). Adequate amount of oxygen and nutrient must be available for the maximum growth of degrading microorganisms (Tahri et al., 2013). Availability of the oxygen is an important factor that affects the biodegradation process. Aerobic and anaerobic degradation may take place depending upon the availability of the oxygen. The anaerobic degradation process is slower than the aerobic process (Bagi, 2018; Quecholac-Piña et al., 2020). Other important factors are temperature and pH, since they influence the optimum growth of microorganisms and the activity of enzymes involved in the degradation process. Acidic and alkaline conditions of the environment also affect the degradation process. Biodegradation may take place at a wide range of pH, range 6.5–8.5 is considered to be optimal for the biodegradation taking place in the aquatic and terrestrial ecosystems (Tahri et al., 2013; Urbanek et al., 2021).

The type of soil and its organic matter content also plays an important role in biodegradation, because the pollutants that enter the soil binds to the mineral and organic matter through absorption or adsorption (Megharaj et al., 2011; Tahri et al., 2013). The availability of the pollutant to the microbial degradation process depends upon the ability of the soil to release the pollutants. Sometimes microorganisms can attack the adsorbed pollutant. But mostly they act upon dissolved fractions. Volatility and aqueous solubility of the pollutants influences its bioavailability to the microorganisms (Megharaj et al., 2011). Hydrophobic poly aromatic hydrocarbons or high molecular weight aliphatic hydrocarbons can bind to sediment particles thereby limiting their degradation process (Vogt and Richnow, 2013). Contaminants that strongly adsorbs to soil may become resistant to degradation and will remain in the soil for a very long time. Direct degradation at the adsorbed state is possible by the cell attachment and the production of extracellular enzymes. In some cases, biosurfactants help to release the pollutants from the adsorbed phase, making them available to microbial uptake (Ren et al., 2018).

Biodegradability of a polymer is essentially determined by some physical and chemical characteristics, such as the availability of functional groups that increase hydrophobicity, the molecular weight and density of the polymer, the morphology (amount of crystalline and amorphous regions), structural complexity, presence of easily breakable bonds, molecular composition, the nature and physical form of the polymer and the hardness of the polymer (Alshehrei, 2017).

6. Future perspectives

The increasing development of technologies for the degradation of the plastic wastes is a key factor to reduce the accumulation of plastics in the environment. Crucial attention is needed for the microplastics that accumulate in activated sludge during wastewater treatment processes, as this water is then further used in agriculture or discharged into streams (Matjašič et al., 2021; Serrano-Ruiz et al., 2021; Urbanek et al., 2021; Zhong et al., 2020).

Biodegradation is a promising approach, and it can provide energy-efficient and cost-efficient technologies, although only if this is carried out under controlled conditions. The understanding of the mechanisms involved in biodegradation is an important issue (Bahl et al., 2021; Matjašič et al., 2021; Quecholac-Piña et al., 2020).

Plastic wastes management solutions should involve technologies for efficient recycling of resources without the generation of hazardous byproducts or toxic substances to the environment (Gallo et al., 2018; Ganesh Kumar et al., 2020; Yang et al., 2018). The strategy for controlling plastics pollution should focus on source reduction, improved reuse, recycle and recovery of plastic via improved separation efficiency from waste streams and subsequently the development of cost-effective clean-up and bioremediation technologies (Ahmed et al., 2018; Yang et al., 2018).

It is necessary to increase the use of biodegradable polymers and bio-based additives to obtain sustainable materials and bioplastics. Polymeric blending, chemical and physical modifications, have been strategies to enhance the properties and performance of polymeric materials (Agustin-Salazar et al., 2018; Angelini et al., 2014; Panzella et al., 2016; Sánchez-Acosta et al., 2019). Contaminated sites should be studied as a promising source of biodegrading microorganisms.

Finally, considering pretreatments, such as thermal and photo oxidation, and/or by mechanical deterioration and the synergism with biodegradation, could be another way to reduce the environmental impact due to plastic pollution (Agustin-Salazar et al., 2020; Agustin-Salazar et al., 2014; Panzella et al., 2016).

References

Abdel-Motaal, F. F., El-Sayed, M. A., El-Zayat, S. A. and Ito, S. (2014). Biodegradation of poly (ε-caprolactone) (PCL) film and foam plastic by *Pseudozyma japonica* sp. nov., a novel cutinolytic ustilaginomycetous yeast species. *3 Biotech*, 4(5): 507–512. https://doi.org/10.1007/s13205-013-0182-9.

Agustin-Salazar, S., Gamez-Meza, N., Medina-Juárez, L. Á., Soto-Valdez, H. and Cerruti, P. (2014). From nutraceutics to materials: effect of resveratrol on the stability of polylactide. *ACS Sustainable Chemistry & Engineering*, 2(6): 1534–1542. https://doi.org/10.1021/sc5002337.

Agustin-Salazar, S., Cerruti, P., Medina-Juárez, L. Á., Scarinzi, G., Malinconico, M., Soto-Valdez, H. and Gamez-Meza, N. (2018). Lignin and holocellulose from pecan nutshell as reinforcing fillers in poly (lactic acid) biocomposites. *International Journal of Biological Macromolecules*, 115: 727–736. https://doi.org/10.1016/j.ijbiomac.2018.04.120.

Agustin-salazar, S., Cerruti, P. and Scarinzi, G. (2020). 9 Biobased structural additives for polymers. *In*: *Sustainability of Polymeric Materials*. https://doi.org/10.1515/9783110590586-009.

Ahmed, T., Shahid, M., Azeem, F., Rasul, I., Shah, A. A., Noman, M., Hameed, A., Manzoor, N., Manzoor, I. and Muhammad, S. (2018). Biodegradation of plastics: current scenario and future prospects for environmental safety. *Environmental Science and Pollution Research*, 25(8): 7287–7298. https://doi.org/10.1007/s11356-018-1234-9.

Alshehrei, F. (2017). Biodegradation of synthetic and natural plastic by microorganisms. *Journal of Applied & Environmental Microbiology*, 5(1): 8–19. https://doi.org/10.12691/jaem-5-1-2.

Álvarez-Chávez, C. R., Sánchez-Acosta, D. L., Encinas-Encinas, J. C., Esquer, J., Quintana-Owen, P. and Madera-Santana, T. J. (2017). Characterization of extruded poly(lactic acid)/pecan nutshell biocomposites. *International Journal of Polymer Science*, 2017: 1–12. https://doi.org/10.1155/2017/3264098.

Angelini, S., Cerruti, P., Immirzi, B., Santagata, G., Scarinzi, G. and Malinconico, M. (2014). From biowaste to bioresource: Effect of a lignocellulosic filler on the properties of poly(3-hydroxybutyrate). *International Journal of Biological Macromolecules*, 71: 163–173. https://doi.org/10.1016/j.ijbiomac.2014.07.038.

Angelini, S., Cerruti, P., Immirzi, B., Scarinzi, G. and Malinconico, M. (2016). Acid-insoluble lignin and holocellulose from a lignocellulosic biowaste: Bio-fillers in poly(3-hydroxybutyrate). *European Polymer Journal*, 76: 63–76. https://doi.org/10.1016/j.eurpolymj.2016.01.024.

Bagi, A. (2018). Microbial degradation in the aquatic environment. pp. 205–232. *In*: Pampanin, D. M. and Sydnes, M. O. (eds.). *Petrogenic Polycyclic Aromatic Hydrocarbons in the Aquatic Environment: Analysis, Synthesis, Toxicity and Environmental Impact* (Issue January 2017). Bentham Science Publishers. https://doi.org/10.2174/978168108 42751170101.

Bahl, S., Dolma, J., Jyot, J. and Sehgal, S. (2020). Materials Today: Proceedings Biodegradation of plastics: A state of the art review. *Materials Today: Proceedings*, *xxxx*: 4–7. https://doi.org/10.1016/j.matpr.2020.06.096.

Bahl, S., Dolma, J., Jyot Singh, J. and Sehgal, S. (2021). Biodegradation of plastics: A state of the art review. *Materials Today: Proceedings*, 39: 31–34. https://doi.org/10.1016/j.matpr.2020.06.096.

Briassoulis, D., Mistriotis, A., Mortier, N. and Tosin, M. (2020). A horizontal test method for biodegradation in soil of bio-based and conventional plastics and lubricants. *Journal of Cleaner Production*, 242: 118392. https://doi.org/10.1016/j.jclepro.2019.118392.

Chia, W. Y., Ying Tang, D. Y., Khoo, K. S., Kay Lup, A. N. and Chew, K. W. (2020). Nature's fight against plastic pollution: Algae for plastic biodegradation and bioplastics production. *Environmental Science and Ecotechnology*, 4: 100065. https://doi.org/10.1016/j.ese.2020.100065.

Chinaglia, S., Tosin, M. and Degli-Innocenti, F. (2018). Biodegradation rate of biodegradable plastics at molecular level. *Polymer Degradation and Stability*, 147(December 2017): 237–244. https://doi.org/10.1016/j.polymdegradstab.2017.12.011.

Diez, M. C. (2010). Biological aspects involved in the degradation of organic pollutants. *Journal of Soil Science and Plant Nutrition*, 10(3): 244–267. https://doi.org/10.4067/S0718-95162010000100004.

Eskander, S. and Saleh, H. E. M. (2017). Biodegradation: process mechanism. *Biodegradation and Bioremediaton*, 8(January): 1–31.

Europe, P. (2021). *Circular Economy*. https://www.plasticseurope.org/en.

Folino, A., Karageorgiou, A., Calabrò, P. S. and Komilis, D. (2020). Biodegradation of wasted bioplastics in natural and industrial environments: a review. *Sustainability*, 12(15): 6030. https://doi.org/10.3390/su12156030.

Gallo, F., Fossi, C., Weber, R., Santillo, D., Sousa, J., Ingram, I., Nadal, A. and Romano, D. (2018). Marine litter plastics and microplastics and their toxic chemicals components : the need for urgent preventive measures. *Environmental Sciences Europe*. https://doi.org/10.1186/s12302-018-0139-z.

Ganesh Kumar, A., Anjana, K., Hinduja, M., Sujitha, K. and Dharani, G. (2020). Review on plastic wastes in marine environment—Biodegradation and biotechnological solutions. *Marine Pollution Bulletin*, 150(November 2019): 110733. https://doi.org/10.1016/j.marpolbul.2019.110733.

Giacomucci, L., Raddadi, N., Soccio, M., Lotti, N. and Fava, F. (2019). Polyvinyl chloride biodegradation by *Pseudomonas citronellolis* and *Bacillus flexus*. *New Biotechnology*, 52(April): 35–41. https://doi.org/10.1016/j.nbt.2019.04.005.

Giacomucci, L., Raddadi, N., Soccio, M., Lotti, N. and Fava, F. (2020). Biodegradation of polyvinyl chloride plastic films by enriched anaerobic marine consortia. *Marine Environmental Research*, 158(December 2019): 104949. https://doi.org/10.1016/j.marenvres.2020.104949.

Iwańczuk, A., Kozłowski, M., Łukaszewicz, M. and Jabłoński, S. (2015). Anaerobic biodegradation of polymer composites filled with natural fibers. *Journal of Polymers and the Environment*, 23(2): 277–282. https://doi.org/10.1007/s10924-014-0690-7.

Janczak, K., Dabrowska, G., Raszkowska-Kaczor, A., Kaczor, D., Hrynkiewicz, K. and Richert, A. (2020). Biodegradation of the plastics PLA and PET in cultivated soil with the participation of microorganisms and plants. *International Biodeterioration & Biodegradation* 155(March). https://doi.org/10.1016/j.ibiod.2020.105087.

Li, F., Yu, D., Lin, X., Liu, D., Xia, H. and Chen, S. (2012). Biodegradation of poly(ε-caprolactone) (PCL) by a new *Penicillium oxalicum* strain DSYD05-1. *World Journal of Microbiology and Biotechnology*, 28(10): 2929–2935. https://doi.org/10.1007/s11274-012-1103-5.

Li, J., Kim, H. R., Lee, H. M., Yu, H. C., Jeon, E., Lee, S. and Kim, D. H. (2020). Rapid biodegradation of polyphenylene sulfide plastic beads by *Pseudomonas* sp. *Science of the Total Environment*, 720: 137616. https://doi.org/10.1016/j.scitotenv.2020.137616.

Li, Z., Wei, R., Gao, M., Ren, Y., Yu, B., Nie, K., Xu, H. and Liu, L. (2020). Biodegradation of low-density polyethylene by Microbulbifer hydrolyticus IRE-31. *Journal of Environmental Management*, 263: 110402. https://doi.org/10.1016/j.jenvman.2020.110402.

Matjašič, T., Simčič, T., Medveščček, N., Bajt, O., Dreo, T. and Mori, N. (2021). Critical evaluation of biodegradation studies on synthetic plastics through a systematic literature review. *Science of The Total Environment*, 752: 141959. https://doi.org/10.1016/j.scitotenv.2020.141959.

Meereboer, K. W., Misra, M. and Mohanty, A. K. (2020). Review of recent advances in the biodegradability of polyhydroxyalkanoate (PHA) bioplastics and their composites. 22(17): 5519–5558. https://doi.org/10.1039/d0gc01647k.

Megharaj, M., Ramakrishnan, B., Venkateswarlu, K., Sethunathan, N. and Naidu, R. (2011). Bioremediation approaches for organic pollutants: A critical perspective. *Environment International*, 37(8): 1362–1375. https://doi.org/10.1016/j.envint.2011.06.003.

Méité, N., Konan, L. K., Tognonvi, M. T., Doubi, B. I. H. G., Gomina, M. and Oyetola, S. (2021). Properties of hydric and biodegradability of cassava starch-based bioplastics reinforced with thermally modified kaolin. *Carbohydrate Polymers*, 254(October 2020): 117322. https://doi.org/10.1016/j.carbpol.2020.117322.

Mi, R., Chen, C., Keplinger, T., Pei, Y., He, S., Liu, D., Li, J., Dai, J., Hitz, E., Yang, B., Burgert, I. and Hu, L. (2020). Scalable aesthetic transparent wood for energy efficient buildings. *Nature Communications*, 11(1): 1–9. https://doi.org/10.1038/s41467-020-17513-w.

Nchedo Ariole, C. and George-West, O. (2020). Bioplastic degradation potential of microorganisms isolated from the soil. *American Journal of Chemical and Biochemical Engineering*, 4(1): 1. https://doi.org/10.11648/j.ajcbe.20200401.11.

Panzella, L., Cerruti, P., Ambrogi, V., Agustin-Salazar, S., D'Errico, G., Carfagna, C., Goya, L., Ramos, S., Martín, M. A., Napolitano, A. and D'Ischia, M. (2016). A superior all-natural antioxidant biomaterial from spent coffee grounds for polymer stabilization, cell protection, and food lipid preservation. *ACS Sustainable Chemistry & Engineering*, 4(3): 1169–1179. https://doi.org/10.1021/acssuschemeng.5b01234.

Peng, B., Li, Y., Fan, R., Chen, Z., Chen, J., Brandon, A. M., Criddle, C. S., Zhang, Y. and Wu, W. (2020). Biodegradation of low-density polyethylene and polystyrene in superworms, larvae of Zophobas atratus (Coleoptera : Tenebrionidae): Broad and limited extent depolymerization*. *Environmental Pollution*, 266: 115206. https://doi.org/10.1016/j.envpol.2020.115206.

Polman, E. M. N., Gruter, G. M., Parsons, J. R. and Tietema, A. (2021). Comparison of the aerobic biodegradation of biopolymers and the corresponding bioplastics: A review. *Science of The Total Environment*, 753: 141953. https://doi.org/10.1016/j.scitotenv.2020.141953.

Quecholac-Piña, X., Hernandez-Berriel, M. C., Mañon-Salas, M. C., Espinosa-Valdemar, R. M. and Vazquez-Morillas, A. (2020). Degradation of plastics under anaerobic conditions: a short review. *Polymers*, 1–18.

Ren, X., Zeng, G., Tang, L., Wang, J., Wan, J., Liu, Y., Yu, J., Yi, H., Ye, S. and Deng, R. (2018). Sorption, transport and biodegradation—An insight into bioavailability of persistent organic pollutants in soil. *Science of The Total Environment*, 610–611: 1154–1163. https://doi.org/10.1016/j.scitotenv.2017.08.089.

Salazar-Sánchez, M. del R., Campo-Erazo, S. D., Villada-Castillo, H. S. and Solanilla-Duque, J. F. (2019). Structural changes of cassava starch and polylactic acid films submitted to biodegradation process. *International Journal of Biological Macromolecules*, 129: 442–447. https://doi.org/10.1016/j.ijbiomac.2019.01.187.

Sánchez-Acosta, D., Rodriguez-Uribe, A., Álvarez-Chávez, C. R., Mohanty, A. K., Misra, M., López-Cervantes, J. and Madera-Santana, T. J. (2019). Physicochemical characterization and evaluation of pecan nutshell as biofiller in a matrix of poly(lactic acid). *Journal of Polymers and the Environment*, 27(3): 521–532. https://doi.org/10.1007/s10924-019-01374-6.

Serrano-Ruiz, H., Martin-Closas, L. and Pelacho, A. M. (2021). Biodegradable plastic mulches: Impact on the agricultural biotic environment. *Science of The Total Environment*, 750: 141228. https://doi.org/10.1016/j.scitotenv.2020.141228.

Singh, R., Singh, P. and Sharma, R. (2014). Microorganism as a tool of bioremediation technology for cleaning environment: A review. *Proceedings of the International Academy of Ecology and Environmental Sciences*, 4(1): 1–6.

Song, Y., Qiu, R., Hu, J., Li, X., Zhang, X. and Chen, Y. (2020). Science of the total environment biodegradation and disintegration of expanded polystyrene by land snails *Achatina fulica*. *Science of the Total Environment*, 746: 141289. https://doi.org/10.1016/j.scitotenv.2020.141289.

Tahri, N., Bahafid, W., Sayel, H. and El Ghachtouli, N. (2013). Biodegradation: involved microorganisms and genetically engineered microorganisms. *In*: *Biodegradation - Life of Science*. InTech. https://doi.org/10.5772/56194.

Tanjung, F. A., Arifin, Y. and Husseinsyah, S. (2018). Enzymatic degradation of coconut shell powder–reinforced polylactic acid biocomposites. *Journal of Thermoplastic Composite Materials*, 1: 1–17. https://doi.org/10.1177/0892705718811895.

Urbanek, A. K., Strzelecki, M. C. and Mirończuk, A. M. (2021). The potential of cold-adapted microorganisms for biodegradation of bioplastics. *Waste Management*, 119: 72–81. https://doi.org/10.1016/j.wasman.2020.09.031.

Vogt, C. and Richnow, H. H. (2013). *Bioremediation via in situ Microbial Degradation of Organic Pollutants* (pp. 123–146). https://doi.org/10.1007/10_2013_266.

Walker, T. R. and McKay, D. C. (2021). Comment on five misperceptions surrounding the environmental impacts of single-use plastic. *Environmental Science & Technology*, 55(2): 1339–1340. https://doi.org/10.1021/acs.est.0c07842.

Wilkes, R. A. and Aristilde, L. (2017). Degradation and metabolism of synthetic plastics and associated products by *Pseudomonas* sp.: capabilities and challenges. *Journal of Applied Microbiology*, 123(3): 582–593. https://doi.org/10.1111/jam.13472.

Yang, J., Ching, Y. C. and Chuah, C. H. (2019). Applications of lignocellulosic fibers and lignin in bioplastics: A review. *Polymers*, 11(5): 1–26. https://doi.org/10.3390/polym11050751.

Yang, S. S., Brandon, A. M., Xing, D. F., Yang, J., Pang, J. W., Criddle, C. S., Ren, N. Q. and Wu, W. M. (2018). Progresses in polystyrene biodegradation and prospects for solutions to plastic waste pollution. *IOP Conference Series: Earth and Environmental Science*, 150(1): 012005. https://doi.org/10.1088/1755-1315/150/1/012005.

Zhong, Y., Godwin, P., Jin, Y. and Xiao, H. (2020). Advanced industrial and engineering polymer research biodegradable polymers and green-based antimicrobial packaging materials: A mini-review. *Advanced Industrial and Engineering Polymer Research*, 3(1): 27–35. https://doi.org/10.1016/j.aiepr.2019.11.002.

Chapter 2

Use of Renewable Source in Biodegradable Polymer

J.J. Cedillo-Portillo,[1] *J.D. Flores-Valdes,*[1] *W.Y. Villastrigo-López,*[1] *D.W. González-Martínez,*[1] *A.O. Castañeda-Facio,*[1] *S.C. Esparza-Gonzalez,*[2] *R.I. Narro-Céspedes*[1] and *A. Sáenz-Galindo*[1,*]

1. Introduction

Today the use of a renewable source in biodegradable polymers, is a viable alternative to reduce the pollution that is in there on planet earth. Polymeric materials made with this type of natural sources have various applications in different areas of interest such as nanotechnology, materials, food, food packaging, among others. Biodegradable polymers are defined as that material that can be completely degraded by the environment directly or indirectly, thus reducing the environmental impact that it can produce as an adverse effect on the environment and become pollution. Biodegradable polymers are classified according to their origin, as natural, modified natural, synthetic or a combination of natural and synthetic. These types of biodegradable materials have different advantages, where it is highlighted that dependence on oil as raw material for its production is avoided, also that it is susceptible to completely degrade in a relatively short period of time, when compared with synthetic polymer, degradation products are usually water, carbon dioxide and biomass. The purpose of this chapter is to offer an overview of different natural sources that are used as biodegradable polymeric materials, both of natural and synthetic origin.

2. Biodegradable polymers

The growing demand for the production and consumption of plastics has endangered life on planet earth because it is assumed that the accumulation of these residues enters persistent pollutants, almost half of the plastic produced is used as packaging in products. Basically used in supermarkets, after the use of these materials by consumers only 48% is collected and recycled, so these types of materials represents a source of contamination for different biota, although the focus that has risen in recent years of reducing, recycling and reusing, that could offer a solution to the problem, though is not enough to stop the excessive use of plastics in society and conjunction with the environment,

[1] Facultad de Ciencias Química, Universidad Autónoma de Coahuila, Ing. J. Cárdenas Valdez S/N, Col. República C.P. 25280. Saltillo Coahuila, México.

[2] Facultad de Odontología, Universidad Autónoma de Coahuila. Ave. Doctora Cuquita Cepeda de Dávila sin número. Col. Adolfo López mateos. C.P. 25125, Saltillo Coahuila, México.

* Corresponding author: aidesaenz@uadec.edu.mx

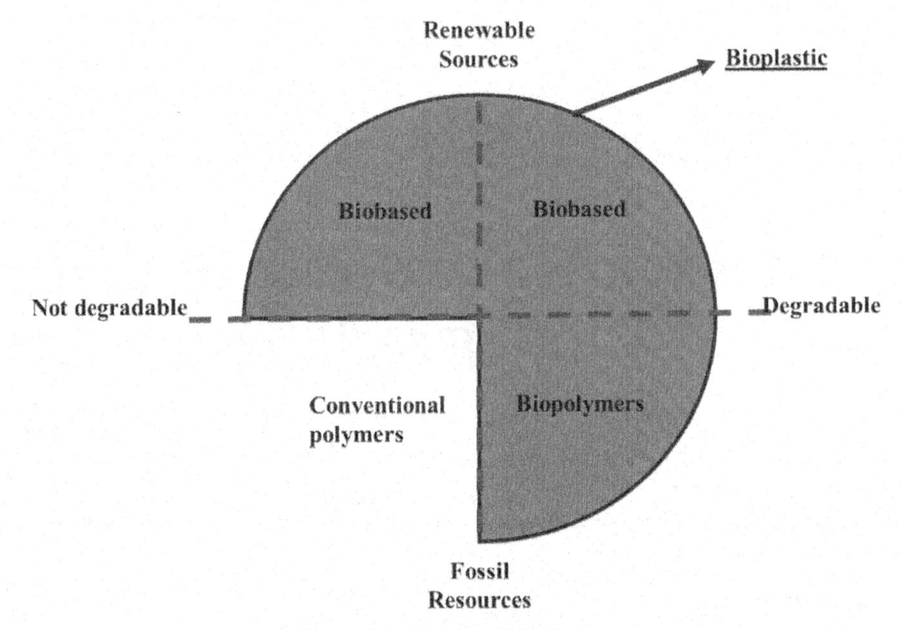

Figure 2.1. Diagram of bioplastics.

some common materials such as wood, glass and metal could be a viable option for the replacement of some plastic materials, however a great disadvantage of these materials is their low durability. The high manufacturing cost, and the inability to adapt to some conditions of use, which is why different researchers have focused on the study of plastic materials with very important characteristics such as biodegradability and compatibility with the environment (Dodande and Vilvalam, 1998; Rai et al., 2021).

Bioplastics are plastic materials that are made partially or totally of natural or synthetic polymers and may or may not be biodegradable, they are mainly composed of polysaccharides, proteins and fibers, the European organization of bioplastic defines them as plastics that are biobased, biodegradable or that meet both characteristics, contrary to what is believed, bioplastics are not something new, they were used by Henry Ford in the manufacture of the T car and are widely used in medical applications, however, what has aroused so much interest in recent years is the use of these to mitigate the environmental impact caused by conventional plastics (Rai et al., 2021; Benhabiles et al., 2012; Zhong et al., 2020). Figure 2.1 Diagram of bioplastics.

2.1 Chitosan

Chitosan is the second most abundant polysaccharide in nature after cellulose, it was described for the first time in 1859 by Rouget, who described it as a linear copolymer composed of β-(1-4)-2-acetamido-D units-glucose and β-(1-4)-2-amino-D-glucose, is an important biopolymer which is characterized by having great properties such as its toxicity which is obtained through the deacetylation of chitin, this being the main component of some skeletons of insects and marine invertebrates such as crabs and shrimps, chitin is generally found in waste from the food industry, in addition to being biodegradable and non-toxic, chitosan has demonstrated different biological activities such as: antimicrobial, anti-inflammatory, antioxidant effects, wound healing and has antitumor activities, all this is mainly due to the functional groups found in its chain, due to which, chitosan is used in applications that have a direct contact with humans (Wu et al., 2021; Hamedi et al., 2018).

The degree of deacetylation of chitosan is one of the most important structural properties since the higher the degree of acetylation, its solubility is decreased in acidic solvents, as well as its antimicrobial activity, which increases when deacetylating chitin (Gan et al., 2021).

Although chitosan has become very popular in recent years in different areas of science, it has several limitations such as its low mechanical stability, its ease of adsorption of water and its easy solubility, mainly in acidic environments, which in some cases limits its use. Due to the above, several authors have reported the modification of chitosan to overcome these adversities and have materials based on this biopolymer more resistant both mechanically and chemically, with crosslinking being one of the best options for modifying this biopolymer (Brás et al., 2020).

2.2 Biodegradable natural polymers

Polymers are long chains of molecules (macromolecules) that have a high molecular weight, these are linked by small units of low molecular weight through covalent bonds. Natural polymers, also called biopolymers, are those that are present or are created by living organisms and are available in large quantities in nature, for example: cellulose, collagen, chitosan, starch, among others (Nazin, 2013). There are two types of natural polymers, the first are those created by living matter and the second are those that need to be polymerized, but they come from renewable sources (Clark, 2013).

Natural polymers can be classified depending on their origin, which are (Villada et al., 2007):

- Animal origin (collagen, gelatin)
- Marine origin (chitin, chitosan)
- Agricultural origin (proteins and polysaccharides)
- Microbial origin (PLA)

Some of the characteristics exhibited by these types of polymers, which make them attractive for multiple applications, are:

- Biodegradability
- Low toxicity
- Low cost
- Availability

One of their drawbacks is that there is the possibility of microbial contamination when exposed to an external environment (Harsha et al., 1981). In addition, natural polymers are grouped based on their method of formation as addition and condensation polymers. Most of these are condensation polymers are formed because of the combination of monomeric units and generally form water as a by-product. Natural polymers can be roughly divided into three groups (Gooch, 2011):

- Proteins and peptides
- Polysaccharides
- Nucleic acids

Figure 2.2, summarizes the classification of natural polymers.

According to different standards ASTM D-5488-94d and EN 13432, a polymer is named "biodegradable" if it can decompose into gases (CO_2, CH_4), water, inorganic salts and biomass (Papadopouluo and Chrissafis, 2017). Biodegradation is produced by the action of enzymes and/or chemical deterioration associated with living organisms, commonly bacteria, fungi and algae, some of the important bacteria in the biodegradation process are *Bacillus*, *Pseudomonas*, *Streptomyces*, *Escherichia*, *Azotobabter*, among others (Leja and Lewandwicz, 2010).

This event occurs in two steps, the first is the fragmentation of the polymer into species of lower molecular weight through abiotic reactions (photodegradation, oxidation or hydrolysis) or biotic (degradation by microorganisms), also called depolymerization and the second step consists

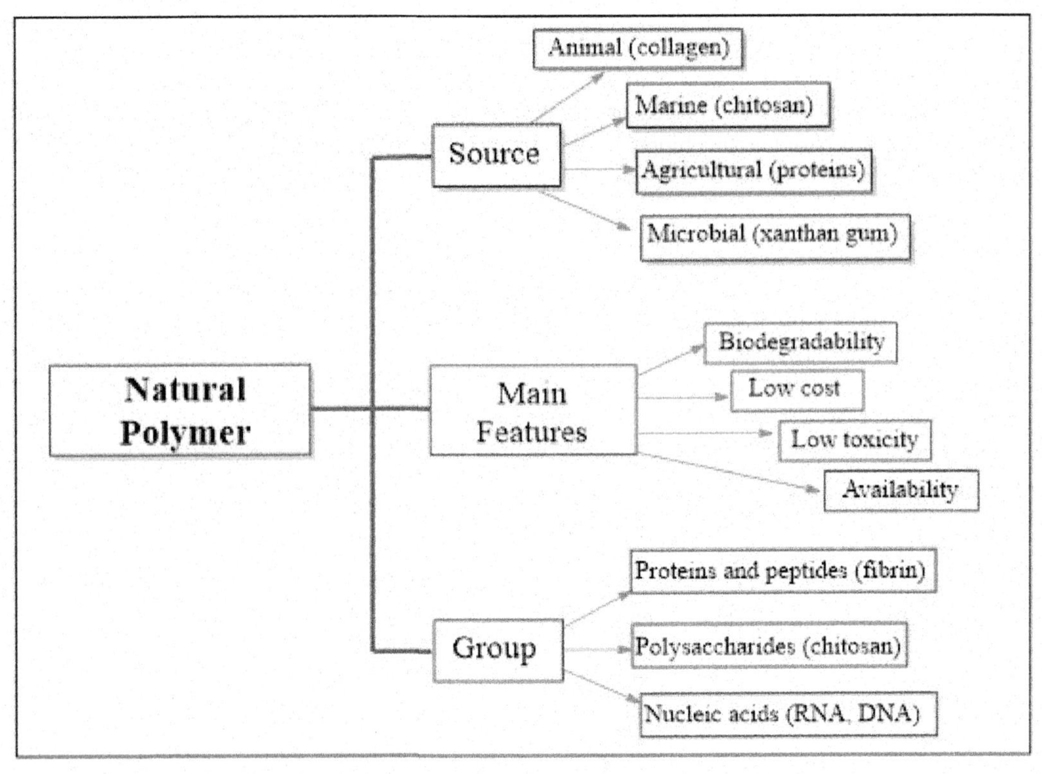

Figure 2.2. General classification of natural polymers.

in the bioassimilation of biopolymer fragments by microorganisms and their mineralization (Bastioli, 2005).

Natural polymers are degraded in biological systems such as oxidation and hydrolysis. The biodegradation process can be divided into aerobic and anaerobic, when there is the presence of oxygen, aerobic biodegradation takes place and CO_2 is produced. And if there is no oxygen, anaerobic degradation occurs and CH_4 is produced. Figure 2.3, shows the degradation process.

The conversion of a natural biodegradable polymer to gas, water, inorganic salts and biomass is called mineralization, and this is complete until all the biodegradable material is consumed and all the carbon is converted to CO_2 (Vroman and Tighzert, 2009). The degradation time is determined by different factors that can affect it, among these are (Vroman and Tighzert, 2009; Witko, 2003; Zapata et al., 2012):

- Characteristics of the polymer:
  - Molecular weight
  - Chemical structure
  - Porosity
  - Elasticity
  - Morphology
- Characteristics of the environment:
  - Humidity
  - Temperature
  - pH

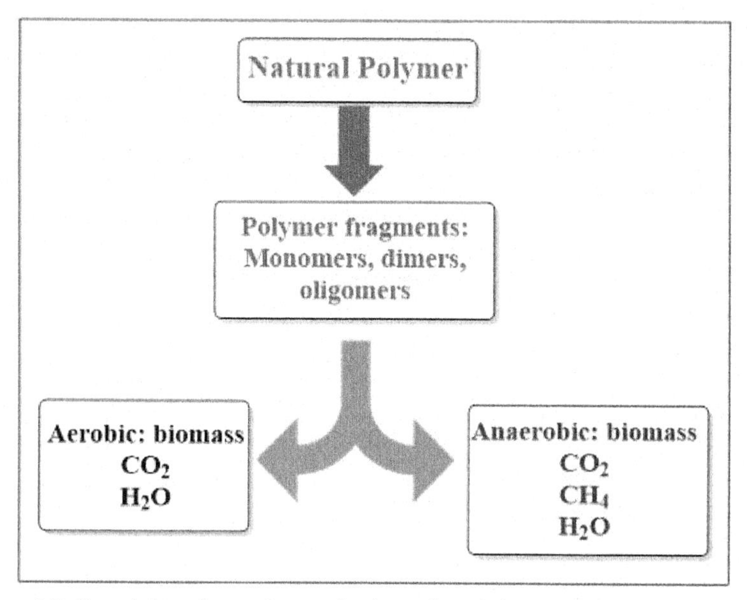

Figure 2.3. Degradation scheme of a natural polymer through the aerobic and anaerobic system.

- Characteristics of microorganisms:
  - Source
  - Quantity
  - Variety
  - Activity

3. Biodegradable synthetic polymers

Synthetic polymers are those that are obtained artificially by synthesis from monomers, such as nylon, polyolefins, polyesters, adhesives, among others. These synthetic polymers are created for specific functions and are obtained from two polymerization methods according to the composition or structure of the polymers. The persistence of synthetic polymers in the environment is dangerous for the human community and wildlife because they generate pollution. This created an urgent need in the implementation of the use of biodegradable plastics, since they can be completely degraded by the environment, thus reducing the environmental impact that these plastics produce (Odian, 2004).

One of the methods is by addition of polymerization consisting of three stages: initiation, propagation and termination. Polymerization is usually a radical pathway derived from vinyl monomers, in which the loss of small molecules does not occur and results from chain reactions comprising some type of reactive center. In the initiation stage the initiator molecule is broken and added to the first monomer molecule, in the propagation the free radical is generated and transmitted from one molecule to another and at the stage of termination the chains of macroradicals are united to finish the reaction. On the other hand, condensation polymerization is obtained from bifunctional or trifunctional monomers and there is loss of small molecules, conventional polymers such as polyethylene (PE), polypropylene (PP), polystyrene (PS), polyvinyl chloride (PVC), etc., widely used in daily life are not biodegradable. However, with population growth and industrialization, there is now greater awareness of the impact of these non-biodegradable polymer materials on the environment (Frackowiak et al., 2018).

Synthetic polymers are used because of their versatility, ease of processing, low density and excellent mechanical properties. Despite these properties, these materials cause pollution and are a big problem for the environment. Therefore, many efforts are now being directed towards the development of several biodegradable polymers. However, despite considerable work on the production of new biodegradable polymers, only a few biodegradable products have been placed on the market. New biodegradable polymers have a wide range of applications and are not only used as substitutes for conventional polymers, but are also used in biomedical applications (Kaur, 2017).

One of the main requirements for a material to be biodegradable is that it must have groups in the main chain that can be easily broken by the action of external agents of a physical or chemical nature, when biodegradable polymeric materials are synthesized it is intended that the properties of that material are maintained and that, when a change in its chemical structure is made, its components are compatible with the environment when decomposed (Tokiwa and Calabia, 2015).

In a polymer, physical or chemical changes may occur because of degradation, there may be discoloration, loss of surface gloss, formation of cracks, and loss of properties such as tensile strength, while chemicals consist of chain breakage or cross-linking reactions of crosslinked, the lower the molecular weight of the polymer the faster the degradation (Pathank and Navneet, 2017).

4. Degradation methods

Photodegradation: when a macro-radical is formed in the amorphous regions of the polymeric substrate, this radical reacts quickly with oxygen to form the peroxide radical, extracting a hydrogen atom from the main polymer chain to produce a hydroperoxide group, this group binds strongly so that highly reactive radicals are produced that allow the chain degradation cycle to continue in the polymer and the degradation cycle ends when two radicals are combined, there are two ways to obtain photodegradable polymers, one is by introducing functional groups sensitive to ultraviolet radiation, such as carbonyl groups, by modifying the polymer, on the other hand, photosensitive additives can be introduced, such as catalysts and peroxidants to accelerate the degradation process (Dinoop et al., 2021).

Thermal degradation: is a process in which, when applying high temperatures to a polymer, it undergoes chemical changes in its structure, usually this degradation occurs together with oxidative degradation if the material is in the presence of oxygen. Thermal degradation is based on the breakdown of covalent bonds because of increased temperature. Thermo degradation causes changes in the glass transition temperature, molecular weight and polymer fluidity index; thus, when a polymer is degraded by breaking the main chain, its molecular weight and viscosity are reduced and the fluidity index increases, two degradation mechanisms of polymer chains are directly related to temperature. The first mechanism is the elimination of the lateral groups forming low molecular weight groups that volatilize at the degradation temperature. The second mechanism consists of depolymerization with the consequent reduction of molecular weight. This mechanism predominates at high temperatures (Thomason and Fernandez, 2021).

Hydrolytic degradation: this occurs when the material is placed in an aqueous medium, by penetrating the water in the polymeric matrix generates expansion, causing the breakdown of intermolecular hydrogen bonds and finally the hydrolysis of weak bonds, this method occurs by two different mechanisms: massive degradation and surface erosion. In the case of mass degradation, water diffusion in a polymer matrix is faster than the rate of hydrolysis. Hydrolytically cleaved bonds in the amorphous parts of the matrix can degrade as water molecules become available due to rapid diffusion. Therefore, the molecular weight of polymers decreases. On the other hand, in the case of surface erosion, water diffusion in the polymer is slower than the rate of degradation of the macromolecule. Hydrolysis is limited to a thin layer on the surface, while the molecular weight of the polymer in the mass remains unchanged. In this mechanism, the rate of degradation depends on the shape of the polymer, hydrolytic degradation takes place in polymers that possess hydrophilic functional groups such as hydroxyl, esters, amide and carbonates. Polymers obtained by condensation reactions such as polyesters, polyamides, polycarbonates and some polyurethanes are particularly sensitive to degradation (Shockley and Muliand, 2020).

When hydrolysis takes place in groups that are in the main polymer chain, the polymer is reduced to oligomers; however, if hydrolysis takes place inside chains, other properties such as solubility are affected. However, the most obvious change that occurs in a polymer after hydrolysis is its mechanical strength, the degree of crystallinity of the polymer also influences hydrolytic degradation. If the polymer is amorphous, the chains are arranged randomly in space, the water can penetrate the matrix more easily than if it has a crystalline structure. One way to accelerate hydrolytic degradation is the addition of different substances (acidic, basic, plasticizer) that increase the rate of hydrolysis of the polymer (Barren and Sparks, 2020).

Oxodegradation: this process begins with the ultraviolet light of the solar radiation, manages to decrease the molecular weight of the polymer due to the breaking of its molecular chains, thereby reducing their molecular weight and being exposed to developing a process of biodegradation over time (Laycock et al., 2017).

5. Biodegradable polymers

Biodegradable polymers, capable of developing aerobic or anaerobic decomposition by the action of microorganisms such as fungi and bacteria, are degraded by the enzymatic action of microorganisms under normal environmental conditions (Zhong et al., 2020).

The biodegradation of a material can be partial (where the chemical structure is altered and some physical properties of the material are lost) or total, where complete mineralization of the material is produced with the production of simpler molecules such as water, mineral salts, CO_2 (under aerobic conditions) or methane (anaerobic conditions) (Dake, 2015).

Microorganisms break down compounds in a simpler way through biochemical transformation, this degradation occurs in sequential steps: biodeterioration (alteration of the physical and chemical properties of the polymer), biofragmentation (decomposition of the polymer in a simpler form by enzymatic excision), assimilation (absorption of molecules by microorganisms) and mineralization (production of oxidized metabolites, after degradation). Mineralization takes place both in anaerobic and aerobic conditions, in aerobic CO_2 and H_2O are formed, while in anaerobic CO_2, CH_4 and H_2O are produced. The biodegradability of synthetic polymers with chemical groups susceptible to microbial attack can be achieved by polycaprolactone, poly-β-hydroxyalkanoates and oil-based polymers (Huang et al., 2020).

Biodegradable polymers can come from petroleum such as polycaprolactone (PCL) which is an aliphatic or biodegradable polyester that does not require a previous photodegradation process (Sanchez, 2020).

Synthetic aliphatic polyesters are synthesized from diols and dicarboxylic acids via condensation polymerization and are completely biodegradable in soil and water. Many polyesters are mixed with starch polymers, to give biodegradable properties with more economical costs (Siddhi and Dilip, 2020).

The biodegradability of plastic does not depend only on the raw materials used for its manufacture; its chemical structure must also be considered. The rate of degradation of polymers depends on the environment (temperature, humidity, pH), as well as the specific characteristics of the polymer: the presence of chemical bonds susceptible to hydrolysis, hydrophobicity, stereochemistry, molecular weight, crystallinity, glass transition temperature, melting temperature and micro-organisms used for their degradation (Yin and Yang, 2020).

Biodegradable synthetic polymers have various applications in medicine: as they are used in tissue engineering (they help in the regeneration of bone tissues, cartilage, tendons and corneas), in the controlled release of drugs and may also have been used on an industrial scale, in the manufacture of food packaging (Tian et al., 2012).

For each application to be given to biodegradable synthetic polymers a set of specific properties is required, for example, mechanical properties such as degradability. With an increasing number of potential applications, a greater variety of materials with different combinations of properties is required. The fields of modern medicine, for example, regenerative medicine, require materials with a variety of functionalities combined in one material. For the successful development of new biodegradable materials, it is necessary to have a solid knowledge of the polymers that are to be used and to consider all their properties and characteristics, especially the degradation property (Rai et al., 2021).

5.1 Aliphatic polyesters

Polyesters today constitute the main family of biodegradable synthetic polymers used in biomedical applications. These polymers have hydrolyzable ester bonds in their structure. Aliphatic polyesters can be classified into two types, according to the mode of binding of the constituent monomers: poly (alkane dicarboxylates) which are synthesized by diol and dicarboxylic polycondensation and polyhydroxyalkanoates, which are polymers of hydroxylic acids (Romatowska et al., 2020).

5.2 Poly (α-hydroxy acids)

This class of polyester has applications in biomedicine since it is possible to adjust its degradation and have excellent biocompatibility. In this category are polymers such as Poly (Glycolic Acid, PGA), Poly (Lactic Acid, PLA) and a range of their copolymers such as Poly (Lactic Acid-co-Glycolic, PLGA). These materials are synthesized by means of polymerization by ring-opening or by condensation, depending on the monomers with which the polymerization is initiated, the first biodegradable synthetic suture approved by the FDA was obtained from polyglycolic acid, which was developed in 1970. This acid has high crystallinity, low solubility in organic solvents and good mechanical properties, so it is used in orthopedic applications, polymerization of lactic acid isomers produces crystalline polymers, while polymerization of racemic mixtures produces amorphous polymers with less mechanical resistance. Poly (L-lactic acid) and poly (DL-lactic acid) are used for biomedical applications. The rate of degradation depends on the molecular weight of the polymer, crystalline and porosity of the polymer matrix (Nottelet et al., 2015).

5.3 Polylactones

It is a semi-crystalline polymer that is soluble in organic solvents, so it is easily processed. Polylactone can be broken down by microorganisms, hydrolytic or enzymatically. However, its degradation is slow, making it less attractive for general applications in tissue engineering, but on the contrary, it has advantages for use in long-lasting implants and drug release (Polyak et al., 2019).

5.4 Polyanhydrides

The low hydrolytic stability of these polymers together with their nature makes them ideal candidates for applications in the controlled release of short-term drugs. Diacids are the most widely used monomers for the synthesis of poly anhydride by activated condensation of acetic anhydride. These are broken down by the hydrolysis of the anhydride bond which is hydrolytically very labile. The rate of degradation can be tailored to the application required, by means of small changes in the polymer structure. Aliphatic polyanhydrides break down at higher rates than aromatic polyanhydrides and aliphatics are soluble in organic solvents, while aromatics are not, they are biocompatible and degrade *in vivo* giving as dialed by-products that are not toxic and because of that can be released by the body. The most studied polyanhydride is poly [(carboxyfenoxyl) propane-sebacic acid], used as a matrix for the controlled release of carmustine, a drug used in chemotherapy against brain cancer (Gijsman et al., 2021).

5.5 Polyurethanes

These polymers belong to a class of synthetic thermoplastic polymers that have been extensively studied for the development of long-lasting implants. Various biodegradable polyurethanes have been obtained, due to mechanical properties and good properties such as biocompatibility, synthetic versatility and biological activity. They are synthesized from three monomers: a diisocyanate, a diol or a diamine extender chain and a long-chain diol. The polymerization reaction of equimolar quantities of diisocyanate and diol produces biodegradable polyurethanes whose composition is decisive in the rate of degradation. The degradation product obtained (diamine obtained after polyurethane hydrolysis) determines the choice of the starting diisocyanate. In other words, aliphatic diisocyanates produce fewer toxic diamines than aromatic diisocyanates. For example, hexamethylenedisocianate and 1,4-butanediocyanate are among the most used dissociates in the formulation of biodegradable polyurethanes, while diphenyl or toluendisocyanate are not used because their breakdown products are toxic. The fact that the chemical and mechanical properties of polyurethanes can be adapted to needs is what explains why they are found in various fields such as the regeneration of neurons, cartilage and bones (Kundys et al., 2018).

As time goes by, polymers have been integrated into the daily life of human beings, however their volume of production and use has increased during the last decades, for which viable alternatives such as biopolymers have been sought have been applied in different areas which are described below:

6. Nanotechnology

Nanotechnology is a science field that studies the control and fabrication of the matter at the molecular level in a scale in to 1 and 100 nm; this is primarily employed in an energic industry, electronic, biomedicine and biotechnology, particularly in chemotherapy and drug administration and textile industry, as well as the alimentary and agricultural (Samrot et al., 2020).

The investigation in nanotechnology focuses in creating materials with unique proprieties of daily use such is the case of the products or self-cleaning fabrics, water and oil repellency, UV protect, comfort in synthetic-based fabrics among others, which are classified in nanofibers, nano coatings, nanocomposites (Wujcik and Monty, 2013).

In 2021 Kim and collaborators studied the synthesis of graphene nanoparticles coated with tannic acid, which was used as non-halogen flame retardant; however, the nanoparticles were electrospun with polyurethane producing multifunctional nanofibers, which showed flame retardant properties as well as antimicrobial activity and mechanical resistance, having potential application in the automotive, construction and medical areas (Kim et al., 2021).

The use of these biopolymers has been used in antimicrobial application, as it is in 2021 where Sackey and collaborators reported the synthesis of silver nanoparticles which were incorporated into polycarbonate membranes (Sackey et al., 2019).

Others application of the biopolymers in nanotechnology, is the textile industry such as was incorporated in medical applications, defense, aerospace, among other (Joshi, 2008).

7. Food

The packaging of foods has an important role in their preservation since it protects against mechanical damage and contamination during their commercialization, also offering a useful, fresh and extended life (Carvalho and Conte Junior, 2020).

The films using for packaging are extracted from biopolymers that include: starch, cellulose and derived monomers such as polylactic acid, these are produced by lamination, coextrusion or coating processes form raw polymer, biopolymers and biodegradable materials (Chisenga et al., 2020).

In 2017 Ivankovic informed that there are three stages of generation of biodegradable polymers, from which films for food packaging are manufactured:

- *First generation*: consists in synthetic polymers as an example, low density polyethylene films are presented, which were applied in shopping bags with a proportion of 5 to 15% of starch and auto-oxidant additives, however this type of polymers decompose into molecules smaller in size that are not biodegradable.
- *Second generation*: it is a generation of biomaterials that comprises a mixture of pregelatinized starch (40–70%) and low density polyethylene incorporating copolymers such as: ethylene acrylic acid, polyvinyl alcohol and vinyl acetate which are used to compact.
- *Third generation*: are produced from biomaterials that are classified as:
  a) *Polymers extracted by biomass*: starch, chitin, chitosan, vegetable proteins and soy.
  b) *Polymers synthesized from monomers*: polylactic acid
  c) *Biomonomers*: natural polymers or genetically modified (Ivankovic et al., 2017; Chiellini, 2008).

Food packaging waste represent a significant part of the urban solid waste increasing environmental risks, whereby they have been developed as biodegradable polymers capable of decomposing in the presence of CO_2, water, inorganic compounds and biomass (Rydz et al., 2018).

In 2020 Mohamad and collaborators mentioned that the polylactic acid is a biobased polymer which has exhibited excellent barrier properties for food preservation compared to conventional petroleum-based films; Mohamed reported fabrication of PLA films which incorporated: timol (T), kesum (K) and curry (C) to 10% in weight. This biomaterial carried out permeability tests and antimicrobial activity, showing positive results against gram-positive bacteria (*Staphylococcus aureus*) as opposed to Gram-negative bacteria (*Escherichia coli*) which did not show inhibition (Mohamad et al., 2020).

T the polycaprolactone (PCL) is a biodegradable synthetic polymer which is used in drug delivery devices and medical devices, just as PLA is considered one of the most promising materials for food packaging, therefore in the year 2018 Xie and collaborators incorporated TiO_2 nanoparticles to these biopolymers with the aim of having bactericidal activity obtaining good results for their application (Xie and Hung, 2018).

8. Automotive

In recent years, the need has increased to improve the properties of biopolymers in terms of strength, stiffness, density and lower cost due to their various applications.

Generally, composite materials, whether natural or synthetic, are used in areas such as: construction, mechanical, automotive, aerospace, biomedical and marine.

In the automotive industry, significant amounts of polymers have been used for the production of both electric and hybrid automobiles, as well as batteries; improving mass reduction and driving anatomy.

The use of these in the automotive industry has been applied mainly in interior linings, in addition to boxes, storage systems, cabins for medium and heavy duty as well as compounds that absorb energy during shock, just as the use of these materials contributes to the cost reduction, production and maintenance extending the useful life of automobiles (Muhammad et al., 2021).

Today the interior parts of cars such as: dashboard, carpet, linings, door molding, burette and seats are made exclusively with polymers, on the other hand, biopolymers, in addition to being light, are profitable, easy to mold and recycle since these have shown greater resistance compared to aluminum, carbon and other metals.

The use of biopolymers in the automotive industry has been employed without compromising quality standards; An example of this is PLA (polylactic acid) and PBS which are applied in seat covers, foams and fabrics, it should be mentioned that pure PLA does not compete in terms of properties with conventional non-biodegradable polymers, however when adding additives as fibers, superior resistance is obtained, which is why it has been chosen as a replacement for conventional polymers.

Companies such as Mazda, Mitsubishi and Ford have pioneered the use of biopolymers for car interiors instead of fossil-oil-based parts (Bhagabati, 2020).

In 2017 Mazda corporation in association with Tejin Fibers Limited developed biodegradable polymers based on PLA, which are biotech with greater mechanical resistance and heat resistance capacity; on the other hand, from the year 2016 Mitsubishi Chemical Corporation with PTT Global Chemical Public Limited developed biodegradable heat resistant PBS (Cooper and Cooper, 2018).

As Beardmore mentioned in 1986, polymeric compounds reinforced with different fillers such as glass fibers, carbon fibers, nylon, carbon black and clay are more effective in high performance applications such as load capacity, chemical and environmental resistance (Company, 1986).

9. Medicine

The application of various materials in medical procedures has grown rapidly. Currently, biomaterials have been used in the treatment of various diseases through the use of biodegradable polymer compounds.

The most commonly used materials in the human body are: metals, ceramics and polymers, however biocomposites are manufactured by combining two or more materials with the aim of achieving improved biocompatibility and biomechanical properties for specific applications.

Biodegradable polymeric materials can be classified into natural and synthetic polymers; where natural polymers have shown problems of instability, immunogenicity and low biodegradability, however synthetic polymers such as polyurethanes (PUR) and poly (co-glycolic acid) (PLGA) offer excellent biocompatibility, being used in wound treatment, orthopedic devices, dental, cardiovascular, drug delivery and tissue engineering applications.

Synthetic polymers such as: polylactic acid (PLA), polyether ether ketone (PEEK) and polymethylmethacrylate (PMMA) are the polymers frequently used in medical applications (Alizadeh-osgouei et al., 2020).

PLA, being a biodegradable polymer, has been used in tissue engineering techniques and regenerative medicine treatment, due to its ability to promote tissue growth in bone graft procedures (DeStefano et al., 2020), such as, it was reported by Bae and collaborators in 2011, which showed that PLA combined with hydroxyapatite has shown to be a promising material in the field of orthopedics (Bae et al., 2011).

On the other hand, Fu et al. reported in 2021 the manufacture of composites with polyester/hydroxyapatite which, like PLA, has shown adequate mechanical resistance, biodegradability, favorable biological properties, good adhesion, and cellular interactions that also allow tissue repair, drug release. Various polyester/HA-based compounds made into microparticles, microspheres, membranes, scaffolds and lumps have been confirmed to have been applied in bone tissue repair, drug delivery and implant fixation.

It should be noted that the aforementioned polymers do not present toxicity in the human body since they gradually degrade within it, without residues, without stimulation and without toxic side effects for the tissues, for which they are approved by the FDA (Drug Administration and food) (Fu et al., 2021).

Likewise, in 2015 Maitz reported the application of biopolymers: Polyamide (PA), polycarbonate (PC), polyester and polymethylmethacrylate (PMMA) in membranes for hemodialysis; Maitz reported that this type of biomaterials has antifouling properties and blood compatibility (Maitz, 2015).

10. Blends of natural and synthetic polymer

Polymer blend is defined as the combination of two or more different polymers. It is necessary to know how to differentiate them from composites, since these are established as those systems that contain polymeric and non-polymeric materials. The mixture of polymers is also known as polyblends or merely blends (Asano, 2017). Synthetic and natural polymer blends is a class of polymeric materials that have not received due attention in the past (Giusti et al., 1995). Today the study of the interactions of natural and synthetic polymer mixtures is more evident due to their potential applications in a wide variety of fields (Maiti and Bidinger, 1981).

Natural polymers have excellent properties, among them is biodegradability which, in addition to its other characteristics, makes them capable of being used in a wide variety of applications, however, they have poor mechanical properties and compared to performance, processability and the price with those polymers obtained from petrochemical sources makes their use limited (Kadla and Kubo, 2004). Therefore, the mixture of natural and synthetic polymers is a convenient method for the development of new polymeric materials for specific purposes (Sionkowska, 2011; Avraam, 2010). Thus, polymer blending is one of the simplest ways to obtain a variety of desired physical and chemical properties by combining the constituent polymers, which is why polymer blending has become a traditional method for producing new ones. High-performance polymeric materials (Geetha and Ramesh, 2007). Since the resulting material could combine the appropriate mechanical correspondence of the synthetic polymer with the biocompatibility of the natural polymer (Cascone et al., 1995).

Polyblends are relatively inexpensive, light weight, easy to manufacture, have high resistance to corrosion, oxidation and the like. These types of polymeric materials are widely used in industrial and engineering applications around the world (Radovskiy and Teltayerv, 2018). The properties of the polymeric system can be modified either chemically or by mixing polymers, making the final material have better technological properties. In polymer blends there is a specific polymer-polymer interaction, which results in different properties than polymers alone. So, it is a suggested method to provide new, or improved material properties (Utracki, 1995). In addition, the properties of polymers are determined by the nature of the substituent, there are different reasons why the mixture of polymers is considered (Bakr and El-Kady, 1996):

- It is a simple way to combine the properties of different polymers.
- They offer attractive engineering applications.
- They are a good option in solving the problem of polymer residues.
- Properties can be maximized.
- In addition to offering an alternative to adapt biodegradability.

The mixture of polymers is one of the most important materials in polymer science, from a theoretical-practical point of view the mixtures are made to take advantage of technical advantages to obtain specific physical properties (Kim et al., 1996). Some of the factors that affect the properties of polymer blends are (Boucher et al., 1996):

- Geometry and size of polymer chains.
- Molecular weight.
- Branching of chains.
- Repeat units in the chain.

The physicochemical properties of polymer blends depend on multiple factors based on the polymer matrix present (Bucknall, 1977):

- Concentration and distribution.
- Composition of the mixture.

- Temperature at which it works.
- Relaxation characteristics.

The physical properties of the polymer mixture depend on the morphology and this in turn depends on the continuous and discontinuous phases, the degree of order of the crystalline or amorphous phases, considering that the microstructure affects the rheological properties of the mixture, therefore that the final properties are also affected. Then the microstructure becomes a key factor in the polymer mixture, consequently, the final morphology of the mixture is based on (Ibrahim Khan and Mazumdar, 2018):

- The characteristics of the materials:
  - Interfacial tension.
  - Elasticity.
  - Molecular weight.
  - Viscosity.
- Processing conditions
  - Mixing temperature.
  - Mixing time.
  - Mixer speed and type.
- Composition of the mixture.

The polymer blend can be homopolymers, copolymers or a mixture of both. Different commercially important mixtures are multicomponent in nature. The final properties of the mixture make it useful in a wide variety of commercial applications and these are strongly influenced by the interface and its morphological development. In mixing systems, the effective modification of properties depends on the miscibility or compatibility of the polymers involved, so the processing of polymer mixtures will depend on the interaction between them (Asano, 2017). There are three types of interaction forces between polymer mixtures, called dispersion, polar and hydrogen bonds (Kramer, 2008):

- *Dispersion*: these types of forces are the result of an atomic dipole formed by a positive nucleus and a negative electron cloud.
- *Polar*: it is divided into two types of permanent and induced dipoles.
- *Hydrogen bridges*: this interaction is due to the existing attraction of a hydrogen atom (positive charge) and an oxygen, fluorine or nitrogen atom (negative charge).

Polymer blends can be classified as follows (Wakabayashi et al., 2008):

- *Miscible mixtures*: polymers interpenetrate each other, symmetric interfaces form miscible mixtures, have good mechanical properties and good cohesion between phases.
- *Immiscible mixtures*: most of these types of mixtures have poor physical properties compared to separate polymers, they can also present different morphologies which depend on the composition, viscosity, elasticity, process conditions, among others.

Those immiscible polymer blends exhibit inferior mechanical properties, and this can be improved by physical or chemical compaction to obtain a better performing material.

Some of the advantages of the polymer blend can be summarized as follows (Utracki and Wilkie, 2014):

- Low cost of production.
- Good properties.

- Combine the properties that make individual polymers unique in multi-component systems.
- Improves mechanical properties such as impact resistance, resistance to cracking, resistance to tension, among others.

10.1 Polycarbonate

Polycarbonate is an aliphatic polyester type polymer with carbonate groups in its structure, which have good biocompatibility and impact resistance; To improve properties such as thermal stability and biodegradation, it is common to combine it with other polymers.

This polymer has a melting point of 100–110°C and its glass transition temperature is 150°C (Balaji et al., 2017).

Aliphatic polycarbonates are optimal materials for the functionalization of biomaterials since they have side chains with functional groups (OH, NH_2, COOH) (Tian et al., 2012).

Bisphenol A (BPA) is the main monomer used for the synthesis of polycarbonate, however it causes deficiencies in the endocrine system when it leaches (Yum et al., 2019).

Some of the applications that this polymer has is the administration of drugs since it has good biocompatibility and non-toxic degradation; polycarbonates degrade *in vivo* by surface erosion, in contrast to the degradation process of aliphatic polyesters. On the other hand, the degradation of this material does not lead to an increase in acidity in the body, which occurs during the degradation of polyester, being dangerous to drugs and healthy tissue.

The design of functionalized cyclic carbonate monomers has received greater interest in polymers and copolymers based on aliphatic polycarbonate that contain hydroxyl, carboxyl and amine groups, the physicochemical properties of these materials that stand out are improved hydrophobicity and biodegradability facilitating the administration of drugs (Chen et al., 2014).

Such is the case of Lv and collaborators in 2021 who carried out the synthesis of polycarbonate with the aim of using it in cardiovascular coatings; The synthesis of a cyclic carbonate monomer containing allyl ester residues, 5-methyl-5-alloxycarbonyl-1 and 3-dioxan-2-one (MAC) was carried out, then the compolymerization of cyclic carbonate with carbonate of 1,3-trimethylene (TMC) with the aim of introducing double bonds, finally 2-methacryloyloxyethylphosphophorylcholine thiolated was grafted to synthesize phospholipids and thus obtain bimimetic polycarbonate obtaining favorable results in contact with blood, being useful in the development of coatings for drugs (Lv et al., 2021).

10.2 Polyamides

A type of nylon that is synthesized from 2-pyrrolidone by ring-opening polymerization this type of polymers is biodegradable from sludge activated by the bacteria *Pseudomonas* sp., which cause the degradation of polyamides through hydrolysis (Yamano et al., 2014).

The biodegradability of nylon is related to the number of carbons in fatty acids; the nylon stockings commonly used are biodegradable polymers in the medical area due to their application in sutures and medical material; on the other hand, polyamides are naturally biodegradable since their GABA (γ-aminobutyric acid) component is a biogenic substance.

Polyamides have physical properties such as: thermo stability and resistance (Yamano et al., 2017).

In 2019, Bastos et al. carried out a conductive fabric used in biosensors using polyamides and natural fabrics such as cotton (Bastos et al., 2019).

10.3 Polylactic acid

PLA is a polyester of the aliphatic type obtained from lactic acid (2-hydroxypropanoic acid) which comes from starch and has application in the pharmaceutical area, which was synthesized in 1932 by Wallace Carother in the Dupont laboratories (Lukachan and Pillai, 2011).

This polymer is a biodegradable, bio-based thermoplastic, which can be synthesized from polycondensation, ring-opening polymerization and direct fermentation of agricultural products, such as starch and sugar (Adesina et al., 2020). The type of polymerizations for this biomaterial is described below:

- *Condensation polymerization*: the first stage of this process is the elimination of water obtained from the production of oligomers from lactic acid; the second stage is the chemical reaction between the oligomers and the catalyst; After this, the last step is the elimination of the water that is produced again by the condensation of the oligomers. Finally, the polymer obtained is cooled below its melting temperature, favoring crystallization.
- *Polymerization by ring opening*: the method used to obtain PLA of high molecular weight, which consists of: polycondensation, depolymerization and polymerization by ring opening (Lim et al., 2008).

The monomer that constitutes PLA is lactic acid (CH_3-CHOHCOOH); PLA can be found in two chiral configurations: L-lactic acid and D-lactic acid.

Biodegradable materials based on polylactic acid are used in biomedicine and tissue engineering due to their degradation and biocompatibility in biological media (Kurzina et al., 2020).

Degradation of PLA occurs through hydrolysis of the polymer chain; the rate of hydrolysis depends on the molecular weight, crystallinity, morphology and diffusion rate of the water. Due to its slow degradation rate, PLA remains *in vivo* for 3–5 years; However, it is possible to accelerate the degradation process, increasing the temperature and acidity, attributing the molecular weight to this process as well.

Among the most important applications for this material are: tissue engineering or regenerative medicine, cardiovascular implants, drug devices as well as tools and medical equipment (DeStefano et al., 2020).

11. Conclusions

There is actually the need in the use of renewable sources as it becomes more elemental and important for the design, obtaining and/or synthesis of polymeric materials, which can be of natural origin or combination of these, natural/synthetic. However, it is important to consider the advantages and disadvantages of the use of these types of materials, always taking care of and respecting the environment. The use of renewable raw materials provides multiple advantages, if anything, its use should be moderate so as not to affect the environment.

Acknowledgements

The authors thank the CONACyT, for the support through the project SEP-CB-2017-2018 A1-S-44977 and Universidad Autónoma de Coahuila, Posgrado en Ciencia y Tecnología de Materiales.

References

Adesina, O. T., Sadiku, E. R., Jamiru, T., Adesina, O. S., Ogunbiyi, O. F., Obadele, B. A. and Salifu, S. (2020). Polylactic acid/graphene nanocomposite consolidated by SPS technique. *Journal of Materials Research and Technology*, 9: 11801–11812.

Alizadeh-osgouei, M., Li, Y. and Wen, C. (2020). Bioactive materials a comprehensive review of biodegradable synthetic polymer-ceramic composites and their manufacture for biomedical applications. *Bioactive Materials*, 4: 22–36.

Asano, A. (2017). Polymer blends and composites. *In*: *Modern Magnetic Resonance*.

Avraam I. Isayev. (2010). *Encyclopedia of Polymer Blends*, Volume 1 Fundamentals.

Bae, J., Won, J., Park, J., Lee, H. and Kim, H. (2011). Improvement of surface bioactivity of poly (lactic acid) biopolymer by sandblasting with hydroxyapatite bioceramic. *Materials Letters*, 65: 2951–2955.

Bakr, N. A. and El-Kady, M. (1996). Mechanical and optical investigations of some polymer blends containing PVC. *Polymer Testing*, 15(3): 281–289.

Balaji, A. B., Pakalapati, H., Khalid, M., Walvekar, R. and Siddiqui, H. (2017). Natural and synthetic biocompatible and biodegradable polymers. Elsevier. pp. 1–30. *In*: *Biodegradable and Biocompatible Polymer Composites: Processing, Properties and Applications*. Malasya.

Barron, A. and Sparks, T. D. (2020). Commercial marine-degradable polymers for flexible packaging. *Iscience* 23: 1–13.

Bastioli, C. (2005). *Handbook of Biodegradable Polymers*. Walter de Gruyter Gmbh, Berlin/Boston.

Bastos, A. R., Pereira da Silva, L., Gomes, V. P., Lopes, P. E., Rodrigues, L. C., Reis, R. L. and Souto, A. P. (2019). Electroactive polyamide/cotton fabrics for biomedical applications. *Organic Electronics*, 77: 105401.

Benhabiles, M. S., Salah, R., Lounici, H., Drouiche, N., Goosen, M. F. A. and Mameri, N. (2012). Antibacterial activity of chitin, chitosan and its oligomers prepared from shrimp shell waste. *Food Hydrocolloids*, 29(1): 48–56.

Bhagabati, P. (2020). Biopolymers and biocomposites-mediated sustainable high-performance materials for automobile applications. pp. 197–2016. *In*: Elsevier. *Sustainable Nanocellulose and Nanohydrogels from Natural Sources*. India.

Brás, T., Rosa, D. and Gonçalves, A.C. (2020). Development of bioactive films based on chitosan and Cynara cardunculus leaves extracts for wound dressings. *Int. J. Biol. Macromol.* 163: 1707–1718.

Boucher, E., Folkers, J. P., Hervet, H., Léger, L. and Creton, C. (1996). Effects of the formation of copolymer on the interfacial adhesion between semicrystalline polymers. *Macromolecules*, 29(2): 774–782.

Bucknall, C. (1977). Toughened Plastics. 48090.

Carvalho, A. P. A. and Conte Junior, C. A. (2020). Green strategies for active food packagings: A systematic review on active properties of graphene-based nanomaterials and biodegradable polymers. *Trends in Food Science and Technology*, 103: 130–143.

Cascone, M. G., Sim, B. and Sandra, D. (1995). Blends of synthetic and natural polymers as drug delivery systems for growth hormone. *Biomaterials*, 16(7): 569–574.

Chen, W., Meng, F., Cheng, R., Deng, C., Feijen, J. and Zhong, Z. (2014). Advanced drug and gene delivery systems based on functional biodegradable polycarbonates and copolymers. *Journal of Controlled Release*, 190: 398–414.

Chiellini, E. (2008). *Environmentally Compatible Food Packaging*. Woodhead publishing limited, Cambridge England 8–10.

Chisenga, S. M., Tolesa, G. N. and Workneh, T. S. (2020). Biodegradable food packaging materials and prospects of the fourth industrial revolution for tomato fruit and product handling. *International Journal of Food Science*, 1–17.

Clark, J. H. (2013). Bio-inspired Polymers Series Editors: Titles in the Series: (Vol. 1).

Company, F. M. (1986). Composite structures for automobiles. *Composite Structures*, 5: 163–176.

Cooper, C. J. and Cooper, C. J. (2018). A biobased multifaceted polymeric material: a case for poly (butylene succinate).

Dake, M. (2015). *Microbial Factories, Biodiversity, Biopolymers, Bioactive Molecules: Biodegradable Polymers: Renewable Nature, Life Cycle, and Application* (volume two). London, Springer.

DeStefano, V., Khan, S. and Tabada, A. (2020). Applications of PLA in modern medicine. *Engineered Regeneration*, 1: 76–87.

Dinoop, L. S., Sunil, J. T., Rajesh, C. and Arun, K. J. (2021). *Accelerated Photodegradation of Solid Phase Polystyrene by Nano Tio_2-Graphene Oxide Composite under Ultraviolet Radiation.* Elsevier, Polymer Degradation and Stability, 184: 1–15.

Dodane, V. and Vilivalam, V. D. (1998). Pharmaceutical applications of chitosan. *Pharmaceutical Science & Technology Today*, 1(6): 246–253.

Frackowiak, S., Ludwiczak, J. and Leluk, K. (2018). Man-made and natural fibres as a reinforcement in fully biodegradable polymer composites: a concise study. *J. Polym. Environ*, 26: 4360–4368.

Fu, Z., Cui, J., Zhao, B., Gf, S. and Lin, K. (2021). An overview of polyester/hydroxyapatite composites for bone tissue repairing. *Journal of Orthopaedic Translation*, 28: 118–130.

Gan, P.G., Sam, S.T., Abdullah, M.F., Omar, M.F. and Tan, W.K. (2021). Water resistance and biodegradation properties of conventionally-heated and microwave-cured cross-linked cellulose nanocrystal/chitosan composite films. *Polym Degrad Stab*., 188: 109563. doi: 10.1016/j.polymdegradstab.2021.109563.

Geetha, D. and Ramesh, P. S. (2007). Ultrasonic studies on polymer blend (natural/synthetic) in strong electrolyte solutions. *Journal of Molecular Liquids*, 136(1–2): 50–53.

Gijsman, P., Hensen, G. and Mak, M. (2021). Thermal initiation of the oxidation of thermoplastic polymers (polyamides, polyesters and UHMwPE). Elsevier, *Polymer Degradation and Stability*, 183: 1–17.

Giusti, P., Lazzeri, L., Cascone, M. G., Seggiani, M., Chimica, I. and Pisa, U. (1995). Macromolecular. *Symposia*, 100: 81–87, 87: 81–87.

Gooch, J. W. (2011). Natural polymers. *In*: *Encyclopedic Dictionary of Polymers*.

Hamedi, H., Moradi, S., Hudson, S. M. and Tonelli, A. E. (2018). Chitosan based hydrogels and their applications for drug delivery in wound dressings: A review. *Carbohydrate Polymers*, 199: 445–460.

Harsha Kharkwal, Bhanu Malhotra and Janaswamy, S. (1981). Natural polymer for drug delivery: an introduction. *Journal of Chemical Information and Modeling*, 53(9): 1689–1699.

Huang, H., Zhang, C., Rong, Q., Li, C., Mao, J., Liu, Y. and Chen, J. (2020). Effect of two organic amendments on atrazine degradation and microorganisms in soil. Elsevier, *Applied Soil Ecology*, 152: 1–8.

Ibrahim Khan, M. A. J. and Mazumder. (2018). Polymer Blends Polymer Blends. In *Introduction to Polymer Compounding: Raw Materials* (Vol 1) (Vol. 1, IssueAugust).

Ivankovic, K. Zeljko, S. Talic, A., Bevanda, M. and Lasic, M. (2017). Biodegradable packaging in the food industry. *Archiv für Lebensmittelhygiene*, 68: 26–38.

Joshi, M. (2008). The impact of nanotechnology on polyesters, polyamides and other textiles. *Polyesters and Polyamides*, 354–415.

Kadla, J. F. and Kubo, S. (2004). Lignin-based polymer blends: Analysis of intermolecular interactions in lignin-synthetic polymer blends. *Composites Part A: Applied Science and Manufacturing*, 35(3): 395–400.

Kaur, G. (2017). Polymers as bioactive materials II: synthetic/biodegradable polymers and composites. *In*: *Bioactive Glasses. Serie in Bioengineering*. Springer, Cham.

Kim, K., Cho, W. and Ha, C. (1996). Properties of dynamically vulcanized EPDM and LLDPE blends. *Journal of Applied Polymer Science*, 59(3): 407–414.

Kim, Y. N., Ha, Y. M., Park, J. E., Kim, Y. O., Jo, J. Y., Han, H., ... Jung, Y. C. (2021). Flame retardant, antimicrobial, and mechanical properties of multifunctional polyurethane nanofibers containing tannic acid-coated reduced graphene oxide. *Polymer Testing*, 93: 107006.

Kramer, E. J. (2008). Effect of End-Anchored Chains on the Adhesion at a Thermoset-Thermoplastic Interface. 1999–2008.

Kundys, A., Bialecka, F. E. and Fabiszewska, A. (2018). Candida antarctica lipase B as catalyst for cyclic esters synthesis, their polymerization and degradation of aliphatic polyesters. *J. Polym. Environ*, 26: 396–407.

Kurzina, I. A., Laput, O. A., Zuza, D. A., Vasenina, I. V., Salvadori, M. C., Savkin, K. P. and Kalashnikov, M. P. (2020). Surface property modification of biocompatible material based on polylactic acid by ion implantation. *Surface and Coatings Technology*, 388: 125529.

Laycock, B., Nikolic, M., Colwell, J. M., Gauthier, E., Halley, P., Bottle, S. and George, G. (2017). Lifetime prediction of biodegradable polymers. Elsevier, *Progress in Polymer Science*, 71: 144–189.

Leja, K. and Lewandowicz, G. (2010). Polymer biodegradation and biodegradable polymers—A review. *Polish Journal of Environmental Studies*, 19(2): 255–266.

Lim, L. T., Auras, R. and Rubino, M. (2008). Processing technologies for poly(lactic acid). *Progress in Polymer Science* (Oxford), 33: 820–852.

Luckachan, G. E. and Pillai, C. K. S. (2011). Biodegradable polymers—a review on recent trends and emerging perspectives. *Journal of Polymers and the Environment*, 19: 637–676.

Lv, D., Li, P., Zhou, L., Wang, R., Chen, H., Li, X. and Huang, N. (2021). Synthesis, evaluation of phospholipid biomimetic polycarbonate for potential cardiovascular stents coating. *Reactive and Functional Polymers*, 163: 104897.

Maiti and Bidinger. (1981). Blends of natural and synthetic polymers: a new route to novel biomaterials. *Journal of Chemical Information and Modeling*, 53(9): 1689–1699.

Maitz, M. F. (2015). Applications of synthetic polymers in clinical medicine. *Biosurface and Biotribology*, 1: 161–176.

Mohamad, N., Mazlan, M. M., Tawakkal, I. S. M. A., Talib, R. A., Kian, L. K., Fouad, H. and Jawaid, M. (2020). Development of active agents filled polylactic acid films for food packaging application. *International Journal of Biological Macromolecules*, 163: 1451–1457.

Muhammad, A., Rahman, M. R., Baini, R. and Bin Bakri, M. K. (2021). Applications of sustainable polymer composites in automobile and aerospace industry. pp. 85–207. *In*: Rahman, R.M. (ed.). *Advances in Sustainable Polymer Composites*. Malaysia.

Nazin Sultana. (2013). *Biodegradable Polymer Based Scaffolds for Bone Tissue Engineering*. Vol. 1. Springer.

Nottelet, B., Darcos, V. and Coudane, J. (2015). Aliphatic polyesters for medical imaging and theranostic applications. Elsevier, *European Journal of Pharmaceutics and Biopharmaceutics*, 97: 350–370.

Odian, G. (2004). *Principles of Polymerization: Polymerization Mechanism* (Fourth Edition). New Jersey, Published by John Wiley & Sons, Inc.

Papadopoulou, E. and Chrissafis, K. (2017). Particleboards from agricultural lignocellulosics and biodegradable polymers prepared with raw materials from natural resources. *Natural Fiber-Reinforced Biodegradable and Bioresorbable Polymer Composites*, 19–30.

Pathak, V. M. and Navneet. (2017). Review on the current status of polymer degradation: a microbial approach. *Bioresources and Bioprocessing*, 4: 1–31.

Polyak, P., Urban, E., Nagy, G. N., Vertessy, B. G. and Pukanszky, B. (2019). The role of enzyme adsorption in the enzymatic degradation of an aliphatic polyester. Elsevier, *Enzyme and Microbial Technology*, 120: 110–116.

Radovskiy, B. and Teltayev, B. (2018). Viscoelastic properties. *Structural Integrity*, 2: 1–22.

Rai, P., Mehrotra, S., Priya, S., Gnansounou, E. and Sharma, S. K. (2021). Recent advances in the sustainable design and applications of biodegradable polymers. Elsevier *Bioresource Technology*, 325: 1–12.

Romatowska, M. P., Haponiuk, J. and Formela, K. (2020). Reactive extrusion of biodegradable aliphatic polyesters in the presence of free-radical-initiators: a review. Elsevier, *Polymer Degradation and Stability*, 182: 1–19.

Rydz, J., Musioł, M., Zawidlak-w, B. and Sikorska, W. (2018). Polymers for food packaging applications. *In*: Alexandru Grumezescu Alina Maria Holban. *Biopolymers for Food Design*.

Sackey, J., Fell, A., Ngilirabanga, J. B., Razanamahandry, L. C., Ntwampe, S. K. O. and Nkosi, M. (2019). Antibacterial effect of silver nanoparticles synthesised on a polycarbonate membrane. *Materials Today: Proceedings*, 36: 336–342.

Samrot, A. V., Sean, T. C., Kudaiyappan, T., Bisyarah, U., Mirarmandi, A., Faradjeva, E., ... Suresh Kumar, S. (2020). Production, characterization and application of nanocarriers made of polysaccharides, proteins, bio-polyesters and other biopolymers: A review. *International Journal of Biological Macromolecules*, 165: 3088–3105.

Sánchez, C. (2020). Fungal potencial for the degradation of petroleum-based polymers: an overview of macro- and microplastics biodegradation. Elsevier, *Biotechnology Advances*, 40: 1–12.

Shockley, M. F. and Muliand, A. H. (2020). Modeling temporal and spatial changes during hydrolytic degradation and erosion in biodegradable polymers. Elsevier, *Polymer Degradation and Stability*, 180: 1–15.

Siddhi, S. P. and Dilip, V. V. (2020). Biodegradable polymeric materials: synthetic approach. *ACS Publication*, 5: 4370–4379.

Sionkowska, A. (2011). Current research on the blends of natural and synthetic polymers as new biomaterials: Review. *Progress in Polymer Science* (Oxford), 36(9): 1254–1276.

Thomason, J. L. and Fernandez, J. L. (2021). Thermal degradation behaviour of natural fibres at thermoplastic composite processing temperatures. Elsevier, *Polymer Degradation and Stability*, 188: 1–10.

Tian, H., Tang, Z., Zhuang, X., Chen, X. and Jing, X. (2012). Biodegradable synthetic polymers: preparation, functionalization and biomedical application. Elsevier *Progress in Polymer Science* (Oxford), 37: 237–280.

Tokiwa, Y. and Calabia, B. P. (2015). *Biodegradable Polymers*. Encyclopedia of Polymeric Nanomaterials. Springer, Berlin.

Utracki, L. A. (1995). History of commercial polymer alloys and blends (from a perspective of the patent literature). *Polymer Engineering & Science*, 35(1): 2–17.

Utracki, L. A. and Wilkie, C. A. (2014). Polymer blends handbook. *In*: *Polymer Blends Handbook*.

Villada, H., Acosta, H. A. and Velasco, R. J. (2007). Biopolymers naturals used in biodegradable packaging. *Journal of the American Chemical Society*, 12(4): 5–13.

Vroman, I. and Tighzert, L. (2009). Biodegradable polymers. *Materials*, 2(2): 307–344.

Wakabayashi, K., Pierre, C., Diking, D. A., Ruoff, R. S., Ramanathan, T., Catherine Brinson, L. and Torkelson, J. M. (2008). Polymer - Graphite nanocomposites: Effective dispersion and major property enhancement via solid-state shear pulverization. *Macromolecules*, 41(6): 1905–1908.

Witko, A. (2003). Edited by Edited by. In World (Vol. 3, Issue February 2004).

Wu, F., Misra, M. and Mohanty, A.K. (2021). Challenges and new opportunities on barrier performance of biodegradable polymers for sustainable packaging. *Prog Polym Sci*. Published online April 20, 101395.

Wujcik, E. K. and Monty, C. N. (2013). Nanotechnology for implantable sensors: Carbon nanotubes and graphene in medicine. Wiley *Interdisciplinary Reviews: Nanomedicine and Nanobiotechnology*, 5: 233–249.

Xie, J. and Hung, Y. C. (2018). UV-A activated TiO_2 embedded biodegradable polymer film for antimicrobial food packaging application. *Lwt*, 96(March): 307–314.

Yamano, N., Kawasaki, N., Oshima, M. and Nakayama, A. (2014). Polyamide 4 with long-chain fatty acid groups - Suppressing the biodegradability of biodegradable polymers. *Polymer Degradation and Stability*, 108: 116–122.

Yamano, N., Kawasaki, N., Ida, S., Nakayama, Y. and Nakayama, A. (2017). Biodegradation of polyamide 4 *in vivo*. *Polymer Degradation and Stability*, 137: 281–288.

Yin, G. Z. and Yang, X. M. (2020). Biodegradable polymers: a cure for the planet, but a long way to go. *Journal of Polymers Research*, 27: 38.

Yum, S., Kim, H. and Seo, Y. (2019). Synthesis and characterization of isosorbide based polycarbonates. *Polymer*, 179: 121685.

Zapata, D., Pujol, R. and Coda, F. (2012). Polímeros biodegradables: una alternativa defuturo a la sostenibilidad del medio ambiente. *Técnica Industrial*, 297: 76–80.

Zhong, Y., Godwin, P., Jin, Y. and Xiao, H. (2020). Biodegradable polymers and green-based antimicrobial packaging materials: A mini-review. *Adv. Ind. Eng. Polym. Res.*, 3(1): 27–35.

Zhong, Y., Godwin, P., Jin, Y. and Xiao, H. (2020). Biodegradable polymers and green-based antimicrobial packaging materials: a mini-review. *Advanced Industrial and Engineering Polymer Research*, 3: 27–35.

Chapter 3

Plastics Technology

Marlene Lariza Andrade-Guel,[1] *Alma Berenice Jasso-Salcedo,*[2]
Diana Iris Medellín-Banda,[1] *Marco Antonio De Jesus-Tellez*[1] and
Christian Javier Cabello-Alvarado[1,3,*]

1. Introduction

The synthetic polymer industry has contributed enormously to the development of many areas, science, life, health care, technology in the last years. Polymers are used as the fundamental materials to prepare many different products in a wide range of applications. They are ubiquitous because of their excellent properties ranging from light weight, flexibility, moisture resistance, high impact strength, moldability, corrosion and chemicals resistance and, relatively low cost. However, and contrast to their durable nature, their persistence in the environment after disposal has created serious environmental problems because of their resistance to degradation, particularly biodegradation.

Today, the consumer comes into daily contact with all kinds of polymer-based materials, most of which are of the non-biodegradable type. Plastic is the most prevalent type of material debris that exists in the world today and is found in all environments including oceans, beaches, drainage systems, rivers, forests, in sparsely inhabited to overpopulated lands. Lately, scientists, authorities, world leaders and people are concerned about ways to deal with solid plastics waste management. Despite all the efforts to resolve the problem of contamination, such as incineration, recycling, filling procedures, source reduction, and so on, the trend does not seem to be abating. Even when plastics are recycled, they never truly leave the environment but are present as smaller pieces called microplastics are five millimeters or less in length.

The aim of this chapter will be focused on biodegradable polymers in terms of their relevance and to summarize the main mechanisms of reactions of their preparation and degradation and provide some current applications. To produce biodegradable polymers at an industrial level, it is necessary to develop an adequate manufacturing process, which must be affordable, efficient and timely. One such method of preparation is polymer extrusion. It is a widely used technique because in the process high viscosity materials can be managed without solvents, saving costs and eliminating volatiles. Reactive extrusion processes offer natural means for polymerization, chemical

[1] Centro de Investigación en Química Aplicada (CIQA), Departamento de Materiales Avanzados, Saltillo, Coahuila, México.
[2] CONACYT-Centro de Investigación en Química Aplicada (CIQA), Departamento de Biociencias y Agrotecnología, Saltillo, Coahuila, México.
[3] CONACYT-Centro de Investigación y de innovación del Estado de Tlaxcala (CITLAX), Tlaxcala de Xicoténcatl, Tlaxcala, México.
* Corresponding author: christian.cabello@ciqa.edu.mx

crosslinking and grafting through the inherent ability to stage reactants. Additionally, in recent decades, membranes made from biodegradable polymers have attracted a lot of attention, because their preparative technology can be chemically used in a wide variety of applications.

The rediscovering of 3-dimensional architectures made by biopolymers in nature has changed the paradigm of polymers structuring in 2-dimensions and proposed a way to overcome the upscaling limitations. This chapter also highlights the advances in the manufacturing of monoliths and scaffolds of biodegradable polymers by 3D printing techniques such as fused deposition melting and selective laser sintering; and the advantages of greenest Deep Eutectic Solvents (DES) and Ionic Liquids (ILs) for the preparation of 3-dimensional foams and aerogels.

2. Background principles

Biodegradable Polymers (BPs) are an excellent alternative with which to replace polymers and plastic, materials that undergo extremely slow degradation that leads to continuous and unwanted damage to the environment (Tokiwa and Calabia, 2007). BPs allow useful biomaterials that possess much shorter life cycles to avoid such deleterious environmental effects. The development of polymers non-resistance to degradation mainly in areas like food packaging (Sorrentino et al., 2007; Seppälä et al., 2004), hygiene products, agricultural uses (Serrano-Ruiz et al., 2021) and biomedical applications such as surgery, drug-controlled release, prosthetics, tissue engineering are at an all-time high (Luckachan and Pillai, 2011; Bei et al., 1997; Bruck, 1981; Kariduraganavar et al., 2014). Biodegradable polymers belong to the family of polymers, that are made of long chains of repeating molecular units and that mostly contain functional groups such acetal, amide, anhydride, azo, carbonates, ester, urethanes, phosphates and glycoside to name a few (Pandey et al., 2019; Ulery et al., 2011).

Generally, biodegradable polymers are classified based on the source of their origin (Sorrentino et al., 2007).

- Those extracted from biomass such as proteins, polypeptides polynucleotides and polysaccharides (Becker and Wurm, 2018; Wróblewska-Krepsztul et al., 2018). These types of polymers are normally degraded in biological systems by hydrolysis followed by oxidation degradation (Hocking, 1992).
- Those produced from the activity of microorganisms such as polyhydroxyalkanoates, polyhydroxy butyrate. Unlike non-biodegradable polymers, the long chains of biodegradables polymers are a source of carbon for useful microorganisms (Maddever and Chapman, 1989).
- Biodegradation of macromolecules from microorganisms takes place through the action of bacteria, fungi or algae leaves no toxic residues (Gross and Kalra, 2002; Amass et al., 1998; Sperling and Carraher, 1985).
- Those chemically synthesized using renewable biobased monomers such as polycaprolactone and polyglycolic acid (Mangaraj et al., 2019). Biodegradation processes occur because of hydrolytic and/or enzymatic chains scission leading to hydroxyacids (Brannigan and Dove, 2017).
- Those chemically synthesized from bio-derived monomers. Among the most biodegradable synthetic polymers are Poly Lactic Acid (PLA), Poly Glycolic Acid (PGA), poly(butylene succinate), polycaprolactone (PCL), and polydioxanone (PDO) (Ncube et al., 2020; Brannigan and Dove, 2017).

3. Synthesis of biodegradable polymers

BPs are mainly synthesized employing techniques such as polycondensation and Ring-Opening Polymerization (ROP); each methodology possesses its advantages and disadvantages (Tschan et al., 2012; Tian et al., 2012). For example, condensation reactions are achieved between diols and

Table 3.1. Classes of biodegradable polymers, chemical structure and their acronyms.

Polymer class	Repeating unit	Biopolymer	Applications	Ref.
Polyacetals	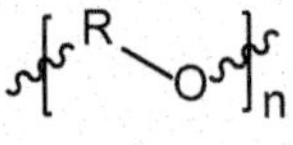	Poly formaldehyde	Engineering plastic. Delivery of vaccines, prostheses	(Murthy et al., 2003; Brin et al., 2008; Kipper et al., 2006; Salman et al., 2009)
Polyanhydrides	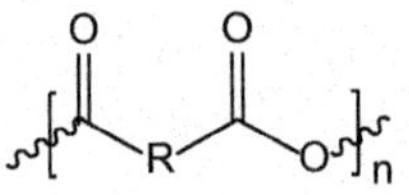	PCPX	Delivery of chemotherapeutics, antibiotics and vaccines	(Carrillo-Conde et al., 2010; Tamayo et al., 2010; Leong et al., 1987; Vasanthan, 2009; Kluin et al., 2009; Asikainen et al., 2005; Brin et al., 2008; Kipper et al., 2006; Salman et al., 2009)
Polyamides	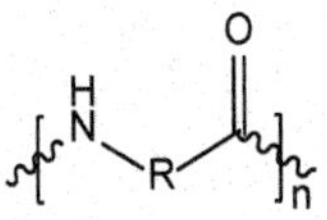	Hydroxylated nylon	Textile applications	(Vasanthan, 2009)
Policarbonates	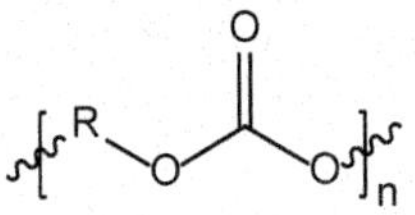	PTMC PDTE	Delivery of angiogenic agents and antibiotics. Tissue engineering of bone, vasculature, and muscle	(Roldughin et al., 2016; Asikainen et al., 2006; Briggs et al., 2009; Pospíšil et al., 1998; Johnson et al., 2010; Kluin et al., 2009; Asikainen et al., 2005)
Polyesters	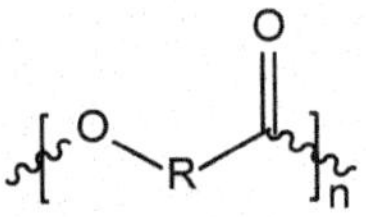	PGA PLLA PLGA PHB PCL PPF	Tissue regeneration, drug delivery, stent deliver of chemotherapeutics, bone screws and plates. Filling bone defects	(Wang et al., 2010; Dunne et al., 2010; Pihlajamäki et al., 2010; Erggelet et al., 2007; Mahmoudifar and Doran, 2010; Andrady and Neal, 2009; Pihlajamäki et al., 2007; Xu et al., 2010; Lu et al., 2008; Lensen et al., 2010; Betancourt et al., 2007; Liu et al., 2010; Darney et al., 1989; Christenson et al., 2007; Young et al., 2009)
Poly iminocarbonates	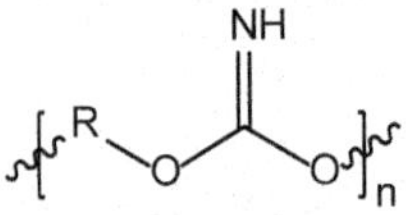	Poly(BPA-iminocarbonate)	Food packaging	(Pulapura et al., 1990; Li and Kohn, 1989)

Table 3.1 contd. ...

...Table 3.1 contd.

Polymer class	Repeating unit	Biopolymer	Applications	Ref.
Polyorthoesters	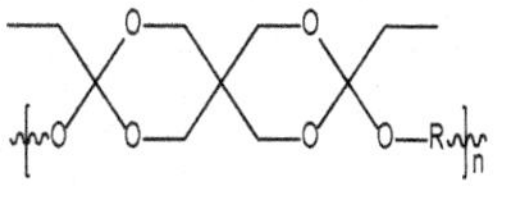	DETOSHU-HD	Delivery of analgesics, DNA vaccines and antiproliferative drugs	(Qi et al., 2008; Wang et al., 2004; Nguyen et al., 2008; Polak et al., 2008)
Polyphosphazenes	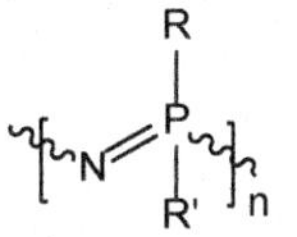	Poly [bis(trifluoroethoxy) phosphazene]	Stent coating	(Radeleff et al., 2008)
Polyphosphoesters	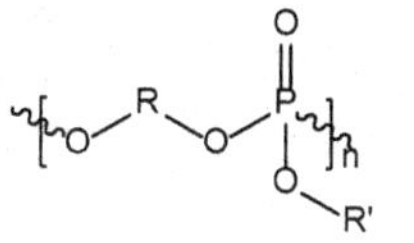	Poly (propylene phosphonate)	Drug delivery system	(Brosse et al., 1989)
Polyurethanes	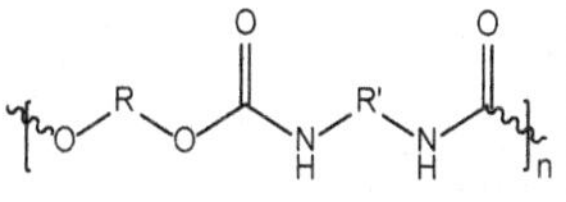	Pellethane®	Tissue engineering	(Ulery et al., 2011)

dicarboxylic acid derivatives or bifunctional molecules (i.e., HO-R-COOH). These reactions can be achieved with or without the presence of the catalyst and do not require anhydrous reaction conditions or detailed purification of the raw materials. Nevertheless, low values in molar mass are obtained with a wide range of dispersity (Đ); moreover, the extraction of condensation subproducts is required to avoid the equilibrium displacement to reagent (Edlund and Albertsson, 2003; Nair and Laurencin, 2007). On the other hand, ROP is perhaps the most employed methodology to obtain biodegradable polymers because it can be and is applied to a wide range of cyclic monomers (i.e., lactides, carbonates, lactams, phosphoesters, phosphazenes, lactones and derivatives). These polymers are synthesized from low to high values of molar mass and near Đ in short periods with the use of an organic or organometallic catalyst (Santoro et al., 2020; Toshikj et al., 2020). It is worth mentioning that the presence of oxygen and humidity in ROP plays an unfavorable role in affecting the deactivation of the catalyst or low reaction yield. Therefore, care should be taken in the handling of some catalysts by using a glovebox and deoxygenated solvents and reagents dried before the reaction to achieve an acceptable conversion of the monomers (Thomas et al., 2012). Table 3.1 summarizes the different polymer classes of BPs along with their applications.

4. Mechanisms of degradation of biodegradable polymers

A short overview of the principle mechanisms of polymer degradation is covered here. Degradation mechanisms refer to the transformation of a product, generally into other products with noticeable changes in its properties and the end of its useful lifetime as the original material. One can find in literature several terms related to phenomena such as degradation, decomposition, disintegration, erosion and resorption associated with the biodegradation of polymeric materials.

Degradation in polymers is defined as "a deleterious change in the chemical structure, physical properties, or appearance of a polymer, which may result from chemical cleavage of the macromolecules forming a polymeric item, regardless of the mechanism of chain cleavage" (Pospíšil et al., 1998).

An illustration of degradation routes and mechanisms is shown in Fig. 3.1.

Polymeric properties, as well as processing, can influence degradation behavior, including chemical structure, chemical composition, molecular weight, molecular weight distribution, degree

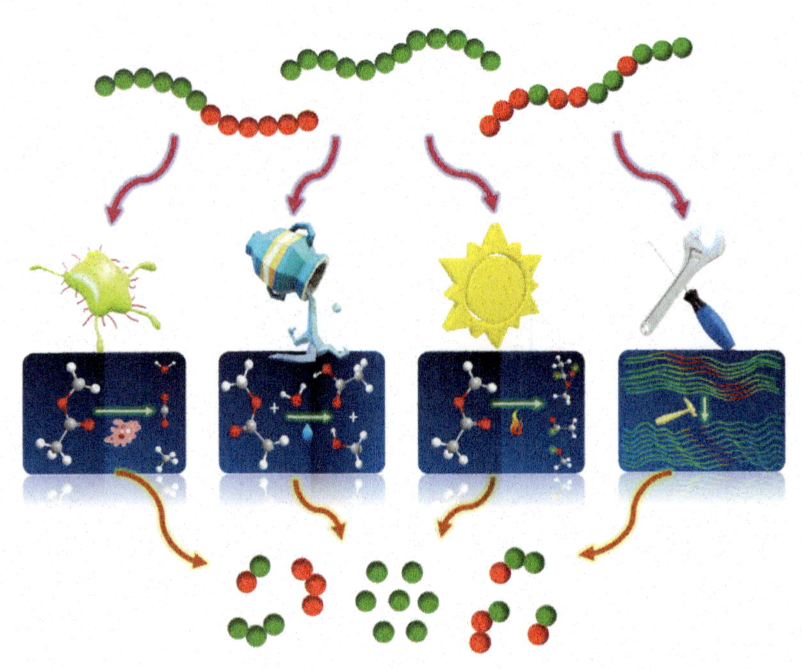

Figure 3.1. Schematic illustration of degradation paths and mechanisms (Authors).

of polymerization, the hydrophilicity-hydrophobicity balance, water-solubility, crystallinity, glass transition temperature, functional groups (ionic), configuration structure, surface area, processing conditions (additives, charges, stabilizers) porosity and water diffusion (Singh and Sharma, 2008; Pandey et al., 2019; Becker and Wurm, 2018). Also, initial molecular weight significantly affects degradation (Gleadall et al. 2014). The degradation process takes place in acidic or basic conditions and is influenced by the polymer's surroundings such as climatic conditions, soil characteristics, presence and population of microorganisms, humidity, temperature, solar radiation, soil nutrients, and so on. A combination of these may be cumulative. Herein, the biodegradation of polymers refers to the process by which organic substances are broken down into compounds by living microbial organisms. Some biodegradable polymers go through bulk degradation while others tend to degrade by surface erosion. Deterioration generally leads to waste microplastics. Bulk degradation behavior is characterized by the hydrolysis of chemical bonds in the polymer chain. The material shows compositional changes, which can result in the formation of empty shells called carapaces. This process entails mass loss throughout the only with the material however maintaining its size (Li, 1999). Polylactides and aliphatic polyesters exhibit bulk degradation (Woodard and Grunlan, 2018). Erosion degradation usually is limited to the surface of a polymer. The process starts with chain cleavage, which results in the production of oligomers and monomers; the polymer undergoes a mass loss and size reduction (Tamada and Langer, 1993). This phenomenon occurs in polymers with functional groups that hydrolyze in an aqueous environment where water is consumed mainly on the surface of the polymers by hydrolysis. Poly(anhydrides) and poly(orthoesters) (Heller, 1994) are examples of surface-eroding biodegradable polymers and are mainly used in biomedical applications pharmacology, orthopedic, etc. The degree of crystallinity of polymers plays an important role in the degradation process, crystalline with regions being impermeable to water; the presence of amorphous regions allows for the diffusion of water (Woodard and Grunlan, 2018). Once water enters into the amorphous parts, hydrolytic bond cleavage occurs to form water-soluble fragments; thereafter, moisture penetrates into the crystalline cells of the polymer. These types of polymer degradation are illustrated in Fig. 3.2.

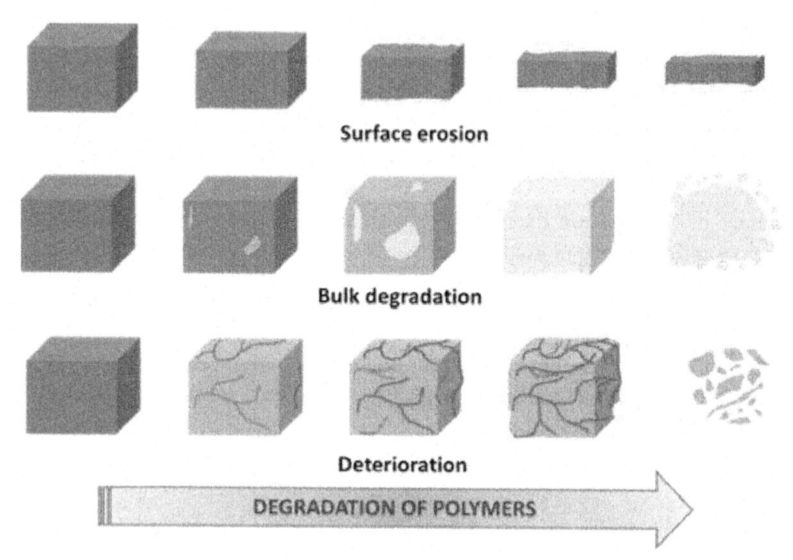

Figure 3.2. Schematic illustration of three types of polymer degradation surface erosion, bulk degradation and deterioration (Authors).

4.1 Degradation by biological activity

4.1.1 Biodegradation

A biodegradable polymer can be defined as a polymer whose labile chemical structure undergoes chain fragmentation resulting in molecules of low molecular mass, until its complete degradation to biomass and metabolic products under suitable conditions. Polymer biodegradation polymers processes consist of achieving changes in the chemical structure of the macromolecule into low molecular mass (Jayasekara et al., 2005). The first stage of the process consists of scission of the long chains of the polymer backbone, a process called depolymerization, which involves complex biotic and abiotic phenomena to degrade the organic matter. Chain fragmentation can be initiated by hydrolysis or enzyme-catalyzed hydrolysis (hydrolases) (Monsigny et al., 2018; Brannigan and Dove, 2017). Once the fragmentation of the chains has been achieved, products are subjected to digestion by microorganisms (bacteria, fungi or algae) (Song et al., 2011; Conato and Sumera, 2012; Bastioli, 2014). Short chains like oligomers and monomers are more easily degraded and mineralized.

Biodegradation takes place in two different conditions depending upon the presence of oxygen in the medium.

1) Microbiological activity decomposes and minerals to eventually yield carbon dioxide and water under aerobic conditions besides microbial biomass (Amann and Minge, 2012; Jayasekara et al., 2005). Notably, the outcome of the complete process is the transformation into carbon dioxide (CO_2).
2) Anaerobic biodegradation is the process of degradation of polymeric material by biological activity in the absence of oxygen resulting in carbon dioxide, water, methane, organic acids and biomass (Letcher, 2020). Both these degradations can occur in the environment under certain circumstances.

a) Aerobic biodegradation

Polymeric material + $O_2 \rightarrow CO_2 + H_2O$ + Biomass

b) Anaerobic biodegradation

Polymeric material $\rightarrow CO_2 + CH_4 + H_2O$ + Biomass

To know the degree of biodegradation in the materials of interest, and the efficiency of the process, one considers the time taken for mineralization, which is equivalent to the measurement of the oxygen consumed in the first stages of the process (Laycock et al., 2017). There are several methods to estimate the level of biodegradability of polymeric materials, which is a function of the oxygen uptake during the first stage of the process under aerobic conditions (Jayasekara et al., 2005).

The efficiency of degradation by microorganisms is a function of the adsorption and rate of hydrolysis reaction and these are undisputedly affected by various factors:

1) Biotic factors include the nutritional requirements of microorganisms, enzyme activity and the proliferation process (Joutey, 2013; Eve Riser-Roberts, 1998). Physicochemical properties of the substrate, for example, include molecular mass, chemical composition, crystallinity, surface area, hydrophilicity and processing conditions (stabilizers, additives) (Makadia and Siegel, 2011; Reis and Román, 2004).
2) The medium conditions include temperature, pH and moisture (Kliem et al., 2020).

4.1.2 Enzymatic degradation

Enzymatic degradation is based on an enzyme-catalyzed scission of bonds of polymers. This process is comprised of four mechanistic steps (Göpferich, 1996; Lyu and Untereker, 2009; Uhrich et al., 1999; Pitt et al., 1981; Zhang et al., 2006; Heller et al., 2002; Muggli et al., 1999; Ghaffar et al., 2014; Burkersroda et al., 2002):

1) Diffusion of the enzyme from the bulk solution to the solid surface;
2) Adsorption of the enzyme on the substrate, resulting in the formation of the enzyme–substrate complex;
3) Catalysis of the hydrolysis reaction, and
4) Diffusion of the soluble degradation products from the solid substrate to the solution.

4.2 Abiotic involvement

Polymers can undergo irreversible changes in their properties under environmental conditions such as light, solar radiation, ambient air, oxygen, temperature and moisture. All of these environmental factors have an influence on polymer performance that can result in discoloration, crazing, cracking, phase separation, erosion, etc. Among other types of degradation mechanisms, although where biological activity is not involved are the phenomena such as hydrolysis, oxidation, photothermal and physical degradation (Amass et al., 1998).

4.2.1 Hydrolytic degradation

Polymers with hydrolyzable linkages in the backbone are very useful in a range of degradable materials. Hydrolyzable covalent bonds are present in chemical functionalities such as amides, anhydrides, carbamates, carbonates, esters, ortho-esters and ureas which exhibit a higher susceptibility to hydrolytic degradation under physiologic conditions (Larson and Weber, 1994).

Hydrolysis is a kind of chemical degradation and can be either acid or base-catalyzed. The rate of hydrolytic degradation depends on many other factors including the polymer chain length, the hydrophilicity/hydrophobicity nature crystallinity, molecular weight distribution, swellability, size, shape and geometry of polymer matrix, surface pretreatment, porosity, pore size and distribution, pore geometry, overall device dimensions, processing conditions and water diffusivity in the polymer matrix (Reddy et al., 2013; Lucas et al., 2008; Vroman and Tighzert, 2009; Laycock et al., 2017). As mentioned earlier, the crystallinity of the polymers is the main focus if one refers to hydrolytic degradation. Polymers with higher levels of crystallinity show a lower rate of degradation because hydrolysis tends to be faster in amorphous domains where the scission of bonds in backbone chains occurs in shorter periods (Zong et al., 1999; Vieira et al., 2010).

4.2.2 Oxidative degradation

Oxidative degradation, like the hydrolytic process, occurs as a result of the scission of susceptible bonds; in this case, chain fragmentation occurs by radical attack supported by peroxide (Nair and Laurencin, 2007). The most powerful agent provoking the degradation are atmospheric forms of oxygen (O_2, O_3), which attack covalent bonds, creating free radicals. The degradation rate depends on the polymer structure and unsaturated links or branched chains. Polyolefins, polyethers vinyl polymers, polyurethanes and polyamines, are generally more susceptible, they are generally more susceptible to undergo oxidative degradation (Brannigan and Dove, 2017; Lyu and Untereker, 2009; Burkersroda et al., 2002).

4.2.3 Photothermal degradation

In this type of degradation, polymers deteriorate by the action of light or heat. This is common in non-biodegradable polymer materials in outdoor applications (PE, PP, PMMA)/(Nagai et al., 2005). When the polymer is exposed to solar radiation, oxidation or cleavage C-C can be induced by high-energy radiation (Platzer, 1986). The most damaging effects on polymers are the visual effect, like optical properties after the loss of mechanical properties. It refers to the deterioration mechanism and the outcome is the degradation of the material. First changes are manifested in its physical properties, later its mechanical and optical properties are altered (Martin et al., 2003). Thermogravimetric analysis (*TGA*) is a useful technique for determining the degree of oxidative and thermal degradation in polymers (Singh and Sharma, 2008).

4.2.4 Physical degradation

This type of degradation refers to the exterior and interior deterioration of polymers caused by wear, exposure to the elements such as sunlight and wind storms, and in general, by mechanical stresses (Booth, 1963) which cause scission of chemical bonds in the polymeric main chain. The polymer degradation may be caused by one of the mechanisms mentioned above or a combination of these.

5. Degradation estimation techniques

Degradation studies are carried out by measuring any change in relevant physical properties, such as appearance, strength, thermal and electrical conductivity, and so on. Optical and electron microscopy analyses, for example, can be used to observe the erosion level of a polymer surface. Changes in the molar mass, as another example, can be recorded by Size Exclusion Chromatography (SEC). Differential Scanning Calorimetry (DSC), as well as X-Ray Diffraction (XRD) may be used to evaluate the degree of polymer crystallization. Monitoring of chemical subproducts, such as hydrolysis, oxidation, organic volatile compounds, etc., can be determined by many spectroscopic methods (Laycock et al., 2017). Kinetic measurements of hydrolysis, for example, can be determined by NMR Methods (Baran and Penczek, 1995; Singh and Sharma, 2008). These techniques as well as others offer a vast array of methods available for the detection and estimation of the degree and kind of degradation that may occur in polymers and plastics.

6. Blends of biodegradable polymers

A large number of different types of polymers are used for packaging applications (Andrady and Neal, 2009) because of their better physical and chemical properties, such as their strength, lightness, resistance to water and most water-borne microorganisms. However, and despite the advantages of biodegradable polymers, their oxygen barrier and mechanical properties are critical factors that determine a biomaterial's viability for packaging application purposes. A strategy to improve these properties and reduce the production cost of biodegradable polymers is to use blends and compatibilizers (Muthuraj et al., 2018). Different types of biodegradable polymers are often blended with other biodegradable and non-biodegradable polymers mainly for food packaging applications. Starch, for example, is often used because of its abundance, low cost and availability (Wang et al., 2004). Muthuraj and collaborators have summarized a variety of biodegradable polymer blends compatibilized by melt processing with a focus on *ex situ* and *in situ* compatibilization strategies. They enlist biodegradable polymer blends with different compatibilizers for application in the packaging industry (Spiegel, 2018).

7. Biodegradable polymer processing

7.1 Extruder design and operation

In recent years, the melt mixing method for polymer manufacture has proven to be a versatile technique compared to in situ polymerization and solution polymerization. Melt mixing is considered more flexible for the formulation of mixtures of different polymers and polymers with additives, this technique has been economical and compatible to carry out processes on an industrial scale, benefiting the obtaining of some commercial products and managing to produce high volumes of plastic products. Some products such as tubes, frames, plastic sheets and films are obtained by melt mixing (Zhang et al., 2017).

The most commonly used devices in melt polymerization are the extruder and roller mixer, which are based on shear flow.

The concept extrude is used when a material is forced through a restricted hole at a specific temperature, this variable is adopted according to the melting point of each polymer. The restriction through which the molten polymer is often passed is called the given. The movement with which the polymer is moved to the smallest hole is caused by an endless screw or spindle. An extruder can be composed of one or two rotating screws mounted on a barrel, the function of the screws is to progressively increase the pressure and push the melt mixture forward through a die. The thermomechanical phenomenon that occur in the barrel contribute to what is called texturization of the product (initially in granular form), while the expansion in the melt or matrix is responsible for the conformation of the final product (Sauceau et al., 2011).

The extruder generally consists of one or two rotating screws, either rotating or counter-rotating, within a stationary cylindrical (barrel). The barrel is often manufactured in sections to shorten the residence time of the molten materials. The sectioned parts of the barrel are screwed or clamped together. An end plate die is connected to the end of the barrel which is determined according to the shape of the extrudates.

In the most advanced twin-screw systems, the extrusion of materials is carried out using a configuration or arrangement of mixing and transport parts in the screw, this type of configuration can help the mixing and the better dispersion of the additives that are added to the polymer. Depending on the equipment, the rotating or counter-rotating spindle can be disassembled. The extruder must be able to rotate the screw at a predetermined speed while compensating for the torque and shear generated, both by the material that is being extruded and by the screws that are used, this influences the type of process and its complexity. A typical extrusion process consists of a motor that acts as a drive unit, an extrusion barrel, a rotating screw and an extrusion die (polymer with or without an additive). A central electronic control unit is connected to the extrusion unit to control process parameters such as screw speed, temperature and pressure (Maniruzzaman et al., 2012).

Single screw extrusion primarily helps distributive mixing, in this process performance depends on screw speed and pressure profile. Twin-screw extrusion supports the adjustment of several independent process variables, including feed rate, screw rotations. Variables such as the screw speed profile and screw temperature along the screw axis lead to high process flexibility and potential for optimization (Uitterhaegen and Evon, 2017).

The operating conditions for biopolymers change increasing screw speed (from 50 to 200 rpm) and decreasing flow rate (from 8.1 to 1.9 kg/hr), the result is sometimes disrupted extrudates, this aspect requires a thermomechanical treatment that involves modifying the conditions according to the nature of the biopolymers. Various studies have been developed regarding the conditions used for the transformation of these biopolymers, the most studied variants are temperature, processing, screw rotation speed, feeding, etc. One of the most evaluated polymers is PLA, when processing this material changes in mechanical and thermal properties are observed after being subjected to multiple twin-screw extrusion, the results indicate an increase in the degree of crystallinity and the flow index (MFI). The increase in these parameters indicates the degradation of the material and the fragmentation of the polymeric chains ("Thermoplastic Processing of Protein-based Bioplastics: Chemical Engineering Aspects of Mixing, Extrusion and Hot Molding" 2003).

Studies have been carried out using the twin-screw extrusion process to obtain hydrophobic compounds, using maltodextrin and a compatibilizer such as a pea protein isolate and Hi-cap 100. The operating conditions of the extruder were at a temperature of 50°C and specific mechanical energy between 120 and 370 $Wh.kg^{-1}$, the encapsulation efficiency rate of the hydrophobic compound was 96.3%. Melt extrusion is a technique that can be used for the encapsulation of compounds using biopolymers such as maltodextrin, modified starch and proteins, resulting in homogeneous mixtures with good porosity (Castro et al., 2020).

This polymer processing technique facilitates the handling of small inorganic charges such as some nanoparticles (Cabello-Alvarado et al., n.d.). Some biopolymers such as proteins, plasticized starch and other polysaccharides, exhibit a change in their properties after being extruded, showing non-Newtonian rheological behavior, this type of behavior is important for the determination of the process conditions (Emin and Schuchmann, 2017).

7.2 Reactive extrusion

Reactive extrusion is considered a method where individual components are chemically modified during the process. Twin-screw extruders are commonly used for this type of extrusion since they allow better control of mixing and residence time. There must be surveillance in feed administered to the extruder, this must be of high precision for an adequate stoichiometry relationship between the components. The aim is to achieve with this extrusion method adequate compatibilization between the compounds based on polysaccharides, it is also easily scalable at an industrial level for the esterification of the polysaccharides (Raquez et al., 2008; Imre et al., 2019).

There is research related to modified starches in a continuous process in a single step, these materials were obtained in a twin-screw extruder. The results indicate that this system offers good mixing conditions and is a very useful tool for processing highly viscous fluids, such as gelatinized starch. The twin-screw extruder shows good heat transfer and the different geometries in the screw design provide adequate control over residence times, as well as having the opportunity to add a greater amount of reagents and/or additives that act as adjuvants and process stabilizers (Moad, 2011)

For the production of modified starch, the systems used for batch reactions involve long residence times of between 2 to 24 hours, when using reactive extrusion this can be improved many times due to the following conditions:

- There is a homogeneous reaction medium.
- Starch concentrations can reach 60 to 80%.
- Higher temperatures depend on starch.
- A mix is carried out efficiently (de Graaf et al., 1995).

Reactive extrusion has been used for the modification of corn starches for encapsulation purposes. This encapsulation is prepared based on emulsions with modified starch using a minimum of surfactant materials since they are toxic and costs are reduced. In a recent study, the extraction of starch from amaranth seeds was reported. The modification of this starch was carried out by extrusion, to later prepare the emulsions to encapsulate compounds. The modification of the starches was carried out employing the crosslinking method by chemical reactions in extrusion. Crosslinking implied the formation of hemiacetal bonds caused by the two carbonyl groups of the crosslinking agent, in this case, the bond was between glyoxal and the hydroxyl groups of the starch molecule. First, the starch was mixed with the crosslinking agent, water and glycerol, obtaining a mixture with 39% humidity, then the material was subjected to an extrusion process in a single screw extruder at 110°C and 240 rpm, the samples at different times were taken. Modified amaranth starch at 80 s exit time from the extruder was the material with optimal properties (García-Armenta et al., 2021).

Currently, research on BPs has focused on the creation of green compounds or biodegradable characteristics, where biopolymers are reinforced with natural fibers and/or particles as inorganic fillers or fillers to increase the mechanical properties of biopolymers. Reactive extrusion is the main method for the preparation of these green compounds. Wu et al. obtained green compounds based on a mixture of poly (butylene succinate), PBS, and poly (butylene-adipate-co-terephthalate PBAT), using natural fibers as reinforcement Miscanthus fiber and Oat hull using reactive extrusion. The mixture of PBS/PBAT with hydrogen peroxide was added to the first feeder of the extruder and the fillers were introduced with an alternating feeder at a speed of 100 rpm and at a temperature of 180°C from the feeding zone to die. The green compounds showed an improvement in stiffness and impact resistance (Wu et al., 2020).

8. Membrane preparation methods

The commercially available separation membranes are usually made of synthetic polymers, the number of discarded membranes used for reverse osmosis increases more every day, only in 2015 12,000 tons of them were discarded, this generates a waste problem because the polymer can break down into microplastics that contaminate soil and water. Due to this, new alternatives for the manufacture of membranes based on other materials such as chitosan with silver nanoparticles have been recently studied, as a result, biodegradability and antibacterial tests were carried out. The silver nanoparticles were incorporated uniformly into the BPs network, this helped in an important way to create a membrane with good properties (Shi et al., 2020).

Some of the membrane preparation methods are mentioned below (Table 3.2):

8.1 Phase inversion

Most commercial membranes are manufactured by this method because it is flexible, inexpensive and can be industrially scaled up. In the 80s, the preparation of membranes by thermally induced phase separation was introduced, a polymer solution is formed at high temperature and cooled to induce phase separation and polymer solidification. Porous membranes are obtained after the removal of the diluent. This technique is normally used for crystalline polymers (Liu et al., 2017).

8.2 Vapor-induced phase separation

This process consists of the preparation of a homogeneous polymer solution, made by evaporating the appropriate solvent, then the solution is poured onto a substrate of the desired thickness, and placed in a steam chamber to induce phase separation. Finally, the polymer is immersed in a non-solvent bath and the membrane is dried. Two of the conditions with which care must be taken to control membrane porosity and properties are exposure time to non-solvent vapors and relative humidity (Venault et al., 2013).

8.3 Non-solvent induced phase separation

This system uses a ternary composition that includes a polymer, a solvent and a non-solvent. The process precipitation by immersion. It begins with the preparation of a homogeneous solution between the polymer and a solvent, then the polymer solution is poured onto a support as a thin film

Table 3.2. Some examples of the biodegradable membrane.

Biopolymer	**Method**	**Membrane structure**	**Application**	**Ref.**
PLA	Phase inversion	Possesses smooth and compact surfaces and fingerlike cross-sections, which is a typical asymmetric structure	Hemodialysis	(Tomietto et al., 2020)
Methylcellulose/ PLA	Vapor-Induced Phase Separation	A homogeneous and dense membrane	------------------	(Gao et al., 2014)
PLA/TiO_2	Non-solvent induced phase separation	Nano-TiO_2 particles inlayed in the hierarchical texture.	Oil/water Separation	(Guo et al., n.d.)
Cellulose/starch	Solution casting process	Smooth surface and starch granules were not observed on the surface and cross-section of the membrane	Plastic packaging	(Xiong et al., 2017)
PLLA/PHB	Electrospinning	Smooth surfaces and relatively uniform fiber	Drug delivery	(Soni et al., 2020)
Polyhydroxy butyrate/poly-3-caprolactone (PHB/PCL)	Electrospinning	Fibrous membranes surfaces	Packaging materials	(Cao et al., 2018)

or extruded through a matrix to generate shapes such as hollow fiber or flat sheets. A coagulation bath that contains a non-solvent and therefore phases separation takes place when solvents are exchanged and precipitation occurs in the polymer solution (Wang and Lai, 2013).

8.4 Pickering emulsions

It is an innovative method for the manufacture of porous gelatin membranes, which are stabilized with hexagonal boron nitride nanosheets, in several steps. Hexagonal boron nitride nanosheets 2% by weight are first suspended in water and sonicated in an ultrasound device for 1 hour to ensure good dispersion of the nanoparticles. Then the solution was heated to 60°C with mechanical stirring and gelatin powder (20% by weight) was added slowly to avoid lumps, the solution was left under these conditions for 2 hours. Reine oil is added to the suspension and it is sonicated for 7 minutes. A whitish emulsion was obtained. Films were prepared at 37°C by pouring the solution dry onto a flat surface. Gelatin membranes were cured at 20°C under 45 ± 5% relative humidity for different drying times (1, 3 and 5 hours). This "curing time" allows the system to set. Then, the membrane was washed with ethanol and water (Mateur et al., 2020).

Studies have been carried out to produce a homogeneous electrospinning mixture with nanocellulose, chitosan and PHA using ultrasound, the method used was named Pickering emulsion-electrospinning method. The PHA biocomposite with nanocellulose and chitosan was evaluated by adsorption of the Congo red dye, the adsorption isotherm to which it was adjusted was that of Langmuir, presenting pseudo-second order kinetics with a chemisorption nature. This bi-compound is proposed to be used as a sewage filter for the removal of colorants (Soon et al., 2019).

8.5 Electrospinning

In 1897 Raleigh first introduced the term electrospinning (Bhardwaj and Kundu, 2010). This method can create ultrafine fibers with diameters ranging from microns to nanometers, it is widely used to prepare nano-fibrous membranes. These membranes have high gravimetric porosity, low density, pore interconnectivity, controllable thickness and good mechanical resistance, and can be used in different applications such as membrane membranes for water and air filter (Kim et al., 2015).

Electrospinning can be carried out both in solution and in the molten state, the advantages of electrospinning in the melt are productivity, since the process can be easily scalable, one of the advantages is that it does not use toxic solvents, the process is clean and less costly.

The electrospinning process is carried out by making a solution of the polymer or else a melt is placed in a capillary, then a strong electric field is applied between a spinneret and a grounded collector. When the applied voltage overcomes the surface tension of the polymer fluid, the strong electric field causes a droplet shape to deform into a conical structure. Then the polymer is expelled in the form of drops. The solvent is evaporated due to continuous movement and elongation by electrostatic repulsion. Finally, the solidification of the fluid filament forms an electrospun membrane (Bhardwaj and Kundu, 2010).

One of the applications of membranes obtained by electrospinning is a distillation and wastewater treatment because, in the electrospinning process parameters such as morphology, pore size and shape, their distribution and hydrophobicity can be controlled. In the case of using a polymer solution in the electrospinning technique, certain parameters in the solution must be taken into account, the most important being the selection of the solvent.

9. 3D printing of biodegradable polymers

BPs of 3-dimensional architectures appeared in the early 2000s in the medical field because 3D print processing deals with shape versatility, reproducibility and upscaling. The adaptation of PLA, PCL, PHA and other BPs to the existing technology requires the adjustment of parameters like working temperature, print speed, surface modification, plasticizers, etc. This section highlights advances in green additives like cellulose to improve the 3D printing processability of biopolymers.

Fused deposition modeling (FDM) 3D printing. FMD is extensively used on thermoplastics in which a filament of biopolymer is heated to build the 3D structure. The biopolymer "ink" with the lowest working temperature is PCL < PHBH < PLA limiting it to few biopolymers. The incorporation of cellulose nanocrystals (CNC) or nanofibrils (CNF) is challenging but can improve the mechanical properties, rheology, processability and biodegradability of the ink. The strategies addressing this problem through a combination of chemical and physical homogenization are presented here.

For instance, Wang et al. (2020) reported the mechanical homogenization of CNFs and their dry mix with biopolymer to FMD 3D print biodegradable 1.75 mm smooth filaments using a Wellzoom desktop single screw extruder at 170°C. First, a mechanically homogenized cellulose suspension was freeze-dried to obtain a high surface area 3-dimensional network. The resultant fine powder was dry dispersed with PLA and PEG600. A loading of 2.5 wt.% CNFs increased the tensile strength from 45 MPa for PLA to 57 MPa for PLA/PEG/CNF. On the other hand, Guibilini et al. (2020) experimented with a wet mixing of CNCs and biopolymer to create complex designs like finger splints. The wet acetylated CNCs were dissolved in with poly (3-hydroxybutyrate-co-3-hydroxyhexanoate) (PHBH) in a "one-pot reaction" which is a green procedure that avoids strong solvents like dimethylformamide or pyridine. The obtained films were used to print 1.75 mm filament at 145°C. The high CNCs loadings (5 to 20 wt.%) particularly increased thermal stability from 220 to 265°C as well as the viscosity of PHBH from 253 Pa s to 487 Pa s for the 10 wt.% CNC/PHBH.

Many PLA composites present a delayed biodegradability in the presence of additives that defeats the purpose of biopolymers, both authors demonstrated that this is not the case.

Direct ink writing (DIW) 3D printing. A variation of the FDM 3D printing technique that uses viscous or gel-ish "inks" at room temperature is referred to as Direct Ink Writing (DIW). A series of natural polymers like lignocellulose, starch, chitin and protein-based biopolymers used for DIW 3D printing have been extensively revised recently (Gauss et al., 2021; Liu et al., 2019). Here we highlight a few reports where nanocellulose can be blended with other polysaccharides for the preparation of monoliths and scaffolds using 3D printing techniques.

For example, Sultan et al. (2019) combined carboxylated CNFs (TOCNF) and alginate to prepare smooth and DIW printable "ink" and print scaffolds using a 3D printer Ultimaker 2+ of 410 μm diameter nozzle at room temperature. The biobased scaffolds were easily stabilized by ionic crosslinking using $CaCl_2$. Indeed stable 100% microcellulose and bacterial cellulose 3D printed cellular scaffolds can be prepared without additives (Siqueira et al., 2017). Silk Fibroin (SF) extracted from cocoons of silkworm crosslinked with 10 wt.% hydroxypropyl methyl cellulose (HPMC) can be processed as a hydrogel for 3D printing using nozzles of 260 μm diameter at room temperature (Zhong et al., 2019). The structural integrity of such a scaffold was achieved using the aforementioned freeze-drying technique.

The disadvantage of DIW 3D printing is the loss of the CNCs surface area that can be avoided in 3D foaming processing using supercritical fluids are reviewed next.

9.1 Foaming of biodegradable polymers

The major advantage of using supercritical fluids is the incorporation of temperature-sensitive compounds like biopolymers apart from the reduction of organic solvents, and the low density of the resultant 3D structure. Recent advances in green solvents are highlighted here.

Biopolymer foaming using supercritical fluids. A three-step foaming process includes (1) dispersion of the components that are submitted to a CO_2 atmosphere at high pressure (20 MPa, 40°C, and 1 to 6 hours), (2) gelation, and (3) drying by depressurization. For instance, 1-butyl-3-methylimidazolium acetate ([BMIM]Ac) and 1-butyl-3-methylimidazolium chloride ([BMIM]Cl) are ionic liquids (ILs) that dissolve both PLA and starch to produce self-standing 3D foams (Martins et al., 2014). However, ILs may present some toxicity so it has been replaced by the next generation of solvents called natural Deep Eutectic Solvents (DES).

The biodegradability testing of the processed foams is scarce though fundamental interacting with living beings. Due to DES natural origin it is easier to accomplish fully biodegradability. However, nothing better than non-solvent biopolymer mixtures processed in supercritical conditions. For instance, Alvarez et al. (2020) used ethyl lactate to prepare poly(lactide-co-glycolide) (PLGA) foams. Furthermore, Markočič et al. (2015) produced PLGA foams on PBS media buffer solution.

Foaming of cellulose and other biopolymers. A new generation of fully biodegradable, light and flexible 3D foams are based on low concentration aqueous dispersion (0.5–2 wt.% for nanocellulose and 10–15 wt.% for other cellulose forms). Novel dispersions may include low amounts of polar solvents like tert-butyl alcohol, methanol and ethanol/HCl (Sakai et al., 2016). Even the concentration may be reduced to 0.3 wt.% by crosslinking cellulose nanofibrils using γ-glycidoxypropyltrimethoxysilane (GPTMS) and branched polyethyleneimine (PEI Mw = 600) in acetone (Li et al., 2019).

A good choice of solvents and drying conditions may contribute to keeping nanocellulose surface area. For instance, TOCNF foams prepared in water vs tert-butyl alcohol showed an increase of an order magnitude of surface area (20 to 160 m^2/g) (Frerich, 2015). A freeze-dried foam displayed a super-high surface area of 2330 m^2/g compared with oven-dried hybrid foam (900 m^2/g) (Sultan et al., 2019). Even algae can be foam-processed into algae aerogels (174 m^2/g) (Alnaief et al., 2018), as well as a series of natural polymers such as alginate, chitin, pectin, starch, vegetal and animal proteins (Nita et al., 2020; Chen et al., 2021).

One major challenge is still the energy consumption at a large scale of the over 100 hours foaming process. Nonetheless, these studies can inspire future research on biodegradable materials processing as represented in Fig. 3.3.

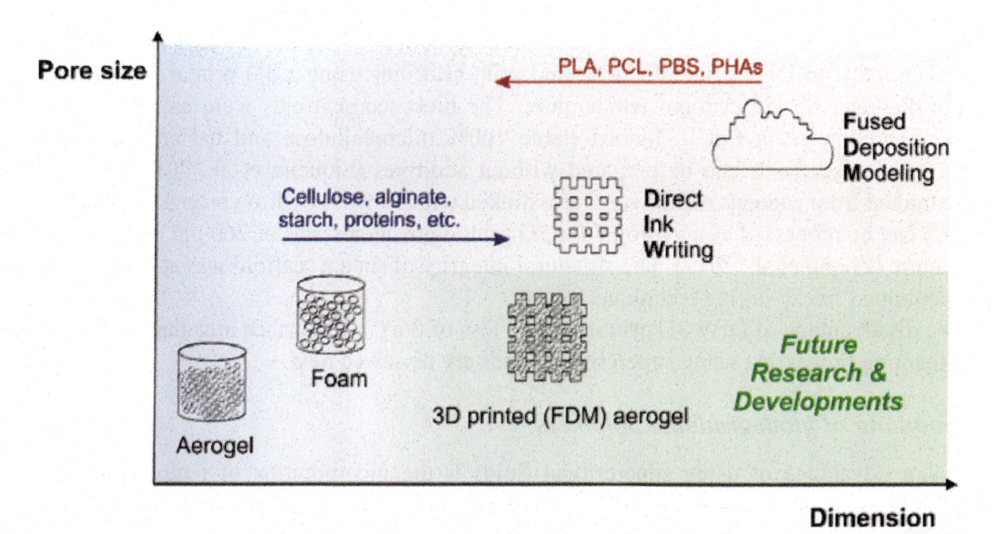

Figure 3.3. Schematic representation of 3D structures processed using BPs like poly (lactic acid) (PLA), poly(caprolactone) (PCL), Poly(Butylene Succinate) (PBS) and poly(hydroxyalcanoates) (PHAs). Future research may explore large 3D printed foams of high porosity and surface area using supercritical fluids (green colored section) (Authors).

Conclusions

BPs provide great opportunities to develop novel materials. There are now a variety of BPs, having different chemical structures and different properties, and they offer a large field of applications. In the last 10 years, the market for BPs has grown significantly due to the interest of society in caring for the environment and avoiding water pollution with microplastics. However, the cost of production and processing is still high, due to this, new synthesis techniques, processing and applications with BPs are required. To provide added value to biodegradable polymers, some advanced technologies have been applied. Improving the 3D printing processability of biopolymers. Recently different studies have been reported concerning the use of metal nanoparticles with biodegradable polymers, especially with silver. Nano-biopolymers are under investigation. Consequently, biodegradable polymers are the topics of much research.

Acknowledgements

The MLAG, DIMB and CJCA appreciate financial support from CIQA project 6605 and 6572. MAJT (CVU: 334512/Id. Project: 76219) acknowledges the financial support from Consejo Nacional de Ciencia y Tecnología (Conacyt) through the Postdoctoral grants program "Estancias postdoctorales por México". MLAG kindly acknowledge the scholarship postdoctoral provided by CIQA.

References

Alnaief, Mohammad, Rana Obaidat and Hadeia Mashaqbeh. (2018). Effect of processing parameters on preparation of carrageenan aerogel microparticles. *Carbohydrate Polymers,* 180(January): 264–75. https://doi.org/10.1016/j.carbpol.2017.10.038.

Álvarez, I., Gutiérrez, C., Rodríguez, J. F., de Lucas, A. and García, M. T. (2020). Production of biodegradable plga foams processed with high pressure CO_2. *Journal of Supercritical Fluids,* 164. https://doi.org/10.1016/j.supflu.2020.104886.

Amann, Manfred and Oliver Minge. (2012). Biodegradability of poly(vinyl acetate) and related polymers bt - synthetic biodegradable polymers. pp. 137–72. *In*: Bernhard Rieger, Andreas Künkel, Geoffrey W. Coates, Robert Reichardt, Eckhard Dinjus and Thomas A. Zevaco (eds.). *Synthetic Biodegradable Polymers.* Berlin, Heidelberg: Springer Berlin Heidelberg. https://doi.org/10.1007/12_2011_153.

Amass, Wendy, Allan Amass and Brian Tighe. (1998). A review of biodegradable polymers: uses, current developments in the synthesis and characterization of biodegradable polyesters, blends of biodegradable polymers and recent advances in biodegradation studies. *Polymer International,* 47(2): 89–144. https://doi.org/https://doi.org/10.1002/(SICI)1097-0126(1998100)47:2<89::AID-PI86>3.0.CO;2-F.

Andrady, Anthony L. and Mike A. Neal. (2009). Applications and societal benefits of plastics. *Philosophical Transactions of the Royal Society B: Biological Sciences,* 364(1526): 1977–84. https://doi.org/10.1098/rstb.2008.0304.

Asikainen, A. J., Noponen, J., Mesimäki, K., Laitinen, O., Peltola, J., Pelto, M., Kellomäki, M., Ashammakhi, N., Lindqvist, C. and Suuronen, R. (2005). Tyrosine derived polycarbonate membrane is useful for guided bone regeneration in rabbit mandibular defects. *Journal of Materials Science: Materials in Medicine,* 16(8): 753–58. https://doi.org/10.1007/s10856-005-2613-6.

Asikainen, Antti J., Jukka Noponen, Christian Lindqvist, Mika Pelto, Minna Kellomäki, Hanne Juuti, Harri Pihlajamäki and Riitta Suuronen. (2006). Tyrosine-derived polycarbonate membrane in treating mandibular bone defects. An experimental study. *Journal of The Royal Society Interface,* 3(10): 629–35. https://doi.org/10.1098/rsif.2006.0119.

Baran, Jolanta and Stanislaw Penczek. (1995). Hydrolysis of polyesters of phosphoric acid. 1. kinetics and the ph profile. *Macromolecules,* 28(15): 5167–76. https://doi.org/10.1021/ma00119a002.

Bastioli, Catia. (2014). *Handbook of Biodegaradable Polymers.* Edited by Catia Bastioli. Second. Smithers Rapra.

Becker, Greta and Frederik R. Wurm. (2018). Functional biodegradable polymers via ring-opening polymerization of monomers without protective groups. *Chemical Society Reviews,* 47(20): 7739–82. https://doi.org/10.1039/C8CS00531A.

Bei, Jian-Zhong, Jian-Ming Li, Zhi-Feng Wang, Jia-Chang Le and Shen-Guo Wang. (1997). Polycaprolactone–poly(ethylene-glycol) block copolymer. IV: Biodegradation behavior *in vitro* and *in vivo*. *Polymers for Advanced Technologies,* 8(11): 693–96. https://doi.org/https://doi.org/10.1002/(SICI)1099-1581(199711)8:11<693::AID-PAT702>3.0.CO;2-B.

Betancourt, Tania, Brandon Brown and Lisa Brannon-Peppas. (2007). Doxorubicin-loaded PLGA nanoparticles by nanoprecipitation: preparation, characterization and *in vitro* evaluation. *Nanomedicine,* 2(2): 219–32. https://doi.org/10.2217/17435889.2.2.219.

Bhardwaj, Nandana and Subhas C. Kundu. (2010). Electrospinning: a fascinating fiber fabrication technique. *Biotechnology Advances*. Elsevier. https://doi.org/10.1016/j.biotechadv.2010.01.004.

Booth, C. (1963). The mechanical degradation of polymers. *Polymer,* 4: 471–78. https://doi.org/https://doi.org/10.1016/0032-3861(63)90060-0.

Brannigan, Ruairí P. and Andrew P. Dove. (2017). Synthesis, properties and biomedical applications of hydrolytically degradable materials based on aliphatic polyesters and polycarbonates. *Biomaterials Science,* 5(1): 9–21. https://doi.org/10.1039/C6BM00584E.

Briggs, Tonye, Matthew D. Treiser, Paul F. Holmes, Joachim Kohn, Prabhas V. Moghe and Treena Livingston Arinzeh. (2009). Osteogenic differentiation of human mesenchymal stem cells on poly(ethylene glycol)-variant biomaterials. *Journal of Biomedical Materials Research. Part A,* 91(4): 975–84. https://doi.org/10.1002/jbm.a.32310.

Brin, Yaron S., Jacob Golenser, Boaz Mizrahi, Guy Maoz, Abraham J. Domb, Shyamal Peddada, Shmuel Tuvia, Abraham Nyska and Meir Nyska. (2008). Treatment of osteomyelitis in rats by injection of degradable polymer releasing gentamicin. *Journal of Controlled Release,* 131(2): 121–27. https://doi.org/https://doi.org/10.1016/j.jconrel.2008.07.022.

Brosse, Jean-Claude, Daniel Derouet, Laurent Fontaine and Supachat Chairatanathavorn. (1989). Fixation of pharmacologically active amines on polyphosphonates, 2. Application to benzocaïne and phenethylamine. *Die Makromolekulare Chemie,* 190(9): 2339–45. https://doi.org/https://doi.org/10.1002/macp.1989.021900935.

Bruck, S. O. (1981). Biomedical Polymers—Polymeric Materials and Pharmaceuticals for Biomedical Use. Goldberg, E.P. and Nakajima, A. (eds.). Academic Press, New York, 1980, pp. 457, US $32.00.

Burkersroda, Friederike von, Luise Schedl and Achim Göpferich. (2002). Why degradable polymers undergo surface erosion or bulk erosion. *Biomaterials,* 23(21): 4221–31. https://doi.org/https://doi.org/10.1016/S0142-9612(02)00170-9.

Cabello-Alvarado, C., Reyes-Rodríguez, P., Andrade-Guel, M., Cadenas-Pliego, G., Pérez-Alvarez, M., Cruz-Delgado, V. J., Melo-López, L., Quiñones-Jurado, Z. V. and Ávila-Orta, C. A. n.d. Melt-mixed thermoplastic nanocomposite containing carbon nanotubes and titanium dioxide for flame retardancy applications. *Polymers,* https://doi.org/10.3390/polym11071204.

Cao, K., Liu, Y., Olkhov, A. A., Siracusa, V. and Iordanskii, A. L. (2018). PLLA-PHB fiber membranes obtained by solvent-free electrospinning for short-time drug delivery. *Drug Delivery and Translational Research,* 8(1): 291–302. https://doi.org/10.1007/s13346-017-0463-7.

Carrillo-Conde, Brenda, Elise Schiltz, Jing Yu, F. Chris Minion, Gregory J. Phillips, Michael J. Wannemuehler and Balaji Narasimhan. (2010). Encapsulation into amphiphilic polyanhydride microparticles stabilizes yersinia pestis antigens. *Acta Biomaterialia,* 6(8): 3110–19. https://doi.org/https://doi.org/10.1016/j.actbio.2010.01.040.

Castro, Natalia, Vanessa Durrieu, Christine Raynaud and Antoine Rouilly. (2020). Twin-screw extrusion encapsulation of MCT-oil in a maltodextrin matrix using compatibilizing biopolymers. *Colloids and Surfaces B: Biointerfaces,* 195(November): 111267. https://doi.org/10.1016/j.colsurfb.2020.111267.

Chen, Yiming, Lin Zhang, Yang Yang, Bo Pang, Wenhui Xu, Gaigai Duan, Shaohua Jiang and Kai Zhang. (2021). Recent progress on nanocellulose aerogels: preparation, modification, composite fabrication, applications. *Advanced Materials*, February, 2005569. https://doi.org/10.1002/adma.202005569.

Christenson, Elizabeth M., Wafa Soofi, Jennifer L. Holm, Neil R. Cameron and Antonios G. Mikos. (2007). Biodegradable fumarate-based PolyHIPEs as tissue engineering scaffolds. *Biomacromolecules,* 8(12): 3806–14. https://doi.org/10.1021/bm7007235.

Conato, Marlon and Florentino Sumera. (2012). Biodegradable polyesters and polyamides from difunctionalized lauric and coconut fatty acids. *Journal of Polymers and the Environment,* 20(1): 217–23. https://doi.org/10.1007/s10924-011-0397-y.

Darney, Philip D., Scott E. Monroe, Cynthia M. Klaisle and Ana Alvarado. (1989). Clinical evaluation of the capronor contraceptive implant: preliminary report. *American Journal of Obstetrics and Gynecology,* 160(5, Part 2): 1292–95. https://doi.org/https://doi.org/10.1016/S0002-9378(89)80015-8.

Dunne, Nicholas, Valerie Jack, Rochelle O'Hara, David Farrar and Fraser Buchanan. (2010). Performance of calcium deficient hydroxyapatite–polyglycolic acid composites: an *in vitro* study. *Journal of Materials Science: Materials in Medicine* 21(8): 2263–70. https://doi.org/10.1007/s10856-010-4021-9.

Edlund, U. and Albertsson, A.-C. (2003). Polyesters based on diacid monomers. *Advanced Drug Delivery Reviews,* 55(4): 585–609. https://doi.org/https://doi.org/10.1016/S0169-409X(03)00036-X.

Emin, M. A. and Schuchmann, H. P. (2017). A mechanistic approach to analyze extrusion processing of biopolymers by numerical, rheological, and optical methods. *Trends in Food Science and Technology*. Elsevier Ltd. https://doi.org/10.1016/j.tifs.2016.10.003.

Erggelet, Christoph, Katja Neumann, Michaela Endres, Kathrin Haberstroh, Michael Sittinger and Christian Kaps. (2007). Regeneration of ovine articular cartilage defects by cell-free polymer-based implants. *Biomaterials,* 28(36): 5570–80. https://doi.org/https://doi.org/10.1016/j.biomaterials.2007.09.005.

Eve Riser-Roberts. (1998). *Remediation of Petroleum Contaminated Soils Biological, Physical, and Chemical Processes.* 1st ed.

Frerich, Sulamith, C. (2015). Biopolymer foaming with supercritical CO_2—thermodynamics, foaming behaviour and mechanical characteristics. *Journal of Supercritical Fluids,* 96(January): 349–58. https://doi.org/10.1016/j.supflu.2014.09.043.

Gao, Ailin, Fu Liu and Lixin Xue. (2014). Preparation and evaluation of heparin-immobilized poly (lactic acid) (PLA) membrane for hemodialysis. *Journal of Membrane Science,* 452(February): 390–99. https://doi.org/10.1016/j.memsci.2013.10.016.

García-Armenta, Evangelina, Lorenzo A. Picos-Corrales, Gustavo F. Gutiérrez-López, Roberto Gutiérrez-Dorado, Janitzio X. K. Perales-Sánchez, Santiago García-Pinilla, Fernanda Reynoso-García, Martín Martínez-Audelo, J. and Alejandro Armenta-Manjarrez, M. (2021). Preparation of surfactant-free emulsions using amaranth starch modified by reactive extrusion. *Colloids and Surfaces A: Physicochemical and Engineering Aspects,* 608(January): 125550. https://doi.org/10.1016/j.colsurfa.2020.125550.

Gauss, Christian, Kim L. Pickering and Lakshmi Priya Muthe. (2021). The use of cellulose in bio-derived formulations for 3D/4D printing: A review. *Composites Part C: Open Access,* 4(March): 100113. https://doi.org/10.1016/j.jcomc.2021.100113.

Ghaffar, A., Schoenmakers, P. J. and van der Wal, S. j. (2014). Methods for the chemical analysis of degradable synthetic polymeric biomaterials. *Critical Reviews in Analytical Chemistry,* 44(1): 23–40. https://doi.org/10.1080/10408347.2013.831729.

Giubilini, Alberto, Gilberto Siqueira, Frank J. Clemens, Corrado Sciancalepore, Massimo Messori, Gustav Nyström and Federica Bondioli. (2020). 3D-printing nanocellulose-poly(3-hydroxybutyrate-co-3-hydroxyhexanoate) biodegradable composites by fused deposition modeling. *ACS Sustainable Chemistry and Engineering,* 8(27): 10292–302. https://doi.org/10.1021/acssuschemeng.0c03385.

Gleadall, Andrew, Jingzhe Pan, Marc-Anton Kruft and Minna Kellomäki. (2014). Degradation mechanisms of bioresorbable polyesters. Part 2. Effects of initial molecular weight and residual monomer. *Acta Biomaterialia,* 10(5): 2233–40. https://doi.org/https://doi.org/10.1016/j.actbio.2014.01.017.

Göpferich, Achim. (1996). Mechanisms of polymer degradation and erosion. *Biomaterials,* 17(2): 103–14. https://doi.org/https://doi.org/10.1016/0142-9612(96)85755-3.

Graaf, R. A. de, Broekroelofs, G. A., Janssen, L. P. B. M. and Beenackers, A. A. C. M. (1995). The kinetics of the acetylation of gelatinised potato starch. *Carbohydrate Polymers,* 28(2): 137–44. https://doi.org/10.1016/0144-8617(95)00088-7.

Gross, Richard A. and Bhanu Kalra. (2002). Biodegradable polymers for the environment. *Science,* 297(5582): 803 LP–807. https://doi.org/10.1126/science.297.5582.803.

Guo, Panjie, Fen Wang, Tongtong Duo, Zhihong Xiao, Airong Xu, Rukuan Liu and Chaohui Jiang. n.d. Facile fabrication of methylcellulose/PLA membrane with improved properties. https://doi.org/10.3390/coatings10050499.

Heller, J. (1994). Use of poly(ortho esters) and polyanhydrides in the development of peptide and protein delivery systems. *In*: *Formulation and Delivery of Proteins and Peptides*, 567: 17–292. ACS Symposium Series. American Chemical Society. https://doi.org/doi:10.1021/bk-1994-0567.ch017.

Heller, Jorge, John Barr, Steven Y. Ng, Khadija Schwach Abdellauoi and Robert Gurny. (2002). Poly(ortho esters): synthesis, characterization, properties and uses. *Advanced Drug Delivery Reviews,* 54(7): 1015–39. https://doi.org/https://doi.org/10.1016/S0169-409X(02)00055-8.

Hocking, Philippa J. (1992). The classification, preparation, and utility of degradable polymers. *Journal of Macromolecular Science, Part C,* 32(1): 35–54. https://doi.org/10.1080/15321799208018378.

Imre, Balázs, Lidia García, Debora Puglia and Francisco Vilaplana. (2019). Reactive compatibilization of plant polysaccharides and biobased polymers: review on current strategies, expectations and reality. *Carbohydrate Polymers*. Elsevier Ltd. https://doi.org/10.1016/j.carbpol.2018.12.082.

Jayasekara, Ranjith, Ian Harding, Ian Bowater and Greg Lonergan. (2005). Biodegradability of a selected range of polymers and polymer blends and standard methods for assessment of biodegradation. *Journal of Polymers and the Environment,* 13(3): 231–51. https://doi.org/10.1007/s10924-005-4758-2.

Johnson, Patrick A., Arnold Luk, Aleksey Demtchouk, Hiral Patel, Hak-Joon Sung, Matthew D. Treiser, Simon Gordonov et al. (2010). Interplay of anionic charge, poly(ethylene glycol), and iodinated tyrosine incorporation within tyrosine-derived polycarbonates: effects on vascular smooth muscle cell adhesion, proliferation, and motility. *Journal of Biomedical Materials Research Part A,* 93A(2): 505–14. https://doi.org/https://doi.org/10.1002/jbm.a.32544.

Joutey, Nezha Tahri. (2013). Biodegradation: involved microorganisms and genetically engineered microorganisms. In, edited by Wifak Bahafid, Ch. 11. Rijeka: IntechOpen. https://doi.org/10.5772/56194.

Kariduraganavar, Mahadevappa Y., Arjumand A. Kittur and Ravindra R. Kamble. (2014). Chapter 1 - Polymer synthesis and processing. pp. 1–31. *In*: Sangamesh G. Kumbar, Cato T. Laurencin and Meng, B. T. (eds.). *Natural and Synthetic Biomedical Polymers Deng.* Oxford: Elsevier. https://doi.org/https://doi.org/10.1016/B978-0-12-396983-5.00001-6.

Kim, Jeong F., Ji Hoon Kim, Young Moo Lee and Enrico Drioli. (2015). Thermally induced phase separation and electrospinning methods for emerging membrane applications: a review. *American Institute of Chemical Engineers AIChE J.,* 62: 461–90. https://doi.org/10.1002/aic.15076.

Kipper, Matt J., Jennifer H. Wilson, Michael J. Wannemuehler and Balaji Narasimhan. (2006). Single dose vaccine based on biodegradable polyanhydride microspheres can modulate immune response mechanism. *Journal of Biomedical Materials Research Part A,* 76A(4): 798–810. https://doi.org/https://doi.org/10.1002/jbm.a.30545.

Kliem, Silvia, Marc Kreutzbruck and Christian Bonten. (2020). Review on the biological degradation of polymers in various environments. *Materials.* https://doi.org/10.3390/ma13204586.

Kluin, Otto S., Henny C. van der Mei, Henk J. Busscher and Daniëlle Neut. (2009). A surface-eroding antibiotic delivery system based on poly-(trimethylene carbonate). *Biomaterials,* 30(27): 4738–42. https://doi.org/https://doi.org/10.1016/j.biomaterials.2009.05.012.

Larson, R. A. and Weber, E. J. (1994). *Reaction Mechanisims in Environmental Chemistry,* CRC Press.

Laycock, Bronwyn, Melissa Nikolić, John M. Colwell, Emilie Gauthier, Peter Halley, Steven Bottle and Graeme George. (2017). Lifetime prediction of biodegradable polymers. *Progress in Polymer Science,* 71: 144–89. https://doi.org/https://doi.org/10.1016/j.progpolymsci.2017.02.004.

Lensen, Dennis, Kevin van Breukelen, Dennis M. Vriezema and Jan C. M. van Hest. (2010). Preparation of biodegradable liquid core PLLA microcapsules and hollow PLLA microcapsules using microfluidics. *Macromolecular Bioscience,* 10(5): 475–80. https://doi.org/https://doi.org/10.1002/mabi.200900404.

Leong, K. W., Simonte, V. and Langer, R. (1987). Synthesis of polyanhydrides: melt-polycondensation, dehydrochlorination, and dehydrative coupling. *Macromolecules,* 20(4): 705–12. https://doi.org/10.1021/ma00170a001.

Letcher, Trevor M. (2020). Chapter 1 - Introduction to plastic waste and recycling. pp. 3–12. *In*: Trevor, M. B. T. (ed.). *Plastic Waste and Recycling Letcher*. Academic Press. https://doi.org/https://doi.org/10.1016/B978-0-12-817880-5.00001-3.

Li, Chun and Joachim Kohn. (1989). Synthesis of poly(iminocarbonates): degradable polymers with potential applications as disposable plastics and as biomaterials. *Macromolecules,* 22(5): 2029–36. https://doi.org/10.1021/ma00195a001.

Li, Suming. (1999). Hydrolytic degradation characteristics of aliphatic polyesters derived from lactic and glycolic acids. *Journal of Biomedical Materials Research,* 48(3): 342–53. https://doi.org/https://doi.org/10.1002/(SICI)1097-4636(1999)48:3<342::AID-JBM20>3.0.CO;2-7.

Li, Yingzhan, Nathan Grishkewich, Lingli Liu, Chang Wang, Kam C. Tam, Shanqiu Liu, Zhiping Mao and Xiaofeng Sui. (2019). Construction of functional cellulose aerogels via atmospheric drying chemically cross-linked and solvent exchanged cellulose nanofibrils. *Chemical Engineering Journal,* 366(January): 531–38. https://doi.org/10.1016/j.cej.2019.02.111.

Liu, Jie, Zhiye Qiu, Shenqi Wang, Lei Zhou and Shengmin Zhang. (2010). A modified double-emulsion method for the preparation of daunorubicin-loaded polymeric nanoparticle with enhancedin vitroanti-tumor activity. *Biomedical Materials,* 5(6): 65002. https://doi.org/10.1088/1748-6041/5/6/065002.

Liu, Jun, Lushan Sun, Wenyang Xu, Qianqian Wang, Sujie Yu and Jianzhong Sun. (2019). Current advances and future perspectives of 3D printing natural-derived biopolymers. *Carbohydrate Polymers.* Elsevier Ltd. https://doi.org/10.1016/j.carbpol.2018.11.077.

Liu, Zhe, Zhenyu Cui, Yuwei Zhang, Shuhao Qin, Feng Yan and Jianxin Li. (2017). Fabrication of polysulfone membrane via thermally induced phase separation process. *Materials Letters,* 195(May): 190–93. https://doi.org/10.1016/j.matlet.2017.02.070.

Lu, Jianjun, John K. Jackson, Martin E. Gleave and Helen M. Burt. (2008). The preparation and characterization of anti-VEGFR2 conjugated, paclitaxel-loaded PLLA or PLGA microspheres for the systemic targeting of human prostate tumors. *Cancer Chemotherapy and Pharmacology,* 61(6): 997–1005. https://doi.org/10.1007/s00280-007-0557-x.

Lucas, Nathalie, Christophe Bienaime, Christian Belloy, Michèle Queneudec, Françoise Silvestre and José-Edmundo Nava-Saucedo. (2008). Polymer biodegradation: mechanisms and estimation techniques—a review. *Chemosphere,* 73(4): 429–42. https://doi.org/https://doi.org/10.1016/j.chemosphere.2008.06.064.

Luckachan, Gisha E. and Pillai, C. K. S. (2011). Biodegradable polymers—a review on recent trends and emerging perspectives. *Journal of Polymers and the Environment,* 19(3): 637–76. https://doi.org/10.1007/s10924-011-0317-1.

Lyu, SuPing and Darrel Untereker. (2009). Degradability of polymers for implantable biomedical devices. *International Journal of Molecular Sciences.* https://doi.org/10.3390/ijms10094033.

Mahmoudifar, Nastaran and Pauline M. Doran. (2010). Chondrogenic differentiation of human adipose-derived stem cells in polyglycolic acid mesh scaffolds under dynamic culture conditions. *Biomaterials,* 31(14): 3858–67. https://doi.org/https://doi.org/10.1016/j.biomaterials.2010.01.090.

Makadia, Hirenkumar K. and Steven J. Siegel. (2011). Poly lactic-co-glycolic acid (PLGA) as biodegradable controlled drug delivery carrier. *Polymers.* https://doi.org/10.3390/polym3031377.

Mangaraj, S., Ajay Yadav, Lalit M. Bal, Dash, S. K. and Naveen K. Mahanti. (2019). Application of biodegradable polymers in food packaging industry: a comprehensive review. *Journal of Packaging Technology and Research,* 3(1): 77–96. https://doi.org/10.1007/s41783-018-0049-y.

Maniruzzaman, Mohammed, Joshua S. Boateng, Martin J. Snowden and Dennis Douroumis. (2012). A review of hot-melt extrusion: process technology to pharmaceutical products. *ISRN Pharmaceutics,* 2012: 1–9. https://doi.org/10.5402/2012/436763.

Markočič, Elena, Tanja Botić, Sabina Kavčič, Tonica Bončina and Željko Knez. (2015). *In vitro* degradation of poly(d, l -lactide-co-glycolide) foams processed with supercritical fluids. *Industrial and Engineering Chemistry Research,* 54(7): 2114–19. https://doi.org/10.1021/ie504579y.

Martin, Jonathan W., Joannie W. Chin and Tinh Nguyen. (2003). Reciprocity law experiments in polymeric photodegradation: a critical review. *Progress in Organic Coatings,* 47(3): 292–311. https://doi.org/https://doi.org/10.1016/j.porgcoat.2003.08.002.

Martins, M., Rita Craveiro, Alexandre Paiva, Ana Rita C. Duarte and Rui L. Reis. (2014). Supercritical fluid processing of natural based polymers doped with ionic liquids. *Chemical Engineering Journal,* 241: 122–30. https://doi.org/10.1016/j.cej.2013.11.080.

Mateur, Molka Nafti, Danae Gonzalez Ortiz, Dorra Jellouli Ennigrou, Karima Horchani-Naifer, Mikhael Bechelany, Philippe Miele and Céline Pochat-Bohatier. (2020). Porous gelatin membranes obtained from pickering emulsions stabilized with H-BNNS: application for polyelectrolyte-enhanced ultrafiltration. *Membranes,* 10(7). https://doi.org/10.3390/membranes10070144.

Moad, Graeme. (2011). Chemical modification of starch by reactive extrusion. *Progress in Polymer Science (Oxford).* Elsevier Ltd. https://doi.org/10.1016/j.progpolymsci.2010.11.002.

Monsigny, Louis, Jean-Claude Berthet and Thibault Cantat. (2018). Depolymerization of waste plastics to monomers and chemicals using a hydrosilylation strategy facilitated by brookhart's iridium(III) catalyst. *ACS Sustainable Chemistry & Engineering,* 6(8): 10481–88. https://doi.org/10.1021/acssuschemeng.8b01842.

Muggli, Dina Svaldi, Amy K. Burkoth and Kristi S. Anseth. (1999). Crosslinked polyanhydrides for use in orthopedic applications: degradation behavior and mechanics. *Journal of Biomedical Materials Research,* 46(2): 271–78. https://doi.org/https://doi.org/10.1002/(SICI)1097-4636(199908)46:2<271::AID-JBM17>3.0.CO;2-X.

Murthy, Niren, Mingcheng Xu, Stephany Schuck, Jun Kunisawa, Nilabh Shastri and Jean M. J. Fréchet. (2003). A macromolecular delivery vehicle for protein-based vaccines: acid-degradable protein-loaded microgels. *Proceedings of the National Academy of Sciences,* 100(9): 4995 LP–5000. https://doi.org/10.1073/pnas.0930644100.

Muthuraj, Rajendran, Manjusri Misra and Amar Kumar Mohanty. (2018). Biodegradable compatibilized polymer blends for packaging applications: a literature review. *Journal of Applied Polymer Science,* 135(24): 45726. https://doi.org/https://doi.org/10.1002/app.45726.

Nagai, Yasutaka, Daisuke Nakamura, Tomoyuki Miyake, Hitoshi Ueno, Naomi Matsumoto, Atsushi Kaji and Fujio Ohishi. (2005). Photodegradation mechanisms in poly(2,6-butylenenaphthalate-co-tetramethyleneglycol) (PBN–PTMG). I: influence of the PTMG content. *Polymer Degradation and Stability,* 88(2): 251–55. https://doi.org/https://doi.org/10.1016/j.polymdegradstab.2004.10.016.

Nair, Lakshmi, S. and Cato T. Laurencin. (2007). Biodegradable polymers as biomaterials. *Progress in Polymer Science,* 32(8): 762–98. https://doi.org/https://doi.org/10.1016/j.progpolymsci.2007.05.017.

Ncube, Lindani K., Albert U. Ude, Enoch N. Ogunmuyiwa, Rozli Zulkifli and Isaac N. Beas. (2020). Environmental impact of food packaging materials: a review of contemporary development from conventional plastics to polylactic acid based materials. *Materials.* https://doi.org/10.3390/ma13214994.

Nguyen, David N., Shyam S. Raghavan, Lauren M. Tashima, Elizabeth C. Lin, Stephen J. Fredette, Robert S. Langer and Chun Wang. (2008). Enhancement of poly(orthoester) microspheres for dna vaccine delivery by blending with poly(ethylenimine). *Biomaterials,* 29(18): 2783–93. https://doi.org/https://doi.org/10.1016/j.biomaterials.2008.03.011.

Nita, Loredana Elena, Alina Ghilan, Alina Gabriela Rusu, Iordana Neamtu and Aurica P. Chiriac. (2020). New trends in bio-based aerogels. *Pharmaceutics,* 12(5): 499.

Pandey, Sharad Prakash, Tripti Shukla, Vinod Kumar Dhote, Dinesh K. Mishra, Rahul Maheshwari and Rakesh K. Tekade. (2019). Chapter 4 - Use of polymers in controlled release of active agents. pp. 113–72. *In*: *Advances in Pharmaceutical Product Development and Research*, edited by Rakesh K B T - Basic Fundamentals of Drug Delivery Tekade. Academic Press. https://doi.org/https://doi.org/10.1016/B978-0-12-817909-3.00004-2.

Pihlajamäki, Harri, Olli Tynninen, Pertti Karjalainen and Pentti Rokkanen. (2007). The impact of polyglycolide membrane on a tendon after surgical rejoining. a histological and histomorphometric analysis in rabbits. *Journal of Biomedical Materials Research Part A,* 81A(4): 987–93. https://doi.org/https://doi.org/10.1002/jbm.a.31144.

Pihlajamäki, Harri K., Sari T. Salminen, Olli Tynninen, Ole M. Böstman and Outi Laitinen. (2010). Tissue restoration after implantation of polyglycolide, polydioxanone, polylevolactide, and metallic pins in cortical bone: an experimental study in rabbits. *Calcified Tissue International,* 87(1): 90–98. https://doi.org/10.1007/s00223-010-9374-z.

Pitt, G. G., Gratzl, M. M., Kimmel, G. L., Surles, J. and Sohindler, A. (1981). Aliphatic polyesters II. The degradation of poly (dl-lactide), poly (ε-caprolactone), and their copolymers *in vivo*. *Biomaterials,* 2(4): 215–20. https://doi.org/https://doi.org/10.1016/0142-9612(81)90060-0.

Platzer, N. (1986). Encyclopedia of Polymer Science and Engineering. Mark, H.F., Bikales, N.M., Overberger, C.G. and Menges, G. (eds.). Wiley-Interscience, New York, 1985, 720 pp.

Polak, Marianne Berdugo, Fatemeh Valamanesh, Olivia Felt, Alicia Torriglia, Jean-Claude Jeanny, Jean-Louis Bourges, Patrice Rat et al. (2008). Controlled delivery of 5-chlorouracil using poly(ortho esters) in filtering surgery for glaucoma. *Investigative Ophthalmology & Visual Science,* 49(7): 2993–3003. https://doi.org/10.1167/iovs.07-0919.

Pospíšil, Jan, Zdenêk Horák, Zdenêk Kruliš and Stanislav Nešpůrek. (1998). The origin and role of structural inhomogeneities and impurities in material recycling of plastics. *Macromolecular Symposia,* 135(1): 247–63. https://doi.org/https://doi.org/10.1002/masy.19981350127.

Pulapura, Satish, Chun Li and Joachim Kohn. (1990). Structure-property relationships for the design of polyiminocarbonates. *Biomaterials,* 11(9): 666–78. https://doi.org/https://doi.org/10.1016/0142-9612(90)90025-L.

Qi, Mingbo, Xiaohong Li, Ye Yang and Shaobing Zhou. (2008). Electrospun fibers of acid-labile biodegradable polymers containing ortho ester groups for controlled release of paracetamol. *European Journal of Pharmaceutics and Biopharmaceutics,* 70(2): 445–52. https://doi.org/https://doi.org/10.1016/j.ejpb.2008.05.003.

Radeleff, Boris, Heidi Thierjung, Ulrike Stampfl, Sibylle Stampfl, Ruben Lopez-Benitez, Christof Sommer, Irina Berger and Goetz M. Richter. (2008). Restenosis of the CYPHER-Select, TAXUS-Express, and Polyzene-F nanocoated cobalt-chromium stents in the minipig coronary artery model. *CardioVascular and Interventional Radiology,* 31(5): 971–80. https://doi.org/10.1007/s00270-007-9243-y.

Raquez, Jean-Marie, Yogaraj Nabar, Ramani Narayan and Philippe Dubois. (2008). *In situ* compatibilization of maleated thermoplastic starch/polyester melt-blends by reactive extrusion. *Polymer Engineering & Science,* 48: 1747–54. https://doi.org/10.1002/pen.21136.

Raquez, Jean Marie, Yogaraj Nabar, Madhusudhan Srinivasan, Boo Young Shin, Ramani Narayan and Philippe Dubois. (2008). Maleated thermoplastic starch by reactive extrusion. *Carbohydrate Polymers,* 74(2): 159–69. https://doi.org/10.1016/j.carbpol.2008.01.027.

Reddy, Murali M., Singaravelu Vivekanandhan, Manjusri Misra, Sujata K. Bhatia and Amar K. Mohanty. (2013). Biobased plastics and bionanocomposites: current status and future opportunities. *Progress in Polymer Science,* 38(10): 1653–89. https://doi.org/https://doi.org/10.1016/j.progpolymsci.2013.05.006.

Reis, Rui L. and Julio San Román. (2004). *Biodegradable Systems in Tissue Engineering and Regenerative Medicine.* Edited by Rui L. Reis. 1st ed. CRC PRESS. https://doi.org/10.1201/9780203491232.

Roldughin, Vjacheslav I., Olga A. Serenko, Elena V. Getmanova, Natalia A. Novozhilova, Galina G. Nikifirova, Mikhail I. Buzin, Sergey N. Chvalun, Alexander N. Ozerin and Aziz M. Muzafarov. (2016). Effect of hybrid nanoparticles on glass transition temperature of polymer nanocomposites. *Polymer Composites,* 37(7): 1978–90. https://doi.org/10.1002/pc.23376.

Sakai, Koh, Yuri Kobayashi, Tsuguyuki Saito and Akira Isogai. (2016). Partitioned airs at microscale and nanoscale: thermal diffusivity in ultrahigh porosity solids of nanocellulose. *Scientific Reports,* 6(1): 1–7. https://doi.org/10.1038/srep20434.

Salman, Hesham H., Juan M. Irache and Carlos Gamazo. (2009). Immunoadjuvant capacity of flagellin and mannosamine-coated poly(anhydride) nanoparticles in oral vaccination. *Vaccine,* 27(35): 4784–90. https://doi.org/https://doi.org/10.1016/j.vaccine.2009.05.091.

Santoro, Orlando, Xin Zhang and Carl Redshaw. (2020). Synthesis of biodegradable polymers: a review on the use of schiff-base metal complexes as catalysts for the ring opening polymerization (ROP) of cyclic esters. *Catalysts.* https://doi.org/10.3390/catal10070800.

Sauceau, Martial, Jacques Fages, Audrey Common, Clémence Nikitine and Elisabeth Rodier. (2011). New challenges in polymer foaming: a review of extrusion processes assisted by supercritical carbon dioxide. *Progress in Polymer Science (Oxford).* Pergamon. https://doi.org/10.1016/j.progpolymsci.2010.12.004.

Seppälä, Jukka V., Antti O. Helminen and Harri Korhonen. (2004). Degradable polyesters through chain linking for packaging and biomedical applications. *Macromolecular Bioscience,* 4(3): 208–17. https://doi.org/https://doi.org/10.1002/mabi.200300105.

Serrano-Ruiz, Hadaly, Lluis Martin-Closas and Ana M. Pelacho. (2021). Biodegradable plastic mulches: impact on the agricultural biotic environment. *Science of The Total Environment,* 750: 141228. https://doi.org/https://doi.org/10.1016/j.scitotenv.2020.141228.

Shi, Shuyu, Xin Liu, Weiyi Li, Zhuo Li, Guoquan Tu, Baolin Deng and Chongxuan Liu. (2020). Tuning the biodegradability of chitosan membranes: characterization and conceptual design. *ACS Sustainable Chemistry and Engineering,* 8(38). https://doi.org/10.1021/acssuschemeng.0c04585.

Singh, Baljit and Nisha Sharma. (2008). Mechanistic implications of plastic degradation. *Polymer Degradation and Stability,* 93(3): 561–84. https://doi.org/https://doi.org/10.1016/j.polymdegradstab.2007.11.008.

Siqueira, Gilberto, Dimitri Kokkinis, Rafael Libanori, Michael K. Hausmann, Amelia Sydney Gladman, Antonia Neels, Philippe Tingaut, Tanja Zimmermann, Jennifer A. Lewis and André R. Studart. (2017). Cellulose nanocrystal inks for 3D printing of textured cellular architectures. *Advanced Functional Materials,* 27(12): 1604619. https://doi.org/10.1002/adfm.201604619.

Song, Fei, Dao-Lu Tang, Xiu-Li Wang and Yu-Zhong Wang. (2011). Biodegradable soy protein isolate-based materials: a review. *Biomacromolecules,* 12(10): 3369–80. https://doi.org/10.1021/bm200904x.

Soni, Raghav, Taka Aki Asoh and Hiroshi Uyama. (2020). Cellulose nanofiber reinforced starch membrane with high mechanical strength and durability in water. *Carbohydrate Polymers,* 238(June): 116203. https://doi.org/10.1016/j.carbpol.2020.116203.

Soon, Chu Yong, Norizah Abdul Rahman, Yee Bond Tee, Rosnita A. Talib, Choon Hui Tan, Khalina Abdan and Eric Wei Chiang Chan. (2019). Electrospun biocomposite: nanocellulose and chitosan entrapped within a poly(hydroxyalkanoate) matrix for congo red removal. *Journal of Materials Research and Technology,* 8(6): 5091–5102. https://doi.org/10.1016/j.jmrt.2019.08.030.

Sorrentino, Andrea, Giuliana Gorrasi and Vittoria Vittoria. (2007). Potential perspectives of bio-nanocomposites for food packaging applications. *Trends in Food Science & Technology,* 18(2): 84–95. https://doi.org/https://doi.org/10.1016/j.tifs.2006.09.004.

Sperling, L. H. and Carraher, C. E. (1985). Encyclopedia of polymer science and engineering. *In*: *Encyclopedia of Polymer Science and Engineering*, 12th ed. wiley.

Spiegel, Stefan. (2018). Synthetic biodegradable polymers. *Journal of Applied Polymer Science,* 135(24): 46279. https://doi.org/https://doi.org/10.1002/app.46279.

Sultan, Sahar, Hani Nasser Abdelhamid, Xiaodong Zou and Aji P. Mathew. (2019). CelloMOF: Nanocellulose enabled 3D printing of metal–organic frameworks. *Advanced Functional Materials,* 29(2): 1805372. https://doi.org/10.1002/adfm.201805372.

Tamada, J. A. and Langer, R. (1993). Erosion kinetics of hydrolytically degradable polymers. *Proceedings of the National Academy of Sciences,* 90(2): 552 LP–556. https://doi.org/10.1073/pnas.90.2.552.

Tamayo, I., Irache, J. M., Mansilla, C., Ochoa-Repáraz, J., Lasarte, J. J. and Gamazo, C. (2010). Poly(anhydride) nanoparticles act as active Th1 adjuvants through toll-like receptor exploitation. *Clinical and Vaccine Immunology,* 17(9): 1356 LP–1362. https://doi.org/10.1128/CVI.00164-10.

Thermoplastic Processing of Protein-based Bioplastics: Chemical Engineering Aspects of Mixing, Extrusion and Hot Molding. (2003). *In Macromolecular Symposia* 197(1): 207–18. https://doi.org/10.1002/masy.200350719.

Thomas, Coralie, Frédéric Peruch and Brigitte Bibal. (2012). Ring-opening polymerization of lactones using supramolecular organocatalysts under simple conditions. *RSC Advances,* 2(33): 12851–56. https://doi.org/10.1039/C2RA22535B.

Tian, Huayu, Zhaohui Tang, Xiuli Zhuang, Xuesi Chen and Xiabin Jing. (2012). Biodegradable synthetic polymers: preparation, functionalization and biomedical application. *Progress in Polymer Science,* 37(2): 237–80. https://doi.org/https://doi.org/10.1016/j.progpolymsci.2011.06.004.

Tokiwa, Yutaka and Buenaventurada P. Calabia. (2007). Biodegradability and biodegradation of polyesters. *Journal of Polymers and the Environment,* 15(4): 259–67. https://doi.org/10.1007/s10924-007-0066-3.

Tomietto, Pacôme, Maewenn Carré, Patrick Loulergue, Lydie Paugam and Jean Luc Audic. (2020). Polyhydroxyalkanoate (PHA) based microfiltration membranes: tailoring the structure by the non-solvent induced phase separation (NIPS) process. *Polymer,* 204(September): 122813. https://doi.org/10.1016/j.polymer.2020.122813.

Toshikj, Nikola, Jean-Jacques Robin and Sebastien Blanquer. (2020). A simple and general approach for the synthesis of biodegradable triblock copolymers by organocatalytic ROP from poly(lactide) macroinitiators. *European Polymer Journal,* 127: 109599. https://doi.org/https://doi.org/10.1016/j.eurpolymj.2020.109599.

Tschan, Mathieu J.-L., Emilie Brulé, Pierre Haquette and Christophe M. Thomas. (2012). Synthesis of biodegradable polymers from renewable resources. *Polymer Chemistry,* 3(4): 836–51. https://doi.org/10.1039/C2PY00452F.

Uhrich, Kathryn E., Scott M. Cannizzaro, Robert S. Langer and Kevin M. Shakesheff. (1999). Polymeric systems for controlled drug release. *Chemical Reviews,* 99(11): 3181–98. https://doi.org/10.1021/cr940351u.

Uitterhaegen, Evelien and Philippe Evon. (2017). Twin-screw extrusion technology for vegetable oil extraction: a review. *Journal of Food Engineering.* Elsevier Ltd. https://doi.org/10.1016/j.jfoodeng.2017.06.006.

Ulery, Bret D., Lakshmi S. Nair and Cato T. Laurencin. (2011). Biomedical applications of biodegradable polymers. *Journal of Polymer Science Part B: Polymer Physics,* 49(12): 832–64. https://doi.org/https://doi.org/10.1002/polb.22259.

Vasanthan, N. (2009). 7 - Polyamide fiber formation: structure, properties and characterization. *In*: *Woodhead Publishing Series in Textiles*, edited by S. J. Eichhorn, J. W. S. Hearle, M. Jaffe, and T. B. T. - Handbook of Textile Fibre Structure Kikutani, 1: 232–56. Woodhead Publishing. https://doi.org/https://doi.org/10.1533/9781845696504.2.232.

Venault, Antoine, Yung Chang, Da-Ming Wang and Denis Bouyer. (2013). A review on polymeric membranes and hydrogels prepared by vapor-induced phase separation process. *Polymer Reviews,* 53(4): 568–626. https://doi.org/10.1080/15583724.2013.828750.

Vieira, A. C., Vieira, J. C., Guedes, R. M. and Marques, A. T. (2010). Degradation and viscoelastic properties of PLA-PCL, PGA-PCL, PDO and PGA fibres. *Materials Science Forum,* 636–637(January): 825–32. https://doi.org/10.4028/www.scientific.net/MSF.636-637.825.

Vroman, Isabelle and Lan Tighzert. (2009). Biodegradable polymers. *Materials.* https://doi.org/10.3390/ma2020307.

Wang, Chun, Qing Ge, David Ting, David Nguyen, Hui-Rong Shen, Jianzhu Chen, Herman N. Eisen, Jorge Heller, Robert Langer and David Putnam. (2004). Molecularly engineered poly(ortho ester) microspheres for enhanced delivery of DNA vaccines. *Nature Materials,* 3(3): 190–96. https://doi.org/10.1038/nmat1075.

Wang, Da Ming and Juin Yih Lai. (2013). Recent advances in preparation and morphology control of polymeric membranes formed by nonsolvent induced phase separation. *Current Opinion in Chemical Engineering.* Elsevier Ltd. https://doi.org/10.1016/j.coche.2013.04.003.

Wang, Limin, Nathan H. Dormer, Lynda F. Bonewald and Michael S. Detamore. (2010). Osteogenic differentiation of human umbilical cord mesenchymal stromal cells in polyglycolic acid scaffolds. *Tissue Engineering Part A,* 16(6): 1937–48. https://doi.org/10.1089/ten.tea.2009.0706.

Wang, Qianqian, Chencheng Ji, Lushan Sun, Jianzhong Sun and Jun Liu. (2020). Cellulose nanofibrils filled poly(lactic acid) biocomposite filament for FDM 3D printing. *Molecules,* 25(10): 2319. https://doi.org/10.3390/molecules25102319.

Wang, Xiu-Li, Ke-Ke Yang and Yu-Zhong Wang. (2004). Properties of starch blends with biodegradable polymers. *Journal of Macromolecular Science® Part C—Polymer Reviews Vol. C43* No. 3(January): 385–409. https://doi.org/10.1081/MC-120023911.

Wayne, J., Maddever, W. and Chapman, G. M. (1989). Modified starch-based biodegradable plastics. *Plastics Engineering,* 31–34.

Woodard, Lindsay N. and Melissa A. Grunlan. (2018). Hydrolytic degradation and erosion of polyester biomaterials. *ACS Macro Letters,* 7(8): 976–82. https://doi.org/10.1021/acsmacrolett.8b00424.

Wróblewska-Krepsztul, Jolanta, Tomasz Rydzkowski, Gabriel Borowski, Mieczysław Szczypiński, Tomasz Klepka and Vijay Kumar Thakur. (2018). Recent progress in biodegradable polymers and nanocomposite-based packaging materials for sustainable environment. *International Journal of Polymer Analysis and Characterization,* 23(4): 383–95. https://doi.org/10.1080/1023666X.2018.1455382.

Wu, Feng, Manjusri Misra and Amar K. Mohanty. (2020). Sustainable green composites from biodegradable plastics blend and natural fibre with balanced performance: synergy of nano-structured blend and reactive extrusion. *Composites Science and Technology,* 200(November): 108369. https://doi.org/10.1016/j.compscitech.2020.108369.

Xiong, Zhu, Haibo Lin, Yun Zhong, Yan Qin, Tiantian Li and Fu Liu. (2017). Robust superhydrophilic polylactide (PLA) membranes with a TiO_2 nano-particle inlaid surface for oil/water separation. *J. Mater. Chem. A,* 5(14): 6538–45. https://doi.org/10.1039/C6TA11156D.

Xu, Liang, Dejun Cao, Wei Liu, Guangdong Zhou, Wen Jie Zhang and Yilin Cao. (2010). *In vivo* engineering of a functional tendon sheath in a hen model. *Biomaterials,* 31(14): 3894–3902. https://doi.org/https://doi.org/10.1016/j.biomaterials.2010.01.106.

Young, Simon, Zarana S. Patel, James D. Kretlow, Matthew B. Murphy, Paschalia M. Mountziaris, L. Scott Baggett, Hiroki Ueda et al. (2009). Dose effect of dual delivery of vascular endothelial growth factor and bone morphogenetic protein-2 on bone regeneration in a rat critical-size defect model. *Tissue Engineering Part A,* 15(9): 2347–62. https://doi.org/10.1089/ten.tea.2008.0510.

Zhang, Guizhen, Ting Wu, Wangyang Lin, Yongbin Tan, Rongyuan Chen, Zhaoxia Huang, Xiaochun Yin and Jinping Qu. (2017). Preparation of polymer/clay nanocomposites via melt intercalation under continuous elongation flow. *Composites Science and Technology,* 145(June): 157–64. https://doi.org/10.1016/j.compscitech.2017.04.005.

Zhang, Zheng, Roel Kuijer, Sjoerd K. Bulstra, Dirk W. Grijpma and Jan Feijen. (2006). The *in vivo* and *in vitro* degradation behavior of poly(trimethylene carbonate). *Biomaterials,* 27(9): 1741–48. https://doi.org/https://doi.org/10.1016/j.biomaterials.2005.09.017.

Zhong, Nongping, Tao Dong, Zhongchun Chen, Yongwei Guo, Zhengzhong Shao and Xia Zhao. (2019). A novel 3D-printed silk fibroin-based scaffold facilitates tracheal epithelium proliferation *in vitro. Journal of Biomaterials Applications,* 34(1): 3–11. https://doi.org/10.1177/0885328219845092.

Zong, Xin-Hua, Zhi-Gang Wang, Benjamin S. Hsiao, Benjamin Chu, Jack J. Zhou, Dennis D. Jamiolkowski, Eugene Muse and Edward Dormier. (1999). Structure and morphology changes in absorbable poly(glycolide) and poly(glycolide-co-lactide) during *in vitro* degradation. *Macromolecules,* 32(24): 8107–14. https://doi.org/10.1021/ma990630p.

Chapter 4
Analysis and Testing of Biopolymers

María C. García-Castañeda,[1] *Kassandra T. Ávila-Alvarez,*[2] *Marco A. García-Lobato,*[2] *Anna Ilyina*[2] and *Rodolfo Ramos-González*[3,*]

1. Thermal analyses

1.1 Thermogravimetric analysis

Thermogravimetric or Thermal Gravimetric Analysis (TGA) is an analytic method of thermal analysis in which the mass of a sample is measured as a function of the temperature or time. The results of the measurements in TGA are displayed as a thermogram curve in which percent mass (or mass in milligrams) is plotted against temperature, time or both (Bottom, 2008). In the biomaterials and biopolymers fields, the TGA technique is mainly applied to analyze the content and/or the thermal stability of these materials.

These days, the course of the Science and Technology (S&T) is moving into renewable resources. In this setting, biopolymers have been evaluated as renewable resources and as materials with several applications in a great diversity of areas (Bahri et al., 2021). Das et al. (2021) recently synthesized a new guar gum indole acetate ester for applications in tissue engineering. They realized the study of the thermal properties of the biopolymer, and the film prepared with it, by TGA. The analyses were carried on from 30°C to 500°C with a heating rate of 10°/min under nitrogen atmosphere. The TGA results allowed to know the moist content and the decomposition temperature of the materials. Hanauer et al. (2021), synthesized a pectin-based biohydrogels reinforced with *Eucalyptus globulus* sawdust. The pectin biohydrogels obtained were employed as β-D-galactosidase enzyme immobilization. And a possible application in food processing is described. The TGA experiments were performed from ambient temperature to 500°C with a heating rate of 20°/min and an N_2 atmosphere. The thermal analyses corroborate the results obtained by FTIR and demonstrate that the pectin-based biohydrogel reinforced with *E. globulus* sawdust has higher crosslinking densities. On the other hand, an adhesive from hydrolyzed lignin, epoxidized soybean oil and soy protein have been recently developed by Chen et al. (2021). The thermogravimetric analysis serves to characterize the thermal properties of the material. The analysis was performed

[1] CONACYT – Universidad de Guanajuato, Loma del Bosque 103, León, Guanajuato 37150, Mexico.
[2] Facultad de Ciencias Químicas, Universidad Autonoma de Coahuila, Blvd. V. Carranza & J. Cárdenas Valdés, Saltillo, Coahuila 25280, Mexico.
[3] CONACYT – Universidad Autonoma de Coahuila, Blvd. V. Carranza & J. Cárdenas Valdés, Saltillo, Coahuila 25280, Mexico.
* Corresponding author: rodolfo.ramos@uadec.edu.mx

in the range of 25°C–600°C at a heating rate of 10°/min in an N_2 atmosphere (flow of 40 mL/min). The results obtained by TGA demonstrated the thermal stability of the biopolymer-based adhesive.

One of the most versatile analysis techniques is TGA. Using this technique, it is possible to analyze both inorganic, organic and natural samples. Providing important information on the compositional and thermal stability of both compounds and finished products. The obtained TGA thermograms bring information about moisture or volatile constituents present in the sample, oxidative decomposition of organic samples, carbonization and pyrolysis of organic compounds and heterogeneous chemical reactions (Bottom, 2008).

1.2 Differential scanning calorimetry

Differential Scanning Calorimetry (DSC) is one of the most widely used thermal analysis techniques. It provides a quick and easy method to use obtaining a large amount of information about a sample, both natural and synthetic. DSC is used in various areas such as polymers and plastics, food and pharmaceuticals, proteins and life science materials (Gabbott, 2008). DSC is a thermal analytical technique that measures the difference of energy required to increase or decrease the temperature of the analyzed sample and a reference material (generally air) (Gabbott, 2008; Sindhu et al., 2015). The most important properties that can be analyzed by differential scanning calorimetry are crystallization and melting behavior, glass transition temperature (Tg) and kinetics and reaction enthalpies. Differential scanning calorimetry is also a suitable thermal analytical technique for the study of the hydration of biopolymers (Dehabadi et al., 2016).

Carboxymethyl cellulose is a biopolymer with a wide application in diverse areas of S&T, such as tissue engineering, food packaging, tissue engineering and drug-controlled delivery system (Klunklin et al., 2020). Yaradoddi et al. (2020) developed a carboxymethyl cellulose-based packaging material. DSC analysis allows the evaluation of the glass transition, crystallization and melting temperatures of the films. Klunklin et al. (2020) synthesized carboxymethyl cellulose from *Asparagus officinalis* stalk. The thermal characterization was conducted by DSC, specifically, the authors evaluated the effect of NaOH concentration, in the synthesis of the biopolymer, on its the melting temperature. The analyses were conducted from 40°C to 450°C with a heating rate of 10°/min and under an N_2 atmosphere.

Moreover, differential scanning calorimetry is a suitable technique for the study of biomolecules stability (Bowers and Markova, 2019). Vaskoska et al. (2021) studied the changes in secondary structure of bovine muscles proteins. DSC analysis allowed to characterize and differentiate between specific proteins by analyzing the denaturalization temperature. On the other hand, the analyses of lipids, in terms of health and food technology, are usually performed by conventional methods and techniques that are hazardous, tedious and time-consuming. An alternative technique, as DSC, have been used to evaluate and characterize the oxidative stability of lipids (Tengku-Rozaina and Birch, 2019). Differential scanning calorimetry is also an adequate technique to study and characterize polysaccharides as starch, cellulose, methyl cellulose, polyacrylamide, chitin, among others (Zhorin and Kiselev, 2020). Mlčoch and Kučerík (2013) used DSC to study the hydration and drying processes of various biopolymers such as carboxymethyl cellulose, cellulose, chitosan, hyaluronan and schizophyllan.

1.3 Mechanical thermal analyses

As part of the new strategies to reduce single-use plastics, such as packaging, disposables, water containers, etc., as well as the search for new industrial and technological applications for bio-based polymers such as biomedical and dental prostheses, wound healing materials, automotive and electrical components, sporting goods and fibers, among others; new developments based on biodegradable or environmentally friendly materials, should have good mechanical performance. Among the tests carried out by the best known thermal analysis methods are Thermomechanical

Analysis (TMA) and Dynamic Mechanical Analysis (DMA), which allow studies of a mechanical nature as a function of temperature using small amounts of material.

1.3.1 Thermomechanical analysis

Thermomechanical analysis is a simple technique that allows monitoring the glass transition, the coefficient of thermal expansion of materials and modulus with the possibility of working as a function of time or with different programmed temperature ramps (Duncan, 2008). This analysis can use different geometries that keep the subject under constant stress, compression or tension; from different probe types such as compression, penetration, tension and volumetric (Price and Duncan, 2016).

In recent decades, efforts have been made to improve the mechanical properties of polyhydroxyalkanoates, and it has been determined that the key to achieving this is through the modification of the morphology of the crystalline phase. Scalioni et al. (2017), worked with poly(hydroxybutyrate) (PHB), by using Curaua fibers and triethyl citrate as an eco-friendly plasticizer. The thermal expansion coefficient measurements made it possible to directly relate this value to the degrees of freedom of the polymer chains, contributing to the determination of the PHB morphology.

Likewise, the use of natural reinforcements in Poly(Lactic Acid) (PLA) has improved its mechanical resistance without impairing the biodegradability of this material, making it attractive for the packaging sector. An analysis carried out from 30 to 120°C in an inert atmosphere with a ramp of 3°C min^{-1} showed a reduction in the expansion of PLA, making evident the decrease in its stiffness as a consequence of the change from glass to a rubber state when using different contents of cellulose (Espinach et al., 2018).

Ashaduzzaman et al. (2020) prepared thermoplastic starch films with different kaolin content and, through dilatometry studies with TMA, indicated that the filler played a role in the crystalline phase that prevents thermal expansion in the range of 37–95°C by increasing the interfacial interaction with the polymeric matrix. Moreover, Price and Duncan (2016), pointed out that dimensional changes in polymers can also be attributed to frozen stresses resulting from cooling, which generate a type of residual stress in the material.

1.3.2 Dynamic mechanical analysis

Dynamic mechanical analysis tests make it possible to measure the viscoelastic behavior of polymeric materials subjected to sinusoidal deformation or oscillating stress as a function of temperature, time or frequency. Additionally, the DMA can provide information about changes in stiffness, damping and thermal transitions of polymeric systems (De Nardo and Farè, 2017). Moreover, the modulus obtained by this technique does not correspond to the classical stress-strain curve; but to a complex shear module (E*), defined by the sum of the elastic (storage) and viscous (loss) moduli (E' and E'', respectively) (Dyamenahalli et al., 2015). Compared with other thermomechanical techniques, DMA has a high sensitivity since it monitors the mechanical property at the molecular level that allows to detect secondary thermal transitions or other types of reactions such as curing, crosslinking and degradation. Similarly, this type of analysis is also helpful to study the efficiency of reinforcements in nanocomposites as a function of temperature (Gan et al., 2020). In the analysis of membranes or films, the viscoelastic properties can be related to the relaxation of the amorphous phase of the bio-based polymers (Tomoda et al., 2020). On the other hand, DMA has different geometric arrangements such as simple shear, tension, clamped bending and three-point bending that can be adapted according to the stiffness of the materials (Duncan, 2008).

Biodegradability studies have been monitored using the film claw tension with a frequency of 1 Hz, 0.15 N preload, 400 mm amplitude, and a ramp of 3°C min^{-1} from 40 to 270°C for cellulose acetate films for 6 months where the results showed that the amorphous region of the polymer tends to degrade first, leaving more significant crystalline regions that increased the storage modulus during the first months of the study (Freitas and Botaro, 2018). On the other hand, a similar study was carried out to monitor the crosslinking of cellulose acetate films with DMSO, using a frequency of 1 Hz, 4 mm amplitude, 0.25 N preload of 25–260°C at 3°C min^{-1} (Kaschuk et al., 2021). In this case, changes in the glass transition were associated with reducing the rotational mobility of single bonds in the non-crystalline region to crosslinked specimens. While an analysis of the Tand curves allowed to identify two secondary transitions of the Tg corresponding to a non-crosslinked region and at higher temperatures a region of short segments between crosslinks.

2. Mechanical properties

One of the objectives of biopolymers is to have properties equal to or similar to consumer commodities, and it is for this reason, mechanical properties frequently limit and determine the final applications of bio-based plastics (Tomoda et al., 2020). These materials are subjected to tests that are generally already standardized under specific norms that provide information related to the preparation and conditioning of the samples and the operating conditions of the equipment.

2.1 Tensile test

Tensile tests are a valuable tool to monitor the mechanical and structural properties of bio-based polymers and bio-composites. It is a destructive method in which the test specimens are subjected to regular forces (pulling) in order to know Young's modulus through the slope of the stress-strain curve; and it is possible to know how rigid material is (tensile strength) and how much it can be elongated (yield strength) to its breaking point (Aaliya et al., 2021). There are standardized methods in which the design and dimension of the specimens are described and the conditioning of the specimens before testing. In the United States, the methods are conducted by the American Society for Testing and Materials (ASTM); the methods in Germany are described by the Deutsches Institut für Normung (DIN) and in the United Kingdom by the British Standards Institution (BSI), among others (Tomoda et al., 2020).

Currently, new biopolymer mixtures are being developed that seek to provide a more sustainable alternative with better properties for the 3D printing industry with applications in medicine. Tensile strength tests under the ISO 527-2 standard for mixtures of poly(lactic) acid/poly(ε-caprolactone) (PLA/PCL) showed that a low content of PCL favored a ductile behavior and on the contrary, brittle behavior can be seen with high PCL content. At the same time, Young's modulus suggested the presence of a phase inversion between both polymers as a consequence of the high cutting speed during the injection process (Jeantet et al., 2021). On the other hand, the effect of crosslinking in cellulose acetate was also followed by tensile tests, where a substantial improvement in their toughness could be observed in the reticulated samples (Kaschuk et al., 2021).

2.2 Impact resistance

Similar to tensile tests, impact resistance is subjected to a destructive process. It is carried out under standardized norms and methods where the characteristics of the specimens (shape and dimensions) are described in detail and the conditioning of temperature and humidity. The main methods for testing are Izod and Charpy, both of which measure the energy required to fracture a sample into two or more pieces mechanically. The main differences lie in the position of the specimen and the hitting direction. In addition, the test can be carried out with or without a notch, so the energy required to fracture the samples without a notch can be one or two orders of magnitude higher than the specimens that are notched (DeArmitt, 2017). This effect can be seen in the studies carried out by Burzic et al. (2019), in which the Charpy tests performed un-notched yielded higher values concerning those that did have a notch for PLA/PHA mixtures. On the other hand, it is typical for the addition of fillers to reduce impact resistance, so in the development of formulations, the dispersion and content of the filler in the polymeric matrix must be taken care of during the formulation of compounds (Aaliya et al., 2021).

3. Degree of crystallinity

The degree of crystallinity of biopolymers such as cellulose is an important parameter to understand and characterize because the presence of crystalline regions in cellulose molecules determines their chemical, physical and mechanical properties (Agarwal et al., 2018; Ju et al., 2015). The Crystallinity Index (CrI) is determined to quantify the content of crystalline cellulose in cellulosic materials, it has also been applied to evaluate the changes in the cellulose structure after been submitted to biological or physicochemical treatments (Ju et al., 2015). Segal et al. (1959) proposed a method to estimate the degree of crystallinity of cellulose by X-ray diffractometry. By this method, the ratio of the height of the reflection of the plane (2 0 0) and the minimum height between the peaks of the planes (2 0 0) and (1 1 0) is used. Ju et al. (2015) proposed an improved method for cellulose crystallinity determination by X-ray diffractometry. Agarwal et al. (2018) developed a new method to estimate the crystallinity of cellulose by Raman spectroscopy. Meanwhile, Agarwal et al. (2013) proposed a method to calculate the crystallinity using near-IR FT-Raman spectroscopy. As can be seen, there are several methods and techniques to the estimation of the crystallinity and all of them possess the advantage of using specialized equipment with high accuracy.

The estimation of CrI is useful to evaluate the treatments and pretreatments of lignocellulosic materials. Aguirre-Fierro et al. (2020) confirmed the effect of high-pressure agave bagasse pretreatment by the estimation of CrI by X-Ray diffractometry. Moreover, the estimation of the crystallinity by XRD is useful to corroborate the sonication time of treatment of chitin to obtain chitosan (Vallejo-Domínguez et al., 2021). On the other hand, poly-3-hydroxybutyrate (PHB) is a biopolymer that is an alternative to the use of polypropylene, this is due to their similar physical properties. PHB has a high crystallinity (about 50–70%) and the determination of the CrI is a rapid and precise manner to evaluate the biosynthesis of this biopolymer (Sirohi et al., 2021).

4. Crosslinking

Biopolymers have been used in diverse areas as delivery devices. Specifically the cross-linked biopolymers networks, are also known as hydrogels. In the medical and pharmaceutical fields, these biomaterials are used as controlled drug delivery systems. In the agro-industrial field they are employed as water delivery systems. In the environmental field, these materials are employed as absorbents of contaminants. Biopolymer-based hydrogels have very interesting properties, such as swelling behavior, absorption capacity, permeability, bioactivity, among others (Younis et al., 2018). Specifically, the swelling index or swelling degree is an important test to analyze in hydrogels to know about the amount (in mass) of solvent (water) that the hydrogel can absorb, in other words, the swelling index is defined as the fractional mass increase of the hydrogel due to the amount of water absorbed (Park et al., 2009).

Younis et al. (2018) synthesized alginate – chitosan – Arabic gum hydrogels as drug release systems. Specifically, optimization of swelling in the acidic and neutral medium was performed. Meng et al. (2019) synthesized "super-swelling" hydrogels utilizing a residual subproduct of the paper industry and acrylic acid, this waste subproduct is rich in lignosulphonate and polysaccharides. The obtained hydrogels exhibit a super-swelling ratio of 280 g/g (mass of water per gram of dry gel) and a slow water release behavior, making it a suitable candidate for applications in agriculture for water retention. On the other hand, Zhang et al. (2020) proposed the sieve filtration method to measure the swelling index of hydrogels. Their method is easy to achieve and is suitable for different types of hydrogels and aqueous solutions. Also large amounts of the sample are not needed.

On the other hand, the crosslinking density is defined as the density of chemical chains (segments) that interconnect two parts of the polymeric network (Wool, 2005). The pore size of a hydrogel can be controlled by adjusting or changing the crosslinking density in the hydrogel network. The higher the crosslinking density, the smaller the pore size, and vice versa (Kopač et al., 2020). It is important to know this property as depending on the pore size its possible applications can be addressed. Kopač et al. (2020) synthesized hydrogels of alginate and 2,2,6,6-tetramethylpiperidine-1-oxyl-oxidized cellulose nanofibers, and calcium chloride as a crosslinker agent. The study aimed to manipulate the crosslinking density by changing the concentrations of the polymer and the crosslinker agent. By the analysis of the crosslinking density, a mathematical model was developed and was useful for predicting the release rate. Crosslinking density normally is measured by swelling methods accompanied by the Flory-Rehner equation. The use of analytical techniques for measuring the crosslinking density is also suitable. Nuclear Magnetic Resonance (NMR) spectroscopy has been used to measure the crosslinking density of vulcanized natural rubbers (bin Ahmad and bin Amu, 1998; Son and Choi, 2019).

5. Degradation

The degradation of bio-based polymers, in general, depends mainly on the nature of their composition and structure, since these characteristics are subject to various thermal, chemical (oxidation, hydrolysis, cyclizations, etc.), and photodestructive processes and the loss of mechanical properties, which limit the usefulness of the polymer and its ability to be reprocessed (Myasoedova, 2017).

Materials such as poly (lactic acid) (Shojaeiarani et al., 2019), polybutylene succinate and polyhydroxyalkanoate blends (Resch-Fauster et al., 2017) have difficulties to be reprocessed due to the ease they have to thermally degrade due to the generation of hydrolysis reactions, consequently reducing their mechanical properties. Other materials derived from glucose chains such as cellulose and starch decompose, releasing CO, CO_2 and H_2O in oxidative atmosphere and carbon in the presence of nitrogen (Aggarwal et al., 1997). The photodegradation of biopolymers is carried out by the exposure of these materials to sources of UV radiation. Mobile panels for weathering tests designed to monitor the amount of UV light coming from the sun have been built, and these tests can usually last weeks or months of exposure in the open. On the other hand, accelerated aging tests allow degradation studies to be carried out under controlled conditions and less time of UV light exposure. These tests use simulated environmental factors such as humidity, rain, sunlight, wind, etc. Studies of biocomposites of PLA and agave fibers have been carried out accelerating aging tests in which the samples are exposed to 4 hours cycles of UV light followed by cycles of water condensation according to the ASTM G154 standard (Martín del Campo et al., 2021).

In general, the monitoring of the degradation of these materials is mainly given by thermal, mechanical and spectroscopic analyzes. However, other techniques such as the whiteness and yellowness index (WI and YI, respectively) are helpful in the industry since the measurements are carried out quickly and easily using a manual colorimeter or spectrophotometers. Three coordinates can describe all colors, L*, with values from 0–100 (indicating a scale from black to white, respectively); positive and negatives values of a* (redness and greenness, respectively); and b* values (positive and negative values corresponding to yellowness and blueness, respectively), such that changes in those coordinates could be an indicator of the degradation of the materials (Croitoru, 2018; Shahabi-Ghahfarrokhi et al., 2019).

Conclusion

Over the years, a large number of valuable techniques have been utilized to analyze and test biopolymers. This review showed the state-of-the-art, most frequently used testing techniques employed to analyze natural polymers and biopolymers. The described methods were classified into five properties that can be analyzed in these biomaterials. These categories were thermal analyses, mechanical tests, crystallinity degree tests, crosslinking, and degradation tests. Also, current applications and existing methods were briefly outlined and can serve as a first approach for the interested researcher to characterize natural polymers or biopolymer samples.

References

Aaliya, B., Sunooj, K. V. and Lackner, M. (2021). Biopolymer composites: a review. *International Journal of Biobased Plastics*, 3(1): 40–84. https://doi.org/10.1080/24759651.2021.1881214.

Agarwal, U. P., Reiner, R. R. and Ralph, S. A. (2013). Estimation of cellulose crystallinity of lignocelluloses using near-IR FT-raman spectroscopy and comparison of the raman and segal-WAXS methods. *Journal of Agricultural and Food Chemistry*, 61(1). https://doi.org/10.1021/jf304465k.

Agarwal, U. P., Ralph, S. A., Reiner, R. S. and Baez, C. (2018). New cellulose crystallinity estimation method that differentiates between organized and crystalline phases. *Carbohydrate Polymers*, 190. https://doi.org/10.1016/j.carbpol.2018.03.003.

Aggarwal, P., Dollimore, D. and Heon, K. (1997). Comparative thermal analysis study of two biopolymers, starch and cellulose. *Journal of Thermal Analysis*, 50(1–2): 7–17. https://doi.org/10.1007/BF01979545.

Aguirre-Fierro, A., Ruiz, H. A., Cerqueira, M. A., Ramos-González, R., Rodríguez-Jasso, R. M., Marques, S. and Lukasik, R. M. (2020). Sustainable approach of high-pressure agave bagasse pretreatment for ethanol production. *Renewable Energy*, 155. https://doi.org/10.1016/j.renene.2020.04.055.

Ashaduzzaman, M., Saha, D. and Mamunur Rashid, M. (2020). Mechanical and thermal properties of self-assembled kaolin-doped starch-based environment-friendly nanocomposite films. *Journal of Composites Science*, 4(2): 38. https://doi.org/10.3390/jcs4020038.

Bahri, F., Shadi, M., Mohammadian, R., Javanbakht, S. and Shaabani, A. (2021). Cu-decorated cellulose through a three-component Betti reaction: An efficient catalytic system for the synthesis of 1,3,4-oxadiazoles via imine C H functionalization of N-acylhydrazones. *Carbohydrate Polymers*, 265. https://doi.org/10.1016/j.carbpol.2021.118067.

Bin Ahmad, A. and Bin Amu, A. (1998). Estimation of crosslink density by solid-state NMR spectroscopy. *In*: Tinker, A. J. and Jones, K. P. (eds.). *Blends of Natural Rubber* (1st ed.). Springer Netherlands. https://doi.org/10.1007/978-94-011-4922-8_4.

Bottom, R. (2008). Thermogravimetric analysis. *In*: Gabbott, P. (ed.). *Principles and Applications of Thermal Analysis* (1st ed.). Blackwell Publishing Ltd. https://doi.org/10.1002/9780470697702.ch3.

Bowers, K. and Markova, N. (2019). Value of DSC in characterization and optimization of protein stability. *In*: Ennifar, E. (ed.). *Microcalorimetry of Biological Molecules. Methods in Molecular Biology* (1st ed., Vol. 1964). Humana Press. https://doi.org/10.1007/978-1-4939-9179-2_3.

Burzic, I., Pretschuh, C., Kaineder, D., Eder, G., Smilek, J., Másilko, J. and Kateryna, W. (2019). Impact modification of PLA using biobased biodegradable PHA biopolymers. *European Polymer Journal*, 114: 32–38. https://doi.org/10.1016/j.eurpolymj.2019.01.060.

Chen, S., Chen, Y., Wang, Z., Chen, H. and Fan, D. (2021). Renewable bio-based adhesive fabricated from a novel biopolymer and soy protein. *RSC Advances*, 11(19). https://doi.org/10.1039/D1RA00766A.

Croitoru, C. (2018). A durability assessment and structural characterization of biopolymer-impregnatedwood. Drewno: Prace Naukowe, Doniesienia, Komunikaty, 61(202): 129–143. https://doi.org/10.12841/wood.1644-3985.232.01.

Das, A., Das, A., Basu, A., Datta, P., Gupta, M. and Mukherjee, A. (2021). Newer guar gum ester/chicken feather keratin interact films for tissue engineering. *International Journal of Biological Macromolecules*, 180. https://doi.org/10.1016/j.ijbiomac.2021.03.034.

De Nardo, L. and Farè, S. (2017). Dynamico-mechanical characterization of polymer biomaterials. pp. 203–232. *In*: Tanzi, M. C. and Faré, S. (eds.). *Characterization of Polymeric Biomaterials*. Elsevier. https://doi.org/10.1016/B978-0-08-100737-2.00009-1.

DeArmitt, C. (2017). Functional fillers for plastics. pp. 517–532. *In*: Kutz, M. (ed.). *Applied Plastics Engineering Handbook*. Elsevier. https://doi.org/10.1016/B978-0-323-39040-8.00023-7.

Dehabadi, L., Udoetok, I. A. and Wilson, L. D. (2016). Macromolecular hydration phenomena. *Journal of Thermal Analysis and Calorimetry*, 126(3). https://doi.org/10.1007/s10973-016-5673-6.

Duncan, J. (2008). Principles and applications of mechanical thermal analysis. pp. 119–163. *In*: Gabbott, P. (ed.). *Principles and Applications of Thermal Analysis*. Oxford, UK: Blackwell Publishing Ltd. https://doi.org/10.1002/9780470697702.ch4.

Dyamenahalli, K., Famili, A. and Shandas, R. (2015). Characterization of shape-memory polymers for biomedical applications. pp. 35–63. *In*: Yahia, L. (ed.). *Shape Memory Polymers for Biomedical Applications* (Elsevier). Elsevier. https://doi.org/10.1016/B978-0-85709-698-2.00003-9.

Espinach, F. X., Boufi, S., Delgado-Aguilar, M., Julián, F., Mutjé, P. and Méndez, J. A. (2018). Composites from poly(lactic acid) and bleached chemical fibres: Thermal properties. *Composites Part B: Engineering*, 134: 169–176. https://doi.org/10.1016/j.compositesb.2017.09.055.

Freitas, R. R. M. and Botaro, V. R. (2018). Biodegradation behavior of cellulose acetate with DS 2.5 in simulated soil. World Academy of Science, Engineering and Technology, Open Science Index 139. *International Journal of Chemical and Molecular Engineering*, 12(7): 347–351. https://doi.org/https://doi.org/10.5281/zenodo.1340536.

Gabbott, P. (2008). A practical introduction to differential scanning calorimetry. *In*: Gabbott, P. (ed.). *Principles and Applications of Thermal Analysis* (1st ed.). Blackwell Publishing Ltd. https://doi.org/10.1002/9780470697702.ch1.

Gan, P. G., Sam, S. T., Abdullah, M. F. bin and Omar, M. F. (2020). Thermal properties of nanocellulose-reinforced composites: A review. *Journal of Applied Polymer Science*, 137(11): 48544. https://doi.org/10.1002/app.48544.

Hanauer, D. C., de Souza, A. G., Cargnin, M. A., Gasparin, B. C., Rosa, D. dos S. and Paulino, A. T. (2021). Pectin-based biohydrogels reinforced with eucalyptus sawdust: Synthesis, characterization, β-D-Galactosidase immobilization and activity. *Journal of Industrial and Engineering Chemistry*, 97. https://doi.org/10.1016/j.jiec.2021.02.022.

Jeantet, L., Regazzi, A., Taguet, A., Pucci, M. F., Caro, A.-S. and Quantin, J.-C. (2021). Biopolymer blends for mechanical property gradient 3D printed parts. *Express Polymer Letters*, 15(2): 137–152. https://doi.org/10.3144/expresspolymlett.2021.13.

Ju, X., Bowden, M., Brown, E. E. and Zhang, X. (2015). An improved X-ray diffraction method for cellulose crystallinity measurement. *Carbohydrate Polymers*, 123. https://doi.org/10.1016/j.carbpol.2014.12.071.

Kaschuk, J. J., Borghei, M., Solin, K., Tripathi, A., Khakalo, A., Leite, F. A. S., … Rojas, O. J. (2021). Cross-linked and surface-modified cellulose acetate as a cover layer for paper-based electrochromic devices. *ACS Applied Polymer Materials*, 3(5): 2393–2401. https://doi.org/10.1021/acsapm.0c01252.

Klunklin, W., Jantanasakulwong, K., Phimolsiripol, Y., Leksawasdi, N., Seesuriyachan, P., Chaiyaso, T., Insomphun, C., Phongthai, S., Jantrawut, P., Sommano, S. R., Punyodom, W., Reungsang, A., Ngo, T. M. P. and Rachtanapun, P. (2020). Synthesis, characterization, and application of carboxymethyl cellulose from asparagus stalk end. *Polymers*, 13(1). https://doi.org/10.3390/polym13010081.

Kopač, T., Ručigaj, A. and Krajnc, M. (2020). The mutual effect of the crosslinker and biopolymer concentration on the desired hydrogel properties. *International Journal of Biological Macromolecules*, 159. https://doi.org/10.1016/j.ijbiomac.2020.05.088.

Martín del Campo, A. S., Robledo-Ortíz, J. R., Arellano, M., Rabelero, M. and Pérez-Fonseca, A. A. (2021). Accelerated weathering of polylactic acid/agave fiber biocomposites and the effect of fiber–matrix adhesion. *Journal of Polymers and the Environment*, 29(3): 937–947. https://doi.org/10.1007/s10924-020-01936-z.

Meng, Y., Liu, X., Li, C., Liu, H., Cheng, Y., Lu, J., Zhang, K. and Wang, H. (2019). Super-swelling lignin-based biopolymer hydrogels for soil water retention from paper industry waste. *International Journal of Biological Macromolecules*, 135. https://doi.org/10.1016/j.ijbiomac.2019.05.195.

Mlčoch, T. and Kučerík, J. (2013). Hydration and drying of various polysaccharides studied using DSC. *Journal of Thermal Analysis and Calorimetry*, 113(3). https://doi.org/10.1007/s10973-013-2946-1.

Myasoedova, V. V. (2017). Agro-polymers and biopolyesters. *Russian Journal of General Chemistry*, 87(6): 1357–1363. https://doi.org/10.1134/S1070363217060378.

Park, H., Guo, X., Temenoff, J. S., Tabata, Y., Caplan, A. I., Kasper, F. K. and Mikos, A. G. (2009). Effect of swelling ratio of injectable hydrogel composites on chondrogenic differentiation of encapsulated rabbit marrow mesenchymal stem cells *in vitro*. *Biomacromolecules*, 10(3). https://doi.org/10.1021/bm801197m.

Price, D. M. and Duncan, J. C. (2016). Thermomechanical, dynamic mechanical and dielectric methods. pp. 164–213. *In*: Gaisford, S., Kett, V. and Haines, P. (eds.). *Principles of Thermal Analysis and Calorimetry*: 2nd Edition (2nd ed.). The Royal Society of Chemistry.

Resch-Fauster, K., Klein, A., Blees, E. and Feuchter, M. (2017). Mechanical recyclability of technical biopolymers: Potential and limits. *Polymer Testing*, 64: 287–295. https://doi.org/10.1016/j.polymertesting.2017.10.017.

Scalioni, L. V., Gutiérrez, M. C. and Felisberti, M. I. (2017). Green composites of poly(3-hydroxybutyrate) and curaua fibers: Morphology and physical, thermal, and mechanical properties. *Journal of Applied Polymer Science*, 134(14): 44676. https://doi.org/10.1002/app.44676.

Segal, L., Creely, J. J., Martin, A. E. and Conrad, C. M. (1959). An empirical method for estimating the degree of crystallinity of native cellulose using the X-ray diffractometer. *Textile Research Journal*, 29(10). https://doi.org/10.1177/004051755902901003.

Shahabi-Ghahfarrokhi, I., Goudarzi, V. and Babaei-Ghazvini, A. (2019). Production of starch based biopolymer by green photochemical reaction at different UV region as a food packaging material: Physicochemical characterization. *International Journal of Biological Macromolecules*, 122: 201–209. https://doi.org/10.1016/j.ijbiomac.2018.10.154.

Shojaeiarani, J., Bajwa, D. S., Rehovsky, C., Bajwa, S. G. and Vahidi, G. (2019). Deterioration in the physico-mechanical and thermal properties of biopolymers due to reprocessing. *Polymers*, 11(1): 58. https://doi.org/10.3390/polym11010058.

Sindhu, R., Binod, P. and Pandey, A. (2015). Microbial poly-3-hydroxybutyrate and related copolymers. *In*: Pandey, A., Höfer, R., Taherzadeh, M., Nampoothiri, K. M. and Larroche, C. (eds.). *Industrial Biorefineries & White Biotechnology* (1st ed.). Elsevier. https://doi.org/10.1016/B978-0-444-63453-5.00019-7.

Sirohi, R., Pandey, J. P., Tarafdar, A., Agarwal, A., Chaudhuri, S. K. and Sindhu, R. (2021). An environmentally sustainable green process for the utilization of damaged wheat grains for poly-3-hydroxybutyrate production. *Environmental Technology & Innovation*, 21. https://doi.org/10.1016/j.eti.2020.101271.

Son, C. E. and Choi, S.-S. (2019). Analytical techniques for measurement of crosslink densities of rubber vulcanizates. *Elastomers and Composites*, 54(3): 209–219.

Tengku-Rozaina, T. M. and Birch, E. J. (2019). Thermal analysis for lipid decomposition by DSC and TGA. *In*: Melton, L., Shahidi, F. and Varelis, P. (eds.). *Encyclopedia of Food Chemistry* (1st ed., Vol. 2). Elsevier. https://doi.org/10.1016/B978-0-08-100596-5.21674-0.

Tomoda, B. T., Yassue-Cordeiro, P. H., Ernesto, J. V., Lopes, P. S., Péres, L. O., da Silva, C. F. and de Moraes, M. A. (2020). Characterization of biopolymer membranes and films: Physicochemical, mechanical, barrier, and biological properties. pp. 67–95. *In*: Agostini de Moraes, M., Ferreira da Silva, C. and Silveira Vieira, R. (eds.). *Biopolymer Membranes and Films*. Elsevier. https://doi.org/10.1016/B978-0-12-818134-8.00003-1.

Vallejo-Domínguez, D., Rubio-Rosas, E., Aguila-Almanza, E., Hernández-Cocoletzi, H., Ramos-Cassellis, M. E., Luna-Guevara, M. L., Rambabu, K., Manickam, S., Siti Halimatul Munawaroh, H. and Loke Show, P. (2021). Ultrasound in the deproteinization process for chitin and chitosan production. *Ultrasonics Sonochemistry*, 72. https://doi.org/10.1016/j.ultsonch.2020.105417.

Vaskoska, R., Vénien, A., Ha, M., White, J. D., Unnithan, R. R., Astruc, T. and Warner, R. D. (2021). Thermal denaturation of proteins in the muscle fibre and connective tissue from bovine muscles composed of type I (masseter) or type II (cutaneous trunci) fibres: DSC and FTIR microspectroscopy study. *Food Chemistry*, 343. https://doi.org/10.1016/j.foodchem.2020.128544.

Wool, R. P. (2005). Properties of triglyceride-based thermosets. *In*: Wool, R. P. and Sun, X. S. (eds.). *Bio-Based Polymers and Composites* (1st ed.). Elsevier. https://doi.org/10.1016/B978-012763952-9/50008-3.

Yaradoddi, J. S., Banapurmath, N. R., Ganachari, S. v., Soudagar, M. E. M., Mubarak, N. M., Hallad, S., Hugar, S. and Fayaz, H. (2020). Biodegradable carboxymethyl cellulose based material for sustainable packaging application. *Scientific Reports*, 10(1). https://doi.org/10.1038/s41598-020-78912-z.

Younis, M. K., Tareq, A. Z. and Kamal, I. M. (2018). Optimization of swelling, drug loading and release from natural polymer hydrogels. *IOP Conference Series: Materials Science and Engineering*, 454. https://doi.org/10.1088/1757-899X/454/1/012017.

Zhang, K., Feng, W. and Jin, C. (2020). Protocol efficiently measuring the swelling rate of hydrogels. *MethodsX*, 7. https://doi.org/10.1016/j.mex.2019.100779.

Zhorin, V. A. and Kiselev, M. R. (2020). A DSC study of the endothermic process, associated with hydrogen bonding. *In*: Polysaccharides after High-Pressure Plastic Deformation. *High Energy Chemistry*, 54(4). https://doi.org/10.1134/S0018143920040141.

Chapter 5

Structure and Morphology of Biodegradable Polymers

Felipe Avalos Belmontes,[1,*] *Francisco J. González,*[1]
Mónica Esmeralda Contreras Camacho,[1] *Aidé Sáenz-Galindo*[1] and
Rodrigo Ortega Toro[2]

1. Introduction

The percentage of an ordered or disordered material within the total mass of a sample of a semicrystalline polymer is very important, as properties generally depend on the amorphous or crystalline concentration. For example, processing and transformation differ for an amorphous polymer, where only the temperature makes the polymer soft, while a semicrystalline polymer presents a real melting temperature. Besides, it occurs with other properties such as mechanical or gas permeability, those which depend on the morphology developed during the cooling process when finished pieces are fabricated.

In the case of biodegradable polymers, order and structural regularity are crucial, especially in the disordered phase where enzymatic degradation occurs, due to the higher free volume present on this phase, and to the relatively lower thermodynamic equilibrium compared with the crystalline phase (Chen et al., 2021; DelRe et al., 2021; Hall et al., 2010; Houfani et al., 2020; Wang et al., 2020a; Zaaba and Jaafar, 2020).

The formation of ordered entities in crystalline reticles can be achieved by manipulating the cooling conditions from the molten state (see Fig. 5.1a). Furthermore, it is possible that the formation of a specific type of crystals, when starting with a polymer at room temperature and then heated, in this case, a recrystallization process may occur. This phenomenon is presented in highly semicrystalline polymers such as Nylon of polyethylene terephthalate (PET), but it can also occur in biodegradable polymers (Furushima et al., 2020; Hall et al., 2021; Jog, 2020; Tonelli, 2020; Vyazovkin, 2020; Zhang et al., 2017).

In this context, when semicrystalline polymers are used to fabricate one finished piece, transformation conditions through the extrusion die or mold filling in an injection process can produce crystallization from the molten state under a shear rate. In this case, crystals formation

[1] Departamento de Ciencia y Tecnología de Polímeros, Facultad de Ciencias Químicas, Universidad Autónoma de Coahuila, 25280 Saltillo, Coahuila, México.

[2] Programa de Ingeniería de Alimentos, Food Packaging and Shelf Life Research Group and Research Group in Complex Fluids Engineering and Food Rheology (IFCRA), Facultad de Ingeniería, Universidad de Cartagena, Carrera 6 # 36-100, Cartagena de Indias D.T y C, Colombia.

* Corresponding author: favalos@uadec.edu.mx

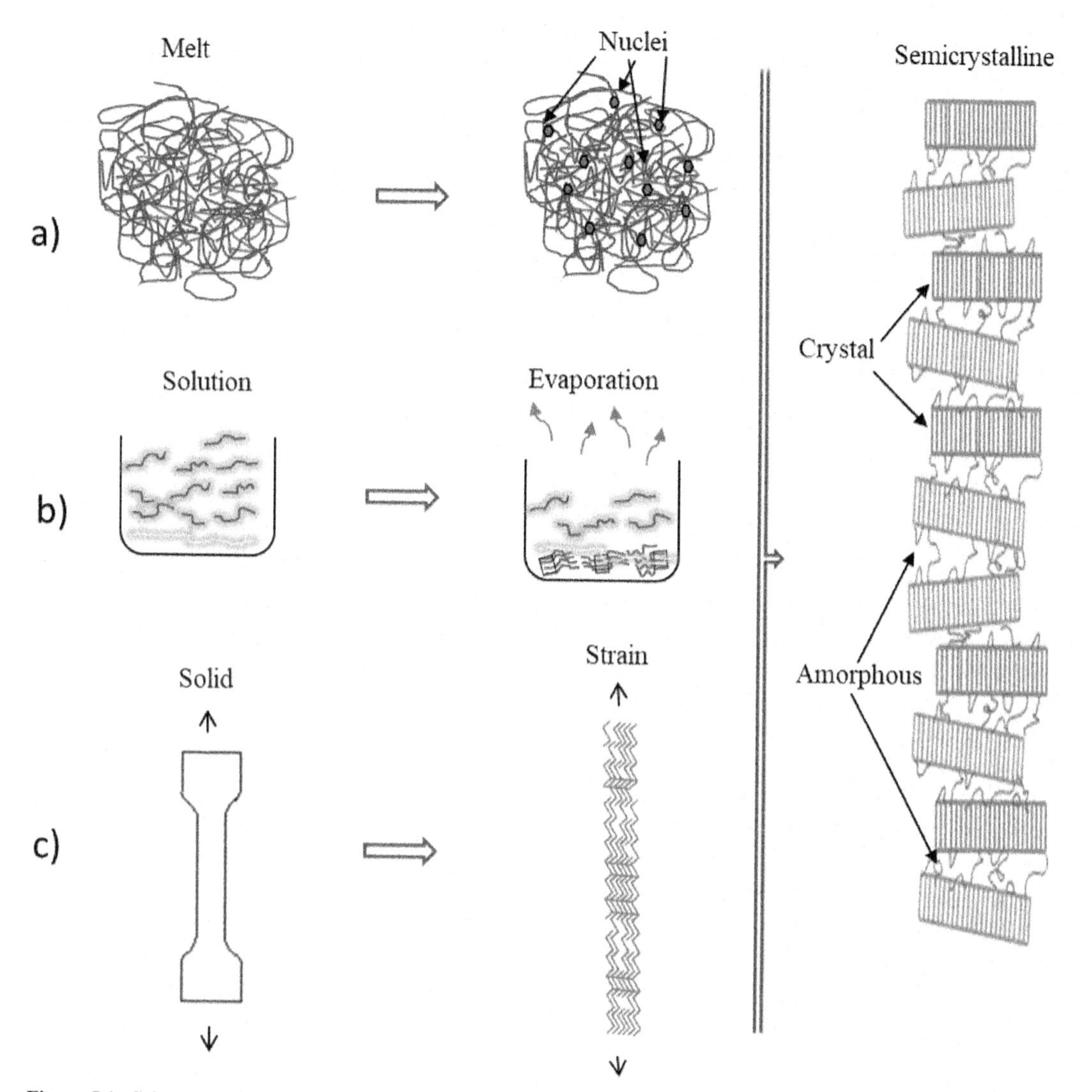

Figure 5.1. Schematic representation of polymeric semi-crystallization process. (a) From melt; (b) from solution; (c) from strain stress.

may be favored due to different factors such as solidification, extrusion and injection velocities, the cooling process and the induced orientation during processing, which leads to an increase in the mechanical resistance of pieces. This phenomenon is known as flow-induced crystallization (Chu et al., 2020a; Kotula and Migler, 2021; Liu et al., 2020a; Wang et al., 2016; Zhao et al., 2020).

Furthermore, it is feasible to obtain polymeric crystals through a solution process (see Fig. 5.1b). While the solvent is evaporating and the macromolecules are desolvating, a molecular creeping occurs, which allows the arrangement of specific molecular sections to form ordered packings. Desolvation is a typical process of biodegradable polymers such as Poly Lactic Acid (PLA) and thermoplastic starch (TPA), those which tend to separate from plasticizers used during the formulation process, giving as a result, a higher crystallinity percentage due to aging (Deng et al., 2021; Feng, 2020; Gu et al., 2018; Welch and Muthukumar, 2001; Zhang and Lucia, 2020; Zhang et al., 2017).

A third way to achieve crystals formation is by applying longitudinal stress over finished pieces (see Fig. 5.1c). While molecules can be deformed due to the applied stress, polymer chains are aligned and joined, forming as a result spatially ordered packings. In biodegradable polymers, this phenomenon has been observed for cellulose obtained from plants during the formation of primary cell walls, when the cell growth occurs (Aygün Klinge, 2020; Chu et al., 2020b; Forestier et al., 2020; Zhang et al., 2017).

The process of molecular packing for crystal formation depends not only on the thermodynamics conditions present but also on the polymer molecular weight, as a molecule with a lower size will have a higher creeping capability than a bigger molecule (Mao et al., 2020; Yousefzade et al., 2020). Moreover, crystal formation also depends on the monomer chemical nature, and on the ability of the polymer chain side groups to be arranged on the space, associated with their stereoisomerism, which leads to crystallization in zig-zag formation or in helicoidal form (Li et al., 2020; Wang et al., 2020b). Under these conditions, a higher packing of their side groups is sometimes reached with *trans* dispositions than with *cis*, forming zig-zag packings. Nevertheless, a higher packing can be reached with *cis* dispositions when helicoidal crystals are present, since the main chain side groups tend to accommodate spatially outside (Hu et al., 2021; Kochervniskii et al., 2020; Nguyen et al., 2021; Staub and Li, 2020; Zheng and Pan, 2020).

From a thermodynamic point of view, despite more than one crystalline phase may coexist on the same polymer, one among the others is the most stable, and the rest are metastable forms at specific pressures and temperatures. Metastable forms are formed due to a kinetic-thermodynamic mechanism, which provides a favorable route for the polymer to fall on a minimum of free local and punctual energy. In general, when a polymer crystallizes from a molten state, stable phases are formed at high crystallization temperatures and low cooling velocities, while metastable phases are formed at high cooling velocities and lower temperatures than stable phases (Furushima et al., 2020; Hall et al., 2021; Jog, 2020; Tonelli, 2020; Vyazovkin, 2020; Zhang et al., 2017).

Crystalline phases can be formed from melt initiating with a crystalline nucleus, which can be the material itself when specific thermodynamic conditions are reached. Once the nucleus is formed, small spatially ordered sections of macromolecules start to deposit over it, to form the crystallites as first-order spatial structures. These crystallites are arranged, in a second stage, to form lamellar structures, those which are organized to finally form spherulitic or dendritic structures (Lai et al., 2020; Raimo, 2021; Song et al., 2020; Staub and Li, 2020; Xuzhen et al., 2021; Woo et al., 2020). At other times, the nucleation points may be additive particles conventionally joined to the polymer formulations, for example, pigments or inorganic fillers. Additionally, organic fillers can act as nucleation points if they present a solidification temperature higher than processing polymer temperature, in case they melt (Aliotta et al., 2017; Barletta and Pizzi, 2021; Liu et al., 2020b; Choudhary et al., 2021; Zhang et al., 2021).

Allotropic polymorphism is defined as the existence of more than one type of crystalline structures in a polymer with the same chemical composition. It describes the macromolecules packing in different spatial arrangements to form crystalline cells. Crystalline phase morphology may present different arrangements, which means that various types of crystals may be present in the same polymer. Diverse crystalline forms can be found in a semicrystalline polymer, then depending on the spatial arrangement and on the degrees of freedom of side groups to the main chain, more than one crystalline form may coexist. This fact may be implicated in (1) the stereoisomerism of precursor monomers, for example, the case of racemic mixtures of lactic acid used for polylactic acid synthesis; (2) processing and transformation conditions; and (3) to the presence of certain additives added to the polymer formulations (Qin et al., 2021; Zheng and Pan, 2020; Raimo, 2021; Hsieh et al., 2020; Righetti et al., 2020; Pan and Inoue, 2009; Wang et al., 2021a; Wu et al., 2020).

The amorphous and crystalline phase of polymers also depends on polymerization conditions, and on the use or not of stereo-specific catalysts. In the case of biodegradable polymers, it may also depend on the monomer source used to obtain that material. In addition to the amorphous and crystalline phases present in semicrystalline polymers, an intermediate phase among them is reported, conventionally known as mesophase. Moreover, some authors propose the existence of amorphous subphases for polymers like polylactic acid (Monnier et al., 2020; Righetti, 2017; Yu et al., 2020).

Considering the revealed points, and the importance which molecular packing has in the synthesis-structure-processing-properties interrelation of polymers, this chapter reviews the characteristics and crystallization forms of most used biodegradable polymers in finished products, in addition to the most common techniques for the morphological characterization of these materials.

2. Polymer morphological characterization techniques

Several techniques are used for the morphological characterization of polymers, for instance, spectroscopic, microscopic, thermal, mechanical, radiation and densimetry techniques. Some peripheric systems can be coupled to those mentioned techniques like heating plates or mechanical stress systems, to observe morphological changes as a function of variables like temperature or strain level. Some of the most employed techniques for this purpose are then revealed.

Among the diverse techniques used for direct characterization of polymer morphology, X-Ray Diffraction (XRD), is probably the one that provides the most information. This technique allows, not only to differentiate between the amorphous and the crystalline phase, but also to identify the different types of packing by analyzing the diffraction patterns of samples. Using this technique and applying Bragg's Law, it is possible to identify, through the position of the angle on which the diffraction band appears, the distance between the different crystalline planes, and with this information, the dimensions of the crystalline lattice and unit cell can be determined. This characterization technique, combined with other complementary ones, allows performing a more rigorous analyses of materials with excellent results (Chrysafi et al., 2021; Doumeng et al., 2021; Hong et al., 2021; Kotula and Migler, 2021; Mondal et al., 2021).

Alternatively, Fourier-transform infrared spectroscopy (FTIR), allows identifying, with the position of the emission bands of sample spectra, the type of packing of semicrystalline materials. With this information it is possible to determine the relative crystallinity percentage, through the ratio of the absorption coefficients between the crystalline and the amorphous bands. Moreover, this technique can be coupled with a heating plate to perform crystallization kinetics studies. This characterization technique, combined with other complementary ones, allows performing more rigorous analyses of materials with excellent results (Acik, 2020; Chrysafi et al., 2021; Hong et al., 2021; Malz et al., 2021; Mondal et al., 2021; Sintim et al., 2020).

The use of Nuclear Magnetic Resonance (NMR) for morphological analysis is based on the different relaxation times between the crystalline and amorphous polymer domains. Protons located in the crystalline or rigid domains return much faster to the equilibrium state than protons in mobile domains. The measurement of decay signal in two characteristics times provides a relation that corresponds to the sample's crystallinity. This characterization technique, combined with other complementary ones, allows performing more rigorous analyses of materials with excellent results (Acik, 2020; Malz et al., 2021; Sintim et al., 2020).

Raman spectroscopy is used for the structural characterization of polymers. In this case, one small portion of monochromatic light interacts with the sample and is inelastically dispersed, showing small frequency changes which are characteristics of the analyzed material which depend on material composition and its physical state. This characterization technique, combined with other complementary ones, allows performing more rigorous analyses of materials with excellent results (Cosby et al., 2021; Doumeng et al., 2021; Kotula and Migler, 2021).

By using microscopy techniques, it is possible to observe different morphologies of semicrystalline materials. Particularly, Optical Microscopy (OM), Scanning Electronic Microscopy (SEM), Transmission Electronic Microscopy (TEM) and Atomic Force Microscopy (AFM) can be used. When OM with polarized light is used, it is possible to detect changes in the different phases of a polymer, this is because refraction index of crystalline phase differs from the amorphous phases, and at macro level, spherulitic or dendritic structures can be observed. This technique can also be applied to carry out kinetic studies of *in situ* crystalline growth, which can be correlated with different types of crystalline growth of polymorphic or allotropic materials. This characterization technique, combined with other complementary ones, allows performing more rigorous analyses of materials with excellent results (Chrysafi et al., 2021; Hong et al., 2021; Kotula and Migler, 2021; Mondal et al., 2021).

Along with XRD, Differential Scanning Calorimetry (DSC) are two of the most used techniques for phase quantification. In the first case, a correlation is made between the amorphous halo and the intensity of crystalline bands to calculate the relative crystallinity percentage. On the other hand, when DSC is used, the heat absorbed in a phase change during the heating process is correlated with the heat necessary for the same phase change if the polymer were theoretically 100% crystalline. Furthermore, it is possible to detect different melting points of the crystalline forms in a single material, resulting in multiple peaks on the calorimetric curve. This technique also allows to carry out kinetic studies of crystalline growth from the molten state. This characterization technique, combined with other complementary ones, allows performing more rigorous analyses of materials with excellent results (Chrysafi et al., 2021; Doumeng et al., 2021; Hong et al., 2021; Kotula and Migler, 2021; Mondal et al., 2021).

3. Biodegradable polymers

A review of the most used biodegradable polymers not only for containers and packaging but also for the biomedical, pharmaceutical, cosmetics and food industry is presented below. This analysis focuses on the morphology, crystallization types and unit cell dimensions. Most of these polymers can be used as a pure component, combined with other polymers with similar characteristics or with non-biodegradable polymers. In the latter case, the objective is to decrease as far as possible the environmental impact caused by polymers that are non-sensitive to microbiological degradation.

3.1 Polylactic acid (PLA)

Polylactic acid (Fig. 5.2), is synthesized from lactic acid, which can be produced from chemical or biotechnological methods. The latter is based on the fermentation of substrates rich in carbohydrates by using fungi or bacteria. Yield and production of enantiomers D+ or L-depend both on biosynthesis conditions of the monomer and the source of the raw material used (Singhvi et al., 2019; Djukić-Vuković et al., 2019).

Polymerization of lactic acid is a condensation process. The polymerization becomes complex due its own nature, as water must be eliminated, while it is generating as a reaction sub-product to make the process irreversible. Due to this, synthesis is normally carried out in two stages. In the first stage, the monomer pre-polymerizes until it reaches oligomer sizes also called lactides; then, a distillation process is applied to separate the water present in the reactor; and finally, these oligomers react using catalysts to obtain macromolecular sizes (Casalini et al., 2019; Di Lorenzo and Androsch, 2018).

From the morphologic point of view, due to conformers re-arrangements, this polymer presents the highest number of allotropic formations. Depending on polymerization conditions and the source

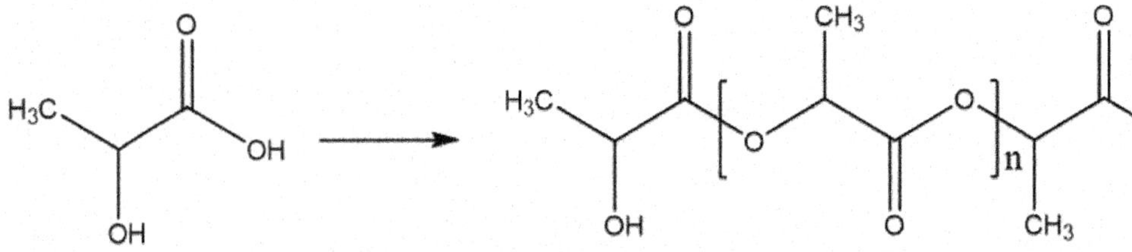

O
H3C
OH
OH
O
CH3
O
H3C
O
O
n
OH
OH
O
CH3

Figure 5.2. Polymer structure of polylactic acid.

used for monomer obtention, it may form crystalline arrangements preferably if the monomer is L-lactic acid type, as the presence of monomer D-Lactic acid affects the crystallinity percentage, as well as its glass transition temperature (Tg) and melting temperature (Tm). At the industrial scale, the percentage of D-Lactic acid is around 2–6% w/w.

The first crystalline form and the thermodynamically most stable is the α form, it can be formed from melt through a helicoidal arrangement 10_3. On the other hand, a helicoidal arrangement 3_1, may give the formation of crystalline structures of β type, which are formed at high strain velocities and high temperatures. It is also possible to find, ϒ type crystalline formations, which grow in an epitaxial layer.

From the molten state and around 100°C, it is possible to form α′ type crystalline structures. These structural arrangements resemble α type crystals, but they are slightly bigger, distorted and thermodynamically metastable, therefore, they end into α type structures at higher temperatures. Furthermore, it is possible to identity other two co-crystallized forms, due to the combination of D-lactic and L-lactic acid, denoted as Stereo complex Crystals (SC) triclinic and SC trigonal (Pan and Inoue, 2009; Zheng and Pan, 2020).

The amorphous phase of PLA has three variants (Monnier et al., 2020; Righetti, 2017; Yu et al., 2020). Two amorphous phases, one with free and the other with restricted movement which vitrify and de-vitrify in the region close to Tg, and a third amorphous rigid phase located in the crystal-amorphous interface which vitrifies and de-vitrifies above Tg, with dimensions of 1–4 nm for this third phase and around 8 nm in the mobile phase.

With respect to the dimensions of different crystalline formations of PLA, Zheng and Pan, in an excellent review about polyesters crystallization (Pan and Inoue, 2009; Zheng and Pan, 2020), assumed the following data of the different unit cells that this material present. Orthorhombic crystalline systems α and α′, obtained from molten state or dissolutions; a = 1.034–1.034 nm, b = 0.597–0.645 nm, c = 2.780–2.880, and a = 1.080 nm, b = 0.620 nm, c = 2.880 nm, respectively. For crystalline systems orthorhombic β, and trigonal β, obtained by solution-spinning, solid-state extrusion, stretching or shearing, the dimensions are a = 1.031–1.041 nm, b = 1.770–1.821 nm, c = 0.880–0.900 nm and a = b = 1.052 nm, c = 0.880 nm, respectively. In the case of the epitaxial crystalline system ϒ, the dimensions are a = 0.995 nm, b = 0.625 nm, c = 0.880 nm.

For Stereocomplex co-Crystallization (SC) obtained when poly(L-lactide) and poly(D-lactide) co crystallize, the dimensions are for the SC triclinic; a = 0.912–0.916 nm, b = 0.913–0.916 nm, c = 0.870–0.930 nm. While that for SC trigonal, according to Zheng and Pan (2020), the dimensions are a = b = 1.498–1.500 nm, c = 0.823–0.870 nm.

3.2 Cellulose

Cellulose is a biopolymer composed of molecules of β-D-glucopyranose bonded covalently through acetal groups. It is the most abundant biomolecule in nature (Fig. 5.3). It is in essence one structural polymer as it forms part of the cell wall of many plant species. In general, it can occupy around 40–50% of the weight of dried plants; however, in cotton, it can be even more than 90% in weight. In addition to its synthesis through plants, this polymer can also be bacteriologically synthesized, in this latter case, the crystalline percentage is much higher due to the high purity of biosynthesized cellulose (Avolio et al., 2012; Hao et al., 2020; Hall et al., 2010; Hernández-Varela et al., 2021; Salari et al., 2019).

Its molecular weight ranges from 800 and up to 10,000 repeated units of β-glucose, and its crystallinity depends mainly on its source. Normally, when it comes from plants, it is possible to find cellulose in the form of a composite combined with lignin molecules, hemicellulose and other carbohydrates (Avolio et al., 2012).

Native cellulose, commercially known as type I, is structurally dimorphic, at it can be arranged in type I_α crystalline packing with a triclinic structure, and in type I_β with a monoclinic structure. In this case, the form I_α is metastable, it is known that it is formed by the strain the cell is exposed

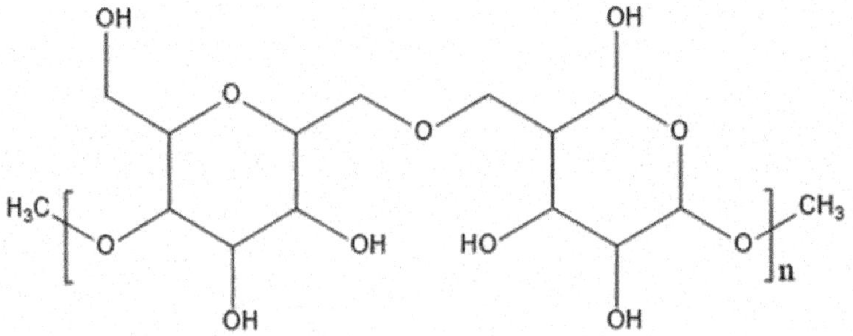

Figure 5.3. Polymer structure of cellulose.

during its growth of the primary wall. Then, once cellular growth is stopped and the stress of the cell *in vivo*, I_{α} is formed. On the other hand, the form I_{β}, more thermodynamically stable, is dominant in the secondary cell walls. Cellulose microfibrils are located densely packed in the secondary cellular walls, while in primary walls, they are in a more disordered form (Wu et al., 2020).

Different types of cellulose exist commercially. Type I is the native cellulose extracted from plants biosynthesized through bacteria. Crystalline form I_{α} presents crystalline unit cell dimensions of a = 0.672 nm, b = 0.596 nm and c = 0.104 nm; and I_{β} form presents dimensions of a = 0.778 nm, b = 0.820 nm and c = 1.03 nm. Cellulose type II with monoclinic structure is a mercerized cellulose and it is thermodynamically more stable since the mercerization process eliminates the disordered amorph phase. The dimensions of the crystalline unit cell are a = 0.810 nm, b = 0.903 nm and c = 1.031 nm. Cellulose type III maintains the monoclinic structure, and it is obtained by a treatment to graft amino groups to celluloses types I and II. It presents crystalline unit cell dimensions a = 0.445 nm, b = 0.785 nm and c = 1.031 nm, while cellulose type IV, is obtained through a glycerol treatment of cellulose type III (Wu et al., 2020).

Due to cellulose biodegradable characteristics, which generally occur through enzymatic hydrolysis, cellulose is generally combined with other polymers for its use in finished pieces, which may be biodegraded completely when the second polymer is also biodegradable or combined with other non-biodegradable polymers to decrease as far as possible its environmental impact. In this case, a milling process is usually carried out during the preparation of these mixtures. It is recommended that the mixing time should not be excessive, as the fibrous structure can be lost and consequently the size of the crystalline domains, having as consequence, a decrease on thermal stability and an increase in the moisture uptake (Arefian et al., 2020; Bhasney et al., 2020; Wang et al., 2020c).

3.3 Starch (Termoplastic starch)

From a chemical point of view, starch is a carbohydrate formed by anhydroglucose units joined by glycosidic bonds. Although it can be obtained from many plants, commercially the main sources are corn, wheat, rice, potatoes and tapioca (see Fig. 5.4). Starch contains two types of macrostructures, one linear, known as amylose, which is joined through α-1-4 glycosidic bonds and one branched structure, amylopectin, which mostly presents α-1-6 glycosidic bonds. The ramifications size depends mainly on the starch source. Amylose molecular weight is around 10^6 and amylopectin molecular weight is much higher, which gives this material a very high viscosity (Castillo et al., 2019; Diyana et al., 2021; Hernández-Medina et al., 2008).

Starch does not present any thermoplastic properties without the additions of plasticizers. When plasticizers are added to starch formulations and applying high temperatures and shear rates, it is possible to obtain a material that easily flows and that can be processed by conventional polymer processing and transformation techniques. During the plastification process, starch is transformed from a granular semicrystalline material to a partially amorphous product. The effect of adding plasticizers like glycerol, sorbitol, water, urea, etc., reduces the interaction among the starch

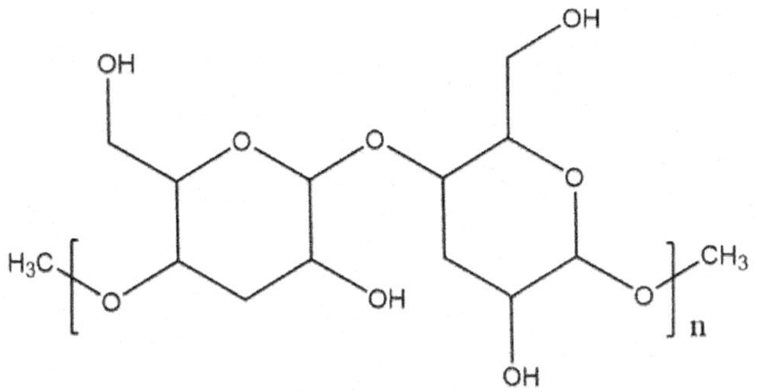

Figure 5.4. Polymer structure of starch.

molecules, having as a result, an increase in free volume, decreasing the material Tg. Properties of plasticized starch, commercially known as thermoplastic starch (TPA), allow it to reach higher elongations just before disintegrating when it is subjected to longitudinal stress, and despite its mechanical resistance decrease, the tenacity in finished pieces increase. Another observable effect in TPA is that its gas permeability increases due to the plastification effect (Abera et al., 2020; Diyana et al., 2021; Gonzalez et al., 2020; Wang et al., 2021b).

One undesirable defect of TPA is that, depending on plasticizer concentration, time and environmental conditions, plastification may experience the retrogradation phenomenon, which is an antiplastification process. Water and other plasticizers are excluded from the amorphous TPA domains, which results on the starch restructuration, forming crystals once more. The presence of plasticizers gives TPA highly hydrophilic properties, which may alter its physicochemical properties due to moisture absorption; however, it is known that this effect can be avoided, by mixing this material with other polymers with hydrophobic characteristics or through chemical modifications of free -OH groups, like esterification, etherification or acetylation. Additionally, hydrophobic inorganic agents are sometimes added in TPA formulations, for example, talc and montmorillonite (Eaton et al., 2021; Fričová et al., 2020; Hernandez and Herminsul, 2021; Zhang and Geng, 2020).

From a morphological point of view, it is known that amylose fraction is the one that predominantly gives starch the crystalline character. Despite amylopectin also forms ordered structures, to date, there are no reports which clearly describe its spatial order. However, one study reported in 2020 (Rodriguez-Garcia et al., 2020), starch from avocado seeds and amaranth using Grazing Incidence X-Ray Diffraction (GIXRD), defines the coexistence of two crystallographic structures for amylose, one type α orthorhombic monoclinic and one type β hexagonal. The same report mentions the different sizes of starch crystalline unit cells reported so far by other researchers. In the case of TPA, it is reported that amylose-glycerol systems are organized in a type V structure with two modalities: hydrated and anhydrous, both denoted as V_h and V_a respectively, whose difference is based on the presence or absence of water molecules inside the starch crystalline lattice (Castillo et al., 2019; Rodriguez-Garcia et al., 2020).

TPA can recrystallize in different polymorphic forms depending on processing, transformation and environmental conditions such as humidity, time and temperature where the material is exposed. As mentioned earlier, plastification breaks the starch crystalline order and transforms it in amorphous, for this reason, this type of materials present two Tg values, the first one due to plasticized domains and the second one due to starch-rich domains (Castillo et al., 2019).

However, there are not many recent references about starch crystalline structures dimensions, Rodriguez-García et al. report dimensions of a = 1.1694 nm, b = 1.7585 nm and c = 1.0659 nm, for orthorhombic amylopectin.

3.4 Chitin (Chitosan)

Chitin is a linear biopolymer formed by a polysaccharide constituted by units of N-Acetylglucosamine linked by β-D (1,4) bonds (see Fig. 5.5). Its chemical structure resembles cellulose, but with an amino group instead of hydroxyl. Its function, like cellulose in plants and collagen in animals is structural. Chitin is easily found in crustacean shells and shellfish, insects and arthropods exoskeleton. It is a low soluble and reactive material, which limits its applications; however, its partial deacetylation using strong alkalis to substitute its acetamido groups with amino groups, allows it to have higher solubility and reactivity when it is transformed into chitosan (El Knidri et al., 2020; Joseph et al., 2021; Parhi, 2020; Santos et al., 2020; Wang et al., 2020d; Zargar et al., 2015).

Chitosan can be easily dissolved in diluted organic acids such as citric, acetic, tartaric, etc. Solubility depends on deacetylation conditions, which can reach up to 90%. Due to the presence of side amino groups on the main chain, chitosan can be chemically modified to carry out enzyme anchor reactions, grafting with other chemical groups or crosslinking reactions to obtain products for specific applications, such as biotechnological, biomedicine, pharmaceutical, food, agricultural, etc. (El Knidri et al., 2020; Parhi, 2020; Ru et al., 2019; Zargar et al., 2015).

Chitin is polymorphic and presents three crystalline forms, α, β and γ. They differ from each other by the number of chains and water molecules located inside the unit cell. Chitin crystalline forms also depend on their source, for instance, if it comes from arthropods exoskeleton, preferentially type α crystals are present, with a great number of polymer chains packed with antiparallel arrangements, having as a consequence an increase of intermolecular attractions. On the other hand, type β, with molecular parallel dispositions, are preferably found in shellfish which are structurally more flexible, such as squids. Finally, ϒ crystalline forms can be found in combination with materials where type α and β crystals exist (Ablouh et al., 2020; Kaya et al., 2017; Ru et al., 2019; Zargar et al., 2015; Zhang et al., 2020).

Each chitin crystalline form has proper characteristics of solubility, for example, type α crystals, are insoluble in water, diluted organic solutions and organic solvents. On the other hand, type β crystals tend to adsorb water and swell, and they are soluble in formic acid. The solubility of both crystalline forms depends not only on their molecular weight but also on the deacetylation degree. Therefore, a chitin deacetylation degree of 28%, makes chitin soluble in acetic acid and 49% makes it soluble in water. Chitin average molecular weight ranges from 1.03×10^6 to 2.5×10^6; however, deacetylation reactions from 5 to 60% to allow chitosan to decrease its molecular weight from 0.15×10^6 to 1.1×10^6. Furthermore, it is reported that crystallinity increases with the deacetylation degree (Facchinatto et al., 2020; Tavares et al., 2020; Wang et al., 2020d; Zargar et al., 2015; Zhang et al., 2021).

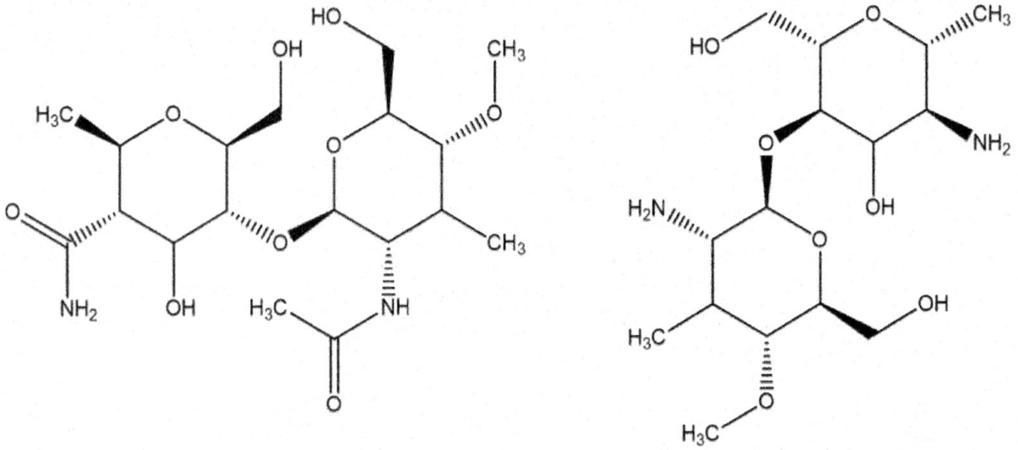

Figure 5.5. Polymeric structure of chitin and chitosane.

Since 1990, some studies have reported on the dimensions of orthorhombic α-chitin in the $P2_12_12_1$ space group, as mentioned by Rubina et al. (2020) and Tian and Liu (2020). These studies report values of a = 0.474–0.890 nm, b = 1.699–1.886 nm and c = 0.9853–1.1030 nm and a = 0.4819 nm, b = 0.9239 nm and c = 1.0384 nm in the $P2_1$ space group for β-chitin. These parameters suggest that the unit cell is composed by four glucosamine sections symmetrically connected in positions $P2_12_12_1$, where the c axis correspond to the fiber axis, with two antiparallel chains per unit cell, each one located on each crystalline axis (Cartier et al., 1990; Okuyama et al., 1997; Raabe et al., 2006; Rubina et al., 2020; Tian and Liu, 2020).

3.5 Polyvinil alcohol (PVOH)

Since vinyl alcohol is unstable due to the isomerization to acetaldehyde, its polymer must be prepared by indirect methods. Polyvinyl alcohol is prepared using polyvinyl acetate as raw material, which reacts with methanol in the presence of a strong base, generally NaOH (Fig. 5.6). This reaction can be controlled up to the point that no acetate groups will be present in the polymer, for this reason, depending on the hydrolysis grade, a single polymer or a copolymer can be present. Therefore, the polymer presents hydrophobic acetate groups but also hydrophilic hydroxyl groups, as a consequence, its properties depend on the polymerization degree and hydrolysis (Aruldass et al., 2019; Aslam et al., 2018; Nagarkar and Patel, 2019).

It is well known that the semicrystalline nature of PVOH is due to hydrogen bonding between polymer chains, in this case, hydroxyl groups are inserted into the crystalline reticulum although the polymer chain is atactic. Assender and Windle, 1998 analyzed variations in the polymer crystalline structure related to tacticity changes using molecular mechanics modeling. They found that crystalline unit cell dimensions may change according to the tacticity grade for a = 0.786–7.88 nm, b = 0.258–0.260 nm, and c = 0.531–0.536 nm. Their results demonstrated good accordance with experimental data, that in this case, they were a = 0.781 nm, b = 0.252 nm, and c = 0.551–0.536 nm. Recently, Tashiro et al., 2020 reported experimental measurements of the unit cell of this polymer, obtaining a = 0.781 nm, b = 0.551 nm and c = 0.255 nm, with angles α = ɣ = 90° and β = 92.2° and it is estimated that these chains adopt a zigzag plane extended configuration.

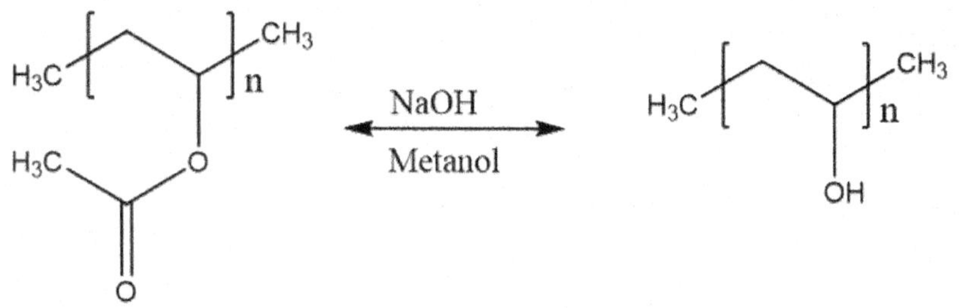

H_3C CH$_3$ n H_3C O O — NaOH / Metanol — H_3C CH$_3$ n OH

Figure 5.6. Polymer structure of polyviniyl alcohol.

3.6 ε-Polycaprolactone

Polycaprolactone is a polymer of the aliphatic family of polyesters, whose monomeric unit is ε-caprolactone. From a chemical point of view, its polymer chain has a sequence of methylene units linked through ester-type groups (see Fig. 5.7). It is a semicrystalline polymer with a melting point exceptionally low for a polymer, 58–60°C and one Tg of –60°C (Casas et al., 2011; Kakroodi et al., 2018). Two technologies have been developed to produce polycaprolactone, the polycondensation of a hydrocarboxylic acid (6-hidroxihexanoic acid) and the Ring-Opening Polymerization (ROP) of lactone, using a catalyzer such as stannous octanoate. The advantage of ROP polymerization is that there is no need to eliminate any subproduct at the end of the reaction, which reduces the number of reactors used for its production (Labet and Thielemans, 2009). There are some reports about enzymatic polymerization of polycaprolactone, the advantage of this approach is that reaction

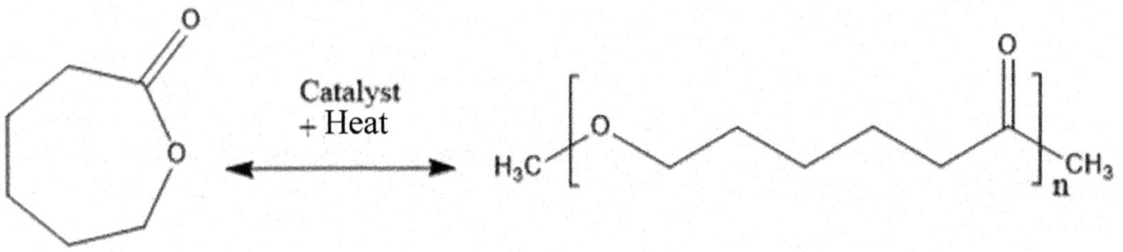

Figure 5.7. Polymer structure of poly(ε-Caprolactone).

conditions can be easily controlled due to the enzymes' catalytic selectivity (Kaplan et al., 2020; Torron et al., 2017). Furthermore, different weights of lactones copolymers of different sizes have been successfully designed using biocatalyst *Candida antarctica lipase B* (Ulker, 2018).

Since 2009 different studies have been reported about unit cell dimensions of ϵ-caprolactone in the $P2_12_12_1$ space group with a complete molecular arrangement in *trans* isomer. Labet Thielemans, 2009 reported values of a = 0.7496 nm, b = 0.4974 nm and c = 1.7297 nm. Other parameters reported by Casas et al. (Casas et al., 2011) are a = 0.747 nm, b = 0.498 nm and c = 1.705 nm, and recently Shkarina et al., 2018 reported unit cell parameters of a = 0.7521 nm, b = 0.4989 nm and c = 1.7164 nm.

4. Conclusions

The importance of identifying the morphology of biodegradable polymers is crucial to establish the synthesis-structure-processing-properties interrelation of these materials. As already mentioned in the chapter, the characteristics of these materials depend on their source and their synthesis process, for example, from bacteriologic fermentation or chemical synthesis. As a consequence, different properties may vary such as purity, density, molecular weight distribution, crystalline packing and crystallinity degree. As these polymers must be subsequently processed to produce finished pieces, it seems evident that the mentioned parameters will influence processing conditions and final properties of finished products.

References

Abera, G., Woldeyes, B., Demash, H. D. and Miyake, G. (2020). The effect of plasticizers on thermoplastic starch films developed from the indigenous Ethiopian tuber crop Anchote (*Coccinia abyssinica*) starch. *International Journal of Biological Macromolecules*, 155: 581–587.

Ablouh, E. H., Jalal, R., Rhazi, M. and Taourirte, M. (2020). Surface modification of α-chitin using an acidic treatment followed by ultrasonication: Measurements of their sorption properties. *International Journal of Biological Macromolecules*, 151: 492–498.

Acik, G. (2020). Preparation of antimicrobial and biodegradable hybrid soybean oil and poly (L-lactide) based polymer with quaternized ammonium salt. *Polymer Degradation and Stability*, 181: 109317.

Aliotta, L., Cinelli, P., Coltelli, M. B., Righetti, M. C., Gazzano, M. and Lazzeri, A. (2017). Effect of nucleating agents on crystallinity and properties of poly (lactic acid) (PLA). *European Polymer Journal*, 93: 822–832.

Arefian, M., Hojjati, M., Tajzad, I., Mokhtarzade, A., Mazhar, M. and Jamavari, A. (2020). A review of Polyvinyl alcohol/Carboxiy methyl cellulose (PVA/CMC) composites for various applications. *Journal of Composites and Compounds*, 2(3): 69–76.

Aruldass, S., Mathivanan, V., Mohamed, A. R. and Tye, C. T. (2019). Factors affecting hydrolysis of polyvinyl acetate to polyvinyl alcohol. *Journal of Environmental Chemical Engineering*, 7(5): 103238.

Aslam, M., Kalyar, M. A. and Raza, Z. A. (2018). Polyvinyl alcohol: A review of research status and use of polyvinyl alcohol based nanocomposites. *Polymer Engineering & Science*, 58(12): 2119–2132.

Assender, H. E. and Windle, A. H. (1998). Crystallinity in poly (vinyl alcohol) 2. Computer modelling of crystal structure over a range of tacticities. *Polymer*, 39(18): 4303–4312.

Avolio, R., Bonadies, I., Capitani, D., Errico, M. E., Gentile, G. and Avella, M. (2012). A multitechnique approach to assess the effect of ball milling on cellulose. *Carbohydrate Polymers*, 87(1): 265–273.

Aygün, S. and Klinge, S. (2020). Continuum mechanical modeling of strain-induced crystallization in polymers. *International Journal of Solids and Structures*, 196: 129–139.

Barletta, M. and Pizzi, E. (2021). Optimizing crystallinity of engineered poly (lactic acid)/poly (butylene succinate) blends: The role of single and multiple nucleating agents. *Journal of Applied Polymer Science*, 138(16): app50236.

Bhasney, S. M., Mondal, K., Kumar, A. and Katiyar, V. (2020). Effect of microcrystalline cellulose [MCC] fibres on the morphological and crystalline behaviour of high density polyethylene [HDPE]/polylactic acid [PLA] blends. *Composites Science and Technology*, 187: 107941.

Cartier, N., Dotard, A. and Chanzy, H. (1990). Single crystals of chitosan. *International Journal of Biological Macromolecules*, 12(5): 289–294.

Casas, M. T., Puiggalí, J., Raquez, J. M., Dubois, P., Córdova, M. E. and Müller, A. J. (2011). Single crystals morphology of biodegradable double crystalline PLLA-b-PCL diblock copolymers. *Polymer*, 52(22): 5166–5177.

Castillo, L. A., López, O. V., García, M. A., Barbosa, S. E. and Villar, M. A. (2019). Crystalline morphology of thermoplastic starch/talc nanocomposites induced by thermal processing. *Heliyon*, 5(6): e01877.

Chen, J., Lin, Y., Chen, Y., Koning, C. E., Wu, J. and Wang, H. (2021). Low-crystallinity to highly amorphous copolyesters with high glass transition temperatures based on rigid carbohydrate-derived building blocks. *Polymer International*, 70(5): 536–545.

Choudhary, S., Dhatarwal, P. and Sengwa, R. J. (2021). Study on crystalline phases and degree of crystallinity of the melt compounded PVA/MMT and PVA/PVP/MMT nanocomposites. *Indian Journal of Pure & Applied Physics* (IJPAP), 59(2): 92–102.

Chrysafi, I., Pavlidou, E., Christodoulou, E., Vourlias, G., Klonos, P. A., Kyritsis, A. and Bikiaris, D. N. (2021). Effects of poly (hexylene succinate) amount on the crystallization and molecular mobility of poly (lactic acid) copolymers. *Thermochimica Acta*, 698: 178883.

Chu, Z., Liu, L., Liao, Y., Li, W., Zhao, R., Ma, Z. and Li, Y. (2020a). Effects of strain rate and temperature on polymorphism in flow-induced crystallization of Poly (vinylidene fluoride). *Polymer*, 203: 122773.

Chu, Z., Liu, L., Lou, Y., Zhao, R., Ma, Z. and Li, Y. (2020b). Flow-induced crystallization of crosslinked poly (vinylidene fluoride) at elevated temperatures: formation and evolution of the electroactive β-phase. *Industrial & Engineering Chemistry Research*, 59(10): 4459–4471.

Cosby, T., Aiello, A., Durkin, D. P. and Trulove, P. C. (2021). Kinetics of ionic liquid-facilitated cellulose decrystallization by Raman spectral mapping. *Cellulose*, 28(3): 1321–1330.

DelRe, C., Jiang, Y., Kang, P., Kwon, J., Hall, A., Jayapurna, I. and Xu, T. (2021). Near-complete depolymerization of polyesters with nano-dispersed enzymes. *Nature*, 592(7855): 558–563.

Deng, Y. F., Zhang, D., Zhang, N., Huang, T., Lei, Y. Z. and Wang, Y. (2021). Electrospun stereocomplex polylactide porous fibers toward highly efficient oil/water separation. *Journal of Hazardous Materials*, 407: 124787.

Di Lorenzo, M. L. and Androsch, R. (eds.). (2018). *Industrial Applications of Poly (lactic acid)* (Vol. 282). Cham: Springer.

Diyana, Z. N., Jumaidin, R., Selamat, M. Z., Ghazali, I., Julmohammad, N., Huda, N. and Ilyas, R. A. (2021). Physical properties of thermoplastic starch derived from natural resources and its blends: a review. *Polymers*, 13(9): 1396.

Djukić-Vuković, A., Mladenović, D., Ivanović, J., Pejin, J. and Mojović, L. (2019). Towards sustainability of lactic acid and poly-lactic acid polymers production. *Renewable and Sustainable Energy Reviews*, 108: 238–252.

Doumeng, M., Makhlouf, L., Berthet, F., Marsan, O., Delbé, K., Denape, J. and Chabert, F. (2021). A comparative study of the crystallinity of polyetheretherketone by using density, DSC, XRD, Raman spectroscopy techniques. *Polymer Testing*, 93: 106878.

Eaton, M. D., Domene-López, D., Wang, Q., Montalbán, M. G., Martin-Gullon, I. and Shull, K. R. (2021). Exploring the effect of humidity on thermoplastic starch films using the quartz crystal microbalance. *Carbohydrate Polymers*, 261: 117727.

El Knidri, H., Laajeb, A. and Lahsini, A. (2020). Chitin and chitosan: chemistry, solubility, fiber formation, and their potential applications. pp. 35–57. *In*: *Handbook of Chitin and Chitosan*. Elsevier.

Facchinatto, W. M., Dos Santos, D. M., Fiamingo, A., Bernardes-Filho, R., Campana-Filho, S. P., de Azevedo, E. R. and Colnago, L. A. (2020). Evaluation of chitosan crystallinity: A high-resolution solid-state NMR spectroscopy approach. *Carbohydrate Polymers*, 250: 116891.

Feng, C., Chen, Y., Shao, J. and Hou, H. (2020). The crystallization behavior of poly (l-lactic acid)/poly (d-lactic acid) electrospun fibers: effect of distance of isomeric polymers. *Industrial & Engineering Chemistry Research*, 59(17): 8480–8491.

Forestier, E., Combeaud, C., Guigo, N., Monge, G., Haudin, J. M., Sbirrazzuoli, N. and Billon, N. (2020). Strain-induced crystallization of poly (ethylene 2, 5-furandicarboxylate). Mechanical and crystallographic analysis. *Polymer*, 187: 122126.

Fričová, O., Hutníková, M., Kovaľaková, M. and Baran, A. (2020). Influence of aging on molecular motion in PBAT-thermoplastic starch blends studied using solid-state NMR. *International Journal of Polymer Analysis and Characterization*, 25(4): 275–282.

Furushima, Y., Toda, A. and Schick, C. (2020). Effect of multi-step annealing above the glass transition temperature on the crystallization and melting kinetics of semicrystalline polymers. *Polymer*, 202: 122712.

Gonzalez, K., Iturriaga, L., Gonzalez, A., Eceiza, A. and Gabilondo, N. (2020). Improving mechanical and barrier properties of thermoplastic starch and polysaccharide nanocrystals nanocomposites. *European Polymer Journal*, 123: 109415.

Gu, Z., Yang, R., Yang, J., Qiu, X., Liu, R., Liu, Y. and Nie, Y. (2018). Dynamic Monte Carlo simulations of effects of nanoparticle on polymer crystallization in polymer solutions. *Computational Materials Science*, 147: 217–226.

Hall, M., Bansal, P., Lee, J. H., Realff, M. J. and Bommarius, A. S. (2010). Cellulose crystallinity—a key predictor of the enzymatic hydrolysis rate. *The FEBS Journal*, 277(6): 1571–1582.

Hall, K. W., Percec, S., Shinoda, W. and Klein, M. L. (2021). Chain-end modification: a starting point for controlling polymer crystal nucleation. *Macromolecules*, 54(4): 1599–1610.

Hao, W., Wang, M., Zhou, F., Luo, H., Xie, X., Luo, F. and Cha, R. (2020). A review on nanocellulose as a lightweight filler of polyolefin composites. *Carbohydrate Polymers*, 116466.

Hernandez, M. and Herminsul, J. (2021). Effect of the incorporation of polycaprolactone (PCL) on the retrogradation of binary blends with cassava thermoplastic starch (TPS). *Polymers*, 13(1): 38.

Hernández-Medina, M., Torruco-Uco, J. G., Chel-Guerrero, L. and Betancur-Ancona, D. (2008). Caracterización fisicoquímica de almidones de tubérculos cultivados en Yucatán, México. *Food Science and Technology*, 28(3): 718–726.

Hernández-Varela, J. D., Chanona-Pérez, J. J., Benavides, H. A. C., Sodi, F. C. and Vicente-Flores, M. (2021). Effect of ball milling on cellulose nanoparticles structure obtained from garlic and agave waste. *Carbohydrate Polymers*, 255: 117347.

Hong, X., Xu, Y., Zou, L., Li, Y. V., He, J. and Zhao, J. (2021). The effect of degree of polymerization on the structure and properties of polyvinyl alcohol fibers with high strength and high modulus. *Journal of Applied Polymer Science*, 138(10): 49971.

Houfani, A. A., Anders, N., Spiess, A. C., Baldrian, P. and Benallaoua, S. (2020). Insights from enzymatic degradation of cellulose and hemicellulose to fermentable sugars—a review. *Biomass and Bioenergy*, 134: 105481.

Hsieh, Y. T., Nozaki, S., Kido, M., Kamitani, K., Kojio, K. and Takahara, A. (2020). Crystal polymorphism of polylactide and its composites by X-ray diffraction study. *Polymer Journal*, 52(7): 755–763.

Hu, Y., Teat, S. J., Gong, W., Zhou, Z., Jin, Y., Chen, H. and Zhang, W. (2021). Single crystals of mechanically entwined helical covalent polymers. *Nature Chemistry*, 1–6.

Jog, J. P. (2020). Crystallization of polymers: polyethylene terephthalate and polyphenylene sulfide. pp. 661–679. *In*: *Handbook of Applied Polymer Processing Technology*. CRC Press.

Joseph, S. M., Krishnamoorthy, S., Paranthaman, R., Moses, J. A. and Anandharamakrishnan, C. (2021). A review on source-specific chemistry, functionality, and applications of chitin and chitosan. *Carbohydrate Polymer Technologies and Applications*, 2: 100036.

Kakroodi, A. R., Kazemi, Y., Rodrigue, D. and Park, C. B. (2018). Facile production of biodegradable PCL/PLA in situ nanofibrillar composites with unprecedented compatibility between the blend components. *Chemical Engineering Journal*, 351: 976–984.

Kaplan, D., Dani, M., Verdoliva, A. and Bellofiore, P. (2020). 12. Enzyme catalysis in the synthesis of biodegradable polymers. pp. 339–392. *In*: Bastioli, C. (ed.). *Handbook of Biodegradable Polymers*. Berlin, Boston: De Gruyter.

Kaya, M., Mujtaba, M., Ehrlich, H., Salaberria, A. M., Baran, T., Amemiya, C. T. and Labidi, J. (2017). On chemistry of γ-chitin. *Carbohydrate Polymers*, 176: 177–186.

Kochervniskii, V. V., Astakhov, V. A., Bedin, S. A., Malyshkina, I. A., Shmakova, N. A., Korlyukov, A. A. and Volkov, V. V. (2020). Peculiarities of structure and dielectric relaxation in ferroelectric vinylidene fluoride-tetrafluoroethylene copolymer at different crystallization conditions. *Colloid and Polymer Science*, 298(9): 1169–1178.

Kotula, A. P. and Migler, K. B. (2021). Percolation implications in the rheology of polymer crystallization. *Polymer Crystallization*, 4(2): e10162.

Labet, M. and Thielemans, W. (2009). Synthesis of polycaprolactone: a review. *Chemical Society Reviews*, 38(12): 3484–3504.

Lai, D., Li, Y., Wang, C., Liu, Y., Li, D. and Yang, J. (2020). Inhibition effect of aminated montmorillonite on crystallization of dendritic polyamide 6. *Materials Today Communications*, 25: 101578.

Lanna, A., Suklueng, M., Kasagepongsan, C. and Suchat, S. (2020). Performance of novel engineered materials from epoxy resin with modified epoxidized natural rubber and nanocellulose or nanosilica. *Advances in Polymer Technology*, 2020.

Li, X., He, Y., Dong, X., Ren, X., Gao, H. and Hu, W. (2020). Effects of hydrogen-bonding density on polyamide crystallization kinetics. *Polymer*, 189: 122165.

Liu, L., Chu, Z., Liao, Y., Ma, Z. and Li, Y. (2020a). Flow-induced crystallization in butene-1/1, 5-hexadiene copolymers: mutual effects of molecular factor and flow stimuli. *Macromolecules*, 53(19): 8476–8486.

Liu, J. H., Cai, J. H., Tang, X. H., Weng, Y. X. and Wang, M. (2020b). Achieving highly crystalline rate and crystallinity in Poly (L-lactide) via *in-situ* melting reaction with diisocyanate and benzohydrazine to form nucleating agents. *Polymer Testing*, 81: 106216.

Malz, F., Arndt, J. H., Balko, J., Barton, B., Büsse, T., Imhof, D. and Brüll, R. (2021). Analysis of the molecular heterogeneity of poly (lactic acid)/poly (butylene succinate-co-adipate) blends by hyphenating size exclusion chromatography with nuclear magnetic resonance and infrared spectroscopy. *Journal of Chromatography A*, 1638: 461819.

Mao, H. I., Chen, C. W. and Rwei, S. P. (2020). Synthesis and nonisothermal crystallization kinetics of poly (butylene terephthalate-co-tetramethylene ether glycol) copolyesters. *Polymers*, 12(9): 1897.

Mondal, S. (2020). Nanocellulose reinforced polymer nanocomposites for sustainable packaging of foods, cosmetics, and pharmaceuticals. pp. 237–253. *In: Sustainable Nanocellulose and Nanohydrogels from Natural Sources*. Elsevier.

Mondal, K., Sakurai, S., Okahisa, Y., Goud, V. V. and Katiyar, V. (2021). Effect of cellulose nanocrystals derived from Dunaliella tertiolecta marine green algae residue on crystallization behaviour of poly (lactic acid). *Carbohydrate Polymers*, 261: 117881.

Monnier, X., Cavallo, D., Righetti, M. C., Di Lorenzo, M. L., Marina, S., Martin, J. and Cangialosi, D. (2020). Physical aging and glass transition of the rigid amorphous fraction in poly (l-lactic acid). *Macromolecules*, 53(20): 8741–8750.

Muthukumar, M. and Welch, P. (2000). Modeling polymer crystallization from solutions. *Polymer*, 41(25): 8833–8837.

Nagarkar, R. and Patel, J. (2019). Polyvinyl alcohol: A comprehensive study. *Acta Scientific Pharmaceutical Sciences*, 3(4): 34–44.

Nguyen, N. Q., Chen, T. F. and Lo, C. T. (2021). Confined crystallization and chain conformational change in electrospun poly (ethylene oxide) nanofibers. *Polymer Journal*, 1–11.

Okuyama, K., Noguchi, K., Miyazawa, T., Yui, T. and Ogawa, K. (1997). Molecular and crystal structure of hydrated chitosan. *Macromolecules*, 30(19): 5849–5855.

Pan, P. and Inoue, Y. (2009). Polymorphism and isomorphism in biodegradable polyesters. *Progress in Polymer Science*, 34(7): 605–640.

Parhi, R. (2020). Drug delivery applications of chitin and chitosan: a review. *Environmental Chemistry Letters*, 18(3): 577–594.

Qin, Y., Litvinov, V., Chassé, W., Zhang, B. and Men, Y. (2021). Change of lamellar morphology upon polymorphic transition of form II to form I crystals in isotactic Polybutene-1 and its copolymer. *Polymer*, 215: 123355.

Raabe, D., Romano, P., Sachs, C., Fabritius, H., Al-Sawalmih, A., Yi, S. B. and Hartwig, H. G. (2006). Microstructure and crystallographic texture of the chitin–protein network in the biological composite material of the exoskeleton of the lobster Homarus americanus. *Materials Science and Engineering: A*, 421(1-2): 143–153.

Raimo, M. (2021). Impact of thermal properties on crystalline structure, polymorphism and morphology of polymer matrices in composites. *Materials*, 14(9): 2136.

Righetti, M. C. (2017). Amorphous fractions of poly (lactic acid). *Synthesis, Structure and Properties of Poly (lactic acid)*, 195–234.

Righetti, M. C., Marchese, P., Vannini, M., Celli, A., Lorenzetti, C., Cavallo, D. and Androsch, R. (2020). Polymorphism and multiple melting behavior of bio-based poly (propylene 2, 5-furandicarboxylate). *Biomacromolecules*, 21(7): 2622–2634.

Rodriguez-Garcia, M. E., Hernandez-Landaverde, M. A., Delgado, J. M., Ramirez-Gutierrez, C. F., Ramirez-Cardona, M. m., Millan-Malo, B. M. and Londono-Restrepo, S. M. (2020). Crystalline structures of the main components of Strach. *Current Opinion in Food Science*.

Ru, G., Wu, S., Yan, X., Liu, B., Gong, P., Wang, L. and Feng, J. (2019). Inverse solubility of chitin/chitosan in aqueous alkali solvents at low temperature. *Carbohydrate Polymers*, 206: 487–492.

Rubina, M. S., Elmanovich, I. V., Shulenina, A. V., Peters, G. S., Svetogorov, R. D., Egorov, A. A. and Vasil'kov, A. Y. (2020). Chitosan aerogel containing silver nanoparticles: From metal-chitosan powder to porous material. *Polymer Testing*, 86: 106481.

Salari, M., Khiabani, M. S., Mokarram, R. R., Ghanbarzadeh, B. and Kafil, H. S. (2019). Preparation and characterization of cellulose nanocrystals from bacterial cellulose produced in sugar beet molasses and cheese whey media. *International Journal of Biological Macromolecules*, 122: 280–288.

Santos, V. P., Marques, N. S., Maia, P. C., Lima, M. A. B. D., Franco, L. D. O. and Campos-Takaki, G. M. D. (2020). Seafood waste as attractive source of chitin and chitosan production and their applications. *International Journal of Molecular Sciences*, 21(12): 4290.

Shkarina, S., Shkarin, R., Weinhardt, V., Melnik, E., Vacun, G., Kluger, P. J. and Surmenev, R. A. (2018). 3D biodegradable scaffolds of polycaprolactone with silicate-containing hydroxyapatite microparticles for bone tissue engineering: High-resolution tomography and *in vitro* study. *Scientific Reports*, 8(1): 1–13.

Singhvi, M. S., Zinjarde, S. S. and Gokhale, D. V. (2019). Polylactic acid: synthesis and biomedical applications. *Journal of Applied Microbiology*, 127(6): 1612–1626.

Sintim, H. Y., Bary, A. I., Hayes, D. G., Wadsworth, L. C., Anunciado, M. B., English, M. E. and Flury, M. (2020). *In situ* degradation of biodegradable plastic mulch films in compost and agricultural soils. *Science of The Total Environment*, 727: 138668.

Song, G., Zhang, J. and Nishiyama, Y. (2020). Twisted pseudo-tetragonal orthorhombic lamellar crystal in cellulose/ionic liquid spherulite. *Cellulose*, 27: 5449–5455.

Staub, M. C. and Li, C. Y. (2020). Towards shape-translational symmetry incommensurate polymer crystals. *Polymer*, 195: 122407.

Tashiro, K., Kusaka, K., Yamamoto, H. and Hanesaka, M. (2020). Introduction of disorder in the crystal structures of atactic poly (vinyl alcohol) and its iodine complex to solve a dilemma between X-ray and neutron diffraction data analyses. *Macromolecules*, 53(15): 6656–6671.

Tavares, L., Flores, E. E. E., Rodrigues, R. C., Hertz, P. F. and Noreña, C. P. Z. (2020). Effect of deacetylation degree of chitosan on rheological properties and physical chemical characteristics of genipin-crosslinked chitosan beads. *Food Hydrocolloids*, 106: 105876.

Tian, B. and Liu, Y. (2020). Chitosan-based biomaterials: From discovery to food application. *Polymers for Advanced Technologies*, 31(11): 2408–2421.

Tonelli, A. E. (2020). Enhancing the melt crystallization of polymers, especially slow crystallizing polymers like PLLA and PET. *Polymer Crystallization*, 3(1): e10095.

Torron, S., Johansson, M. K., Malmström, E., Fogelström, L., Hult, K. and Martinelle, M. (2017). Telechelic polyesters and polycarbonates prepared by enzymatic catalysis. pp. 29–64. *In*: *Handbook of Telechelic Polyesters, Polycarbonates, and Polyethers*. Jenny Stanford Publishing.

Ulker, C. (2018). Enzymatic synthesis and characterization of biodegradable poly (ω-pentadecalactone-co-ε-caprolactone) copolymers. *Journal of Renewable Materials*, 6(6): 591–598.

Vyazovkin, S. (2020). Activation energies and temperature dependencies of the rates of crystallization and melting of polymers. *Polymers*, 12(5): 1070.

Wang, Z., Ma, Z. and Li, L. (2016). Flow-induced crystallization of polymers: Molecular and thermodynamic considerations. *Macromolecules*, 49(5): 1505–1517.

Wang, P., Linares-Pastén, J. A. and Zhang, B. (2020a). Synthesis, molecular docking simulation, and enzymatic degradation of AB-type indole-based polyesters with improved thermal properties. *Biomacromolecules*, 21(3): 1078–1090.

Wang, T., Li, X., Luo, R., He, Y., Maeda, S., Shen, Q. and Hu, W. (2020b). Effects of amide comonomers on polyamide 6 crystallization kinetics. *Thermochimica Acta*, 690: 178667.

Wang, Y., Ying, Z., Xie, W. and Wu, D. (2020c). Cellulose nanofibers reinforced biodegradable polyester blends: Ternary biocomposites with balanced mechanical properties. *Carbohydrate Polymers*, 233: 115845.

Wang, W., Xue, C. and Mao, X. (2020d). Chitosan: Structural modification, biological activity and application. *International Journal of Biological Macromolecules*.

Wang, W., Fenni, S. E., Ma, Z., Righetti, M. C., Cangialosi, D., Di Lorenzo, M. L. and Cavallo, D. (2021a). Glass transition and aging of the rigid amorphous fraction in polymorphic poly (butene-1). *Polymer*, 123830.

Wang, J., Liang, Y., Zhang, Z., Ye, C., Chen, Y., Wei, P. and Xia, Y. (2021b). Thermoplastic starch plasticized by polymeric ionic liquid. *European Polymer Journal*, 148: 110367.

Welch, P. and Muthukumar, M. (2001). Molecular mechanisms of polymer crystallization from solution. *Physical Review Letters*, 87(21): 218302.

Woo, E. M., Lugito, G. and Nagarajan, S. (2020). Dendritic polymer spherulites: birefringence correlating with lamellae assembly and origins of superimposed ring bands. *Journal of Polymer Research*, 27(1): 1–22.

Wu, Q., Xu, J., Wu, Z., Zhu, S., Gao, Y. and Shi, C. (2020). The effect of surface modification on chemical and crystalline structure of the cellulose III nanocrystals. *Carbohydrate Polymers*, 235: 115962.

Xuzhen, Z., Xin, W., Chenmeng, Z., Wenjian, H. and Yong, L. (2021). Defects in polylactide spherulites: Ring line cracks and micropores. *Polymer Degradation and Stability*, 183: 109416.

Yousefzade, O., Jeddi, J., Franco, L., Puiggali, J. and Garmabi, H. (2020). Crystallization kinetics of chain extended poly (L-lactide)s having different molecular structures. *Materials Chemistry and Physics*, 240: 122217.

Yu, T. H., Su, Y. H., Huang, H. H., Tsai, H. J. and Hsu, W. K. (2020). Amorphous fraction controlled mechanical and optical properties of polylactic acid below glass transition temperature. *Polymer Testing*, 91: 106731.

Zaaba, N. F. and Jaafar, M. (2020). A review on degradation mechanisms of polylactic acid: Hydrolytic, photodegradative, microbial, and enzymatic degradation. *Polymer Engineering & Science*, 60(9): 2061–2075.

Zargar, V., Asghari, M. and Dashti, A. (2015). A review on chitin and chitosan polymers: structure, chemistry, solubility, derivatives, and applications. *Chem. Bio Eng. Reviews*, 2(3): 204–226.

Zhang, M. C., Guo, B. H. and Xu, J. (2017). A review on polymer crystallization theories. *Crystals*, 7(1): 4.

Zhang, Y. and Geng, X. 2020. Principle of biopolymer plasticization. pp. 1–19. *In*: *Processing and Development of Polysaccharide-Based Biopolymers for Packaging Applications*. Elsevier.

Zhang, Z. and Lucia, L. (2020). Improved reswelling behaviors and thermal stability of polyvinyl alcohol composite gels assisted by salt. *Materials Letters*, 281: 128743.

Zhang, W., Zhao, Y., Xu, L., Song, X., Yuan, X., Sun, J. and Zhang, J. (2020). Superfine grinding induced amorphization and increased solubility of α-chitin. *Carbohydrate Polymers*, 237: 116145.

Zhang, X., Yang, B., Fan, B., Sun, H. and Zhang, H. (2021). Enhanced nonisothermal crystallization and heat resistance of poly (l-lactic acid) by d-sorbitol as a homogeneous nucleating agent. *ACS Macro Letters*, 10(1): 154–160.

Zhao, R., Chu, Z. and Ma, Z. (2020). Flow-induced crystallization in polyethylene: effect of flow time on development of shish-kebab. *Polymers*, 12(11): 2571.

Zheng, Y. and Pan, P. (2020). Crystallization of biodegradable and biobased polyesters: Polymorphic crystallization, cocrystallization, and structure-property relationship. *Progress in Polymer Science*, 101291.

Chapter 6

Rheology Properties of Biodegradable Polymers

Juan Pablo Castañeda Niño,[2] *José Herminsul Mina Hernandez,*[2]
Heidi Andrea Fonseca Florido,[3] *Leticia Melo López,*[3]
Margarita del Rosario Salazar Sánchez[4] and *Jose Fernando Solanilla Duque*[1,*]

1. Introduction

Generally, three types of tests are performed to study the time-dependent properties of viscoelastic materials, finding the behavior of the creep phenomenon, stress relaxation and the implementation of Creep by multiple cycles. In contrast, the tension test, dynamic analysis and thermo-mechanical analysis are temperature-dependent. The tests can be performed using different geometries to obtain different stresses, strains and moduli (E', E", G' and G"): bending, compression, tension, shear, compressibility modulus, parallel plates, among others. When determining the type of polymeric material to be evaluated, only one of the tests is chosen to describe the changes in the properties over time; however, there is also the option of reporting information on the viscoelastic properties of the polymeric materials in a relatively short time (Menard, 2008; Tajvidi et al., 2006). The following is a description of the test methods that can be used in the DMA. Some of the thermal and rheological properties are shown in Table 6.1.

2. Types of polymer melt processing and their flow behavior

The melt processing of biodegradable polymers helps to achieve unique applications for different industries, however, their flow behavior is an important property to consider in their processing by melting. Biodegradable plastics can be melted by heating, shaped and solidified according to their end use. During the melt processing biodegradable polymers have high viscosities that may or may not improve their behavior in the processing equipment usually used for conventional polymers. The types of biodegradable polymers used by melt processing are classified by: naturally biodegradable as thermoplastic starch (TPS), Cellulose Acetate (CA), polyhydroxyalkanoates (PHA); designed to be biodegradable from renewable sources as polylactic acid (PLA); and from non-renewable sources as polycaprolactone (PCL), polybutylene adipate-co-terephthalate (PBAT) and polybutylene

[1] Departamento de agroindustria, Universidad del Cauca, Popayán, Colombia.
[2] Escuela de Ingeniería de Materiales, Universidad del Valle, Cali, Colombia.
[3] Centro de Investigación en Química Aplicada (CIQA), Coahuila, Mexico.
[4] Departamento de Ciencias Agroindustriales, Facultad de Ingenierías y Tecnológicas, Universidad Popular del Cesar, Cesar, Colombia.
* Corresponding author: jsolanilla@unicauca.edu.co

Table 6.1. Thermal and rheological properties of some biopolymers. Polymer density (ρ, in g/cm³). Thermal properties: glass transition temperature (Tg, in °C), crystallization temperature (Tc in °C) and melt point (Tm, in °C). Consistency index (K), Flow index (n), enthalpies of fusion (ΔHm in J/g), heat capacity (Cp in J/g °C to 25°C).

Polymers		**Thermal properties**						
Molecular mass (kDa)		ρ	*Cp*	*Tm*	*Tc*	*ΔHm*	*Tg*	*Ref.*
PLA		1.21–1.25	1.20	150–162	155–170		45.60–59.2	(Farah et al., 2016)
PLA 3051D-		1.25		146.5–148.4	106.4	27.6–33.4	55–65	(Gordobil et al., 2014)
PLA 2002D				168.2	122.4		58.6–61.37	(Ge et al., 2013)
PLA 4032D				169.14	104.59	24.95	61.55	(Ding et al., 2018)
PLLA	100–300	1.24–1.30		170–200		0.6–36.5	55–65	(Farah et al., 2016)
				174–190		36.5	54	(Baimark and Srihanam, 2015)
PDLLA	70	1.25–1.27		am*			50–60	(Ahlinder et al., 2018)
		1.25–1.27		176		44.8	59	(Baimark and Srihanam, 2015)
PGA		1.50–1.71		220–233			35–45	(Beltrán-García et al., 2001)
PCL	40–80	1.11–1.146		58–65	26.1	39	(–60)–(–65)	(Olewnik-Kruszkowska et al., 2020)
PHB		1.18–q.262		168–182			15.0–5	(Sánchez-Safont et al., 2018)
PBS	112			127.5–146.5	60–110	67.4	(–18)–(–32)	(Šerá et al., 2020a)
PBAT				60.4/126.5	48.4	21.8	(–21.83)–(–33.2)	(de Oliveira et al., 2019)
PCLA	100–500			170–190	84			(Ilyas et al., 2020)
PTMC	14			228.1	177.6	27.0	–30	(Ilyas et al., 2020)
PGLA	40–100			161	120			(Ilyas et al., 2020)
PHA				148.72	107.85	6.9	50.39–64.64	(Georgiopoulos et al., 2014)

Polymers		**Rheological properties**						
Molecular mass (kDa)		σ	*E*	ϵ	σ*	E*	*n*	*Ref.*
PLA		21–60	0.35–3.5	2.5–6	16.8–48.0	0.28–2.80	35 (190°C, 2.16 Kg)	(Park et al., 2020)
PLA 3051D		48	3.64				0.934 (160°C, 2.16 Kg), 10–25 (210°C, 2.16 Kg)	(Lv et al., 2017)
PLA 2002D		69.8 ± 63.2	1.777	5.7 ± 0.3	1.24		6–6.4 (190°C, 2.16 Kg)	(Ali Nezamzadeh et al., 2017)

Table 6.1 contd. ...

...Table 6.1 contd.

Polymers		Rheological properties						
PLA 4032D		58.5 ± 2						(Wang et al., 2019)
PLLA	100–300	15.5–150	2.7–4.14	3.0–10.0	40.0–66.8	2.23–3.85		(Weir et al., 2004)
		40.0–66.8	2.7–4.14	3.0–10	1.24–1.30		60.5 (170°C, 2.16 Kg) –60.8 (250°C, 2.16 Kg)	(Farah et al., 2016)
PDLLA	70	27.6–50	1–3.45	2.0–10.0	22.1–39.4	0.80–2.36		(Aluthge et al., 2013)
							60.5 (170°C, 2.16 Kg) –60.8 (250°C, 2.16 Kg)	(Ahlinder et al., 2018)
PGA		60–99.7	6.0–7.0	1.5–20	40.0–45.1	5.0–4.51		(Zhang et al., 2021)
PCL	40–80	20.7–42	0.21–0.44	300–1000	18.6–36.7	0.19–0.38		(Avella et al., 2021)
PHB		40	3.5–4	5.0–8.0	32.0–33.9	2.80–2.97		(Weinmann and Bonten, 2019)
PBS	112	34–41.47			1.26	0.53	10	(Arabeche et al., 2020)
PBAT		22.0 ± 3.7	0,0557	544.5 ± 113.4			7.10 (190°C, 2.16 Kg); 2.5–4.5 (190°C, 2.16 Kg)	(Ding et al., 2018)
PCLA	100–500							
PTMC	14	59.0	0,013	199 ± 31			30(300°C, 1.2 Kg)	(He et al., 2008; Li et al., 2020; Šnejdrová et al., 2021; Vidyasagar et al., 2017)
PGLA	40–100							
PHA		326.6 ± 44	0,392	116 ± 21				(Tian et al., 2021)

succinate (PBS). The processing of polymers includes extrusion, injection molding, compressing molding, film blowing and fiber spinning.

2.1 Thermoplastic starch (TPS)

Native starch is converted to thermoplastic starch using a plasticizer by solution, extrusion, melt mixing, among others, and include morphological and structure changes of starch granules. The gelatinization, melting and degradation are related with shear and temperature and cause variations in viscosity affecting starch processability, good performance during shaping process and mixing with others polymers. Starch flow conditions and viscous behavior depend on starch botanical origin, type of plasticizer, concentration, processing temperature and previous thermomechanical treatment (Decaen et al., 2020). Nevoralová et al. (2019) studied the mixing of three different types of starches (wheat, corn and tapioca) with polycaprolactone (PCL). From oscillatory shear flow, the tapioca starch showed the lowest elasticity and all TPS exhibited gel-like behavior with the same slope of storage modulus curves. The damping factor ($\tan \delta = G''/G'$) was constant in evaluating the frequency range. From frequency dependencies the values of complex viscosity decreased in order of wheat, corn and tapioca. PCL presented the lowest viscosity value. In dynamic mechanical thermal analysis, wheat and corn TPS/PCL blends showed similar rheological characteristics with the storage modulus higher than the loss modulus, but blends with tapioca starch displayed the opposite behavior suggesting that the PCL-phase has a dominant effect on the rheological properties of the tapioca/PCL blend.

2.2 Cellulose acetate (CA)

CA presents difficulties in processing because high viscosity and the melt processing temperature is very close to its decomposition temperature. To reduce these issues CA could be mixed with other biodegradable polymers or plasticizers (Bendaoud and Chalamet, 2014). Wang et al. (2016) plasticized CA varying amount of polyethylene glycol (PEG 200) and triethyl citrate (TEC). The Melt Flow Index (MFI) of CA plasticized with PEG were much higher than TEC, presenting a lower viscosity, also, MFI grew with increasing content of plasticizers. A non-Newtonian, shear thinning behavior was observed and the melt viscosity of CA decreased with the addition of plasticizers, because the interaction between CA chains is reduced by the interaction with plasticizers. The PEG was more efficient for reduced the viscosity of CA at all test shear rates because it had a very low viscosity and good lubricating property.

2.3 Polylactic acid (PLA)

Measuring the viscoelastic behavior of PLA presents difficulties due to its low thermostability, resulting in degradation during measurement in the melt. It has been shown that melt viscosity of low molecular weight PLAs (~ 40,000 Dalton) shows Newtonian-like behavior at shear rates typical of film extrusion. In high molecular weight PLAs (~ 100,000–300,000 Dalton) the melt viscosity is in the order of 500–1000 Pa. s. at shear rates of 10–50 s^{-1}. The melts of these high molecular weight PLAs behave like a pseudoplastic non-Newtonian fluid and the polymer melt exhibits shear-thinning behavior. This rheological behavior is the consequence of gradual disentanglement of the highly physically entangled molecules of PLA (Lim et al., 2008). Wang et al. (2019) prepared polylactic acid/polybutylene adipate-co-terephthalate (PLA/PBAT) blends by melt blending in the presence of a multifunctional epoxy oligomers as reactive compatibilizer (ADR). PBAT is a more flexible molecular chain that increase the elasticity and melt strength of PLA, however, the addition of ADR to the blends causes a more pronounced shear thinning behavior and changed the Newtonian behavior in the intermediate frequency region and increased the complex viscosity at low frequencies due to the enhancement of interfacial interaction and entanglement between PLA and PBAT chains.

2.4 Polycaprolactone (PCL)

PCL exhibits low melt viscosity and strength, reducing its melt processing by film extrusion or film blowing, and also presents low crystallization rate limiting its use in injection molding. To overcome these drawbacks, PCL is blended with other polymers, nanomaterials and also chemically modified (Avella et al., 2021). Balali et al., 2018 elaborated silk-reinforced PLA/PCL composites with 1–7% of silk fibers by melt mixing. The blends had a typical terminal relaxation behavior from the nature of PLA and PCL, however the addition of silk modifies this behavior especially at low frequencies and both moduli increased their values as silk load was higher, because the polymer chains movement restriction.

2.5 Polybutylene adipate-co-terephthalate (PBAT)

It is known that, an important factor in controlling the blend morphology is the viscosity ratio (the ratio of the viscosity of the dispersed polymer to the viscosity of the continuous polymer), a closer viscosity could result in much finer morphology. Complex viscosity (η^*) and MFI depend on the molecular weight and chain structure and are used to characterize the flow behavior of molten polymers (Lu et al., 2017). Lu et al. (2017) demonstrated that the addition of dicumyl peroxide (DCP) on PBAT increases the η^*, while the MFI decreases. This behavior is explained by the entanglement theory, which establishes that the presence of crosslinking network or long-chain branching structure promotes the tangles of the PBAT chains, and the polymer mobility.

2.6 Polybutylene succinate (PBS)

The PBS has a highly linear chain structure and weak intermolecular interaction resulting in low melt strength and melt viscosity, which reduce its processing by film blowing, spinning and foaming. To increase the chains entanglement and improve the melt strength and melt viscosity many methods could be used like chain extension and crosslinking. The improvement of melt viscosity and strength favors the fabrication of articles as blown bottle, foamed films and fibrillated materials (Li et al., 2015). Marek et al. (2019) elaborated nanocomposites of Mg2Al-based PBS nanocomposites dispersed with inorganic–organic hybrid materials (Layered Double Hydroxides, LDHs), were functionalized with the amino acids L-histidine (HIS) and L-phenylalanine (PHE) to increase the Newtonian zero-shear viscosity to improve the melt viscosity of PBS for the melt processing. Both organo-modified LDHs exhibited a remarkable chain-extension effect for PBS with an outstanding increase in the zero-shear viscosity of almost 90 times in compared to filler-free PBS.

3. Application of rheology in biopolymers

Biodegradable polymers or biopolymers are the best alternatives to replace traditional petroleum-based polymers, which generally lead to persistent waste, once they have completed their function. Among the biodegradable polymers that have been used and/or studied the most to date are polyvinyl alcohol (Sharma, 2020), polylactic acid (Park et al., 2020), polyhydroxyalkanoates (Lemes et al., 2015), polyhydroxybutyrate (Botana et al., 2010), alginate (Nishio et al., 2004), starch (Romero-Bastida et al., 2018) and gelatin (Mu et al., 2013). Although some biodegradable polymers such as polyvinyl alcohol have excellent film-forming properties, good mechanical properties, high solubility in water and high resistance to chemicals (Sharma, 2020), some other biopolymers have disadvantages such as high gas or water vapor permeability, lower mechanical properties (such as gelatin) (Mu et al., 2013) or insufficient properties for injection and extrusion molding such as polylactic acid (Harris and Lee, 2008), compared to synthetic polymers. An option to modify and/or improve the mechanical, physical and thermal properties of biodegradable polymers is through the incorporation of nanoparticles, either by *in situ* polymerizations (Ilsouk et al., 2017), by melt mixing (Botta et al., 2021), or in solution (Feijoo et al., 2020). The improvement of the properties of the polymers can be achieved even when using low concentrations of nanoparticles, as long as they are homogeneously dispersed and that there are polymer-nanoparticle interactions. As nanoparticles have a high surface-to-volume ratio, they present very high van der Waals and electrostatic forces (Ridi et al., 2014) that induce their agglomeration. To avoid agglomeration, nanoparticles can be superficially modified before incorporating them into the polymer and thus achieve a better interfacial interaction, which generates an effective load transfer between polymer and filler (Abraham et al., 2017). Nanoparticle-polymer interactions restrict the long- and/or short-range dynamics of the polymer chains, which is reflected in the change in viscosity of the composite material compared to the neat polymer (Botta et al., 2021). One way to monitor the interaction, orientation and mobility of the polymeric chains in the polymeric nanocomposites is through the rheological methods, which reflect the inner disorders, interpret viscous as well as elastic properties of polymer nanocomposites (Sivaraman et al., 2017), in addition to that they can detect the presence of internal structures (Abraham et al., 2017). From the rheological studies of the nanocomposites, it is possible to identify the concentration of the charge sufficient for the behavior of the nanocomposite to change from liquid-like to solid-like (Jiang et al., 2016). That is, it is possible to know the amount of nanometric filler with which forms a 3D network, which generates an abrupt change of the processing and rheological characteristics, known as rheological percolation threshold of polymer nanocomposite (Park et al., 2020).

When evaluating the rheological properties of biopolymeric nanocomposites, it has been seen that there are behaviors that occur in various systems, these behaviors are mentioned below. In shear flow, as the nanoparticle loading increased, storage modulus enhanced, and the solid-like transition is observed. In addition, shear viscosity and shear thinning behavior also increased. While in extensional flow, strain hardening behavior was observed if the nanoparticle is well dispersed in the system (Lee et al., 2005). On the other hand, when graphing the complex viscosity as a function of the frequency of the neat and filled materials, several effects are observed: (a) the nanocomposites present a non-Newtonian behavior more pronounced than the neat polymer behavior, (b) the viscosity of the nanocomposites is higher than that of the neat matrix, and it expanded with increasing the filler content, (c) at high frequencies, the flow curves of the nanocomposites materials and the flow curve of the neat polymer remain closer among them, although the viscosity values of the filled materials remain higher than the pure polymer. This rheological behavior is attributed to a strong interaction between the dispersed nanoparticle and the polymer that restricts the movements of the polymer chains (Scaffaro et al., 2016).

The rheological studies allow knowing the minimum concentration of nanoparticles to be incorporated, which helps in reducing costs by generating polymeric nanocomposites, using the least necessary amount of nanoparticles. This optimization of resources increases the possibility of using polymeric nanocomposites with superior properties in various applications because their production cost could be competitive with respect to the conventionally used materials. In addition, this type of study has application as a quality control tool that allows reducing batch-to-batch product variations in the industry (Abraham et al., 2017). Table 6.2 shows the application of rheological measurement of some biodegradable polymeric nanocomposites.

The rheological study of nanocomposites based on biodegradable biopolymers allows one to accurately know the interactions that may exist between the nanoparticles and the polymer matrix. With the help of this tool, together with others not mentioned here, it is possible to take decisions to optimize the generation of nanocomposites with the best thermal, mechanical and processing properties. These decisions are based on the understanding of the internal structure of the materials, which are in function of the sizes and shapes of the particles, their compatibility with the polymeric matrix, the way the nanocomposite is prepared and the concentration of nanoparticles, among others. The rheological studies allow employing more accurately the nanotechnology in various applications.

Table 6.2. Application of rheological measurement to biodegradable polymer nanocomposites.

Biodegradable polymer	Nanoparticle and amount evaluated	Preparation	Rheological measurement	Application	Ref.
Bioflex F2110 MaterBi® EF05B MaterBi® EF04P	Cloisite 20A, 5 wt.%; Socal 312; 5 wt.%,	Melt extrusion	Oscillatory rheometer in parallel plate geometry and capillary rheometer	Check the viability of using a bionanocomposites for the production of irrigation pipes	(Botta et al., 2021)
Sodium alginate polymer	Maghemite (γ-Fe_2O_3), 1%	Aqueous solutions	Oscillatory rheometer in cone-plate geometry	To obtain bionancomposites controllables by an external magnetic field with applications for drug delivery, sensoring and tissue engineering	(Feijoo et al., 2020)
Poly(Lactic Acid)	Alkylated Graphene Oxide 0.1–2%	Solution blending	Rotational rheometer with a parallel-plate geometry	Increase the barrier characteristics against gas and water vapor transfer for packaging	(Park et al., 2020)
Poly (vinyl alcohol)	Graphene oxide, 0.5, 1.0 & 1.5 wt%	Solution	Rotational rheometer with a parallel-plate geometry	Potential to be employed in the packaging and coating applications	(Sharma, 2020)
Poly(Lactic Acid)	Graphene, 1.5–9%	mMlt extrusion	Oscillatory rheometer with a parallel-plate geometry	Materials with superior mechanical, thermal, gas barrier, electrical and flame retardant properties	(Ivanova and Kotsilkova, 2019)
Poly(Lactic Acid)	Hydroxyapatite, 5–25 wt%	Melt-compounding	Torque rheometer	Fabrication of biocompatible composites for medical applications	(Backes et al., 2020)
Polymer blend of poly(lactic acid) (PLA) and a copolyester (BioFlex®)	Graphene nanoplatelets and ciprofloxacin	Melt-compounding	Plate–plate rotational rheometer	biopolymer-based nanocomposites with antimicrobial properties suitable for medical device packaging	(Scaffaro et al., 2016)
GREENPOL (polyethylene, aliphatic polyester and starch).	Cloisite 15A and Cloisite 30B, 1–7%	Melt Intercalation Method	Rheometer with a parallel-plate geometry and extensional rheometer	Biodegradable nanocomposites with superior properties to unmodified polymers.	(Lee et al., 2005)

4. Dynamic-mechanical analysis (DMA)

This technique provides information related to first, second and third-order transitions by studying the viscoelastic properties and morphology of different materials. Likewise, the analytical technique allows characterizing the material under different conditions to determine its structure and performance in possible applications (Devi et al., 2011; Menard, 2008; Menard and Menard, 2020; Saba et al., 2016; Sepe, 1998). Other parameters that can be evaluated in polymers and biocomposites are crosslink density, dynamic brittleness, dynamic/complex viscosity, storage and loss modulus conformation, creep and stress relaxation concerning time at a given temperature (Saba et al., 2016). Among the samples that can be evaluated and analyzed in the DMA are polymeric materials such as homopolymers, copolymers, polymeric blends, and composites/biocomposites, exposed to a specific frequency in a temperature range (Das and Chakraborty, 2006). This analytical technique has a high degree of sensitivity by providing oscillating forces, causing sinusoidal stresses in the

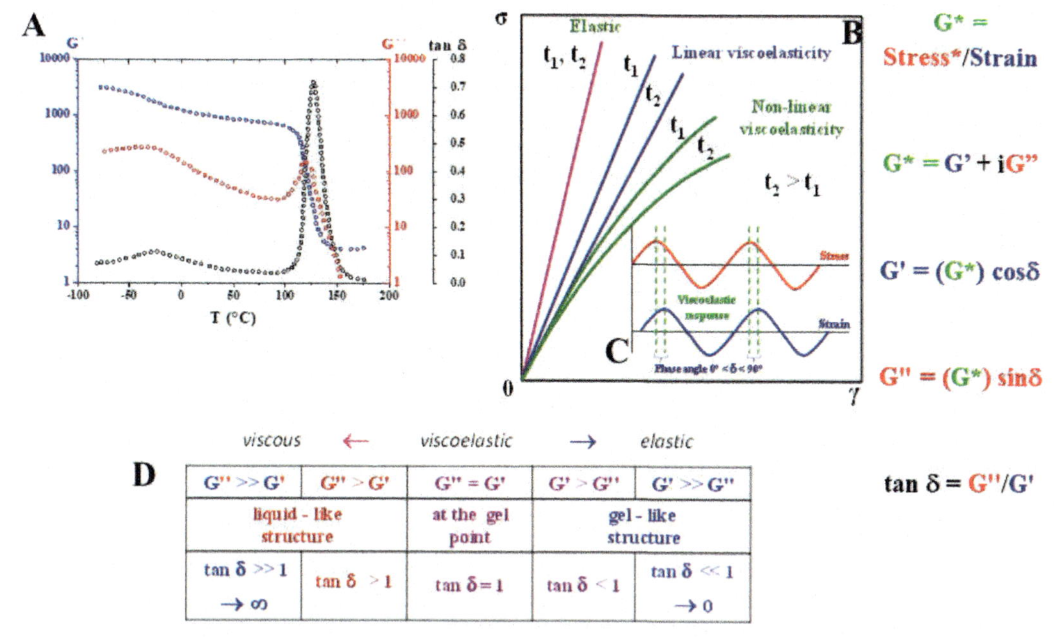

Figure 6.1. Basic fundamentals of Dynamic-Mechanical Analysis (DMA). (A) typical DMA figure, (B) typical Viscoelasticity figure, (C) phase angle figure, (D) viscoelastic, viscous and solid behavior parameters. Complex modulus (G*), elastic modulus (G'), loss modulus (G"), and phase angle (δ). Adapted from (Chan et al., 2017; Lazaridou et al., 2003; Menard and Menard, 2020; Park et al., 2020; Romero-Bastida et al., 2018).

sample, which generate the respective sinusoidal deformations as a response, characterizing the structure as a function of frequency, temperature, time, stress, atmosphere or mixture of the above parameters (Fig. 6.1). The results provided by this analytical technique depend on the physical or structural phases such as the type of polymeric matrix, the type of interface, morphology and nature of the constituents of the evaluated material (Devi et al., 2011; Menard, 2008; Menard and Menard, 2020; Saba et al., 2016; Sepe, 1998). From the above, the properties related to the flow tendency (or viscosity) and the stiffness (modulus) of the material to be evaluated can be calculated (Liang et al., 2007).

The technique determines the complex modulus (E*) through the stress/strain ratio and through trigonometric fundamentals, separates the dynamic mechanical responses of materials into two distinct parts (Fig. 6.1): the elastic component or storage modulus (E' or G') and the viscous component or loss modulus (E" or G"). The elastic process describes the energy stored in the system, while the viscous component describes the energy dissipated as heat during the process (Tajvidi and Falk, 2004; Tsang et al., 2019).

The storage modulus can be related to the modulus of elasticity or Young's modulus, being associated with the stiffness of the evaluated material, while the loss moduli is related to the internal friction of the material, being sensitive to different types of molecular movements, transitions, relaxation processes, morphologies and presence of structural heterogeneities (Wang et al., 2019). The mechanical loss factor or tangent delta (tan δ) is another functional and dimensionless parameter to compare the viscoelastic behaviors of different materials, including the materials' ability in energy dissipation (E"/E') as shown in Fig. 6.1 (Pothan and Thomas, 2003). The presence of high values of the Tan δ in the evaluated sample indicates a higher plastic deformation behavior, while at low values, high elasticity is evidenced. The relationship between Tan δ, storage and loss moduli concerning temperature is shown in Fig. 6.1A, being one of the advantages of the DMA based on obtaining the moduli and the Tan δ in each oscillation, allowing the generation of sweeps or curves in temperature intervals comprising approximately 200°C between 20 to 40 minutes (Menard and Menard, 2020).

Polymers are viscoelastic fluids that act as elastic or viscous materials depending on the flow or deformation ratio. Viscoelastic materials undergo mechanical changes at different temperatures, corresponding to relaxation transitions associated with molecular mechanisms.

The mentioned transitions are the glass transition temperature (Tg) and melting temperature (Tm) in amorphous and crystalline materials, being essential thermal parameters for the application of polymers. Amorphous materials such as polymethylmethacrylate are used below Tg, semi-crystalline polymers such as nylon can be used below or above Tg but without exceeding their Tm, while cross-linked rubbers can be used above their Tg without reaching thermal decomposition. Apart from the polymers' thermal properties, this analytical technique allows characterizing their structure and morphology by determining the degree of polymer's stiffness, taking into account the moduli and Tan δ (Botta et al., 2021; Shukor et al., 2014).

5. Testing methods using the DMA

5.1 Tension test

It is possible to determine the stress, Young's modulus and strain in the polymeric material using a constant temperature and strain rate, a method similar to that performed by a universal testing machine (Domene-López et al., 2019).

5.2 Dynamic analysis

It is the most widely used analysis method, consisting of determining the curves of the elastic or storage modulus (E' or G'), viscous modulus or loss (E" or G") and Tan δ as a function of temperature. From the engineering point of view, the above parameters allow evaluating the performance in the solid state, while the shear is employed for flexible systems such as thermoplastic materials, adhesives, pastes and molten material.

The most commonly used geometries are tension and bending. On the other hand, it is also possible to perform a direct analysis between the elastic modulus as a function of temperature, based on an initial temperature sweep, evaluating the changes of the elastic modulus (G') in the polymeric material. This type of graph makes it possible to identify different transitions related to movements and/or molecular rotations of the material under study before reaching its respective melting (Fig. 6.2). Among the events mentioned above, there are the α transitions, corresponding to the same glass transition temperature (Tg), indicating the existence of polymeric chains' movements due to temperature. The β transitions are associated with movements of functional groups present in the polymeric chains and are often related to the degree of toughness of the polymer. While other transitions, such as γ and δ, represent the bending and stretching of point fractions of the molecules (Ghanbari et al., 2018; Šerá et al., 2020b).

5.3 Thermo-mechanical analysis

The polymeric material is subjected to bending, applying constant stress as the temperature increases. This type of analysis allows determining the glass transition temperature (Tg) with high precision due to the ability to register changes in the free volume of the polymeric structure through variations in its stiffness. The most commonly used geometries are: Flexure, tension, penetration, expansion and dilatometry (Saba et al., 2018).

5.4 Creep phenomenon

It consists of a time-dependent distortion presented in polymeric materials when they are subjected to constant stress or deformation of low magnitude over a long period of time and determines the

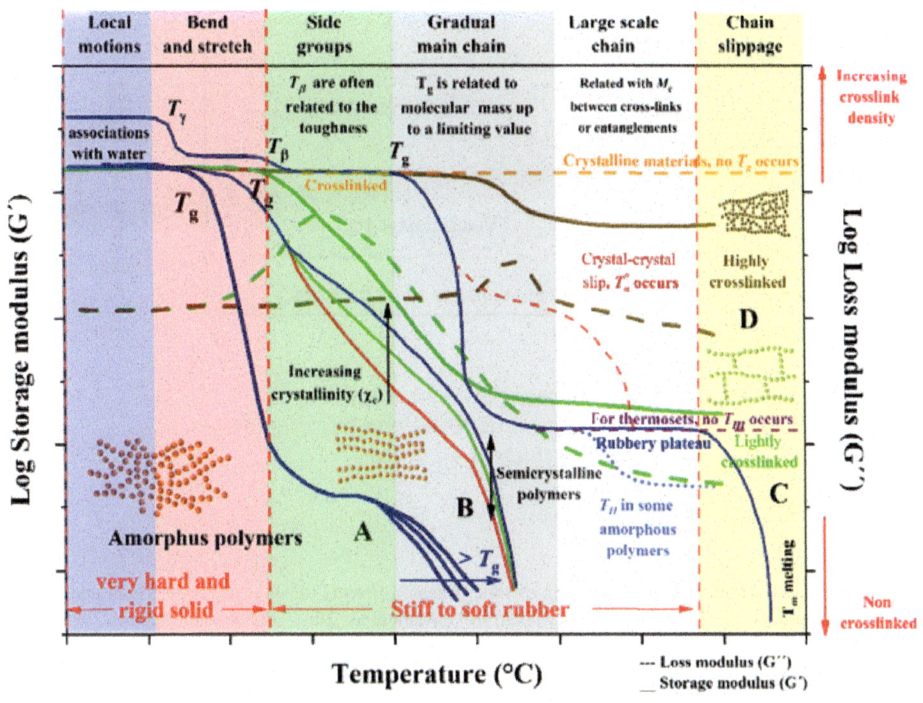

Figure 6.2. Types of polymeric structures and their rheological behaviors. Adapted from (Bayart et al., 2021; Cai et al., 2014; Devi et al., 2011; Frone et al., 2013; Li et al., 2021; Liu et al., 2011; Menard, 2008; Menard and Menard, 2020; Saba et al., 2016; Sepe, 1998).

possible structural, dimensional or mechanical properties changes through the samples' recovery or loss of properties when the stress or deformation established in the test is suppressed. There are three types of creep analysis, which differ in: (a) application of constant stress as a function of time and the strain changes are reported; (b) application of a constant strain as a function of time and the stress changes are reported (stress relaxation) and (c) determination of fatigue. The above alternatives can be performed until the material suffers a fracture, for a defined time or number of cycles using three geometries: bending, tension and compression (Sadasivuni et al., 2020). One of the material properties that change with time is deformation, mainly when the material is evaluated above the glass transition temperature (Tajvidi et al., 2006). Creep experimentation can collect information at low frequencies, while material recovery requires high frequencies through free oscillations (McKeen, 2015; Zhang et al., 2021).

5.4.1 Creep from deformation as a function of time

In this test, a creep characteristic curve is generated (Fig. 6.3), based on three stages: (1) A constant load or stress is applied to the material, inducing an instantaneous deformation; (2) the creep ratio reaches equilibrium; (3) In the final stage, the creep ratio is increased until the fracture is achieved by maintaining the stress on the material (Fig. 6.3) or by suppressing the stress, generating an elastic, viscoelastic recovery and an unrecovered viscous flow (Fig. 6.3) (Ramamoorthy et al., 2019). A quantifiable parameter to characterize creep is compliance, which is related to the inverse of the modulus of elasticity and signifies the degree of molecular relaxation of the polymer (Khalifah, 2021).

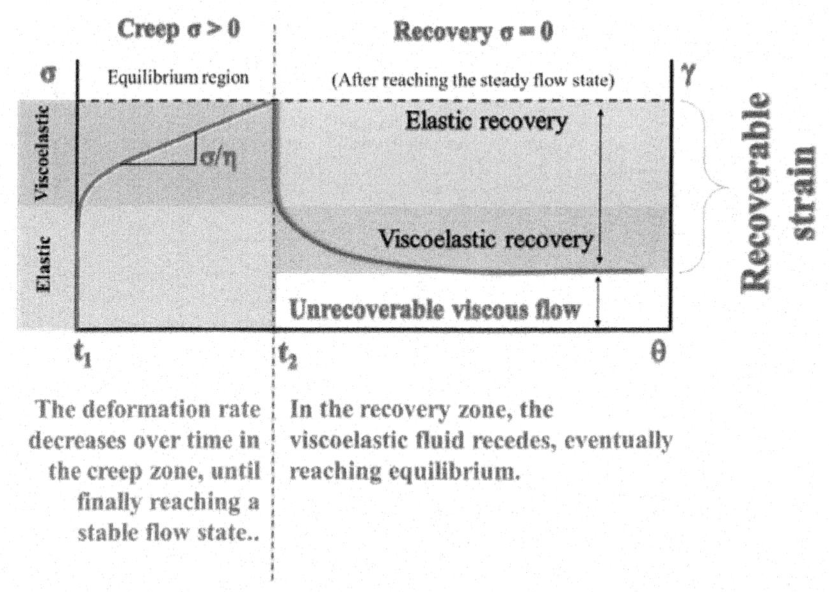

Figure 6.3. Creep curves from strain as a function of time. Adapted from (Colinet et al., 2009; Guan and Hanna, 2005; Kwaśniewska et al., 2020; Ma and Harwalkar, 1991; Mark et al., 2004; Menard and Menard, 2020).

5.4.2 Stress relaxation

A constant strain (value less than 0.25%) of a polymeric sample is defined, and the stress decreases as time progresses, using low frequencies (see Fig. 6.3) (Bashir, 2021; Cristea et al., 2020; Ghanbari et al., 2018). In the equilibrium region of the creep curve, the level of crosslinking between the thermosetting polymer chains can be established, taking into account their deformation capacity. The addition of plasticizers in the polymeric material increases the Creep phenomenon, mainly when the temperature is above the Tg (Liu et al., 2011; Šerá et al., 2020a).

5.4.3 Multiple cycle creep

This methodology allows predicting the long-term performance of the polymeric material by subjecting it to repetitive stresses or deformations that represent a simulation corresponding to a given application. Cycles can be performed for a given time, using a temperature range or using other environmental factors. By comparing the properties in the first and last cycle, the degree of mechanical decomposition can be determined (see Fig. 6.3). A second alternative of analysis consists of using temperature increments as the implementation of constant stress progresses or modifying the exposure atmosphere of the material by using solvents or gases (Villarreal and Iturriaga, 2016; Zhang et al., 2021)

6. Dynamic analysis in polymers

Through the behaviors of the moduli and Tan δ as a function of temperature, it is possible to differentiate the characteristic structures of amorphous, semicrystalline and crosslinked polymers by identifying the magnitude of Tan δ, finding values higher than 1 in amorphous polymers and lower than 1 in semicrystalline and thermosetting polymers. The difference between the latter two is because the elastic modulus curve of the semicrystalline polymer tends to zero as the temperature increases, achieving a complete fusion of its structure, while in thermosetting polymers, its value is reduced as the temperature increases until achieving a constant value greater than 0, due to the existence of crosslinking between the polymeric chains that prevent their fusion (Avella et al., 2021; Awadhiya et al., 2016).

Using the dynamic curve considering the elastic modulus and temperature, it is also possible to differentiate the structures of amorphous, semi-crystalline and thermosetting polymers, having in common the identification of the Tg and in the case of semi-crystalline polymers, their melting is reported. In amorphous polymers, the drop of E' is evidenced at a higher temperature. The molecular weight can be differentiated between this polymer at high temperatures in its viscous and molten state, evidencing higher modulus in those polymers with higher molecular weight. In the case of semi-crystalline polymers above the Tg and before melting, there is a zone that allows determining the degree of crystallinity; the higher the modulus, the higher the crystallinity. In contrast, thermostable polymers, after identifying their Tg, there is a zone that defines the degree of crosslinking, and the lower the drop in modulus after Tg as the temperature increases, the higher the degree of crosslinking. If the temperature of the crosslinked polymers continues to increase, the elastic modulus remains constant, being characteristic of the presence of primary bonds between polymeric chains that prevent their fusion (Lazaridou et al., 2003).

6.1 Dynamic analysis of polymer blends

By using dynamic curves based on the loss modulus and temperature, an increase in the loss modulus is identified as the temperature increases in the polymeric material, indicating an increase in the mobility of the polymeric chains, manifesting itself in the process of molecular relaxation. The most important signal of this curve is the Tg, manifesting itself through the formation of a peak with the highest value of the storage modulus. Another analysis that can be performed corresponds to determining the degree of miscibility between two or more polymers when mixed, identifying miscible, semi-miscible and non-miscible polymers. The miscible mixture occurs when the glass transition temperatures belonging to each polymer of the mixture are displaced and approach each other until a reduced temperature difference is achieved. A homogeneous mixture is generated that does not allow distinguishing the polymers that compose it. In contrast, the semi-immiscible mixture shows a slight displacement or approach between the glass transition temperatures. Finally, no changes in the respective Tg are considered in the immiscible mixture because there is no compatibility between the polymers (Cristea et al., 2020; Menard and Menard, 2020; Nakasone et al., 2016; Šerá et al., 2020b).

6.2 Delta tangent in polymers and biocomposites

It is considered as the viscoelasticity index since, when having a value equal to or less than 0.1, the polymeric material presents a rigid behavior, and values below 0.03 indicate that the material is below the Tg. This parameter allows defining the degree of impact resistance and determines the ductile failure of the polymer at different temperatures. The behavior of Tan δ in two polymers, polyethersulfone (PES) and polyetherimide (PEI), being materials of similar chemical structures. However, PES maintains higher Tan δ between the temperature range from –50 to 150°C, corresponding to the polymer with higher impact resistance, while PEI achieves a slight approach to that given by PES when heated to 96°C. From this analysis, it is also possible to determine the degree of adhesion between the matrix and the reinforcement at the glass transition temperature, identifying that the lower Tan δ, the greater the adhesion between the biocomposite components (Bashir, 2021).

6.3 Dynamic analysis in biocomposites

A complete mechanical and thermal characterization can be performed using temperature and frequency ranges through the concept of dynamic analysis. Its dynamic properties are controlled by the composition of the mixture and the type of interface generated between the different phases that make up the material (Guancha-Chalapud et al., 2020; Homez et al., 2018; Palechor-Trochez et al., 2021). The DMA manages to relate the macroscopic properties, including the interaction of mechanical and thermal properties with the viscoelastic behavior, identifying changes and structural reorganizations at the microscopic level, finding the following general behaviors from the incorporation of fibers in the matrix:

- The increase of Tg in the biocomposite is related to an increase in the interfacial interaction between its components.
- An increase in the storage modulus is generated, indicating the restriction of molecular movements.
- Influence of fiber orientation.
- Identification of the state of interfacial interactions between the reinforcement and the matrix.
- Tan δ values provide information according to fiber type, phase distribution, fiber/matrix ratio, fiber-matrix interactions and identification of voids in the biocomposite. The increase of their values is related to the decrease of interfacial bonds.

From the dynamic properties, the biocomposite can be characterized through the C coefficient and the Cole-Cole diagram (Eq. 1). The first methodology is related to the effect of filler and/or reinforcement on the storage modulus, which is represented as shown in Equation 1.

$$C = \frac{\left(\frac{E'_G}{E'_R}\right)_{composite}}{\left(\frac{E'_G}{E'_R}\right)_{Resin}} \qquad \text{Eq. 1.}$$

where E'G and E'R correspond to the storage modulus in the glassy and viscous regions, respectively, when high values of C are generated, it indicates the reduction of filler and/or reinforcement efficiency in the modulus. The C coefficient represents a relative measure of the modulus drop as the temperature increases, and the material exceeds the region of the Tg (Palechor-Trochez et al., 2021; Salazar-Sánchez et al., 2022).

While the second methodology allows the study of dielectric dispersion, considering the changes in viscoelastic properties through the relationship between the storage modulus and the loss modulus, resulting from the use of a specific treatment in dielectric relaxation at a certain frequency. This type of analysis is specialized to determine structural changes in cross-linked polymeric matrices mixed with fibers. In homogeneous systems, a semicircular behavior is generated in the diagram, while in heterogeneous systems, the formation of imperfect circles is indicated (Arabeche et al., 2020; Saba et al., 2018).

When using the elastic modulus's analysis of a biocomposite as a function of temperature, the increase of the storage modulus is exhibited as the content of the reinforcement in the polymeric matrix increases before and after the Tg and the glass transition temperature increase. The above behaviors can be displayed in a thermosetting resin and four biocomposites made up of different lignocellulosic hemp fibers concentrations (14 to 65%) (Bashir, 2021; Menard and Menard, 2020).

6.4 Creep phenomena in biocomposites

Biocomposites containing high fiber content generate a higher degree of creep resistance; however, when lignocellulosic fibers are used as reinforcement, there is an alteration in creep behavior due to moisture sensitivity. The creep phenomenon depends on several factors when evaluating and analyzing biocomposites, considering the chemical structure, matrix molecular orientation, crystallinity, magnitude of the applied stress, temperature, humidity, fiber volume, fiber aspect ratio, fiber orientation and interfacial adhesion between fibers and matrix, being factors that determine the different mechanical and thermal properties. Among the reasons for originating a fracture in the biocomposite are the breaking of primary and secondary bonds, viscoelastic deformations, fibril breakage, sliding of polymeric chains and formation of voids. Generally, biocomposites that have unidirectional reinforcements subjected to loads in the same direction as the fibers, their creep behavior is similar to that provided by the reinforcement, while in biocomposites containing a randomly oriented reinforcement, their creep behavior is dominated by the matrix (McKeen, 2015; Salazar, 2016; Salazar Sánchez et al., 2020; Sánchez-Safont et al., 2018).

Conclusions

From the bibliographic review related to the Dynamic-Mechanical Analysis with polymers and biocomposites, six tests are identified (Creep phenomenon, stress relaxation, Creep by multiple cycles, tension test, dynamic analysis and thermo-mechanical analysis) to study the thermal, mechanical and structural properties. Some of the structural parameters that allow analysis and differentiation between the polymeric materials evaluated are: molecular weight, crystallinity, degree of crosslinking, determination of the glass transition temperature and melting temperature, degree of miscibility of two or more polymers in a mixture, impact resistance, creep analysis, the influence of reinforcement and other additives in the biocomposite, determination of the C coefficient and the Cole-Cole diagram.

References

Abraham, J., Sharika, T., George, S. C. and Thomas, S. (2017). Rheological percolation in thermoplastic polymer nanocomposites. *Rheol Open Access*, 1: 1–15.

Ahlinder, A., Fuoco, T. and Finne-Wistrand, A. (2018). Medical grade polylactide, copolyesters and polydioxanone: Rheological properties and melt stability. *Polymer Testing*, 72: 214–222. https://doi.org/10.1016/j.polymertesting.2018.10.007.

Ali Nezamzadeh, S., Ahmadi, Z. and Afshari Taromi, F. (2017). From microstructure to mechanical properties of compatibilized polylactide/thermoplastic starch blends. *Journal of Applied Polymer Science*, 134(16): 1–9. https://doi.org/10.1002/app.44734.

Aluthge, D. C., Xu, C., Othman, N., Noroozi, N., Hatzikiriakos, S. G. and Mehrkhodavandi, P. (2013). PLA–PHB–PLA triblock copolymers: synthesis by sequential addition and investigation of mechanical and rheological properties. *Macromolecules*, 46(10): 3965–3974. https://doi.org/10.1021/ma400522n.

Arabeche, K., Abdelmalek, F., Delbreilh, L., Zair, L. and Berrayah, A. (2020). Physical and rheological properties of biodegradable poly(butylene succinate)/Alfa fiber composites. *Journal of Thermoplastic Composite Materials*, 089270572090409. https://doi.org/10.1177/0892705720904098.

Avella, A., Mincheva, R., Raquez, J.-M. and Lo Re, G. (2021). Substantial effect of water on radical melt crosslinking and rheological properties of poly(ε-caprolactone). *Polymers*, 13(4): 491. https://doi.org/10.3390/polym13040491.

Awadhiya, A., Kumar, D. and Verma, V. (2016). Crosslinking of agarose bioplastic using citric acid. *Carbohydrate Polymers*, 151: 60–67. https://doi.org/10.1016/j.carbpol.2016.05.040.

Backes, E. H., Pires, L. D. N., Beatrice, C. A. G., Costa, L. C., Passador, F. R. and Pessan, L. A. (2020). Fabrication of biocompatible composites of poly(lactic acid)/hydroxyapatite envisioning medical applications. *Polymer Engineering & Science*, 60(3): 636–644. https://doi.org/10.1002/pen.25322.

Baimark, Y. and Srihanam, P. (2015). Influence of chain extender on thermal properties and melt flow index of stereocomplex PLA. *Polymer Testing*, 45: 52–57. https://doi.org/10.1016/j.polymertesting.2015.04.017.

Balali, S., Davachi, S. M., Sahraeian, R., Shiroud Heidari, B., Seyfi, J. and Hejazi, I. (2018). Preparation and characterization of composite blends based on polylactic acid/polycaprolactone and silk. *Biomacromolecules*, 19(11): 4358–4369. https://doi.org/10.1021/acs.biomac.8b01254.

Bashir, M. A. (2021). Use of Dynamic Mechanical Analysis (DMA) for Characterizing Interfacial Interactions in Filled Polymers. *Solids*, 2(1): 108–120. https://doi.org/10.3390/solids2010006.

Bayart, M., Foruzanmehr, M. R., Vuillaume, P. Y., Ovlaque, P., Robert, M. and Elkoun, S. (2021). Poly(lactic acid)/flax composites: effect of surface modification and thermal treatment on interfacial adhesion, crystallization, microstructure, and mechanical properties. *Composite Interfaces*, 1–20. https://doi.org/10.1080/09276440.2021.1884470.

Beltrán-García, M. J., Orozco, A., Samayoa, I. and Ogura, T. (2001). Lignin degradation products from corn stalks enhance notably the radial growth of basidiomycete mushroom mycelia. *Revista de La Sociedad Química de México*, 45(2): 77–81. http://www.scielo.org.mx/scielo.php?pid=S0583-76932001000200007&script=sci_arttext.

Bendaoud, A. and Chalamet, Y. (2014). Plasticizing effect of ionic liquid on cellulose acetate obtained by melt processing. *Carbohydrate Polymers*, 108: 75–82. https://doi.org/10.1016/j.carbpol.2014.03.023.

Botana, A., Mollo, M., Eisenberg, P. and Torres Sanchez, R. M. (2010). Effect of modified montmorillonite on biodegradable PHB nanocomposites. *Applied Clay Science*, 47(3–4): 263–270. https://doi.org/10.1016/j.clay.2009.11.001.

Botta, L., La Mantia, F. P., Mistretta, M. C., Oliveri, A., Arrigo, R. and Malucelli, G. (2021). Structure–property relationships in bionanocomposites for pipe extrusion applications. *Polymers*, 13(5): 782. https://doi.org/10.3390/polym13050782.

Cai, J., Xiong, Z., Zhou, M., Tan, J., Zeng, F., Meihuma, Lin, S. and Xiong, H. (2014). Thermal properties and crystallization behavior of thermoplastic starch/poly(É”-caprolactone) composites. *Carbohydrate Polymers*, 102(1): 746–754. https://doi.org/10.1016/j.carbpol.2013.10.095.

Chan, S. Y., Choo, W. S., Young, D. J. and Loh, X. J. (2017). Pectin as a rheology modifier: Origin, structure, commercial production and rheology. *Carbohydrate Polymers*, 161: 118–139. https://doi.org/10.1016/j.carbpol.2016.12.033.

Colinet, I., Dulong, V., Hamaide, T., Le Cerf, D. and Picton, L. (2009). Unusual rheological properties of a new associative polysaccharide in salt media. *Carbohydrate Polymers*, 77(4): 743–749. https://doi.org/10.1016/j.carbpol.2009.03.019.

Cristea, M., Ionita, D. and Iftime, M. M. (2020). Dynamic mechanical analysis investigations of PLA-based renewable materials: how are they useful? *Materials*, 13(22): 5302. https://doi.org/10.3390/ma13225302.

Das, M. and Chakraborty, D. (2006). Influence of mercerization on the dynamic mechanical properties of bamboo, a natural lignocellulosic composite. *Industrial & Engineering Chemistry Research*, 45(19): 6489–6492. https://doi.org/10.1021/ie0603971.

de Oliveira, T. A., de Oliveira Mota, I., Mousinho, F. E. P., Barbosa, R., de Carvalho, L. H. and Alves, T. S. (2019). Biodegradation of mulch films from poly(butylene adipate co-terephthalate), carnauba wax, and sugarcane residue. *Journal of Applied Polymer Science*, 136(47): 1–9. https://doi.org/10.1002/app.48240.

Decaen, P., Rolland-Sabaté, A., Colomines, G., Guilois, S., Lourdin, D., Della Valle, G. and Leroy, E. (2020). Influence of ionic plasticizers on the processing and viscosity of starch melts. *Carbohydrate Polymers*, 230: 115591. https://doi.org/10.1016/j.carbpol.2019.115591.

Devi, L. U., Bhagawan, S. S. and Thomas, S. (2011). Dynamic mechanical properties of pineapple leaf fiber polyester composites. *Polymer Composites*, 32(11): 1741–1750. https://doi.org/10.1002/pc.21197.

Ding, Y., Lu, B., Wang, P., Wang, G. and Ji, J. (2018). PLA-PBAT-PLA tri-block copolymers: Effective compatibilizers for promotion of the mechanical and rheological properties of PLA/PBAT blends. *Polymer Degradation and Stability*, 147: 41–48. https://doi.org/10.1016/j.polymdegradstab.2017.11.012.

Domene-López, D., García-Quesada, J. C., Martin-Gullon, I. and Montalbán, M. G. (2019). Influence of starch composition and molecular weight on physicochemical properties of biodegradable films. *Polymers*, 11(7): 1084. https://doi.org/10.3390/polym11071084.

Farah, S., Anderson, D. G. and Langer, R. (2016). Physical and mechanical properties of PLA, and their functions in widespread applications—A comprehensive review. *Advanced Drug Delivery Reviews*, 107: 367–392. https://doi.org/10.1016/j.addr.2016.06.012.

Feijoo, A. V., Lopez-Lopez, M. T., Galindo-Gonzalez, C., Stange, S., Nguyen, T. T., Mammeri, F., Ammar-Merah, S. and Ponton, A. (2020). Rheological investigation of magnetic sensitive biopolymer composites: effect of the ligand grafting of magnetic nanoparticles. *Rheologica Acta*, 59(3): 165–176. https://doi.org/10.1007/s00397-020-01191-y.

Frone, A. N., Berlioz, S., Chailan, J. F. and Panaitescu, D. M. (2013). Morphology and thermal properties of PLA-cellulose nanofibers composites. *Carbohydrate Polymers*, 91(1): 377–384. https://doi.org/10.1016/j.carbpol.2012.08.054.

Ge, H., Yang, F., Hao, Y., Wu, G., Zhang, H. and Dong, L. (2013). Thermal, mechanical, and rheological properties of plasticized poly(<scp>L</scp> -lactic acid). *Journal of Applied Polymer Science*, 127(4): 2832–2839. https://doi.org/10.1002/app.37620.

Georgiopoulos, P., Kontou, E. and Niaounakis, M. (2014). Thermomechanical properties and rheological behavior of biodegradable composites. *Polymer Composites*, 35(6): 1140–1149. https://doi.org/10.1002/pc.22761.

Ghanbari, A., Tabarsa, T., Ashori, A., Shakeri, A. and Mashkour, M. (2018). Thermoplastic starch foamed composites reinforced with cellulose nanofibers: Thermal and mechanical properties. *Carbohydrate Polymers*, 197(June): 305–311. https://doi.org/10.1016/j.carbpol.2018.06.017.

Gordobil, O., Egüés, I., Llano-Ponte, R. and Labidi, J. (2014). Physicochemical properties of PLA lignin blends. *Polymer Degradation and Stability*, 108: 330–338. https://doi.org/10.1016/j.polymdegradstab.2014.01.002.

Guan, J. and Hanna, M. a. (2005). Selected morphological and functional properties of extruded acetylated starch—Polylactic acid foams. *Industrial & Engineering Chemistry Research*, 44(9): 3106–3115. https://doi.org/Doi 10.1021/Ie049786q.

Guancha-Chalapud, M. A., Gálvez, J., Serna-Cock, L. and Aguilar, C. N. (2020). Valorization of Colombian fique (Furcraea bedinghausii) for production of cellulose nanofibers and its application in hydrogels. *Scientific Reports*, 10(1): 1–10. https://doi.org/10.1038/s41598-020-68368-6.

Harris, A. M. and Lee, E. C. (2008). Improving mechanical performance of injection molded PLA by controlling crystallinity. *Journal of Applied Polymer Science*, 107(4): 2246–2255. https://doi.org/10.1002/app.27261.

He, C., Kim, S. W. and Lee, D. S. (2008). *In situ* gelling stimuli-sensitive block copolymer hydrogels for drug delivery. *Journal of Controlled Release*, 127(3): 189–207. https://doi.org/10.1016/j.jconrel.2008.01.005.

Homez, A. K., Daza, L. D., Solanilla, J. F. and Váquiro, H. A. (2018). Effect of temperature, starch and plasticizer concentrations on color parameters of ulluco (Ullucus tuberosus Caldas) edible films. *IOP Conference Series: Materials Science and Engineering*, 437(1). https://doi.org/10.1088/1757-899X/437/1/012003.

Ilsouk, M., Raihane, M., Lahcini, M., Meri, R. M., Zicāns, J., Cimdina, L. B. and Kharas, G. B. (2017). Bionanocomposites poly(ϵ-caprolactone)/organomodified Moroccan beidellite clay prepared by *in situ* ring opening polymerization: Characterizations and properties. *Journal of Macromolecular Science, Part A*, 54(4): 201–210. https://doi.org/10.1080/10601325.2017.1282229.

Ilyas, R. A., Sapuan, S. M., Kadier, A., Kalil, M. S., Ibrahim, R., Atikah, M. S. N., Nurazzi, N. M., Nazrin, A., Lee, C. H., Faiz Norrahim, M. N., Sari, N. H., Syafri, E., Abral, H., Jasmani, L. and Ibrahim, M. I. J. (2020). Properties and characterization of PLA, PHA, and other types of biopolymer composites. pp. 111–138. *In*: *Advanced Processing, Properties, and Applications of Starch and Other Bio-Based Polymers*. Elsevier. https://doi.org/10.1016/B978-0-12-819661-8.00008-1.

Ivanova, R. and Kotsilkova, R. (2019). Investigation of rheological and surface properties of poly (lactic) acid polymer/carbon nanofiller nanocomposites and their future applications. *Industry 4.0*, 4(1): 19–23.

Jiang, Z., Zhang, H., Han, J., Liu, Z., Liu, Y. and Tang, L. (2016). Percolation model of reinforcement efficiency for carbon nanotubes dispersed in thermoplastics. *Composites Part A: Applied Science and Manufacturing*, 86: 49–56. https://doi.org/10.1016/j.compositesa.2016.03.031.

Khalifah, K. M. (2021). The effect of creep rate on polymeric composites reinforced by nanoclays and their comparison. *International Journal of Nanoscience*, 20(03): 2150027. https://doi.org/10.1142/S0219581X21500277.

Kwaśniewska, A., Chocyk, D., Gładyszewski, G., Borc, J., Świetlicki, M. and Gładyszewska, B. (2020). The influence of kaolin clay on the mechanical properties and structure of thermoplastic starch films. *Polymers*, 12(1): 73. https://doi.org/10.3390/polym12010073.

Lazaridou, A., Biliaderis, C. G. and Kontogiorgos, V. (2003). Molecular weight effects on solution rheology of pullulan and mechanical properties of its films. *Carbohydrate Polymers*, 52(2): 151–166.

Lee, S. K., Seong, D. G. and Youn, J. R. (2005). Degradation and rheological properties of biodegradable nanocomposites prepared by melt intercalation method. *Fibers and Polymers*, 6(4): 289–296.

Lemes, A. P., Montanheiro, T. L. A., Passador, F. R. and Durán, N. (2015). Nanocomposites of polyhydroxyalkanoates reinforced with carbon nanotubes: chemical and biological properties. pp. 79–108. *In*: Thakur, V. and Thakur, M. (eds.). *Eco-friendly Polymer Nanocomposites: Processing and Properties* (Firts). Springer, New Delhi. https://doi.org/10.1007/978-81-322-2470-9_3.

Li, H., Luo, Y., Qi, R., Feng, J., Zhu, J., Hong, Y., Feng, Z. and Jiang, P. (2015). Fabrication of high-viscosity biodegradable poly(butylene succinate) (PBS)/solid epoxy (SE)/carboxyl-ended polyester (CP) blends. *Journal of Applied Polymer Science*, 132(27): n/a-n/a. https://doi.org/10.1002/app.42193.

Li, X., Becquart, F., Taha, M., Majesté, J.-C., Chen, J., Zhang, S. and Mignard, N. (2020). Tuning the thermoreversible temperature domain of PTMC-based networks with thermosensitive links concentration. *Soft Matter*, 16(11): 2815–2828. https://doi.org/10.1039/C9SM01882D.

Li, Y., Yao, S., Han, C. and Cheng, H. (2021). Miscibility, crystallization and mechanical properties of poly[(3-hydroxybutyrate)-co-(4-hydroxyvalerate)]/poly(propylene carbonate)/poly(vinyl acetate) ternary blends. *Polymer International*, 70(10): 1544–1553. https://doi.org/10.1002/pi.6235.

Liang, L., Ren, S., Zheng, Y., Lan, Y. and Lu, M. (2007). Cure kinetics and mechanical properties of liquid crystalline epoxides with long lateral substituents cured with anhydride. *Polymer Journal*, 39(9): 961–967. https://doi.org/10.1295/polymj.PJ2006219.

Lim, L.-T., Auras, R. and Rubino, M. (2008). Processing technologies for poly(lactic acid). *Progress in Polymer Science*, 33(8): 820–852. https://doi.org/10.1016/j.progpolymsci.2008.05.004.

Liu, H., Chaudhary, D., Yusa, S. I. and Tadé, M. O. (2011). Glycerol/starch/Na+-montmorillonite nanocomposites: A XRD, FTIR, DSC and 1H NMR study. *Carbohydrate Polymers*, 83(4): 1591–1597. https://doi.org/10.1016/j.carbpol.2010.10.018.

Lu, X., Zhao, J., Yang, X. and Xiao, P. (2017). Morphology and properties of biodegradable poly (lactic acid)/poly (butylene adipate-co-terephthalate) blends with different viscosity ratio. *Polymer Testing*, 60: 58–67. https://doi.org/10.1016/j.polymertesting.2017.03.008.

Lv, S., Gu, J., Tan, H. and Zhang, Y. (2017). The morphology, rheological, and mechanical properties of wood flour/starch/poly(lactic acid) blends. *Journal of Applied Polymer Science*, 134(16): 1–9. https://doi.org/10.1002/app.44743.

Ma, C. Y. and Harwalkar, V. R. (1991). Thermal analysis of food proteins. *Advances in Food and Nutrition Research*, 35(C): 317–366. https://doi.org/10.1016/S1043-4526(08)60067-4.

Marek, A. A., Verney, V., Taviot-Gueho, C., Totaro, G., Sisti, L., Celli, A. and Leroux, F. (2019). Outstanding chain-extension effect and high UV resistance of polybutylene succinate containing amino-acid-modified layered double hydroxides. *Beilstein Journal of Nanotechnology*, 10: 684–695. https://doi.org/10.3762/bjnano.10.68.

Mark, J., Ngai, K., Graessley, W. and Mandelkern, L. (2004). *Physical properties of polymers*. https://books.google.com/books?hl=es&lr=&id=ueX1mpDpXTMC&oi=fnd&pg=PR10&dq=Physical+Properties+of+Polymers,&ots=5rakP5twHX&sig=VtDKTtfH9xcNAsdK9A-kAD9gtAE.

McKeen, W. (2015). *The Effect of Creep and Other Time Related Factors on Plastics and Elastomers* (W. McKeen (Ed.); Third). Elsevier. https://doi.org/10.1016/C2013-0-19368-3.

Menard, K. P. (2008). *Dynamic Mechanical Analysis: A Practical Introduction* (K. P. Menard (Ed.); 2nd ed.). CRC Press. https://doi.org/10.1201/9781420053135.

Menard, K. P. and Menard, N. R. (2020). *Dynamic Mechanical Analysis* (K. P. Menard & N. R. Menard (Eds.); 3rd ed.). CRC Press. https://doi.org/10.1201/9780429190308.

Mu, C., Li, X., Zhao, Y., Zhang, H., Wang, L. and Li, D. (2013). Freezing/thawing effects on the exfoliation of montmorillonite in gelatin-based bionanocomposite. *Journal of Applied Polymer Science*, 128(5): 3141–3148. https://doi.org/10.1002/app.38511.

Nakasone, K., Ikematsu, S. and Kobayashi, T. (2016). Biocompatibility evaluation of cellulose hydrogel film regenerated from sugar cane bagasse waste and its *in vivo* behavior in mice. *Industrial and Engineering Chemistry Research*, 55(1): 30–37. https://doi.org/10.1021/acs.iecr.5b03926.

Nevoralová, M., Ujčić, A., Kodakkadan, Y. N. V. and Starý, Z. (2019). Rheological characterization of starch-based biodegradable polymer blends. *AIP Conference Proceedings*, 050005. https://doi.org/10.1063/1.5109511.

Nishio, Y., Yamada, A., Ezaki, K., Miyashita, Y., Furukawa, H. and Horie, K. (2004). Preparation and magnetometric characterization of iron oxide-containing alginate/poly(vinyl alcohol) networks. *Polymer*, 45(21): 7129–7136. https://doi.org/10.1016/j.polymer.2004.08.047.

Olewnik-Kruszkowska, E., Burkowska-But, A., Tarach, I., Walczak, M. and Jakubowska, E. (2020). Biodegradation of polylactide-based composites with an addition of a compatibilizing agent in different environments. *International Biodeterioration and Biodegradation*, 147(March 2019): 104840. https://doi.org/10.1016/j.ibiod.2019.104840.

Palechor-Trochez, J. J., Ramírez-Gonzales, G., Villada-Castillo, H. S. and Solanilla-Duque, J. F. (2021). A review of trends in the development of bionanocomposites from lignocellulosic and polyacids biomolecules as packing material making alternative: A bibliometric analysis. *International Journal of Biological Macromolecules*, 192: 832–868. https://doi.org/10.1016/j.ijbiomac.2021.10.003.

Park, I. H., Lee, J. Y., Ahn, S. J. and Choi, H. J. (2020). Melt rheology and mechanical characteristics of poly(lactic acid)/alkylated graphene oxide nanocomposites. *Polymers*, 12(10): 2402. https://doi.org/10.3390/polym12102402.

Pothan, L. A. and Thomas, S. (2003). Polarity parameters and dynamic mechanical behaviour of chemically modified banana fiber reinforced polyester composites. *Composites Science and Technology*, 63(9): 1231–1240. https://doi.org/10.1016/S0266-3538(03)00092-7.

Ramamoorthy, S. K., Åkesson, D., Rajan, R., Periyasamy, A. P. and Skrifvars, M. (2019). Mechanical performance of biofibers and their corresponding composites. pp. 259–292. *In*: Jawaid, M., Thariq, M. and Saba, N. (eds.). *Mechanical and Physical Testing of Biocomposites, Fibre-Reinforced Composites and Hybrid Composites* (First). Elsevier. https://doi.org/10.1016/B978-0-08-102292-4.00014-X.

Ridi, F., Bonini, M. and Baglioni, P. (2014). Magneto-responsive nanocomposites: Preparation and integration of magnetic nanoparticles into films, capsules, and gels. *Advances in Colloid and Interface Science*, 207: 3–13. https://doi.org/10.1016/j.cis.2013.09.006.

Romero-Bastida, C. A., Chávez Gutiérrez, M., Bello-Pérez, L. A., Abarca-Ramírez, E., Velazquez, G. and Mendez-Montealvo, G. (2018). Rheological properties of nanocomposite-forming solutions and film based on montmorillonite and corn starch with different amylose content. *Carbohydrate Polymers*, 188: 121–127. https://doi.org/10.1016/j.carbpol.2018.01.089.

Saba, N., Jawaid, M., Alothman, O. Y. and Paridah, M. T. (2016). A review on dynamic mechanical properties of natural fibre reinforced polymer composites. *Construction and Building Materials*, 106: 149–159. https://doi.org/10.1016/j.conbuildmat.2015.12.075.

Saba, N., Jawaid, M. and Sultan, M. T. H. (2018). An overview of mechanical and physical testing of composite materials. *In*: *Mechanical and Physical Testing of Biocomposites, Fibre-Reinforced Composites and Hybrid Composites*. Elsevier Ltd. https://doi.org/10.1016/B978-0-08-102292-4.00001-1.

Sadasivuni, K. K., Saha, P., Adhikari, J., Deshmukh, K., Ahamed, M. B. and Cabibihan, J. (2020). Recent advances in mechanical properties of biopolymer composites: a review. *Polymer Composites*, 41(1): 32–59. https://doi.org/10.1002/pc.25356.

Salazar-Sánchez, M. del R., Immirzi, B., Solanilla-Duque, J. F., Zannini, D., Malinconico, M. and Santagata, G. (2022). Ulomoides dermestoides Coleopteran action on Thermoplastic Starch/Poly(lactic acid) films biodegradation: a novel, challenging and sustainable approach for a fast mineralization process. *Carbohydrate Polymers*, 279: 118989. https://doi.org/10.1016/j.carbpol.2021.118989.

Salazar, M. Á. H. (2016). Viscoelastic performance of biocomposites. *Composites from Renewable and Sustainable Materials*, 303.

Salazar Sánchez, M. D. R., Cañas Montoya, J. A., Villada Castillo, H. S., Solanilla Duque, J. F., Rodríguez Herrera, R. and Avalos Belmotes, F. (2020). Biogenerated polymers: an enviromental alternative. *DYNA*, 87(214): 75–84. https://doi.org/10.15446/dyna.v87n214.82163.

Sánchez-Safont, E. L., Aldureid, A., Lagarón, J. M., Gámez-Pérez, J. and Cabedo, L. (2018). Biocomposites of different lignocellulosic wastes for sustainable food packaging applications. *Composites Part B: Engineering*, 145: 215–225. https://doi.org/10.1016/j.compositesb.2018.03.037.

Scaffaro, R., Botta, L., Maio, A., Mistretta, M. and La Mantia, F. (2016). Effect of graphene nanoplatelets on the physical and antimicrobial properties of biopolymer-based nanocomposites. *Materials*, 9(5): 351. https://doi.org/10.3390/ma9050351.

Sepe, M. P. (1998). Dynamic mechanical analysis for plastic engineers. *PDL Handbook Series, Plastic Design Library*.

Šerá, J., Serbruyns, L., De Wilde, B. and Koutný, M. (2020a). Accelerated biodegradation testing of slowly degradable polyesters in soil. *Polymer Degradation and Stability*, 171. https://doi.org/10.1016/j.polymdegradstab.2019.109031.

Šerá, J., Serbruyns, L., De Wilde, B. and Koutný, M. (2020b). Accelerated biodegradation testing of slowly degradable polyesters in soil. *Polymer Degradation and Stability*, 171. https://doi.org/10.1016/j.polymdegradstab.2019.109031.

Sharma, B. (2020). Viscoelastic investigation of graphene oxide grafted PVA biohybrid using ostwald modeling for packaging applications. *Polymer Testing*, 91: 106791. https://doi.org/10.1016/j.polymertesting.2020.106791.

Shukor, F., Hassan, A., Hasan, M., Islam, M. S. and Mokhtar, M. (2014). PLA/Kenaf/APP biocomposites: effect of alkali treatment and ammonium polyphosphate (APP) on dynamic mechanical and morphological properties. *Polymer-Plastics Technology and Engineering*, 53(8): 760–766. https://doi.org/10.1080/03602559.2013.869827.

Sivaraman, A., Ganti, S. S., Nguyen, H. X., Birk, G., Wieber, A., Lubda, D. and Banga, A. K. (2017). Development and evaluation of a polyvinyl alcohol based topical gel. *Journal of Drug Delivery Science and Technology*, 39: 210–216. https://doi.org/10.1016/j.jddst.2017.03.021.

Šnejdrová, E., Martiška, J., Loskot, J., Paraskevopoulos, G., Kováčik, A., Regdon Jr., G., Budai-Szűcs, M., Palát, K. and Konečná, K. (2021). PLGA based film forming systems for superficial fungal infections treatment. *European Journal of Pharmaceutical Sciences*, 163: 105855. https://doi.org/10.1016/j.ejps.2021.105855.

Tajvidi, M. and Falk, R. H. (2004). Dynamic mechanical analysis of compatibilizer effect on the mechanical properties of wood flour-high-density polyethylene composites. *International Journal of Engineering*, 17(1 (Transactions B: Applications)): 95–104. https://www.sid.ir/en/Journal/ViewPaper.aspx?ID=2270.

Tajvidi, M., Falk, R. H. and Hermanson, J. C. (2006). Effect of natural fibers on thermal and mechanical properties of natural fiber polypropylene composites studied by dynamic mechanical analysis. *Journal of Applied Polymer Science*, 101(6): 4341–4349. https://doi.org/10.1002/app.24289.

Tian, J., Zhang, R., Wu, Y. and Xue, P. (2021). Additive manufacturing of wood flour/polyhydroxyalkanoates (PHA) fully bio-based composites based on micro-screw extrusion system. *Materials & Design*, 199: 109418. https://doi.org/10.1016/j.matdes.2020.109418.

Tsang, Y. F., Kumar, V., Samadar, P., Yang, Y., Lee, J., Ok, Y. S., Song, H., Kim, K.-H., Kwon, E. E. and Jeon, Y. J. (2019). Production of bioplastic through food waste valorization. *Environment International*, 127: 625–644. https://doi.org/10.1016/j.envint.2019.03.076.

Vidyasagar, A., Ku, S. H., Kim, M., Kim, M., Lee, H. S., Pearce, T. R., McCormick, A. V., Bates, F. S. and Kokkoli, E. (2017). Design and characterization of a PVLA-PEG-PVLA thermosensitive and biodegradable hydrogel. *ACS Macro Letters*, 6(10): 1134–1139. https://doi.org/10.1021/acsmacrolett.7b00523.

Villarreal, M. E. and Iturriaga, L. B. (2016). Viscoelastic properties of amaranth starch gels and pastes. Creep compliance modeling with Maxwell model. *Starch/Staerke*, 68(11–12): 1073–1083. https://doi.org/10.1002/star.201600065.

Wang, B., Chen, J., Peng, H., Gai, J., Kang, J. and Cao, Y. (2016). Investigation on changes in the miscibility, morphology, rheology and mechanical behavior of melt processed cellulose acetate through adding polyethylene glycol as a plasticizer. *Journal of Macromolecular Science, Part B*, 55(9): 894–907. https://doi.org/10.1080/00222348.2016.1217185.

Wang, X., Peng, S., Chen, H., Yu, X. and Zhao, X. (2019). Mechanical properties, rheological behaviors, and phase morphologies of high-toughness PLA/PBAT blends by *in-situ* reactive compatibilization. *Composites Part B: Engineering*, 173: 107028. https://doi.org/10.1016/j.compositesb.2019.107028.

Weinmann, S. and Bonten, C. (2019). Thermal and rheological properties of modified polyhydroxybutyrate (PHB). *Polymer Engineering & Science*, 59(5): 1057–1064. https://doi.org/10.1002/pen.25075.

Weir, N. A., Buchanan, F. J., Orr, J. F. and Dickson, G. R. (2004). Degradation of poly-L-lactide. Part 1: *In vitro* and *in vivo* physiological temperature degradation. *Proceedings of the Institution of Mechanical Engineers, Part H: Journal of Engineering in Medicine*, 218(5): 307–319. https://doi.org/10.1243/0954411041932782.

Zhang, R., Zhang, S., Jiang, G., Gan, L., Xu, Z. and Tian, Y. (2021). Optimization of fermentation conditions, purification and rheological properties of poly (γ-glutamic acid) produced by *Bacillus subtilis* 1006-3. *Preparative Biochemistry & Biotechnology*, 1–9. https://doi.org/10.1080/10826068.2021.1941103.

Zhang, Y., Liu, X., Yin, B. and Luo, W. (2021). A nonlinear fractional viscoelastic-plastic creep model of asphalt mixture. *Polymers*, 13(8): 1278. https://doi.org/10.3390/polym13081278.

Chapter 7

Structural Analysis of Polymers and Composite Materials

Juan Pablo Castañeda Niño,[1,*] *José Herminsul Mina Hernandez,*[1] *Alex Valadez González*[2] and *Jose Fernando Solanilla Duque*[3]

1. Introduction

From the analysis of rheological parameters such as storage modulus (G'), loss modulus (G") and Tan δ, it is possible to monitor and identify different structural characteristics in polymeric materials and polymer matrix composites. Generally, the G' values are higher as molecular weight, crystallinity and the polymer chains orientation increase. In the case of biocomposites, the presence of the reinforcing fibers also raises the G' values. While with thermosetting polymers, it is possible to estimate the density of crosslinks between polymeric chains. In addition, with the increase of temperature in the previously mentioned tests, the polymer and composite materials reach a reduction in their storage modulus, this decrease being very marked at temperatures higher than the Tg of the material. On the other hand, as an influence of the deformation speed (dε/dt), it is seen that increases in the values of this parameter have repercussions on increases in the material's storage modulus. While incorporating plasticizing additives generates a lower value of G' because they promote higher mobility of the polymeric chains, the incorporation of reinforcements in the polymeric matrix generates increases of G' because they restrict the molecular movement of the polymeric matrix (Menard and Menard, 2020). Next the different behaviors manifested in polymers and polymer matrix biocomposites will be reported.

2. Dynamic and static behavior in biocomposites

Several types of biocomposites have been developed based on the use of lignocellulosic fibers from different botanical sources, these being used as reinforcement in a variety of organic matrices to form new materials that have a biodegradable character in a portion or the totality of their composition. The influence of different types of surface modifications of lignocellulosic fiber and other process additives and the processing techniques used on the mechanical, thermal and viscoelastic properties have been studied using Dynamic-Mechanical Analysis (DMA). Different investigations from DMA studies leading mainly to creep experiments are reported (Menard, 2008).

[1] Grupo Materiales Compuestos. Escuela de Ingeniería de Materiales, Universidad del Valle, Cali, Colombia.
[2] Unidad de Materiales, Centro de Investigación Científica de Yucatán, Mérida, México.
[3] Departamento de agroindustria, Universidad del Cauca, Popayán, Colombia.
* Corresponding author: juancastaneda@unicauca.edu.co

2.1 Dynamic analysis

The dynamic mechanical analysis results obtained in different biocomposites in which the type of matrix used in their preparation varies, which can be derived from petroleum-based polymers, agro-polymers, polymers of microbial origin and biodegradable synthetic polymers.

2.1.1 Petroleum-based polymers

DMA reports corresponding to biocomposites made up of polyolefins, polyester and epoxy resins as matrices, containing lignocellulosic fibers as a reinforcing agent, were studied, indicating changes in storage modulus, maximum loss modulus and Tan δ when evaluating different characteristics related to biocomposite composition (type of matrix, type of fiber, fiber content, filler content and use of coupling agents) and type of processing technique employed (Cristea et al., 2020; Montilla-Buitrago et al., 2021).

Biocomposites based on petroleum-derived synthetic polymers showed a gradual reduction of G' as the temperature increased (see Table 7.1), associating this behavior to higher molecular mobility in the respective matrices, while G" increased up to a maximum value that is related to the glass transition temperature and subsequently decreased for higher temperature values (Gupta and Srivastava, 2015). By mixing high-density polyethylene (HDPE), polypropylene (PP), polyester and epoxy resins matrices with their respective lignocellulosic reinforcements for the elaboration of biocomposites, higher storage moduli were reported that were manifested by the pure matrices. The storage modulus increased as the fiber content increased (Biswal et al., 2011; Gupta and Srivastava, 2015; Palechor-Trochez et al., 2021). However, in the case of a biocomposite of an unsaturated polyester resin matrix blended with 63% alkalinized kenaf fibers, a lower storage modulus was reported in a temperature range between 35 and 160°C that found in a biocomposite consisting of a blend of a polar unsaturated polyester with 56% of alkalinized kenaf fibers. The above behavior is related to the degree of affinity between the matrix and the lignocellulosic reinforcement, allowing the formation of a greater number of secondary bonds between the unsaturated polyester with a higher degree of polarity and the alkalinized fibers (Aziz et al., 2005). The incorporation of fillers and coupling agents in the biocomposite also generated increases in the storage modulus, a behavior identified in the PP biocomposite with 3% clay and grafted with maleic anhydride (Clyne and Hull, 2019) and the biocomposites based on HDPE and wood flour (Solanilla-Duque et al., 2021). The major contribution of maleic anhydride (coupling agent) in increasing the storage modulus is evident from temperatures below 50°C. According to literature among the biocomposites that presented the highest storage modulus, in a temperature range between –100 and 100°C, are those based on HDPE and epoxy resin due to the chemical bonds generated with the maleic anhydride.

According to Gupta and Srivastava (2015), based on the concept of hybrid biocomposites, the type of fiber is another factor that affects the magnitude of the storage modulus in a defined temperature range. Hybridization consists of using two or more types of fibers that allow generating an interaction that increases the reinforcement in the biocomposite, such behavior is evidenced by mixing in equal parts jute fiber and sisal fiber representing 30% of the reinforcement in the epoxy resin matrix.

Regarding the loss modulus (E") and the Tan δ, the identification of the glass transition temperature (Tg) is achieved; however, Akay (1993) recommended using the peak generated in the E", since its values present greater closeness to that granted by the DSC, while the Tg generated in the Tan δ can present deviations between 10 to 20°C above the real value. Taking into account the behavior of Tg (see Table 7.2), by increasing the fiber content, hybridization through the use of two types of lignocellulosic reinforcements, the incorporation of coupling agents and the use of clays in the biocomposites, Tg increases, relating this behavior to the restriction of the mobility of the molecular chains corresponding to the matrix (Kargarzadeh et al., 2017; Lambert and Wagner, 2017; Olewnik-Kruszkowska et al., 2020; Palechor-Trochez et al., 2021). The intensity of Tan δ reveals the degree of interfacial adhesion in biocomposites, finding lower values in those constituted by an

Table 7.1. Storage modulus in biocomposites based on synthetic polymers derived from petroleum.

Biocomposite	Storage Modulus (GPa)					Ref.
	–100°C	–50°C	0°C	50°C	100°C	
HDPE +-* Wood flour (25%)	4.6	3.8	3.0	1.6	0.5	(Tajvidi and Falk, 2004)
HDPE + wood flour (50%)	5.7	4.8	3.9	2.5	1.0	
HDPE + wood flour (50%) + MA (2%)	6.9	5.9	4.7	3.1	1.2	
PP-g-MA + Clay (3%)	4.2	4.1	3.3	1.7	0.8	(Biswal et al., 2011)
PP-g-MA + Clay (3%) + Banana fiber (20%)	4.7	4.5	3.7	2.2	1.0	
PP-g-MA + Clay (3%) + Banana fiber (30%)	4.8	4.6	3.7	2.3	1.3	
Unsaturated polyester + alkalinized kenaf fiber (56%)	-	-	-	2.2	1.1	(Aziz et al., 2005)
Unsaturated polyester (styrene) + Alkalinized kenaf fiber (63%)	-	-	-	1.4	0.6	
Epoxy resin + Sisal fiber (30%)	-	-	-	2.2	0.1	(Babaee et al., 2015)
Epoxy resin + Yute fiber (30%)	-	-	-	2.5	0.4	
Epoxy resin + Yute fiber and Sisal fiber (30%)	-	-	-	3.6	1.4	

Table 7.2. Loss modulus and delta tangent of biocomposites based on synthetic petroleum-derived polymers.

Biocomposite	G'' maximum		Tan δ		Ref.
	T° (°C)	G'' (GPa)	T° (°C)	Intensity	
HDPE + Wood flour (25%)	53.0	0.21	-	-	(Tajvidi and Falk, 2004)
HDPE + Wood flour (50%)	61.0	0.27	-	-	
HDPE + Wood flour (50%) + MA (2%)	62.5	0.31	-	-	
PP-g-MA + Clay (3%)	6.2	0.20	12.5	0.09	(Biswal et al., 2011)
PP-g-MA + Clay (3%) + Banana fiber (20%)	7.5	0.24	9	0.07	
PP-g-AM + Clay (3%) + Banana fiber (30%)	7.0	0.22	9	0.08	
Unsaturated polyester + Alkalinized kenaf fiber (56%)	90.3	0.15	99	0.11	(Aziz et al., 2005)
Unsaturated polyester (styrene) + Alkalinized kenaf fiber (63%)	78.0	0.25	101	0.78	
Epoxy resin + Sisal fiber (30%)	62.6	0.28	71.6	0.55	(Babaee et al., 2015)
Epoxy resin + Yute fiber (30%)	65.8	0.49	70.9	0.46	
Epoxy resin + Yute fiber and Sisal fiber (30%)	84.4	0.52	93.3	0.28	

epoxy resin and two types of fibers (hybridization) (Lila et al., 2019; Ramamoorthy et al., 2019). From the values obtained in the storage and loss moduli, the Cole-Cole diagram determined that the biocomposites elaborated from PP-g-AM mixed with clay and lignocellulosic banana fibers reached a heterogeneous system (Reichert et al., 2020; Shaghaleh et al., 2018).

2.1.2 Agro-polymers

Incorporating natural fibers in thermoplastic starch (TPS) based matrices has the purpose of developing biocomposites with improved mechanical properties, reaching increases of up to four times the maximum tensile strength concerning the unreinforced TPS matrix (Yu et al., 2006). Table 7.2 reports values of G" and Tan δ for investigations related to biocomposites where kenaf and sisal fibers are used, mixed with thermoplastic corn starch.

From the results obtained by DMA, a considerable increase in the storage modulus in a temperature range between –90 to 100°C in TPS-based biocomposites when compared with their respective pure matrices is seen, which was due to the generation of hydrogen bridges and physical anchors between the reinforcement and the TPS [25]. Ghanbari et al., 2018a reported higher modulus at temperatures below –50°C regarding the other studies, due to, among other things, the processing techniques (twin-screw extrusion followed by compression molding) employed, the lower concentration of plasticizer (glycerin), the reinforcing agent and the lower concentration of cellulose nanofibers in the biocomposite. Babaee et al. (Babaee et al., 2015) reported lower modulus values in composites with a higher amount of reinforcing agent in a TPS matrix, and this was probably due to the type of transformation, which was casting. de Freitas et al. (2021), from the mechanical characterization of biocomposites reinforced with Kenaf fibers, also achieved a lower modulus in their material, possibly due to the interaction between the nature of the processing technique employed, the use of short fiber and the presence of bubbles in the biocomposites (see Table 7.3).

Regarding the surface modification of the biocomposite reinforcement, the acetylation (acetylation degree of 0.2) of the nanofibers reduced the storage modulus over the entire temperature range evaluated because in the structure of the acetylated cellulose nanofiber by replacing some hydroxyl groups with acetyl groups, the crystallinity and the polar interactions of the cellulose with the matrix were decreased and promoted a more significant physicochemical interaction with other nanofibers and the matrix, contributing to the formation of gaps or porosities in the biocomposite (Ghanbari et al., 2018b). While with the alkalinization of short fibers, an increase in the storage modulus was reported, revealing a higher adhesion between the modified fiber and the matrix, allowing a biocomposite of higher stiffness concerning the material reinforced with untreated fibers (de Freitas et al., 2021). Regarding the loss modulus, two signals related to molecular movements or α and β transitions are presented in the biocomposites containing 0.5 and 1.5% cellulose nanofibers (see Table 7.4). In the α transition associated with Tg, a slight shift towards higher temperatures is reported, from 58.0 to 60.5°C, probably caused by the increased restriction of the TPS molecular chains and higher thermal stability. However, by increasing the cellulose nanofiber content, a shift to a lower temperature (50.0°C) was generated, a phenomenon that may indicate a possible plasticizing effect. On the other hand, the presence of the β transition in the material was possibly due to vibrations coming from segments of the molecular chains appearing from the TPS, being greater in the biocomposites (–47.0 and –50.0) than in the pure TPS (–56.0). Finally, using the information provided by the Tan δ (temperature and intensity), a higher Tg and interfacial adhesion was found when using unacetylated cellulose nanofibers and the alkalinized fibers. As for the increase of the amount of reinforcement in the matrix, a reduction of Tg and an increase of Tan δ intensity were seen, showing an increase of plastic participation at temperatures above 0°C and lower interfacial adhesion between the reinforcement and the TPS matrix (Ghanbari et al., 2018a; Kargarzadeh et al., 2017; Ramamoorthy et al., 2019).

Table 7.3. Agro-polymer-based biocomposite storage modulus.

Biocomposite	**Storage modulus (GPa)**						**Ref.**
	–90°C	**–50°C**	**–40°C**	**0°C**	**50°C**	**100°C**	
TPS + Cellulose nanofibers (10%)	-	5.4	-	3.4	1.9	0.2	(Babaee et al., 2015)
TPS + Acetylated cellulose nanofibers (10%)	-	1.9	-	1.2	0.5	0.1	
TPS + Cellulose nanofibers (0.5%)	8.2	5.4	-	1.8	0.9	0.1	(Ghanbari et al., 2018b)
TPS + Cellulose nanofibers (1.5%)	12.5	7.8	-	2.2	0.7	0.1	
TPS + Sisal fibers (3%)	-	-	0.15	0.05	0.01	0.01	(de Freitas et al., 2021)
TPS + alkalinizated sisal fibers (3%)	-	-	0.25	0.10	0.02	0.01	

Table 7.4. Modulus of loss and tangent delta in agro-polymer based biocomposites.

Biocomposite	G'' maximum		Tan δ		Ref.
	T° (°C)	G'' (GPa)	T° (°C)	Intensity	
TPS + Cellulose nanofibers	-	-	83	0.23	(Babaee et al., 2015)
TPS + Acetylated cellulose nanofibers	-	-	81	0.29	
TPS + Cellulose nanofibers (0.5%)	–47.0 60.5	0.7 0.18	–37.5 87.5	0.16 0.67	(Ghanbari et al., 2018b)
TPS + Cellulose nanofibers (1.5%)	–50.0 50.0	1.2 0.2	–37.5 80.5	0.18 0.70	
TPS + Sisal fiber (3%)	–40.0	0.035	–43.3 10.0	0.26 0.24	(de Freitas et al., 2021)
TPS + Alkalinized sisal fiber (3%)	–41.0	0.054	–43.9 12.6	0.22 0.22	

2.1.3 Polymers of microbial origin

The blend between some of the biopolyesters that make up the PHA (Polyhydroxyalkanoates) family with the lignocellulosic fibers generate a greater interfacial adhesion compared to that generated in the biocomposites constituted by synthetic polymers derived from petroleum such as polypropylene since PHA can wet the fibers as evidenced in Scanning Electron Microscopy (SEM). However, as the plasticizer content in the biocomposite increases, the interfacial interaction is reduced (Cai et al., 2014; Zaaba et al., 2013). Currently, polyhydroxybutyrate (PHB) (Tokiwa and Calabia, 2007), polyhydroxybutyrate-valerate (PHBV) (Ishigaki et al., 2004) or mixtures of PHB and polyhydroxyvalerate (PHV) (Madbouly, 2021) are used to require less processing energy to obtain the biocomposites since they are alternatives that allow a reduction of the melting temperature without exposing the lignocellulosic fibers to thermal decomposition.

Table 7.5 shows a higher storage modulus, in a temperature range between –50 to 70°C, for biocomposites containing a higher proportion of fibers, in which the generation of primary bonds with the polymeric chains of the matrix is promoted through the use of silanes; highlighting the biocomposite that presents PHB as a matrix. The characteristic behavior of this type of biocomposite is due to its stability at 50°C, maintaining storage moduli between 1.4 to 2.9 GPa, being higher values than those found in polypropylene-based biocomposites (Bhardwaj et al., 2006; Shanks et al., 2004). In the loss modulus, the displacement of the α transition (Tg) towards higher temperatures can be seen by incorporating a second polymer (PHV) in the mixture (hybridization) and using silanes to establish primary bonds between the flax fibers and the matrix, indicating greater rigidity or restriction of molecular mobility in the structure of the biocomposites (see Table 7.6), with silanization being responsible for a greater increase in this transition (Marlina et al., 2018).

Table 7.5. Storage modulus in polyhydroxyalkanoate-based biocomposites.

Biocomposite	Storage modulus (GPa)					Ref.
	–50°C	0°C	30°C	50°C	70°C	
PHBV + Recycled cellulose fiber (15%)	-	-	1.8	1.4	0.9	(Bhardwaj et al., 2006)
PHBV + Recycled cellulose fiber (40%)	-	-	2.9	2.3	1.7	
PHB + Flax fiber (50%)	2.3	2.2	-	1.39	-	(Shanks et al., 2004)
PHB + Silanized flax fiber (50%)	4.76	4.5	-	2.9	-	
PHB + PHV (12%) + Fibra de lino (50%)	4.18	4.26	-	2.71	-	
PHB + PHV (12%) + Silanized flax fiber (50%)	4.67	4.66	-	2.84	-	

Table 7.6. Modulus of loss and tangent delta in polyhydroxyalkanoate-based biocomposites.

Biocomposites	**G′′ maximum**		**Ref.**
	T° (°C)	**G′′ (GPa)**	
PHB + Flax fiber (50%)	16	0.12	(Shanks et al., 2004)
PHB + Silanized flax fiber	23	0.20	
PHB + PHV (12%) + Flax fiber (50%)	18	0.25	
PHB + PHV (12%) Silanized flax fiber (50%)	25	0.31	

2.1.4 Synthetic biodegradable polymers

Several developments have been carried out mainly focused on Poly Lactic Acid (PLA) as a polymeric matrix since it has outstanding mechanical, thermal and physicochemical properties among biodegradable materials (Salazar-Sánchez et al., 2019), allowing to replace some petroleum-derived synthetic polymers in more demanding and stable applications (Jandas et al., 2013). Next, in Tables 7.7 and 7.8, the dynamic analysis in PLA-based biocomposites is shown, including aging studies using parameters such as temperature cycles, environmental conditions and water immersion.

Table 7.7. Storage module in biocomposites based on polylactic acid.

Biocomposite	**Storage modulus (GPa)**					**Ref.**
	–40°C	**30°C**	**50°C**	**70°C**	**90°C**	
PLA + Sisal fiber (30%)	-	-	2.7	0.08	0.39	(Gil-Castell et al., 2016)
PLA + Sisal fiber (30%) + maleic anhydride (2.5%)	-	-	3.1	0.16	0.44	
PLA + Sisal fiber (30%) + Hydrothermal aging at 85°C	-	-	1.3	0.48	0.16	
PLA + Sisal fiber (30%) + maleic anhydride (2.5%) + Hydrothermal aging at 85°C	-	-	0.92	0.38	0.13	
PLA + Fibra reciclada (30%)	-	8.7	8.3	0.9	0.1	(Huda et al., 2005)
PLA + Talc (30%)	-	11.6	10.8	3.2	0.1	
PLA + Plantain fiber (30%)	-	4.5	4.4	0.6	0.1	(Jandas et al., 2012)
PLA + Alkalinized plantain fiber (30%)	-	5.6	5.3	0.8	0.1	
PLA + Plantain fiber (30%) + 3-Aminopropyltriethoxysilane (APS)	-	5.1	4.8	0.6	0.1	
PLA + Plantain fiber (30%) + bis-(3-triethoxy silyl propyl) tetrasulfane (Si-69)	-	4.3	4.1	2.0	0.1	
PLA + Cellulose nanofibers (1%)	-	2.8	-	0.17	-	(Jonoobi et al., 2010)
PLA + Cellulose nanofibers (5%)	-	4.4	-	2.5	-	
PLA + Kenaf Fiber (20%)	6.9	6.2	6.0	-	-	(Palechor-Trochez et al., 2021)
PLA + Nanoclay (5%)	5.4	4.8	4.5	-	-	
PLA + Kenaf Fiber (20%) + Nanoclay (5%)	7.2	6.4	6.1	-	-	
PLA + Cane bagasse fiber (20%)	-	1.0	0.98	0.04	0.01	(Lila et al., 2019)
PLA + Cane bagasse fiber (20%) + thermal cycle 4 weeks	-	0.88	0.85	0.58	0,50	
PLA + Cane bagasse fiber (20%) + thermal cycle 12 weeks	-	0.820	0.80	0.02	0,01	
PLA + Kenaf fiber (25%) + PEG (15%)	-	1.62	0.20	0.01	0,01	(Çokaygil et al., 2014)
PLA + Kenaf fiber (25%) + PEG (15%) + APP (10%)	-	2.25	1.50	0.01	0,01	
PLA + Alkalinized kenaf fiber (25%) + PEG (15%) + APP (10%)	-	5.30	5.30	4.20	2,00	

Table 7.8. Modulus of loss and tangent delta in polylactic acid-based biocomposites.

Biocomposite	**G'' maximum**		**Tan δ**		**Ref.**
	T° (°C)	**G'' (GPa)**	**T° (°C)**	**Intensity**	
PLA + Sisal fiber (30%)	-	-	63.7	0.60	(Gil-Castell et al., 2016)
PLA + Sisal fiber (30%) + maleic anhydride (2.5%)	-	-	63.9	0.16	
PLA + Sisal Fiber (30%)	-	-	71.9	0.48	
PLA + Sisal fiber (30%) + maleic anhydride (2.5%)	-	-	73.9	0.22	
PLA + Recycled fiber (30%)	66	1.88	73.0	1.25	(Huda et al., 2005)
PLA + Talc (30%)	69	2.51	74.0	1.74	
PLA + Plantain fiber (30%)	65	0.75	71.0 97.0	0.70 0.48	(Jandas et al., 2012)
PLA + Alkalinized plantain fiber (30%)	66	0.98	71.0 98.0	0.72 0.54	
PLA + Plantain fiber (30%) + APS	65	0.90	71.0 99.0	0.82 0.55	
PLA + Plantain fiber (30%) + Si-69	70	0.73	76.0 110.0	0.94 0.44	
PLA + Nanocellulose fibers (1%)	-	-	71.0	1.70	(Jonoobi et al., 2010)
PLA + Nanocellulose fibers (5%)	-	-	76.0	1.05	
PLA + Kenaf fiber (20%)	63	0.72	-	-	(Yusoff et al., 2016)
PLA + Clay (5%)	62	0.49	-	-	
PLA + Kenaf fiber (20%) + Clay (5%)	63	0.80	-	-	
PLA + Cane bagasse fiber (20%)	66	0.14	68.0	0.82	(Lila et al., 2019)
PLA + Cane bagasse fiber (20%) + thermal cycle 4 weeks	70	0.08	77.0	0.14	
PLA + Cane bagasse fiber (20%) + thermal cycle 12 weeks	71	0.06	78.0	0.13	
PLA + Kenaf fiber (25%) + PEG (15%)	40	0.95	41	1.4	(Shukor et al., 2014)
PLA + Kenaf fiber (25%) + PEG (15%) + APP (10%)	50	0.57	61	0.70	
PLA + Alkalinized kenaf fiber (25%) + PEG (15%) + APP (10%)	58	0.45	66	0.42	

In the different formulations of PLA-based biocomposites, the same behavior of the storage modulus found in the previous biocomposites is seen (Table 7.7), reporting a reduction of its value as the temperature increases. On the other hand, incorporating a reinforcement and/or filler in the PLA matrix contributes to achieving a higher modulus concerning the pure matrix, even though there is opposition to an adequate dispersion and interfacial adhesion of the PLA lignocellulosic fibers from its polar character. At the same time, PLA tends to present a higher apolarity (Huda et al., 2005). As reported by Kaiser et al. (2013), the incorporation of nanoclays in biocomposites generates an interaction with short kenaf fibers to increase thermal stability as long as a high aspect ratio of fibers and nanoclay is used. Another difference in thermal stability is due to the source of lignocellulosic fibers used, finding a higher modulus in fibers from newspaper recycling, followed by kenaf fibers, fibers from banana pseudostem, Sisal fibers and finally sugarcane bagasse with the lowest values.

Similarly, the implementation of a coupling agent (maleic anhydride or silane), the surface modification of the lignocellulosic reinforcement by alkalinization and the use of cellulose nanofibers contribute to the increase in the storage modulus (see Table 7.7), which is higher than that provided by biocomposites consisting of native lignocellulosic fibers. This is due to increased interfacial adhesion between the lignocellulosic fiber and the PLA, allowing a more effective stress transfer. In the alkalinization of the fibers, hemicellulose and lignin are removed, allowing more significant surface contact with the PLA. At the same time, silanization generates a reduction in molecular mobility due to the formation of covalent bonds and hydrogen bridges between the different molecular chains of the fibers and the PLA, leading to a modification of the structure of the polymer that forms the biocomposite (Jandas et al., 2012). Regarding the use of cellulose nanofibers, an increase in storage modulus was seen as their content in the biocomposite increased, going from 1 to 5% with values of 3.1 and 4.4 GPa, respectively, when exposed to a temperature of 30°C. The above behavior is due to a higher anchorage of the nanofibers in the PLA matrix (Jonoobi et al., 2010).

Shukor et al. (2014) reported the elaboration of biocomposites from PLA, kenaf fibers (reinforcement) in its native and alkalized state, polyethylene glycol (plasticizer) and ammonium polyphosphate (flame retardant) through the use of the twin-screw extrusion process with subsequent compression molding, achieving the highest thermal stability when using 15% plasticizer and 10% flame retardant, when the biocomposite contained alkalized fibers. A reduction in the modulus of elasticity was identified when using a higher content of flame retardant (> 10%), contributing to the generation of free spaces between the polymeric chains, preventing the continuation of the formation of bonds that could stiffen the biocomposite (see Table 7.7). According to the review of PLA biocomposites, greater thermal stability at 70°C was recorded in two formulations, relating the use of talc as filler in the first formulation and alkalinized kenaf fibers mixed with polyethylene glycol and ammonium polyphosphate in the second, achieving a storage modulus of 3.2 and 4.2 GPa, respectively.

In the determination of Tg, most of the authors involved in the review of PLA-based biocomposites used the loss modulus or Tan δ for the determination (see Table 7.8). It can be seen that the increase of the lignocellulosic reinforcement content, the addition of a coupling and filling agent with nanometric dimensions, the surface modification through alkalinization and the effect of thermal aging, contribute to the increase of Tg, indicating a greater molecular restriction in the structure of the PLA-based biocomposite (Gopi et al., 2019; Lila et al., 2019; Otoni et al., 2018; Shukor et al., 2014). However, an different behavior was reported when a plasticizer (polyethylene glycol) was added to the biocomposite, achieving a lower Tg, going from 63 to 71°C to values between 40 and 58°C due to the generation of free volumes between the PLA molecular chains and the lignocellulosic fibers (Ovalle-Serrano et al., 2018). Considering the different formulations identified in the review, the biocomposite consisting of 25% alkalinized kenaf fibers, plasticized with 15% polyethylene glycol and reinforced with 15% ammonium polyphosphate, gives the highest storage modulus in the dynamic analysis and the lowest Tg, being characteristic in materials with higher toughness, achieving an increase in the ability to absorb forces from impacts.

In relating these biocomposites to possible applications, thermal aging studies were applied, using two study media under atmospheric conditions and submerged in water. Lila et al. (2019) processed several PLA biocomposites reinforced with fiber from sugarcane bagasse through the use of extrusion with subsequent injection molding and subjected them to thermal cycles comprised by a temperature variation between –20 to 65°C, developing each cycle in 12 hours, the repetition of the respective cycle was extended for 12 weeks and a dynamic measurement was performed every 4 weeks (see Table 7.8). When monitoring the storage modulus, at temperatures below 70°C, a reduction of the modulus was identified as the thermal cycle time increased, while, at temperatures above 70°C, at week 4, the highest modulus of elasticity was obtained due to increased crystallization. However, at week 12, the former structure is lost in the biocomposite, manifesting itself in reducing the storage modulus.

In a second study, Gil-Castell et al. (2016) reported the thermal aging of a PLA biocomposite reinforced with 30% sisal fibers by immersing it in water at 85°C for 2 hours. In the dynamic analysis, the effect generated by the coupling between the matrix and the fibers through the use of maleic anhydride was taken into account. When evaluating the storage modulus between 50 and

90°C, a pronounced reduction of the modulus was seen from 54°C to 70°C. However, when the biocomposites continued to be heated, the modulus increase was resumed due to the crystallization of some polymeric chains that were previously in an amorphous state. The highest modulus values were found in the coupled biocomposite concerning the uncoupled state. In the case of the biocomposites thermally aged at 85°C, the storage modulus recorded before 70°C was lower concerning the non-aged states, due to the possible hydrolysis or depolymerization of PLA upon exposure in water at high temperature, being a reduction of the modulus with higher pronounced in the coupled biocomposite, since the chemical reaction of maleic anhydride in the presence of dicumyl peroxide, apart from contributing to the generation of primary bonds between the polymeric chains, can generate a reduction in molecular weight and affect surface adhesion between the matrix and the reinforcement when exposed to water at high temperature. In contrast, the storage modulus identified at temperatures close to 70°C was higher concerning non-aged biocomposites due to a higher formation of crystalline structures (de Oliveira et al., 2019).

2.2 Creep phenomenon

This methodology makes it possible to identify the shape memory behavior of polymers and polymer matrix biocomposites after being subjected to constant stress (tension, bending or compression) for a given time. In this sense, Cokaygil et al. (Barmouz and Hossein Behravesh, 2017) elaborated a blend of thermoplastic modified corn starch (54%) with pectin (46%), subsequently added 0.25% of sodium silicate, obtaining a biodegradable material in the form of sheets through the use of extrusion, being compared with a low-density polyethylene film (LDPE) through the use of a creep test at a constant stress of 0.75 MPa for 10 minutes, maintaining a temperature of 23°C and relative humidity of 50%. In the biodegradable film, a maximum deformation of 11.7% was reached in the creep zone, and subsequently, its deformation after recovery was 11.5%, while in the LDPE, its values were 0.43 and 0.06%, respectively. This showed that the starch and pectin mixture reached a higher deformation during the creep phenomenon and a lower elastic recovery capacity concerning LDPE. However, compliance was higher in LDPE (4.81×10^{-9} m^2/N) concerning the biodegradable film (3.71×10^{-9} m^2/N), explaining a greater capacity for deformation and macromolecular relaxation by LDPE (Alves et al., 2007).

In the second type of creep methodology, based on the application of constant stress for 30 minutes, Cyras et al. (2002) used the biodegradable resin Materbi (a mixture of polycaprolactone, starch and other additives) and incorporated short sisal fibers at different concentrations (0 to 40%) to obtain a biocomposite using the twin-screw extrusion processing technique. Initially, the Materbi resin was subjected to creep, exposing it to a stress of 5 MPa at different isotherms (8, 15, 28, 30 and 35°C) for 30 minutes, showing, as expected, a greater deformation capacity at higher temperatures, tending to a region of equilibrium or stabilization as the time evaluated progressed. The resin shows a stable deformation of 1% at 8°C, 2.8% at 28°C, and 4.7% at 35°C. Subsequently, the different biocomposites were evaluated with variations in their sisal fiber content at 28°C, presenting a reduction of the deformation capacity as the fiber content increases, going from 2.8% for the pure resin to 0.8% when incorporating 40% of the sisal reinforcement. The blends of thermoplastic starch with polycaprolactone were characterized by presenting high compliance (Chung, 2010; Menard and Menard, 2020).

In a third development, blends of polylactic acid (PLA) and thermoplastic polyurethane (TPU) were evaluated using PLA/TPU ratios of 90/10, 60/40, and 50/50, respectively. The importance of incorporating TPU into PLA was associated with the generation of higher dimensional resilience of the matrix when blended with cellulose nanofibers and exposed to constant stress. However, since pure PLA acting as a matrix induces high stiffness, a low elastic recovery capacity was identified. The chosen treatment was constituted by 60% PLA and 40% TPU, giving the values of storage and loss modulus with a greater balance to generate greater capacity in the dimensional recovery after subjecting it to stress in a defined time. Initially, creep was performed on the polymer blend with different isotherms (40 to 80°C) at a constant stress of 2 MPa, achieving a maximum deformation in the creep zone of 1.0% at 40°C up to 8.3% at 80°C. However, when adding 4% of cellulose nanofibers in the matrix, the maximum deformation in the creep zone went from 1.5% at 40°C to 4.9% at 80°C, relating this behavior to a restriction of molecular mobility when adding a reinforcement in the matrix constituted by PLA and TPU. As for the viscoelastic recovery, the addition of cellulose nanofibers contributes to higher participation at lower temperatures, while the elastic recovery does not present considerable differences concerning the pure matrix (Park et al., 2020).

Conclusions

The characterization of biocomposites based on different types of a matrix using DMA led to the following conclusions:

- The use of PLA as a matrix for developing a biocomposite, including alkalinized fibers, polyethylene glycol and ammonium polyphosphate, generated greater thermal stability at temperatures above 0°C concerning biocomposites consisting of synthetic polymers derived from petroleum, including thermosetting polymers such as epoxy resin.
- The use of biocomposites containing hybrid reinforcement, coupling agents, the polar or apolar affinity between the lignocellulosic fibers and the matrix, higher content of lignocellulosic fibers, and/or nanoclays and alkalinized fibers contributes to increased thermal stability and more significant interaction between the lignocellulosic fibers and the matrix.
- A limited development of dynamic analysis was found in biocomposites from thermoplastic starches and their respective biocomposites.
- Despite the low thermal stability of thermoplastic starch-based biocomposites compared to those made of petroleum-derived synthetic polymers, polyhydroxyalkanoates and biodegradable synthetic polymers, the highest storage moduli were identified between –90 to 100°C when cellulose nanofibers were used as reinforcement.
- From the high interfacial affinity between polyhydroxyalkanoates and lignocellulosic fibers, moderate thermal stability was identified in their respective biocomposites at 50°C. However, it should be clarified that their thermal resistance is lower than that granted by most PLA-based biocomposites.
- The incorporation of lignocellulosic fibers in the matrix reduces the incidence of the creep phenomenon in the polymeric matrix.
- The elastic and viscoelastic recovery of biocomposites based on thermoplastic starch is lower than that reported for biocomposites based on synthetic polymers derived from petroleum.

References

Akay, M. (1993). Aspects of dynamic mechanical analysis in polymeric composites. *Composites Science and Technology*, 47(4): 419–423. https://doi.org/10.1016/0266-3538(93)90010-E.

Alves, N. M., Saiz-Arroyo, C., Rodriguez-Perez, M. A., Reis, R. L. and Mano, J. F. (2007). Microhardness of starch based biomaterials in simulated physiological conditions. *Acta Biomaterialia*, 3(1): 69–76. https://doi.org/10.1016/j.actbio.2006.07.004.

Aziz, S., Ansell, M., Clarke, S. and Panteny, S. (2005). Modified polyester resins for natural fibre composites. *Composites Science and Technology*, 65(3–4): 525–535. https://doi.org/10.1016/j.compscitech.2004.08.005.

Babaee, M., Jonoobi, M., Hamzeh, Y. and Ashori, A. (2015). Biodegradability and mechanical properties of reinforced starch nanocomposites using cellulose nanofibers. *Carbohydrate Polymers*, 132: 1–8. https://doi.org/10.1016/j.carbpol.2015.06.043.

Barmouz, M. and Hossein Behravesh, A. (2017). Shape memory behaviors in cylindrical shell PLA/TPU-cellulose nanofiber bio-nanocomposites: Analytical and experimental assessment. *Composites Part A: Applied Science and Manufacturing*, 101: 160–172. https://doi.org/10.1016/j.compositesa.2017.06.014.

Bhardwaj, R., Mohanty, A. K., Drzal, L. T., Pourboghrat, F. and Misra, M. (2006). Renewable resource-based green composites from recycled cellulose fiber and poly(3-hydroxybutyrate-co-3-hydroxyvalerate) bioplastic. *Biomacromolecules*, 7(6): 2044–2051. https://doi.org/10.1021/bm050897y.

Biswal, M., Mohanty, S. and Nayak, S. K. (2011). Mechanical, thermal and dynamic-mechanical behavior of banana fiber reinforced polypropylene nanocomposites. *Polymer Composites*, 32(8): 1190–1201. https://doi.org/10.1002/pc.21138.

Cai, J., Xiong, Z., Zhou, M., Tan, J., Zeng, F., Meihuma, Lin, S. and Xiong, H. (2014). Thermal properties and crystallization behavior of thermoplastic starch/poly(É"-caprolactone) composites. *Carbohydrate Polymers*, 102(1): 746–754. https://doi.org/10.1016/j.carbpol.2013.10.095.

Chung, D. D. L. (2010). Composite material structure and processing. pp. 1–34. *In*: Chung, D. D. L. (ed.). *Composite Materials. Engineering Materials and Processes.* (First). Springer, London. https://doi.org/10.1007/978-1-84882-831-5_1.

Clyne, T. W. and Hull, D. (2019). *An Introduction to Composite Materials*. Cambridge University Press. https://doi.org/10.1017/9781139050586.

Çokaygil, Z., Banar, M. and Seyhan, A. T. (2014). Orange peel-derived pectin jelly and corn starch-based biocomposite film with layered silicates. *Journal of Applied Polymer Science*, 131(16): n/a-n/a. https://doi.org/10.1002/app.40654.

Cristea, M., Ionita, D. and Iftime, M. M. (2020). Dynamic mechanical analysis investigations of PLA-based renewable materials: how are they useful? *Materials*, 13(22): 5302. https://doi.org/10.3390/ma13225302.

Cyras, V. P., Martucci, J. F., Iannace, S. and Vazquez, A. (2002). Influence of the fiber content and the processing conditions on the flexural creep behavior of Sisal-PCL-starch composites. *Journal of Thermoplastic Composite Materials*, 15(3): 253–265. https://doi.org/10.1177/0892705702015003454.

de Freitas, R. R. M., do Carmo, K. P., de Souza Rodrigues, J., de Lima, V. H., Osmari da Silva, J. and Botaro, V. R. (2021). Influence of alkaline treatment on sisal fibre applied as reinforcement agent in composites of corn starch and cellulose acetate matrices. *Plastics, Rubber and Composites*, 50(1): 9–17. https://doi.org/10.1080/14658011.2020.1816119.

de Oliveira, T. A., de Oliveira Mota, I., Mousinho, F. E. P., Barbosa, R., de Carvalho, L. H. and Alves, T. S. (2019). Biodegradation of mulch films from poly(butylene adipate co-terephthalate), carnauba wax, and sugarcane residue. *Journal of Applied Polymer Science*, 136(47): 1–9. https://doi.org/10.1002/app.48240.

Ghanbari, A., Tabarsa, T., Ashori, A., Shakeri, A. and Mashkour, M. (2018a). Thermoplastic starch foamed composites reinforced with cellulose nanofibers: Thermal and mechanical properties. *Carbohydrate Polymers*, 197(June): 305–311. https://doi.org/10.1016/j.carbpol.2018.06.017.

Ghanbari, A., Tabarsa, T., Ashori, A., Shakeri, A. and Mashkour, M. (2018b). Preparation and characterization of thermoplastic starch and cellulose nanofibers as green nanocomposites: Extrusion processing. *International Journal of Biological Macromolecules*, 112: 442–447. https://doi.org/10.1016/j.ijbiomac.2018.02.007.

Gil-Castell, O., Badia, J. D., Kittikorn, T., Strömberg, E., Ek, M., Karlsson, S. and Ribes-Greus, A. (2016). Impact of hydrothermal ageing on the thermal stability, morphology and viscoelastic performance of PLA/sisal biocomposites. *Polymer Degradation and Stability*, 132: 87–96. https://doi.org/10.1016/j.polymdegradstab.2016.03.038.

Gopi, S., Amalraj, A., Jude, S., Thomas, S. and Guo, Q. (2019). Bionanocomposite films based on potato, tapioca starch and chitosan reinforced with cellulose nanofiber isolated from turmeric spent. *Journal of the Taiwan Institute of Chemical Engineers*, 96: 664–671. https://doi.org/10.1016/j.jtice.2019.01.003.

Gupta, M. K. and Srivastava, R. K. (2015). Effect of sisal fibre loading on dynamic mechanical analysis and water absorption behaviour of jute fibre epoxy composite. *Materials Today: Proceedings*, 2(4–5): 2909–2917. https://doi.org/10.1016/j.matpr.2015.07.253.

Huda, M. S., Drzal, L. T., Misra, M., Mohanty, A. K., Williams, K. and Mielewski, D. F. (2005). A study on biocomposites from recycled newspaper fiber and poly(lactic acid). *Industrial & Engineering Chemistry Research*, 44(15): 5593–5601. https://doi.org/10.1021/ie0488849.

Ishigaki, T., Sugano, W., Nakanishi, A., Tateda, M., Ike, M. and Fujita, M. (2004). The degradability of biodegradable plastics in aerobic and anaerobic waste landfill model reactors. *Chemosphere*, 54(3): 225–233. https://doi.org/10.1016/S0045-6535(03)00750-1.

Jandas, P. J., Mohanty, S. and Nayak, S. K. (2012). Renewable resource-based biocomposites of various surface treated banana fiber and poly lactic acid: characterization and biodegradability. *Journal of Polymers and the Environment*, 20(2): 583–595. https://doi.org/10.1007/s10924-012-0415-8.

Jandas, P. J., Mohanty, S. and Nayak, S. K. (2013). Surface treated banana fiber reinforced poly (lactic acid) nanocomposites for disposable applications. *Journal of Cleaner Production*, 52: 392–401. https://doi.org/10.1016/j.jclepro.2013.03.033.

Jonoobi, M., Harun, J., Mathew, A. P. and Oksman, K. (2010). Mechanical properties of cellulose nanofiber (CNF) reinforced polylactic acid (PLA) prepared by twin screw extrusion. *Composites Science and Technology*, 70(12): 1742–1747. https://doi.org/10.1016/j.compscitech.2010.07.005.

Kaiser, M., Anuar, H. and Razak, S. (2013). Ductile–brittle transition temperature of polylactic acid-based biocomposite. *Journal of Thermoplastic Composite Materials*, 26(2): 216–226. https://doi.org/10.1177/0892705711420595.

Kargarzadeh, H., Johar, N. and Ahmad, I. (2017). Starch biocomposite film reinforced by multiscale rice husk fiber. *Composites Science and Technology*, 151: 147–155. https://doi.org/10.1016/j.compscitech.2017.08.018.

Lambert, S. and Wagner, M. (2017). Environmental performance of bio-based and biodegradable plastics: The road ahead. *Chemical Society Reviews*, 46(22): 6855–6871. https://doi.org/10.1039/c7cs00149e.

Lila, M. K., Shukla, K., Komal, U. K. and Singh, I. (2019). Accelerated thermal ageing behaviour of bagasse fibers reinforced poly (lactic acid) based biocomposites. *Composites Part B: Engineering*, 156(August 2018): 121–127. https://doi.org/10.1016/j.compositesb.2018.08.068.

Madbouly, S. A. (2021). Bio-based polyhydroxyalkanoates blends and composites. *Physical Sciences Reviews*. https://doi.org/10.1515/psr-2020-0073.

Marlina, D., Sato, H., Hoshina, H. and Ozaki, Y. (2018). Intermolecular interactions of poly(3-hydroxybutyrate-co-3-hydroxyvalerate) (P(HB-co-HV)) with PHB-type crystal structure and PHV-type crystal structure studied by low-frequency Raman and terahertz spectroscopy. *Polymer*, 135: 331–337. https://doi.org/10.1016/j.polymer.2017.12.030.

Menard, K. P. (2008). *Dynamic Mechanical Analysis: A Practical Introduction* (K. P. Menard (ed.); 2nd ed.). CRC Press. https://doi.org/10.1201/9781420053135.

Menard, K. P. and Menard, N. R. (2020). *Dynamic Mechanical Analysis* (K. P. Menard & N. R. Menard (eds.); 3rd ed.). CRC Press. https://doi.org/10.1201/9780429190308.

Montilla-Buitrago, C. E., Gómez-López, R. A., Solanilla-Duque, J. F., Serna-Cock, L. and Villada-Castillo, H. S. (2021). Effect of plasticizers on properties, retrogradation, and processing of extrusion-obtained thermoplastic starch: a review. *Starch - Stärke*, 73(9–10): 2100060. https://doi.org/10.1002/star.202100060.

Olewnik-Kruszkowska, E., Burkowska-But, A., Tarach, I., Walczak, M. and Jakubowska, E. (2020). Biodegradation of polylactide-based composites with an addition of a compatibilizing agent in different environments. *International Biodeterioration and Biodegradation*, 147(March 2019): 104840. https://doi.org/10.1016/j.ibiod.2019.104840.

Otoni, C. G., Lodi, B. D., Lorevice, M. V., Leitão, R. C., Ferreira, M. D., Moura, M. R. d. and Mattoso, L. H. C. (2018). Optimized and scaled-up production of cellulose-reinforced biodegradable composite films made up of carrot processing waste. *Industrial Crops and Products*, 121(November 2017): 66–72. https://doi.org/10.1016/j.indcrop.2018.05.003.

Ovalle-Serrano, S. A., Gómez, F. N., Blanco-Tirado, C. and Combariza, M. Y. (2018). Isolation and characterization of cellulose nanofibrils from Colombian Fique decortication by-products. *Carbohydrate Polymers*, 189(November 2017): 169–177. https://doi.org/10.1016/j.carbpol.2018.02.031.

Palechor-Trochez, J. J., Ramírez-Gonzales, G., Villada-Castillo, H. S. and Solanilla-Duque, J. F. (2021). A review of trends in the development of bionanocomposites from lignocellulosic and polyacids biomolecules as packing material making alternative: A bibliometric analysis. *International Journal of Biological Macromolecules*, 192: 832–868. https://doi.org/10.1016/j.ijbiomac.2021.10.003.

Park, I. H., Lee, J. Y., Ahn, S. J. and Choi, H. J. (2020). Melt rheology and mechanical characteristics of poly(lactic acid)/alkylated graphene oxide nanocomposites. *Polymers*, 12(10): 2402. https://doi.org/10.3390/polym12102402.

Ramamoorthy, S. K., Åkesson, D., Rajan, R., Periyasamy, A. P. and Skrifvars, M. (2019). Mechanical performance of biofibers and their corresponding composites. pp. 259–292. *In*: Jawaid, M., Thariq, M. and Saba, N. (eds.). *Mechanical and Physical Testing of Biocomposites, Fibre-Reinforced Composites and Hybrid Composites* (First). Elsevier. https://doi.org/10.1016/B978-0-08-102292-4.00014-X.

Reichert, C. L., Bugnicourt, E., Coltelli, M. B., Cinelli, P., Lazzeri, A., Canesi, I., Braca, F., Martínez, B. M., Alonso, R., Agostinis, L., Verstichel, S., Six, L., De Mets, S., Gómez, E. C., Ißbrücker, C., Geerinck, R., Nettleton, D. F., Campos, I., Sauter, E., … Schmid, M. (2020). Bio-based packaging: Materials, modifications, industrial applications and sustainability. *In*: *Polymers* (Vol. 12, Issue 7). MDPI AG. https://doi.org/10.3390/polym12071558.

Salazar-Sánchez, M. del R., Campo-Erazo, S. D., Villada-Castillo, H. S. and Solanilla-Duque, J. F. (2019). Structural changes of cassava starch and polylactic acid films submitted to biodegradation process. *International Journal of Biological Macromolecules*, 129: 442–447. https://doi.org/10.1016/j.ijbiomac.2019.01.187.

Shaghaleh, H., Xu, X. and Wang, S. (2018). Current progress in production of biopolymeric materials based on cellulose, cellulose nanofibers, and cellulose derivatives. *RSC Advances*, 8(2): 825–842. https://doi.org/10.1039/c7ra11157f.

Shanks, R. A., Hodzic, A. and Wong, S. (2004). Thermoplastic biopolyester natural fiber composites. *Journal of Applied Polymer Science*, 91(4): 2114–2121. https://doi.org/10.1002/app.13289.

Shukor, F., Hassan, A., Hasan, M., Islam, M. S. and Mokhtar, M. (2014). PLA/Kenaf/APP biocomposites: effect of alkali treatment and ammonium polyphosphate (APP) on dynamic mechanical and morphological properties. *Polymer-Plastics Technology and Engineering*, 53(8): 760–766. https://doi.org/10.1080/03602559.2013.869827.

Solanilla-Duque, J. F., Salazar-Sánchez, M. del R. and Rodríguez Herrera, R. (2021). Potential of lignocellulosic residues from coconut, fique, and sugar cane as substrates for pleurotus and ganoderma in the development of biomaterials. *Environmental Quality Management*. https://doi.org/10.1002/tqem.21826.

Tajvidi, M. and Falk, R. H. (2004). Dynamic mechanical analysis of compatibilizer effect on the mechanical properties of wood flour-high-density polyethylene composites. *International Journal of Engineering*, 17(1 (Transactions B: Applications)): 95–104. https://www.sid.ir/en/Journal/ViewPaper.aspx?ID=2270.

Tokiwa, Y. and Calabia, B. P. (2007). Biodegradability and biodegradation of polyesters. *Journal of Polymers and the Environment*, 15(4): 259–267. https://doi.org/10.1007/s10924-007-0066-3.

Yu, L., Dean, K. and Li, L. (2006). Polymer blends and composites from renewable resources. *Progress in Polymer Science*, 31(6): 576–602. https://doi.org/10.1016/j.progpolymsci.2006.03.002.

Yusoff, R. B., Takagi, H. and Nakagaito, A. N. (2016). Tensile and flexural properties of polylactic acid-based hybrid green composites reinforced by kenaf, bamboo and coir fibers. *Industrial Crops and Products*, 94: 562–573. https://doi.org/10.1016/j.indcrop.2016.09.017.

Zaaba, N. F., Ismail, H. and Jaafar, M. (2013). Effect of peanut shell powder content on the properties of recycled polypropylene (RPP)/peanut shell powder (PSP) composites. *BioResources*, 8(4): 5826–5841. https://doi.org/10.15376/biores.8.4.5826-5841.

Chapter 8
Films

Karla C. Córdova-Cisneros, Paola F. Vera-García,
Karina G. Espinosa-Cavazos, Omar A. Martínez-Anguiano,
Aidé Sáenz-Galindo and *Adali O. Castañeda-Facio**

1. Introduction

A great variety of materials from both renewable sources (biopolymers) and non-renewable sources (synthetic polymers) are used in the manufacture of polymeric films, the vast majority are generally based on non-renewable sources such as petroleum, which are resistant to chemical and physical degradation. Polyethylene (PE) is one of the synthetic polymers that are most used to obtain films in the packaging industry. In addition to PE there are various polymers to get films in very diverse applications such as in the packaging industry, in solar cells, biomedical applications, as well as in dentistry.

Currently, interest has appeared in the manufacture of films based on biopolymers from renewable resources, due to the pollution generated by films derived from non-renewable resources such as petroleum, coupled with this situation, the degradation of synthetic polymers takes a long time to achieve its total decomposition, reaching a critical level in terms of environmental contamination. Biopolymers can be based on cellulose, chitosan, gluten, starch or proteins and are an attractive alternative to replace synthetic polymers, because they are biocompatible, biodegradable, economical and ecological due to the fact that they can reduce the environmental impact.

However, one of the disadvantages of biopolymers is their thermal and mechanical properties, which is why conventional polymers are the most widely used. In this sense, research has been carried out on biopolymers to overcome these limitations. Studies have been made of mixtures of different polymers, both conventional and biopolymers, physical or chemical modifications have also been made or metallic nanoparticles have even been added to obtain reinforced films (nanocomposites) with the intention of improving mechanical, thermal, antimicrobial properties, properties of water vapor/oxygen barrier that have shown improvements in materials for various applications.

Polymeric nanocomposites are materials that are characterized by the homogeneous dispersion of particles in nanometric dimensions (less than 100 nm) within a polymeric matrix. Polymeric nanocomposites present a significant increase in physical, chemical, mechanical, thermal and antimicrobial properties in films, in order to find new applications in areas such as medical and pharmaceutical.

Facultad de Ciencias Químicas, Universidad Autónoma de Coahuila, Ing. J. Cardenas Valdez S/N, Col. República C.P. 25280, Saltillo, Coahuila, México.
* Corresponding author: adali.castaneda@uadec.edu.mx

An example of these nanocomposites is composite or hybrid latexes, made up of organic polymeric particles and inorganic particles (O/I), combining the best characteristics of inorganic solids and the manageability of polymers. Hybrid latexes can be prepared by polymerization in dispersed or heterogeneous media, highlighting the polymerization in suspension, emulsion and miniemulsion, resulting in hybrid latex particles, which require a very strict control of the synthesis parameters of the nanocomposites. The present chapter describes the production of films based on polymethylmethacrylate-co-butylacrylate nanocomposites (PMMA-Co-BuA) and silver nanoparticles (NPAg) obtained from plant extracts.

Within biopolymers, chitosan is a high value-added polysaccharide obtained from fishing industry waste, it is a derivative of chitin, which is the second most abundant polymer in nature. It is a natural polymer that has antimicrobial, anticancer, biodegradable characteristics, among others, that makes it a biocompatible material, for this reason, this chapter presents the obtaining of films based on the mixture of polyvinyl alcohol (PVA) and chitosan, using the casting method.

Among polymeric matrices, polyvinyl alcohol (PVA) is a polymer widely studied for its ease of forming films, its hydrophobicity, good processability, good biocompatibility and for its interesting chemical and physical properties, which when combined with another polymer such as polyvinylpyrrolidone (PVP) can form very resistant films. Furthermore, they have been shown to be polymers of low toxicity and cytotoxicity.

Currently PVA and PVP have been gaining great interest in the pharmaceutical and biomedical areas due to the fact that materials are obtained for applications such as tissue engineering, regenerative medicine, as well as for drug delivery systems. Its other applications are in the textile area, in the pharmaceutical industry, being used as an excipient, as an adhesive and film former for food packaging.

Due to what was described earlier in this chapter, procuring nanocomposites based on PVA/PVP to obtain polymeric films with properties is also discussed. Finally, some applications of films based on synthetic and natural polymers are shown.

2. Films based on PMMA-co-BuA

At present there are various procedures for the elaboration of a nanocomposite, some of these techniques are solution mixing, melt mixing and *in situ* polymerization, among others. *In situ* polymerization is carried out in homogeneous and heterogeneous systems, where nanoparticles are incorporated. Heterogeneous systems one of the most common where emulsion polymerization is the most used in the formation of latex, which when incorporating nanoparticles becomes a nanocomposite or hybrid material, making it a profitable method due to the fact that it achieves a good dispersion of the nanoparticles in the polymeric matrix (Dastjerdi et al., 2017; Donescu et al., 2009; Hübner et al., 2018; Pérez et al., 2017).

Emulsion polymerization is carried out through a chain reaction using free radicals, this polymerization method begins as an oil-in-water type emulsion, polymerization begins when the initiator that will be responsible for generating free radicals is incorporated. The main components of this type of polymerization are the initiator, which has the characteristic of forming free radicals when the molecule decomposes, which are soluble in water, some of the most used are potassium persulfate (K_2SO_4), ammonium persulfate $(NH_4)_2S_2O_8$ and hydrogen peroxide (H_2O_2). On the other hand, the surfactant is composed of two parts: a polar or hydrophilic part and the other part of the surfactant corresponds to the non-polar or hydrophobic part, which is an alkyl-type hydrocarbon chain. These types of molecules are important in polymerization since their main function is to be the nucleation site of the polymer particles and provide colloidal stability to newly formed particles, they can be of the anionic, cationic, amphoteric and ionic types. Regarding monomers, vinyl monomers such as vinyl acetate ($C_4H_6O_2$), acrylic acid ($C_3H_4O_2$), styrene (C_8H_8), butadiene, acrylonitrile are generally used, the most widely used among emulsions being methyl methacrylate

(MMA) and butylacrylate (BuA) (Bhanvase et al., 2014; van Herk, 2010; Hübner et al., 2018; Ovando, 2007; Min et al., 2017; Rios et al., 2013; Zhenqian et al., 2017).

Finally, the dispersion of polymer particles obtained during the resulting polymerization is known as latex and the final properties that are achieved will depend on different parameters such as the form of addition of the monomer, the reaction temperature, initiator and surfactant that affect the molecular weight of the polymer. Nanocomposites obtained from composite latex present better properties, attributed to the properties that nanoparticles provide when incorporated into latex. At present there are various investigations of nanocomposites of a polymeric matrix and metallic nanoparticles as reinforcement as mentioned by Xu et al., in their research where they synthesized nanocomposites based on polymethylmethacrylate (PMMA) and silver nanoparticles (AgNPs) by means of Bi-*in situ* ultrasonic emulsion, they mentioned that the resulting nanocomposite showed a core-shell structure, PMMA as the shell and AgNPs as the core. Demonstrating that there is a chemical interaction between nanoparticles and the polymer chain (Xu et al., 2007). Mamaghani et al., carried out two methods to obtain the nanocomposite of polymethylmethacrylate-butylacrylate-acrylic acid and AgNPs. In the first, the latex was synthesized by emulsion polymerization, once the latex was obtained, the AgNPs were dispersed, and the second method was the *in situ* synthesis of the latex in the presence of AgNPs by polymerization in mini emulsion. Through the results, they were able to show that the compound synthesized by mini emulsion obtained better distribution of NPs and bacterial activity compared to the separate dispersion method (Mamaghani et al., 2011). Muzalev et al., synthesized a polymeric compound based on PMMA and AgNPs by means of high-speed thermal decomposition of silver (Ag) salts in a polymeric solution. Through the results, they showed that the optical properties of nanocomposites can achieve an application in anti-reflective coatings for photocells (Muzalev et al., 2012). Abdelaziz et al. studied the synthesis and evaluation of thermal and optical properties of PMMA and $AgNO_3$ in different concentrations (0–10%) using the casting technique. The results showed that in the thermal and optical analysis they changed as a function of the increase in $AgNO_3$ levels in the polymer, attributed to the sensitivity of the films with $AgNO_3$ for thermal and optical sensor applications (Abdelaziz and Abdelrazek, 2013). On the other hand, An et al., synthesized nanocomposites based on methylmethacrylate-styrene and AgNPs by microemulsion. Through Transmission Electron Microscopy (TEM) and X-Ray Diffraction (XRD) it was observed that the AgNPs show spherical morphology and are uniformly dispersed in the polymer matrix with average particle sizes of 10 nm, in addition, a strong interaction between the copolymer matrix and the AgNPs was observed (An et al., 2015). Berber et al., obtained the nanocomposites of polymethylmethacrylate-co-butylacrylate with graphene oxide (PMMA-co-BuA-FLGO) in different concentrations of FLGO (0–1.5%) synthesized by *in situ* emulsion polymerization. Their results showed that the properties of the compound depended on the amount of FLGO in the copolymer, since the compound with the addition of 1% showed better results in thermal and mechanical properties due to the chemical interaction between PMMA-Co-BuA and FLGO (Berber et al., 2017).

On the other hand, Vera et al., obtained nanocomposites of polymethylmethacrylate-co-butylacrylate (PMMA-Co-BuA) and silver nanoparticles (AgNPs). The nanocomposites were synthesized by free radical polymerization where the most common method is *in situ* emulsion, as monomers a mixture of MMA and BuA was used, as initiator $(NH_4)_2S_2O_8$ and as surfactant Sodium Dodecyl Sulfate (SDS). Subsequently, the AgNPs were added to the reaction system at different concentrations of 0.05–1% with respect to the polymer. Finally, films were formed from the latex by casting, as shown in Fig. 8.1.

The nanocomposites were characterized by different techniques such as ultraviolet-visible spectroscopy (UV-Vis), fourier-transform infrared spectroscopy (FTIR), thermal tests as thermogravimetric analysis (TGA), Differential Scanning Calorimetry (DSC), mechanical stress tests and Scanning Electron Microscopy (SEM). By UV-Vis the characteristic absorbance of the surface plasmon resonance (RSP) of the AgNPs in the nanocomposite was found. Using FTIR, the intermolecular interactions that occur between PMMA-co-BuA and AgNPs were determined,

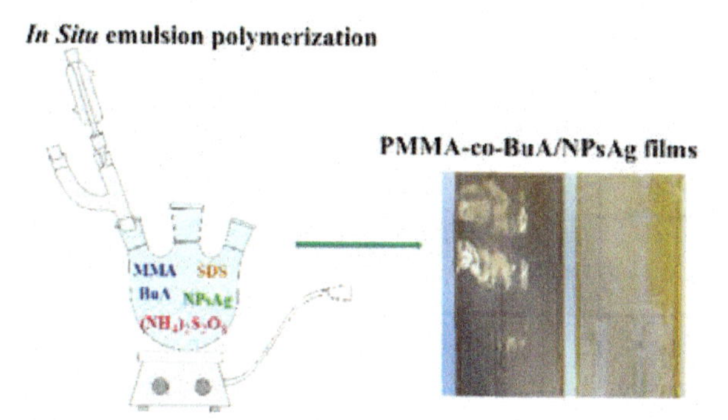

Figure 8.1. Graphic representation of components of *in situ* emulsion polymerization and obtaining the final nanocomposites of PMMA-co-BuA and NPsAg.

with hydrogen bonding being the most likely. Subsequently, in the thermal tests such as TGA, the thermal stability of the nanocomposites was observed, showing that the higher the concentration of nanoparticles the thermal stability decreases. The DSC analysis showed the glass transition temperatures (Tg) of the nanocomposites presenting an increase in the tg attributed to the incorporation of the AgNPs. Then, by means of mechanical stress tests, it was possible to determine the % elongation of the nanocomposites, Young's modulus and tensile strength, obtaining that at a higher concentration of nanoparticles it presented lower mechanical resistance compared to the copolymer without nanoparticles. Finally, by SEM it was confirmed that the AgNPs presented good dispersion in the copolymer and no agglomerates of AgNPs were present (Vera et al., 2019).

Deshmukh et al., similarly investigated the RPS in the PMMA-AgNPs nanocomposites, showing the RPS in the nanocomposite (Deshmukh and Composto, 2007). Mamaghani et al., investigated the nanocomposites based on polymethylmethacrylate-co-butylacrylate-acrylic acid with AgNPs, finding very similar bands between the spectra of the polymer only PMMA-co-BuA and the nanocomposite of PMMA-co-BuA and AgNPs, attributing it to the intermolecular interactions that exist between the copolymer and the nanoparticles (Mamaghani et al., 2011). Buhin et al., determined that by increasing the content of silica nanoparticles in the PMMA-co-BuA polymer, thermal stability increases in the nanocomposites (Buhin et al., 2013). Berber et al., obtained very similar results for the PMMA-co-BuA and FLGO nanocomposite, they observed that when adding the particles, Young's modulus and strength decreased attributing it to the rigid structure of graphene that acts as stress concentration points (Berber et al., 2017).

3. Films of PVA/PVP

In recent years, polymeric composite materials have been under great study because they present improved properties, such as increased mechanical properties (Toh et al., 2021), antimicrobial properties (Guo et al., 2021) and thermal properties (Callister, 1996). Being acceptable for a wide range of applications such as biomedicine, pharmacy and automotive, to mention a few (Zindani and Kumar, 2019). Therefore, for years they were introduced as an area of study. The first polymeric matrix composite materials to be used were those applied in the automotive industry, as in the case of Toyota Research Group in 1993, which introduced a composite material by dispersing layers of montmorillonite silicate (MMT) in a polymer matrix, obtaining a significant increase in properties (LeBaron et al., 1999), later in 1996 the Mercedez-Benz E-class came into the market, presenting an improvement in the door panels, replacing the common material with a composite material of polymeric matrix with epoxy resin and flax/sisal fiber, obtaining an increase in the mechanical properties, as well as a decrease in its weight (Sudell and Evans, 2005).

The great characteristic of these materials is that they present at least two phases, the continuous phase, which refers to a matrix that represents the volume of the material, and the discontinuous phase, which acts as a reinforcement (Toh et al., 2021). The reinforcements in a polymeric compound can allow the increase or improvement of its characteristics such as crystallinity, mechanical resistance, electrical conductivity, thermal conductivity and antimicrobial activity (Yuan et al., 2019). Therefore, the type of reinforcement added to the polymeric matrix is of great importance, until now the most used have been natural fibers (Chandra et al., 2020), synthetic fibers (Mayandi et al., 2020), carbon nanotubes (Mohammed, 2020) ceramic (Alizadeh-Osgouei et al., 2019), polymeric (Li et al., 2020) and metallic nanoparticles, the latter being of great interest due to the characteristics they confer on the matrix, such as antimicrobial properties. The efficiency of polymeric composite materials reinforced with metallic nanoparticles depends a lot on the compatibility of the matrix and the reinforcement, in addition to the dispersion of the NPs in the matrix is the key challenge to achieve that the properties improve in the composite material (Tawfik et al., 2020).

On the other hand, there is a great variety of syntheses for obtaining nanoparticles (NPs), such as chemical methods which are largely studied because they allow adequate control of the size and shape of the NPs (Slistan et al., 2007), however, the addition of stabilizer agents and reducing those that are toxic are needed, generating residues that can harm human health and the environment (Vera et al., 2016). While, physical methods have also been widely used, achieving very good results when obtaining desired characteristics such as the size of the particles, the problem with this type of method is that expensive technologies are needed to carry them out (Boutinguiza et al., 2018), which could generate an increase in the cost when using these types of methodologies. Which is why biological methods have recently begun to be studied to obtain metallic nanoparticles, considered as green methods because bacteria, fungi and plants are used as reducing agents and stabilizers, avoiding the use of toxic chemicals, as well as being simple and low-cost methods (Kobashigawa et al., 2019).

Among the metallic nanoparticles that have been studied, silver nanoparticles (AgNPs) have stood out by exhibiting characteristics that allow attractive applications in various fields, such as optics, biotechnology, electronics and medicine, to name a few (Patil and Kim, 2016). Since NPs act as reinforcements in polymeric matrices, their applications increase considerably. As was the case of Shankar and collaborators in 2018, who obtained films composed of polylactic acid (PLA) and AgNPs to evaluate the barrier properties of water vapor, finding an increase in these and when evaluating the antimicrobial properties, they found a powerful antibacterial activity against *Escherichia coli* and *Listeria monocytogenes* (Shankar et al., 2018).

It has been shown that these types of materials have been dominant among the other existing composite materials, due to the fact that they present good interaction between polymeric matrix and the reinforcing material, in addition to the fact that the polymers that make up the matrices are particularly interesting since they comply essential functions, among which can be highlighted (Besednjak, 2005):

- They transmit the effort to the reinforcement through the interface (boundary between the reinforcement and the matrix).
- They protect the reinforcement from compression efforts.
- They protect the reinforcement from external attacks such as humidity, chemical attack, etc.

Some scientists have used various polymers that act as matrices for these materials, among which polyvinyl alcohol (PVA) and polyvinylpyrrolidone (PVP), being non-toxic polymers, easy to mix and considered suitable materials to withstand various reinforcements can be mentioned. PVA is a biocompatible, hydrophilic polymer with high chemical stability, mechanical resistance and low cost (Atena et al., 2021), as well as good processability to form thin films, while PVP is hydrophilic, it has good stability, good resistance corrosion and high photoabsorption coefficient (Shaoquiang et al., 2017). PVA and PVP are formed by hydrophilic functional groups, being water-soluble polymers, which can form structural networks, related as retic hydrogels (Huang et al., 2017), which can swell in water or biological fluids, considerably increasing their volume and thus avoiding its solubility in polar solvents. Various researchers have studied the crosslinking of these polymers, such as Huang et al. In 2017, they prepared a hydrogel composed of polyvinyl alcohol (PVA) crosslinked with polyvinylpyrrolidone (PVP) and borax, in this study the formation of bonds between PVA and PVP was evidenced, increasing its mechanical properties. Various relationships of PVP with respect to PVA were analyzed, finding that the lower the PVP content the mechanical properties are improved. Huang et al., obtained hydrogels with better mechanical properties for biomedical and industrial applications (Huang et al., 2017). On the other hand, Eisa and collaborators in the same year, obtained PVA/PVP films crosslinked with 2.5% glutaraldehyde (GA), claiming that the mixture functioned as the polymeric matrix that supported NPsAg, they also evaluated the catalytic activity proving to be efficient by degrading the 4-nitrophenol compound and showed that the film could be recycled and used to degrade the compound at least three times, making it more effective (Eisa et al., 2017). Glutaraldehyde has been one of the crosslinking agents studied for the formation of a three-dimensional network between PVA and PVP, allowing the production of films at room temperature.

There are various preparation methods used to obtain polymeric composites, such as melt mixing, in situ polymerization and solution mixing.

- *Melt mixing*: The polymer is brought to a high temperature (up to melting temperature) and mixed with the particles, which are dispersed in the matrix. Very simple techniques such as injection or extrusion are used in this production procedure.
- *Polymerization in situ*: In this method of obtaining, the particles are taken to a monomer solution, where the polymerization is carried out, getting the dispersion of the particles during the reaction, generating a greater interaction between the matrix-particle.
- *Mixed in solution*: It consists of the preparation of a suspension, in which the particles are dispersed in a polar organic solvent, by means of mechanical stirring or ultrasound, subsequently a polymer dissolved in a solvent is added, causing the polymer chains to interact with the particles, subsequently the solvent is evaporated to obtain a film of the composite material. This method is of interest for the realization of nanocomposites (Covarrubias et al., 2013).

Solution mixing in recent years has been widely used for polymers such as PVA/PVP because these can be dissolved in polar solvents and later be cross-linked and add reinforcements to increase their properties.

4. Films of PVA/PVP cross-linked with glutaraldehyde and reinforced with AgNPs

Cordova et al., in 2018 obtained nanocomposites (films) where the following reagents were used: polyvinyl alcohol (Sigma Aldrich Mw ~ 31000–50000). Polyvinylpyrrolidone (Sigma Aldrich Mw ~ 55000) and glutaraldehyde (Sigma Aldrich, grade I, 25% in H_2O). the nanoparticles used as reinforcement were synthesized by biological methods from the extract of *Eucalyptus globulus.*

The methodology consisted in preparing a 10% PVA solution in distilled water, which was kept under constant magnetic stirring at a temperature of 80°C until dissolution, then a 2% PVP solution in water was added. The mixture of the two polymers was kept under constant magnetic stirring at 60°C until a homogeneous mixture was obtained, then the crosslinking agent (GA) was added at 6% with respect to the polymers, the solution was left under constant stirring for 2 hours. Subsequently, the aqueous suspensions of AgNPs (50 μl) were added to the PVA/PVP/GA, maintaining constant magnetic stirring for 30 minutes and then being treated with the ultrasound tip for 15 minutes. Finally, the solution was poured into glass molds and allowed to dry in an oven at a temperature of 40°C for 24 hours. Polymeric matrix composite films were obtained from PVA/PVP reinforced with NPsAg, which were finally characterized.

The crosslinking percentage was evaluated obtaining 97% on the other hand, the films obtained 110% swelling, with which it was determined that in general, the evaluated films show swelling values greater than 100%, which indicates that the films can swell, but do not lose their molecular structure. Regarding the percentage of crosslinking, it showed high percentages that confirm the formation of the crosslinking network. By means of FT-IR it was possible to carry out the structural elucidation of the PVA/PVP compounds cross-linked with GA and reinforced with AgNPs, this due to the fact that the inter or intramolecular interactions have a very important role in the behavior and final properties of the polymers. The mixtures formed by PVA and PVP in which one of the components contains donor groups of protons (OH) and the other acceptor (CO), have spectral modes corresponding to these groups, which tend to be significantly affected both in intensity as in their position when they are part of a hydrogen bond, which could be detected by means of infrared spectroscopy. The PVA/PVP film has characteristic bands of PVA and PVP polymers, important changes were also obtained, one being that the 3200 cm^{-1} band became wider and the 1654 cm^{-1} band intensified. These changes in the bands confirm the high miscibility between the polymers, the crosslinking agent and the silver nanoparticles since they all have similar functional groups, the changes in these bands are due to the formation of hydrogen bridges between the C = O groups and HO (Fig. 8.2) (Choudhary et al., 2018). These results are consistent with the results obtained by Faridi and collaborators (Faridi-Majidi et al., 2017), where in their study they evaluated the mixture of PVA/PVP and hydroxyapatite, obtaining results in FT-IR very similar to the previous ones and mentioned that the mixture between its polymers and hydroxyapatite can be asserted with the changes of the aforementioned bands (Córdova et al., 2018).

Tests were also carried out to measure antibacterial activity against *E. coli*; being a Gram negative bacterium. The method used was disk diffusion, using Potato Dextrose Agar (PDA) as a culture medium for bacterial growth. The behavior of the polymeric composite material containing the AgNPs indicates that the nanocomposite presented a good response to sensitivity against *E. coli* bacteria (Fig. 8.2). This disk diffusion analysis to determine antimicrobial activity was used as a qualitative test since there was no measurement of the inhibition halo (Córdova et al., 2018).

AgNPS reinforced polymeric film formation has been shown to be very effective in increasing antimicrobial properties and can be used in biomedical applications.

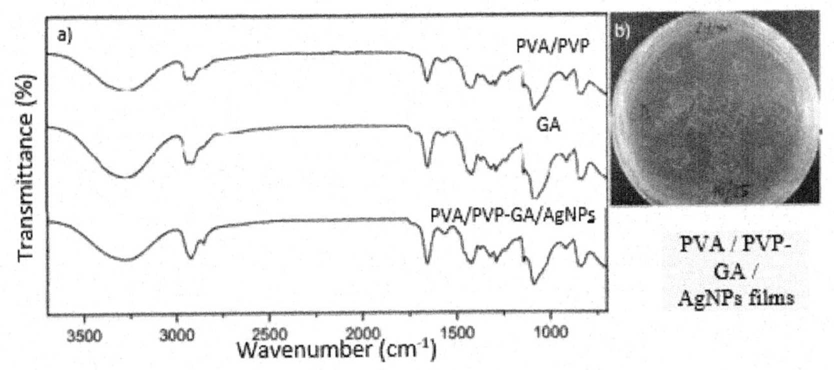

Figure 8.2. (a) FTIR spectra of PVA/PVP-GA/AgNPs films and (b) Antibacterial evaluation (disk diffusion method) of PVA/PVP-GA/AgNPs films against *E. coli* bacteria.

5. Nanocomposites PVA/CS crosslinked with glutaraldehyde and AgNPs synthesized by biosynthesis

Polymeric nanocomposites are materials that are characterized by having homogeneous dispersion of nanoparticles (NPs) with a dimension of less than 100 nm inside the polymeric matrix. Many types of polymeric matrix such as naturals and sintetic polymers exist. Gelatin, gluten, collegen, chitin, cellulose, starch and chitosan are types of natural polymers. On the other hand, there are synthetics such as polyesteramides, polydioxanone, polyanhydro, polyurethanes, polycarbonates and polyvinyl alcohol (Balaji et al., 2018).

The polyvinyl alcohol (PVA) is a semi-crystalline and hydrophilic polymer, have many properties including biocompatibility, excellent chemical resistance, good mechanical properties and biodegradability, has been used for food packaging and medical treatments (Liu et al., 2015). However it presents limitations such as thermal stability, biological activity and resistance to moisture absortption (Yang et al., 2021). The combinations with other polymers has resulted beneficially to improve it properties. There are studies that used PVA with others materials such as natural polymers and metallic nanoparticles.

Chitosan (CS) is a deacetylation product of chitin derived from crustacean, shells, cell wall of fungi and insects. CS is very attractive material for it is non-toxic, biodegradable and biocompatibility (Li et al., 2021). CS is a good material to be used as active packaging materials due to its environmentally friendly nature, biodegradability, biocompatibility, hight-quality, film forming and non-toxicity (Khezrian et al., 2018). CS is a potential material for the production of polymeric films, however, the application of CS film has been limited due to their relatively low antioxidant and antibacterial activities (Liu et al., 2021).

As PVA have hydrophilic characteristics, it has low stability in water, it is necessary for a chemical crosslinking in PVA/QS mixture to ensure a more stable structure.

As PVA is very hydrophilic, has low stability in water, it is necessary use to chemical crosslinking in PVA/QS mixture to ensure a more stable structure and that it does not dissolve in water or physiological fluids for medical applications. Among the chemical crosslinking agents, glutaraldegyde (GA) is by far the most widely used for its efficiency of polymer materials stabilization also is easily accessible, inexpensive and effectively crosslinked (Biji et al., 2001).

Espinosa et al. developed nanocomposite based to PVA/QS in different proportions using GA such as a crosslinking agent to strengthen with AgNPs obtained by biosynthesis using *Geranium* (*Pelargonium* spp.) extract (Espinosa et al., 2019).

It specifically used PVA (Mw ~ 130,000; Sigma Aldrich), CS (Coyotefoods) and glutaraldehyde (25% water solution; Sigma Aldrich) as a crosslinking agent. A PVA solution was prepared with 5 g of PVA were weighed and added to 500 ml of distilled water with mechanical stirring for 2 hours at 80°C. For the CS solution were weighed 2.5 g of CS and added to a 250 ml of acetic acid solution 1% (v/v) with mechanical stirring for 2 hours at 50°C.

Then two solutions were mixed in the selected blend according Table 8.1 and add 2% of glutaraldehyde (GA) as a crosslinking agent, finally were added 2% of NPAg obtained by biosynthesis used Geranium (*Pelargonium* spp.) extract (Espinosa et al., 2022).

The polymeric solution was poured into a dried glass petri dish and was left to dry at room temperature. After drying the film was peeled from the petri dish to obtain a stable film.

Films were characterized by FTIR (Perkin-Elmer Nicolet Nexus 47). A classical gravimetric method was used to evaluate the swelling and gel percentages of the nanocomposite film. Weighed mass of the nanocomposite film (m_1) was soaked in distilled water, then removed from water and the excess on the surface was wiped with a filter paper and weighed (m_2), at the end the film was

Table 8.1. Condition for nanocomposites PVA/CS/GA/AgNPs.

Polivinylalcohol (%)	Chitosan (%)	Glutaraldehyde (%)	AgNPs (%)
100	0	2	2
90	10	2	2
80	20	2	2
70	30	2	2
60	40	2	2
50	50	2	2
0	100	2	2

dried over at 50°C for 24 hours and weigthed (m_3) (Wael et al., 2017). Finally, the next relations were employed:

Swelling % = $m_2/m_1 \times 100$
Gel % = $(m_3/m_1) \times 100$

The thermal analysis of the nanocomposites was carried out by Thermal Gravimetic Analysis (TGA) (DISCOVERY TGA INSTRUMENTS).

The nanocomposites films showed a dark color in all cases due to the increased concentration of CS. The FTIR spectra of PVA/QS/GA/AgNPs composites showed a broad band at 3282.17 cm^{-1} attributed to the stretching mode of –OH groups, also a shift and decrease of the band due to the interactions with PVA was observed, these changes confirmed the high miscibility between both the polymers, crosslinked agent and AgNP. All of the functional groups are similar, each change of the band it is about the hydrogen bond, this behavior was observed by Kumar et al. in 2019, who analyzed the compatibility between QS and PVA and then proposed the formation of hydrogen bond about the interaction on both polymers with the crosslinked agent GA and AgNPs and their functionalization due to obtention method of AgNPs. The band attributed to the vibration of C-H was observed in 2922.11 cm^{-1}, the band corresponding to NH_2 appeared at 1559.28 cm^{-1} and the band assigned to O-C-NH_2 was shown at 1647.16 cm^{-1} (Paipitak et al., 2011).

The QS contain primary amino groups (NH_2), that can be protonated in acetic acid solution, which can increase the solubility in an aqueous medium. On the other hand, PVA contains –OH groups which also makes it soluble in water, for this reason it is important use a crosslinking agent to avoid its solubility since one of the objectives is to obtain structurally stable nanocomposites, for this reason glutaraldehyde has an important role in obtaining crosslinked nanocomposites.

Degrees of swelling were obtained between 180–320%, 320% for PVA/QS/NPs/GA (100:0, PVA:QS respectively as shown in Table 8.1), 187% for PVA/QS/NPs/GA (90:10), 238% for PVA/QS/NPs/GA (80:20), 202% for PVA/QS/NPs/GA (70:30), 285% for PVA/QS/NPs/GA (60:40), 186% for PVA/QS/NPs/GA (50:50) and 207% PVA/QS/NPs/GA (0:100). These values indicate that the films can swell, they do not lose their molecular structure, the formulations are appropriate and can keep its rigidity. The nanocomposite PVA/QS/NPs/GA (60:40) shows the best swelling.

The gel percent obtained is between 96–71%, these high percentages are due to the primary amino groups present in the CS structure that easily react with the -OH groups of GA (Abdeen et al., 2018), creating hydrogen bond type interactions between the –OH group from PVA and –OH from CS, as well as interactions between the functional groups of GA and the functional groups found on the surface of the AgNPs, resulting in three-dimensional networks. Ninety five percent was obtained for the formulation PVA/QS/NPs/GA (100: 0 PVA:QS respectively as shown in Table 8.1), 96% for PVA/QS/NPs/GA (90:10), 71% for PVA/QS/NPs/GA (80:20), 96% for PVA/QS/NPs/GA (70:30), 93% for PVA/QS/NPs/GA (60:40), 95% for PVA/QS/NPs/GA (50:50) and 90% PVA/QS/NPs/GA 0:100.

The thermal analysis was carried out in a temperature range of 30 to 600°C at a heating rate of 20°C/min, with a nitrogen atmosphere. The thermogravimetric analysis was carried out in order to study thermal stability of the composites with the AgNPs, in Fig. 8.3 the thermograms corresponding to PVA/QS/NPs/GA nanocomposites with the different PVA/CS ratios (100:0, 90:10, 80:20, 70:30, 60:40, 50:50). The nanocomposites show a significant weight loss in an approximate range of 240°C and a slight increase in thermal stability is observed in most samples, except PVA and 0% CS formulation. It gives rise to partial intermolecular breakdown of its structure. In general, the behavior observed is that the higher the percentage of CS, the films have greater thermal stability.

Finally, the modulus results were 911 N/mm^2 for PVA/QS/NPs/GA (100:0 PVA:QS respectively as shown in Table 8.1), 1070 N/mm^2 for PVA/QS/NPs/GA (90:10), 1379 N/mm^2 PVA/QS/NPs/GA (80:20), 1634 N/mm^2 for PVA/QS/NPs/GA (70:30), 1903 N/mm^2 for PVA/QS/NPs/GA (60:40), 2587 N/mm^2 for PVA/QS/NPs/GA (50:50) and 3132 N/mm^2 PVA/QS/NPs/GA (0:100).

In general, the characteristic bands of the components could be observed in the FTIR analyzes. The PVA/CS/NPs/GA nanocomposites presented crosslinking percentages of 90–95% and swelling of 180–320%. TGA showed improvement in the thermal stability of the nanocomposites.

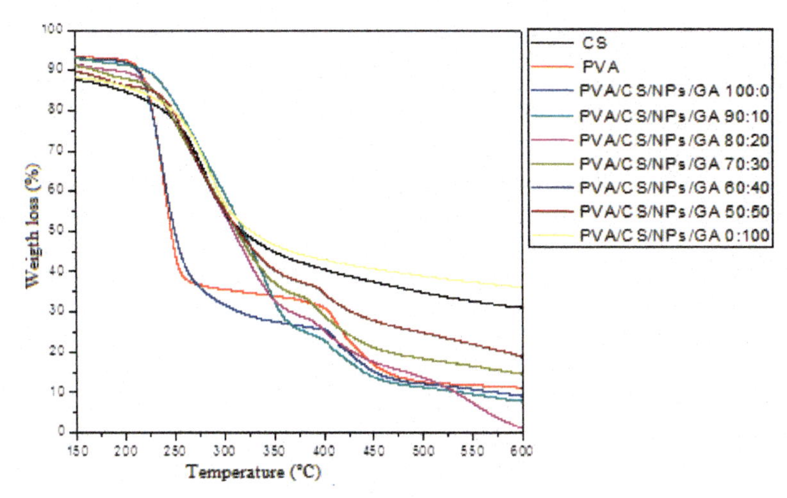

Figure 8.3. TGA curves of the PVA/CS/NPs/GA nanocomposites.

6. Applications

Some of the main applications of polymeric films are in the packaging sector, which is considered an important global industry in the development of countries. Packaging is closely related to the process of conservation and protection of different products, controlling their exposure to some factors such as oxygen, light, humidity and various pollutants. In this way, it seeks to maintain and ensure the quality of the products for a long time. However, the use of synthetic films has led to growing environmental problems due to the lack of biodegradability present in them (Nagaraja et al., 2019).

One of the most prominent and developing biopolymers are polyhydroxyalkanoates (PHA) (Ragaert et al., 2019). PHAs are polyesters of various hydroxyalkanoates synthesized by microorganisms like bacteria, with poly(3-hydroxybutyrate-co-3-hydroxyvalerate) (PHBV) being one of the PHAs with the best properties. In fact, it has similar characteristics to polypropylene (PP) such as low toxicity, biocompatibility, biodegradability and high crystallinity, making PHBV a possible replacement for PP for biodegradable applications despite limitations like the high cost, brittleness and poor thermal stability, these bioplastics find a growing demand especially in the field of sustainable packaging (Varghese et al., 2020).

There are also multiple alternatives to PHBV such as polylactic acid, chitosan, etc., which have shown growing interest due to their low negative impact on the environment to be biodegradable materials. The addition of copolymers increases the miscibility between the components of the films and therefore, represents a novel approach in the field of active packaging without forgetting the antimicrobial properties that many of these polymers present (Moreno-Vásquez et al., 2017).

Chitosan, is a biopolymer capable of forming films with great potential as packaging materials in food because of its natural, non-toxic and biodegradable properties, in addition to its antimicrobial effect. Chitosan is a polysaccharide of natural origin and derived from chitin, one of the most abundant natural polymers, commonly found in the cell walls of fungi and insect exoskeletons (Raveendran et al., 2017). Chitin is partially deacetylated to form chitosan, the molecular weight and degree of deacetylation are essential to evaluate the final characteristics of chitosan (Song et al., 2018).

On the other hand, antibacterial films for food packaging can be prepared by mixtures of Poly (L-Lactic Acid) (PLLA)/PLLA grafted maleic anhydride/epigallocatechin gallate. The miscibility between PLLA and epigallocatechin gallate can be increased by adding PLLA grafted maleic anhydride as a compatibilizing agent (Moreno-Vásquez et al., 2017).

Numerous polymers have also been used for the packaging of drugs and medical devices. In these advanced applications, polymers are not normally present as bulk materials (Huang and Voit,

2013). They are used as coatings on devices in the form of multiple deposited films. However, the adsorption and migration of the bioactive substance to the polymer, changes in pH, oxygen permeation and the release of leachable components must be carefully considered for individual applications. The interaction can affect not only the drug, but also the function of the polymer container (Maitz, 2015).

The polyolefins, high-density polyethylene (HDPE) and PP are the most common polymers for compressible vials, but multilayer packaging is also frequently used to achieve the required properties of inertia, protection against oxygen or UV rays (Jenke, 2014).

Polyvinyl chloride (PVC) containing the phthalate plasticizer diethylhexyl-phthalate (DEHP), is used for many extracorporeal perfusion tubes to provide medicines or also in blood conduction tubes in extracorporeal dialysis or extracorporeal oxygenation. In addition, donations of blood and blood products are typically stored in bags made of this polymer (Maitz, 2015).

Poly (α-esters) on the other hand, contain an aliphatic ester bond in the polymer backbone, are generally hydrophobic and undergo general erosion through an ester bond. Poly (α-esters) can be developed in numerous devices to load a wide range of drug modules such as peptides, proteins, vaccines and macromolecules (García, 2018).

Polymeric films have also been used for biomedical applications, one of these polymers is gelatin, which is widely used in this type of application due to its biocompatibility, biodegradability and versatility in the formation of physically cross-linked hydrogels in physiological environments (Aycan et al., 2019).

Due to the biocompatible and bioactive properties, as well as the demonstrated hemostatic and adhesive behavior, research on gelatin ultimately resulted in the commercialization and clinical use of various biomedical products, including biological glues, topical hemostatic agents, films used in neurological, thoracic and ocular surgery and micro-vehicles for cell culture (Piao et al., 2021).

Another type of polyester elastomers is poly (diol citrate) such as Poly(1,8-Octanediol-co-Citrate) (POC) which has been used over the years in tissue engineering. POC-based films have been compared to other FDA-approved materials with hemocompatible properties. Hemocompatibility tests showed low platelet and protein binding, as well as endothelial cells were able to retain their morphology under flow conditions, presenting good results (Coenen et al., 2018).

The Hyaluronic Acid (HA) is a mucopolysaccharide found naturally in numerous tissues and organs of the human body: epidermis, connective tissue, cartilage, synovial fluid, eyes, etc. The biodegradability, biocompatibility, non-immunogenicity and its viscoelastic properties, makes the HA a great component in applications of nanomedicine, vaccine and mucosal delivery and its application in surgical aid, treatment of arthritis and wound healing (Sharma et al., 2015).

Polycaprolactone (PCL) polyester is a semi-crystalline material in nature and soluble in a wide range of organic solvents, has a low melting point (between 55°C and 60°C) and the ability to form miscible mixtures with a wide range of bioerodible polymers as well as their degradation rate is improved when manufacturing copolymers with lactide or glycolide (Zhang et al., 2018). PCL was explored primarily as a means for controlled drug and vaccine delivery, attributed to its high permeability and non-toxicity to drugs. Subcutaneous insertion of PCL thin film devices for local or systemic administration is due to their adaptability, biodegradability and biocompatibility (Cama et al., 2017).

Nanoparticles (NP) have shown enormous potential for clinical application. In this way, a new discipline called nanomedicine has appeared, which can be defined as the branch of medicine that applies the knowledge and tools of nanotechnology to treat and prevent diseases (Crucho and Barros, 2017). The NPs are defined as particles whose size is smaller at 100 nm, NPs show remarkable bioavailability when administered orally or intravenously. They are capable of transporting drug molecules to tissues or organs, providing more efficient stability and activity (Kassem et al., 2017).

Polymeric nanoparticles (PNPs) due to their small size, have been reported to show distinctive potentials in drug delivery, vaccine delivery, gene delivery and cancer therapy. As theranostics,

PNPs are used for imaging, specific targeting, protection of drug molecules and to improve the therapeutic index (El-Say and El-Sawy, 2017).

Nanogels or hydrogel nanoparticles, are swollen networks of amphiphilic or hydrophilic polyionic polymers (Bose et al., 2019). These materials have the potential to be used as delivery systems due to their unique properties that include multivalent conjugation, high water content and biocompatibility, one of the examples with the lowest environmental impact being chitosan nanoparticles (Yanat and Schroën, 2021).

Nanoparticles of the poly (lactic-co-glycolic acid) copolymer (PLGA) synthesized by the polymerization of Poly (Lactic Acid) (PLA) and Poly (Glycolic Acid) (PGA) have also been employed in medicine. PLGA nanoparticles have been used as core materials mainly in the development of biomimetic nanovaccines. These types of biomimetic nanovaccines employ cell membrane coating technologies, where the entire cancer cell membranes can be used as a source of cell mimicry (Bose et al., 2018).

Dentistry is one of the disciplines that has long been favored through the use of biomaterials in the fields of periodontics and oral surgery, mainly for the regeneration of damaged root structures, dental pulp and dentins (Anju et al., 2020).

Due to their mechanical, physical and biological properties from the polymeric materials and films, there has been an increase in the use of these types of materials in the field of dentistry. Furthermore, these materials can be used for dentin regeneration or as advanced drug delivery systems (Rokaya et al., 2018).

Polymeric films in the area of dentistry can be classified according to their applications, as shown below:

- Prevent the development of biofilms and dental cavities
- Prevent dental erosion
- Restorative dentistry
- Prosthetic dentistry
- Implantology
- Periodontics
- Reduce corrosion
- Reduce friction

Multiple polymeric films have been used for the purpose of protecting teeth against erosion by avoiding contact of the acidic environment of the oral cavity with the teeth. Polymers like propylene glycol alginate, esterified pectin and gum arabic are capable of being absorbed by the teeth, forming a protective layer on the enamel and dentin, reducing the effects of erosion on the teeth as reported by Beyer et al. in 2012.

Natural polymers such as collagen, chitosan, hyaluronic acid, alginate, agar and chitin glycol-based hydrogels were reported to be attractive candidates for the preparation of dental scaffolds. PGA, PLA, PLLA, PLGA and PGA/PLLA are the most common biodegradable synthetic polymers of interest in dental scaffold investigations (Alshehadat et al., 2016).

There are a wide variety of applications for polymeric films, one of which has been reported with the potential to solve or reduce problems related to the water crisis. Wang et al. in 2020, developed polymeric films of sodium maleate grafted polypropylene (PP-g-NaMA) synthesized by solvent-induced self-assembly, which has a hydrophobic part and the other superhydrophilic. These polymeric materials can be used to reduce water evaporation and water collection, both with high efficiency according to researchers, in addition to its low cost, high performance and good stability.

Other applications include the development of applicable materials in sensors, such as humidity or temperature. The use of films based on semiconductor polymers such as PEDOT: PSS (poly (3,4-ethylenedioxythiophene) polystyrene sulfonate) films have been reported as crucial materials to predict changes in electrical properties as a function of environmental conditions (such as, humidity and temperature) for the stability of electrical devices (Kang et al., 2019).

It is possible to use polymeric films for their application in solar cells even with the use of techniques such as plasma deposition of films with the aim of increasing the performance of solar devices (Wong et al., 2016). The use of hybrid semiconductor polymers has also been reported to improve cell efficiency. Yu et al. in 2015 used poly (3-hexylthiophene) (P3HT) films hybridized with multiwalled carbon nanotubes (MWNT), indicating that the ratio of P3HT to MWNT is capable of considerably influencing the performance of solar cells. The efficiency of the solar cells was able to increase based on the ratio of carbon nanotubes, indicating that the optimal ratio of P3HT/MWNT ratio (= 1/1).

For biophotonics applications which include optical imaging optical sensing, and light-activated therapy needs materials with optical properties and these are required to manufacture optical elements such as lenses to transmit, detect and transform light (Shan et al., 2018).

Optical materials are described as those materials with the ability to control or alter electromagnetic radiation in the infrared (IR), ultraviolet (UV) and visible spectrum regions. For these types of applications, the materials must comply with certain optical, mechanical, chemical and biological properties (Lee et al., 2016). Some of the most important properties of optical materials are the degree of transparency and the refractive index, however the mechanical, chemical and biological properties of these types of materials are also fundamental for biophotonic applications (Shabahang et al., 2018). Although conventional silica glasses have favorable optical properties, they are usually toxic and not biocompatible for use in medicine, keeping in mind their mechanical fragility, which poses a risk of injury to the surrounding living tissues. Transparent polymeric biomaterials have advantages over inorganic silica optical materials as these can have mild mechanical properties, biocompatibility and biodegradability, as well as efficient chemical and biological functionalities (Shan et al., 2018).

Some natural polymers, such as cellulose fibers, have been used successfully in biophotonic applications. Cellulose, a polysaccharide of natural origin, has been used to manufacture optical fibers and sensors, due to its high visible light transmittance capacity and its good permeability to water. Microstructured cellulose fiber allows for the integration of optical, microfluidic and drug delivery functionalities. Researchers have also applied cellulose materials as a coating film for microstructured polymer optical fibers with the function of developing optical probes (Shan et al., 2018).

Another widely studied naturally occurring natural polymer is silk. Silks are natural fibers formed by proteins and produced by different types of insects and arachnids such as silkworms, moths, butterflies and spiders. Their hydrophobic B-sheet structures provides a great mechanical resistance. Molecular design and moderate aqueous processing allow them to be engineered with good chemical and biological properties. Films between 20 and 100 μm thick were found to be ideal for optical devices and biophotonic applications. By inducing film crystallization on appropriate substrates, nanopatterned silk optical elements have been developed, including the development of lenses, microlens arrays, diffraction gratings and pattern generators (Applegate et al., 2015).

In summary, polymeric films can be used for a wide range of applications, such as in the packaging sector and biomedical applications, as well as in dentistry and solar cells.

7. Conclusions

Currently biopolymers are gradually replacing synthetic polymers in some specific applications. The use of biopolymers in various areas of the industry has great beneficial and economic potential thanks to recent research carried out to improve mechanical, thermal and chemical properties, among others. Obtaining films based on biopolymers represents a challenge because of the influence it would have on the substitution of synthetic polymers due to the fact that its main advantage is biodegradability, which will allow the reduction of polluting waste. In addition, the use of by-products that are considered waste would be promoted, such as chitosan, which is obtained from shrimp and lobster waste from packing companies. In this way, the reduction of the accumulation of plastic waste is being promoted.

8. Acknowledgements

Thanks to the support of CONACYT for the scholarships awarded, as well as the Faculty of Chemical Sciences of the Autonomous University of Coahuila.

References

Abdeen, Z., Ahmed, F. and Negm, N. (2018). Nanocomposite framework of chitosan/polyvinyl alcohol/ZnO: Preparation, characterization, swelling and antimicrobial evaluation. *Journal of Molecular Liquids*, 250: 335–343.

Abdelaziz, M. and Abdelrazek, E. M. (2013). Thermal-optical properties of polymethylmethacrylate/silver nitrate films. *Journal of Electronic Materials*, 42: 2743–51.

Alizadeh-Osgouei, M., Li, Y. and Wen, C. (2019). A comprehensive review of biodegradable synthetic polymeric-ceramic composites and their manufacture for biomedical applications. *Bioactive Materials*, 4: 22–36.

Alshehadat, S. A., Thu, H. A., Hamid, S. S. A., Nurul, A. A., Rani, S. A. and Ahmad, A. (2016). Scaffolds for dental pulp tissue regeneration: A review. *International Dental & Medical Journal of Advanced Research*, 2: 1–12.

An, J., Luo, Q., Li, M., Wang, D., Li, X. and Yin, R. (2015). A facile synthesis of high antibacterial polymer nanocomposite containing uniformly dispersed silver nanoparticles. *Colloid and Polymer Science*, 293: 1997–2008.

Anju, S., Prajitha, N., Sukanya, V. S. and Mohanan, P. V. (2020). Complicity of degradable polymers in health-care applications. *Materials Today Chemistry*, 16: 100236.

Applegate, M. B., Perotto, G., Kaplan, D. L. and Omenetto, F. G. (2015). Biocompatible silk step-index optical waveguides. *Biomedical Optics Express*, 6: 4221.

Atena, N., Mehri-Saddat, E. and Fatemeh, D. (2021). Enhanced electrocatalitic performance of Pt nanoparticles immobilized on novel electrospun PVA/Ni/NiO/Cu complex bio-nanofiber/Chitosan based on Calotropis procera plant for methanol electro-oxidation. *International Journal of Hydrogen Energy*, 46: 18949–18963.

Aycan, D., Selmi, B., Kelel, E., Yildirim, T. and Alemdar, N. (2019). Conductive polymeric film loaded with ibuprofen as a wound dressing material. *European Polymer Journal*, 109308.

Balaji, A., Pakalapat, H., Khalid, Walvekar, R. and Siddiqui, H. (2018). Natural and synthetic biocompatible and biodegradable polymers. *Biodegradable and Biocompatible Polymer Composites*, 1: 3–32.

Berber, H., Ucar, E. and Sahinturk, U. (2017). Synthesis and properties of waterborne few-layer graphene oxide/ poly(MMA-Co-BuA) nanocomposites by *in situ* emulsion polymerization. *Colloids and Surfaces A: Physicochemical and Engineering Aspects*, 531: 56–66.

Besednjak, A. (2005). Materiales compuestos: Procesos de fabricación de embarcaciones. España: Ediciones UPC.

Beyer, M., Reichert, J., Sigusch, B. W., Watts, D. C. and Jandt, K. D. (2012). Morphology and structure of polymer layers protecting dental enamel against erosion. *Dental Materials*, 28: 1089–1097.

Bhanvase, B. A. and Sonawane, S. H. (2014). Ultrasound assisted *in situ* emulsion polymerization for polymer nanocomposite: a review. *Chemical Engineering and Processing: Process Intensification*, 85: 86–107.

Biji, A., Cojazzi, G., Panzavolta, S., Rubini, K. and Roveri, N. (2001). Mechanical and thermal properties of gelatin films at different degrees of glutaraldehyde crosslinking. *Biomaterials*, 22(8): 763–768.

Bose, R. J., Paulmurugan, R., Moon, J., Lee, S. H. and Park, H. (2018). Cell membrane-coated nanocarriers: the emerging targeted delivery system for cancer theranostics. *Drug Discovery Today*, 23: 891–899.

Bose, R. J., Kim, M., Chang, J. H., Paulmurugan, R., Moon, J. J., Koh, W. G., Lee, S. H. and Park, H. (2019). Biodegradable polymers for modern vaccine development. *Journal of Industrial and Engineering Chemistry*, 77: 12–24.

Boutinguiza, M., Fernández-Arias, J., Del val, J., Buxadera-Palomero, J., Rodriguez, D., Lusquiños F., Gil, F. and Pou, J. (2018). Synthesis and deposition of silver nanoparticles on cp Ti by laser ablation in open air for antibacterial effect in dental implants. *Materials Letters*, 231: 126–129.

Buhin, Z., Blagojevic, S. L. and Leskovac, M. (2013). *In situ* emulsion polymerization and characterization of poly(butyl acrylate-co-methyl methacrylate)/silica nanosystems. *Polymer Engineering and Science*, 1–7.

Callister, W. (2003). Ciencia e Ingeniería de los Materiales. Barcelona: Reverté S. A.

Cama, G., Mogosanu, D. E., Houben, A. and Dubruel, P. (2017). Synthetic biodegradable medical polyesters: Poly-σ-caprolactone. *Science and Principles of Biodegradable and Bioresorbable Medical Polymers: Materials and Properties*, 79–105.

Chandra, S., Ashek-E-Khoda, S., Sayeed, Md., Suruzzaman, Paul, D., Aninda, S. and Grammatikos, S. (2020). On the use of wood charcoal filler to improve the properties of natural fiber reinforced polymer composites. *Materials Today: Proceedings*, 44: 926–929.

Choudhary, S. and Sengwa, R. (2018). ZnO nanoparticles dispersed PVA–PVP blend matrix based high performance flexible nanodielectrics for multifunctional microelectronic devices. *Current Applied Physics*, 18: 1041–1058.

Coenen, A. M. J., Bernaerts, K. V., Harings, J. A. W., Jockenhoevel, S. and Ghazanfari, S. (2018). Elastic materials for tissue engineering applications: Natural, synthetic, and hybrid polymers. *Acta Biomaterialia*, 79: 60–82.

Córdova-Cisneros, K., Castañeda-Facio, A. and Sáenz-Galindo, A. (2018). Obtención de nanocompuestos con nanopartículas de TiO_2 y Ag incorporadas en una matriz de PVA/PVP. (Tesis de maestría). Universidad Autónoma de Coahuila.

Covarrubias, C., Farias, L., Pérez, N. and Hernández, E. (2013). Nanocompuestos a Base de Polímeros Dispersos y Nanofibras de Carbono. *Revista Iberoamericana de Polímeros*, 14: 108–116.

Crucho, C. I. C. and Barros, M. T. (2017). Polymeric nanoparticles: A study on the preparation variables and characterization methods. *Materials Science and Engineering C*, 80: 771–784.

Dastjerdi, Z., Cranston, E. D. and Dubé, M. A. (2017). Synthesis of poly(n-butyl acrylate/methyl methacrylate)/CNC latex nanocomposites via *in situ* emulsion polymerization. *Macromolecular Reaction Engineering,* 11: 1–8.

Deshmukh, R. D. and Composto, R. J. (2007). Surface segregation and formation of silver nanoparticles created *in situ* in poly(methyl methacrylate) films. *Chemistry of Materials,* 19: 745–54.

Donescu, D., Nistor, C. L., Purcar, V., Petcu, C., Serban, S., Corobea, M. C. and Ghiurea, M. (2009). Formation and dissolution of silver nanoparticles. *Optoelectronics and Advanced Materials, Rapid Communications,* 3: 44–48.

Eisa, W., Abdel, T., Mohamed, E. and Mahrous, S. (2017). Crosslinked PVA/PVP supported silver nanoparticles: a reusable and efficient heterogeneous catalyst for the 4-nitrophenol degradation. *J. Inorg. Organomet. Polym*, 27: 1703–1711.

El-Say, K. M. and El-Sawy, H. S. (2017). Polymeric nanoparticles: Promising platform for drug delivery. *International Journal of Pharmaceutics*, 528: 675–691.

Espinosa, K. G. and Castañeda, A. (2019). Síntesis de nanopartículas de plata mediante el uso de extracto de *Geranium* (*Pelargonium* spp.) y su incorporación en las mezclas de PVA/Quitosano para la obtención de películas. Universidad Autónoma de Coahila, Tesis de Maestría.

Espinosa, K. G., Sáenz, A. and Castañeda, A. (2022). Bio-síntesis de nanopartículas de plata mediante el extracto de *Geranium* (*Pelargonium* spp.). *Afinidad*, 79: 595. Accepted article.

Faridi-Majidi, R., Nezafati, N., Pazouki, M. and Hesaraki, S. (2017). The effect of synthesis parameters on morphology and diameter of electrospun hydroxyapatite nanofibers. *Journal of the Australian Ceramic Society*, 53: 225–233.

García, M. C. (2018). Drug delivery systems based on nonimmunogenic biopolymers. *Engineering of Biomaterials for Drug Delivery Systems: Beyond Polyethylene Glycol*, 317–344.

Guo, Z., Poot, A. and Grijpma, D. (2021). Advanced polymer-based composites and structures for biomedical applications. *European Polymer Journal*, 149: 110388.

Herk van, A. M. (2010). Historical overview of (mini) emulsion polymerizations and preparation of hybrid latex particles. pp. 1–52. *In*: Herk van A. M. and Landfester, K. (eds.). *Advanced in Polymer Science: Hybrid Latex Particles.* Germany: Springer.

Huang, M., Hou, Y., Li, Y., Wang, D. and Zhang, L. (2017). High performances of dual network PVA hydrogel modified by PVP using borax as the structure-forming accelerator. *Designed Monomers and Polymers*, 20: 505–513.

Huang, X. and Voit, B. (2013). Progress on multi-compartment polymeric capsules. *Polymer Chemistry*, 4: 435–443.

Hübner, Ch., Fettkenhauer, Ch., Voges, K. and Lupascu, D. C. (2018). *Agglomeration-Free Preparation of Modified Silica Nanoparticles for Emulsion Polymerization - A Well Scalable Process*. 34: 376–83.

Jenke, D. R. (2014). Extractables and leachables considerations for prefilled syringes. *Expert Opinion on Drug Delivery,* 11: 1591–1600.

Kang, T. G., Park, J. K., Kim, B. H., Lee, J. J., Choi, H. H., Lee, H. J. and Yook, J. G. (2019). Microwave characterization of conducting polymer PEDOT:PSS film using a microstrip line for humidity sensor application. *Measurement: Journal of the International Measurement Confederation*, 137: 272–277.

Kassem, M. A., El-Sawy, H. S., Abd-Allah, F. I., Abdelghany, T. M. and El-Say, K. M. (2017). Maximizing the therapeutic efficacy of imatinib mesylate–loaded niosomes on human colon adenocarcinoma using box-behnken design. *Journal of Pharmaceutical Sciences*, 106: 111–122.

Khezrian, A. and Shahbazi, Y. (2018). Application of nanocomposite chitosan and carboxymethyl cellulose films containing natural presercative compounds in minced camel's meat. *International Journal of Biological Macromolecules,* 106: 1146–1158.

Kobashigawa, J., Robles, C., Martínez, M. and Crmarán, C. (2019). Influence of strong bases on the synthesis of silver nanoparticles (AgNPs) using the ligninolytic fungi Trametes trogii. *Sudi Journal of Biological Science*. 26: 1331–1337.

Kumar, S., Krishnakumar, B., Sobral, A. and Koh, J. (2019). Bio-based (Chitosan/PVA/ZnO) nanocomposites film: thermally stable and photoluminescence material for removal of organic dye. *Carbohydrate Polymers*, 1: 559–564.

LeBaron, P. C., Wang, Z. and Pinnavaia, T. J. (1999). Polymer-layered silicate nanocomposites: an overview. *Applied Clay Science*, 15: 11–29.

Lee, K.-S., Andraud, C., Tamada, K., Sokolov, K., Kotz, K. T. and Zheng, G. (2016). Feature issue introduction: biophotonic materials and applications. *Optical Materials Express*, 6: 1747.

Li, F., Gao, Y., Zhang, C., Jin, J., ji, X., Zhang, Y., Zhang, X. and Jiang, W. (2020). Design of high impact thermal plastic polymer composites with balanced toughness and rigidity: Effect of matrix polymer molecular weight, *Polymer*, 208: 122957.

Li, X., Xing, R., Xu, Ch., Liu, S., Qin, Y., Li, K., Yu, H. and Li, P. (2021). Immunostimulatory effect of chitosan and quaternary chitosan: a review of potential vaccine adjuvants. *Carbohydrate Polymers*, 264: 118050.

Liu, W., Xie, J., Li, L., Xue, B., Li, X., Gan, J., Shao, Z. and Sun, T. (2021). Properties of phenolic acid-chitosan composite films and preservatice effect on *Panaeus vannamei*. *Journal of Molecular Structure,* 1239: 130531.

Liu, X., Chen, Q., Lv, L., Feng, X. and Meng, X. (2015). Preparation of transparent PVA/TiO_2 nanocomposite films with enhanced visible-light photocatalytic activity. *Catalysis Communications*, 58: 30–33.

Maitz, M. F. (2015). Applications of synthetic polymers in clinical medicine. *Biosurface and Biotribology*, 1: 161–176.

Mamaghani, M. Y., Pishvaei, M. and Kaffashi, B. (2011). Synthesis of latex based antibacterial acrylate polymer/nanosilver via *in situ* miniemulsion polymerization. *Macromolecular Research,* 19: 243–49.

Mayandi, K., Rajini, N., Ayrilmis, N., Indira, M. P., Siengchin, S., Mohammad, F. and Al-Lohedan, H. (2020). An overview of endurance and ageing performance under various environmental conditions of hybrid polymer composites. *Journal of Materials Research and Technology*, 9: 15962–15988.

Min, T. H. and Choi, H. J. (2017). Synthesis of poly(methyl methacrylate)/graphene oxide nanocomposite particles via pickering emulsion polymerization and their viscous response under an electric field. *Macromolecular Research,* 25: 565–71.

Mohammed, A. (2020). Thermoplastic composite system using polymer blend and fillers. *Journal of King Saud University-Engineering Science*.

Moreno-Vásquez, M. J., Plascencia-Jatomea, M., Sánchez-Valdes, S., Castillo-Yáñez, F. J., Ocaño-Higuera, V. M., Rodríguez-Félix, F., Rosas-Burgos, E. C. and Graciano-Verdugo, A. Z. (2017). Preparation and characterization of films made of poly(l-lactic acid)/poly(l-lactic acid) grafted maleic anhydride/epigallocatechin gallate blends for antibacterial food packaging. *Journal of Plastic Film and Sheeting*, 33: 10–34.

Muzalev, P. A., Kosobudskii, I. D., Kul'Batskii, D. M. and Ushakov, N. M. (2012). Polymer composites based on polymethylmethacrylate with silver nanoparticles, synthesis and optical properties. *Inorganic Materials: Applied Research,* 3: 40–43.

Nagaraja, A., Jalageri, M. D., Puttaiahgowda, Y. M., Raghava Reddy, K. and Raghu, A. V. (2019). A review on various maleic anhydride antimicrobial polymers. *Journal of Microbiological Methods*, 163: 105650.

Ovando, V. M. (2007). Estudio Téorico y Experimental de La Copolimerización de Acetato de Vinilo y Acrilato de Butilo En Microemulsiones Estabilizadas Anionica y Estericamente En Procesos Por Lotes y Semicontinuos. Centro de investigación en química aplicada.

Paipitak, K., Pornpra, T., Mongkontalang, P., Techitdheer, W. and Pecharapa, W. (2011). Characterization of PVA-chitosan nanofibers prepared by electrospinning. *Procedia Engineering*, 8: 101–105.

Patil, M. and Kim, G. (2016). Eco-Friendly approach for nanoparticles synthesis and mechanism behind antibacterial activity of silver and anticancer activity of gold nanoparticles. *Applied Microbiology and Biotechnology*, 101: 79–92.

Pérez, B. T. M., Farías, L. C., Ovando, V. M. M., Asua, J. M., Rosales, L. M. and Tomovska, R. (2017). Miniemulsion copolymerization of (meth)acrylates in the presence of functionalized multiwalled carbon nanotubes for reinforced coating applications. *Beilstein Journal of Nanotechnology,* 8: 1328–37.

Piao, Y., You, H., Xu, T., Bei, H.-P., Piwko, I. Z., Kwan, Y. Y. and Zhao, X. (2021). Biomedical applications of gelatin methacryloyl hydrogels. *Engineered Regeneration*, 2: 47–56.

Ragaert, P., Buntinx, M., Maes, C., Vanheusden, C., Peeters, R., Wang, S., D'hooge, D. R. and Cardon, L. (2019). Polyhydroxyalkanoates for food packaging applications. *Reference Module in Food Science*, 153–177.

Raveendran, S., Rochani, A. K., Maekawa, T. and Kumar, D. S. (2017). Smart carriers and nanohealers: A nanomedical insight on natural polymers. *Materials*, 10. 929.

Rios, L. A., Ocampo, D., Franco, A., Cardona, J. F. and Cardeño, F. (2013). Efecto de Surfactantes Polimerizables En La Distribución de Tamaño de Partícula, PH, Viscosidad, Contenidos de Sólidos y de Monómero Residual de Una Resina Estireno-butilacrilato. *Polimeros Ciência e Tecnologia,* 23: 352–357.

Rokaya, D., Srimaneepong, V., Sapkota, J., Qin, J., Siraleartmukul, K. and Siriwongrungson, V. (2018). Polymeric materials and films in dentistry: An overview. *Journal of Advanced Research*, 14: 25–34.

Shabahang, S., Kim, S. and Yun, S. H. (2018). Light-guiding biomaterials for biomedical applications. *Advanced Functional Materials*, 28: 1–17.

Shan, D., Gerhard, E., Zhang, C., Tierney, J. W., Xie, D., Liu, Z. and Yang, J. (2018). Polymeric biomaterials for biophotonic applications. *Bioactive Materials*, 3: 434–445.

Shankar, S., Rhim, J. W. and Won, K. (2018). Preparation of poly(lactide)/lignin/silver nanoparticles composite films with UV light barrier and antibacterial properties. *International Journal of Biological Macromolecules*, 107: 1724–1731.

Shaoquiang, C., Hongyang, Z., Yuanquing, C., Jianling, Z., Haigang, Y., Long, J. and Yi, D. (2017). Investigation of polypirrole/polyvinyl alcohol-titanium dioxide composite films for photo-catalytic applications. *Applied Surface Science*, 342: 55–63.

Sharma, R., Agrawal, U., Mody, N. and Vyas, S. P. (2015). Polymer nanotechnology based approaches in mucosal vaccine delivery: Challenges and opportunities. *Biotechnology Advances*, 33: 64–79.

Slistan, A., Herrera, R., Rivas, J., Ávalos, M., Castillón, F. and Posada, A. (2007). Synthesis of silver nanoparticles in a polyvinylpyrrolidone (PVP) paste, and their optical properties in a film and in ethylene glycol. *Materials Research Bulletin*, 43: 90–96.

Song, R., Murphy, M., Li, C., Ting, K., Soo, C. and Zheng, Z. (2018). Current development of biodegradable polymeric materials for biomedical applications. *Drug Design, Development and Therapy*, 12: 3117–3145.

Sudell, B. and Evans, W. (2005). Natural fiber composites in automotive. pp. 1–36. *In*: Mohanty, A., Misra, M., Drzal, L. and Selke, S. (eds.). *Natural Fibers, Biopolymers and Biocomposites*. London: CRC Press.

Tawfik, A. S., Nagaraj, P. S., Mahesh, M. S., Kakarla, R. R. and Tejraj, M. A. (2020). Recent trends in functionalized nanoparticles loaded polymeric composites: An energy application. *Materials Science for Energy Technologies*, 3: 515–525.

Toh, H. W., Yee, D. W., Koon, J. C., Ow, V., Lu, S., Tan, L. P., En Hou, P., Venkatraman, S., Huang, Y. and Ying, H. (2021). Polymer blends and polymer composites for cardiovascular implants. *European Polymer Journal*, 146: 110249.

Varghese, S. A., Pulikkalparambil, H., Rangappa, S. M., Siengchin, S. and Parameswaranpillai, J. (2020). Novel biodegradable polymer films based on poly(3-hydroxybutyrate-co-3-hydroxyvalerate) and Ceiba pentandra natural fibers for packaging applications. *Food Packaging and Shelf Life*, 25: 100538.

Vera, D., Jolan, I., Weber, K., Dana, C. M. and Jürgen, P. (2016). *In situ* hydrazine reduced silver colloids synthesis-enhancing SERS reproducibility. *Analytica Chimica Acta*, 946: 73–79.

Vera, P. F., Castañeda, A. O. and Farías. L. (2019). Síntesis de Nanocompuestos de Polimetilmetacrilato-Co-Butilacrilato y Nanopatículas de Plata Mediante Métodos Sustentables. Universidad Autónoma de Coahuila.

Wael, E., Andel-Baset, T., Mahamed, W. and Mahrous, S. (2017). Crosslinked PVA/PVP supported silver nanoparticles: a reusable and efficient heterogeneous catalyst for the 4-Nitrophenol degradation. *Journal Inorganic Organomentallics Polymers,* 27: 1703–1711.

Wang, S., Zhang, X., Jiang, C., Jiang, H. and Qiao, J. (2020). Facile preparation of Janus polymer film and application in alleviating water crisis. *Materials Chemistry and Physics*, 240: 122256.

Wong, W. W. H., Rudd, S., Ostrikov, K., Ramiasa-MacGregor, M., Subbiah, J. and Vasilev, K. (2016). Plasma deposition of organic polymer films for solar cell applications. *Organic Electronics*, 32: 78–82.

Xu, G. C., Xiong, J. Y., Ji, X. L. and Wang, Y. L. (2007). Synthesis of nanosilver/PMMA composites via ultrasonically bi-*in situ* emulsion polymerization. *Journal of Thermoplastic Composite Materials,* 20: 523–33.

Yanat, M. and Schroën, K. (2021). Preparation methods and applications of chitosan nanoparticles; with an outlook toward reinforcement of biodegradable packaging. *Reactive and Functional Polymers*, 161: 104849.

Yang, W., Ding, H., Qi, G., Li, Ch., Xu, P., Zheng, T., Zhu, X., Kenny, J., Puglia, D. and Ma, P. (2021). Highly transparent PVA/nanolignin composite films with excellent UV shielding, anticabterial and andioxidant performance. *Reactive and Functional Polymers*, 162: 104873.

Yu, Y. Y., Chien, W. C., Ko, Y. H., Chen, C. P. and Chang, C. (2015). Preparation of conjugated polymer-based composite thin film for application in solar cell. *Thin Solid Films*, 584: 363–368.

Yuan, S., Shen, F., Kai, Ch. and Zhou, K. (2019). Polymeric composites for powder-based additive manufacturing: Materials and applications. *Progress in Polymeric Science*, 91: 141–168.

Zhang, X., Tan, B. H. and Li, Z. (2018). Biodegradable polyester shape memory polymers: Recent advances in design, material properties and applications. *Materials Science and Engineering C*, 92: 1061–1074.

Zhenqian, Z., Sihler, S. and Ziener, U. (2017). Alizarin Yellow R (AYR) as compatible stabilizer for miniemulsion polymerization. *Journal of Colloid and Interface Science,* 507: 337–43.

Zindani, D. and Kumar, K. (2019). An insight additive manufacturing of fiber reinforced polymer composite. *International Journal of Lightweight Materials and Manufacture*, 2: 267–278.

Chapter 9

Composites and Novel Applications in the Biomedical Field

Claudia Gabriela Cuellar Gaona,[1] *Rosa Idalia Narro Céspedes,*[1,*] *Ricardo Reyna, Martínez,*[1] *Víctor Adán Cepeda Tovar,*[1] *Karina Reyes Acosta*[2] and *Aidé Sáenz Galindo*[1]

I. Introduction

The need for composite materials with improved and tailor-made properties is increasing. Due to their versatility, these composite materials have a wide range of applications. They consist of at least a mixture of two different materials, which can be polymers, ceramics and metals (Mano et al., 2004).

The traditional definition of composite material is that it comprises two or more components whose properties are superior to those of the individual materials. These materials all remain perfectly identifiable in the composite and are made up of at least two phases: continuous and dispersed. The continuous phase is responsible for filling the volume and transferring the charges to the dispersed phase. The dispersed phase is usually reinforcing and is responsible for improving one or more properties of the composite material. Most composite materials are primarily aimed at improving mechanical properties, such as stiffness and strength. However, other properties are required from these materials and are of great interest, such as biocompatibility, biodegradability, biointegration, transport properties (electrical or thermal) or density (Mano et al., 2004). This is because these composite materials can replace and/or repair, as well as cause the regeneration of parts of a living organism and improve the health of the organism.

In the current scenario of the 21st century, plastic waste pollution has reached unprecedented levels, mainly caused by non-biodegradable and single-use materials. This problem needs to implemented in research and initiatives associated with waste material pollution and disposal (Ammala et al., 2011; Ojeda et al., 2009). An important alternative that can solve this problem is the development of biodegradable composite materials. Biodegradability is a critical property in producing these materials since most of the materials labeled as non-biodegradable take a long time

[1] Polymers Department, School of Chemistry, Autonomous University of Coahuila, Postal Code: 25280, Saltillo, Coahuila, Mexico.
[2] Chemical Engineering Department, School of Chemistry, Autonomous University of Coahuila, Postal Code: 25280, Saltillo, Coahuila, Mexico.
* Corresponding author: rinarro@uadec.edu.mx

Table 9.1. Lifetime of materials takes to decompose (Satyanarayana et al., 2009).*

Material	Time to degrade in the environment
1. *Glass and tires*	Uncertain time
2. *Plastic*	450 years
3. *Aluminum can*	200–500 years
4. *Tin can, Rubber boot sole*	50–80 years
5. *Painted wood*	13 years
6. *Natural composites* a) *Wool stocking* b) *Bamboo stick* c) *Cotton, Banana peel, Orange peel* d) *Polycaprolactone-Grating – Maleic Anhydride/starch, Polycaprolactone-starch* e) *Waste Gelatin/Polyvinyl Alcohol, Waste Gelatin/Sugar Cane Bagasse, Conventional copy paper*	 1 year 1–3 year 1–5 year 2 months 1 month
7. *Cotton*	1–5 months
8. *PCL-g-MAH/Starch*	2 months
9. *PCL-Starch*	2 months
10. *WG/PVA*	1 month
11. *WG/SCB*	1 month
12. *WG/WG/SCB/Film*	1 month
13. *PHB-PHB/Starch*	1 month

* *Part reused with permission from Elsevier, License number 5081520439570.*

to decompose (Table 9.1). In general, composite materials can be divided into two main branches: those that are biodegradable and those that are non-biodegradable.

In recent years, considerable efforts have been made to develop composite materials that can replace non-biodegradable synthetic composites. This growing global awareness has led to sustainable and biodegradable materials as an alternative to synthetic composites. Consequently, a great demand for biodegradable alternatives has emerged in different fields and industries to achieve a more environmentally friendly approach to manufacturing and designing composite materials (Rendón-Villalobos et al., 2016; Satyanarayana et al., 2009). Increasingly, degradable alternatives that offer the same functionality as synthetic composites for the required lifetime and at a competitive cost have become essential.

Despite the challenges faced, several industries and areas of science, such as biomedicine, packaging, automation and consumer goods, have been encouraged to apply biodegradable alternatives in their products, taking the necessary preliminary steps towards the commercialization of biodegradable composites.

Products made from these materials are designed to biodegrade at the end of their useful life, thus achieving ecological and sustainable benefits. Current research efforts are directed toward developing composite products that are environmentally friendly and perform better (Zhang et al., 2020).

It is worth noting that biodegradable composites can complement and eventually replace petroleum-based or traditional composites in various applications. The primary motivation for the development of biodegradable composites has been and continues to be to create a new generation of reinforced composite materials that are environmentally compatible in terms of products, use, renewability and biodegradability (Chopparapu et al., 2020; Dziadek et al., 2017).

There is immense opportunity to develop new bio-based products, but the real challenge is to design suitable bio-based products through innovative ideas. Green composites are the future; there are solid prospects for the future, despite the current low production level (Chopparapu et al., 2020; Fernández-Montero et al., 2020; Satyanarayana et al., 2009).

This chapter, therefore, focuses on new research in the field of biodegradable composite materials and their applications in the field of medicine.

2. Importance and classification of composites

The importance of composites is because by joining two materials together, it is possible to achieve the combination of properties that cannot be obtained in the original materials. These properties depend on factors such as the reinforcement content and orientation, the properties of matrix and reinforcement and the production method.

These materials have specific common properties that are mainly linked to their composite nature and the presence of a particular reinforcement: These include high mechanical properties, resistance to corrosion and oxidation, low density, anisotropy depending on the type of reinforcement, physical and mechanical characteristics depending on the constituents and the proportions of these in the composite, possibility of realizing complex geometries, good electrical and dielectric properties among others.

Composites, in general, can be classified in different ways depending on the shape of the reinforcement:

- *Particle-reinforced.* Whose phase immersed in the matrix is composed of particles with shapes approximate to spherical. Particle-reinforced composites are subdivided into large particle-reinforced and dispersion-hardened.
- *Fiber-reinforced.* Fibers are elements in which one dimension is significantly larger than the other two. Within the composite, these reinforcing fibers can be oriented continuously or discontinuously, randomly or aligned. The primary fibers used as reinforcements are glass fibers, carbon fibers, boron fibers, ceramic fibers, metallic fibers, aramid fibers, natural fibers: sisal, hemp, linen. Fibers, in general, can be presented in yarns, mats, tapes or fabrics (Kraus and Trappe, 2021).
- *Structurally reinforced.* A structural composite material is made up of both composite and homogeneous materials, and its properties depend not only on its constituent materials but also on the geometry of the design of the structural elements that make it up. They are classified as laminar, non-laminar or sandwich panels.

Another classification is according to the type of matrix

- *Ceramic matrix composites.* These materials use alumina, hydroxyapatite, CSi, etc. They are essential because they resist high temperatures, their main disadvantage is their brittleness and low resistance to thermal shocks.
- *Metal matrix composites.* These materials are mainly alloying of aluminum, titanium, magnesium, and others. They have high thermal and electrical conductivity, as well as more outstanding durability and wear resistance. Among their advantages is that they do not absorb moisture and their main disadvantage is their high price.
- *Polymer matrix composites.* These materials have low density and parts with very complicated geometries can be obtained. They are currently the most widely used due to their low cost and versatility, and their use has spread to all areas of science. The most critical disadvantage is their fire resistance. These composites are further divided according to the type of organic matrix they use.
  a) *Thermosets*: These materials have a cross-linked structure, forming a network based on covalent bonds. Their main characteristic is that once they have been molded and cooled, they cannot be reprocessed by applying heat or pressure. Their general properties are high thermal stability, rigidity and hardness; they are insoluble and insulating thermally

and electrically. Examples include unsaturated polyesters, epoxy resins, vinyl ester resins, phenols, etc.

b) *Thermoplastics*: Thermoplastic materials are those that, in general, at high temperatures become deformable and melt, so they can acquire the desired shape because when they cool down, they harden, not undergoing chemical changes in the process. Examples are polypropylene, polyamides, polycarbonates, saturated polyesters, polysulfones, to mention a few (Siew et al., 2021).

Finally, composites can also be divided into two main branches: biodegradable and non-biodegradable This study is focused on biodegradable composites and as part of these, biomaterials composites as they are the materials of the future will be discussed.

3. Polymeric, ceramic, metallic matrix

3.1 Biocomposites overview

Nowadays, various composite materials are presented that arise from sources of diverse nature, being those obtained from synthetic sources shown with greater abundance worldwide. An important point about these is their low degradability, which has caused a vast consumption that has led to large batches of disposable products, which are not conductively good in the environment.,On the contrary, they pollute non-renewable resources (soil, water, etc.). An initiative to mitigate this significant problem is using resources that nature itself supplies to man to be taken advantage of and used as different substances. These biomaterials do not have the mechanical, physical and chemical properties they provide obtained from a synthetic or semi-synthetic source. Therefore, techniques have been implemented to combine these materials creating new products that seek to satisfy the needs that society currently demands. The classification and main contributions to the synthesis and end-use (product) of composites designed from a natural or synthetic polymer source are investigated here. There is a growing need for composite materials with improved properties.

One of the areas where composite materials have the most significant impact is in the medical field. To meet the requirements of an application in this field, metals and ceramics can be combined with polymers. Metals have good strength and wear resistance for use in orthopedic implants, while ceramics have good corrosion resistance and good biocompatibility, such as calcium phosphates, which are widely used in the manufacture of bone implants (Brien, 2011; Langer, 2000; Tomba et al., 2014).

3.2 Biodegradable polymer composites, classification, advantages and disadvantages

The biodegradable polymeric compounds that are of interest in this study can be based on natural polymers or synthetic polymers. The material which is briefly described below:

- *Natural polymers*: Natural polymers exist such as biomolecules and compounds that form the body of living beings. They include most proteins, nucleic acids, polysaccharides (complex sugars such as plant cellulose, starch and fungal chitin), rubber, plant gums and polyhydroxyalkanoates.
- *Synthetic polymers*: These polymers can be created in the laboratory by joining specific monomers in a chain, using organic or inorganic inputs, under controlled conditions of temperature, pressure and the presence of catalysts. In this way, a chain or step-by-step reaction is generated, resulting in the generation of the compound. Examples of these polymers are polycaprolactone, polylactic acid, etc.

These composites are formed by the combination of the materials above: Natural and synthetic polymers form these materials. For example: starch and polylactic acid. They have advantages such as low production costs, improved mechanical properties and control of degradation. These materials show significant advantages over traditional plastics as they avoid dependence on petroleum as a raw material for their production. In addition, they degrade entirely in a relatively short time (compared to typical plastics), and their degradation by-products are generally carbon dioxide, water and biomass.

These biomaterials have a significant impact mainly in the area of medical research. Many medical advances and breakthroughs are due to the development of new biodegradable composites that are compatible with the human body and do not leave toxic residues in them. Once these materials have fulfilled their function, they disappear without the need for surgical intervention.

In addition to the medical area, polymer-based composites are available to meet a variety of human needs. These materials are currently being widely used in the automotive industry, in housing, mechanisms of household appliances and electrical devices, among many others. This wide variety of uses is due to their exceptional chemical, mechanical, optical and thermal properties, closely related to their chemical composition and structure. This is due to obtaining unique polymeric compounds, which achieve outstanding properties generally designed for concrete applications (Coreño-Alonso and Méndez-Bautista, 2010).

These extraordinary properties acquired by composites are caused by adding another material that, although in a discontinuous phase within the polymeric matrix, can cause the necessary property for a specific application, such as increased reinforcement or increased conductivity or some other desired characteristic.

However, these materials also have certain disadvantages compared to traditional polymers. One of the most important disadvantages is the higher production cost compared to petroleum-based polymers. However, this disadvantage has diminished over time, allowing the cost gap to become smaller and smaller. In addition, in most cases, biodegradable polymers have lower mechanical properties than typical composites, with lower physical strength.

3.3 Natural and synthetic polymeric biocomposites

Biopolymers naturals are mainly divided into two types: Polysaccharides and proteins. The former contains collagen, alginate, chitosan, dextran, chondroitin sulfate, etc., while proteins contain fibrin, Zeina keratin, and gelatin (Lett et al., 2021).

This variety of natural macromolecules, named biopolymers, can bind, for example, to hydroxyapatite, a ceramic material, because biopolymers naturals are more biodegradable and practical development in medical applications of hydroxyapatite. Although hydroxyapatite is not biodegradable, it meets the European Society for Biomaterials definition. This institution defined biomaterials as those "materials used to assess, cure, correct or replace any tissue, organ or function of the human body."

Polymeric biopolymers have something in common, as they possess qualities as non-toxic and highly biocompatible monomeric units. Biopolymers are commonly manufactured from biological waste, as it is available in abundance in living organisms (Giannuzzi et al., 2005). For example, hydroxyapatite plays an essential role in making it compatible with chitin and chitosan.

On the other hand, synthetic polymers are readily accessible, and most are immeasurably available. Polymers of synthetic origin such as polystyrene (PS) (Anju et al., 2020), Polyvinylpyrrolidone (PVP) (Nasouri et al., 2015), poly-sulfone (PSU) (Cojocaru et al., 2017), poly-caprolactone (PCL) (Khatti et al., 2019), Polyacid-Lactic Acid (PLA) (Naghieh et al., 2017; Xu et al., 2019), polyvinyl alcohol (PVA) (Dattola et al., 2019; Lett et al., 2016; Lett et al., 2020; Swain et al., 2015), polytetrafluoroethylene (PTFE) (Springer et al., 2001), polyethylene terephthalate (PET) (Neves et al., 2005), and other binding agents have been widely used in the synthesis of copolymers as a whole of hydroxyapatite.

These have something in common, as they possess qualities as non-toxic and highly biocompatible monomeric units. Biopolymers are commonly manufactured from biological waste, as it is available in abundance in living organisms (Giannuzzi et al., 2005). Hydroxyapatite plays an essential role in making it compatible with chitin and chitosan.

3.3.1 Natural polymers (Proteins)

Fibrin

Fibrin is one of the main compounds of the extracellular matrix that provides structural integrity and mechanical strength to the tissues of the body (Kim et al., 2017). One of the first fibrin-based products developed was fibrin sealant. Today it has a wide variety of products with adhesive properties. For example, it has been used in tissue sealing during plastic and reconstructive surgeries. It has also been used as a scaffold in adipose tissue, cardiac, muscle, nerve, bone, eye, respiratory, skin, tendons and ligaments. Fibrin has a porous morphology that makes it ideal as a scaffold for splicing and cell proliferation and differentiation, as well as growth factor delivery systems such as basic Fibroblast Growth Factor (bFGF). The only limitation is the rapid degradation rate and weak mechanical strength (Song et al., 2018). Kim et al. in 2017 made fibrin and collagen-based composite, which was in the form of hydrogel and was compressed. Compression helps stiffen the composite because an uncompressed fibrin network may be too soft to withstand high forces. The hydrogel exhibited better synergistic properties than isolated collagen and fibrin matrices; it can be used as a sealant, sponges and tissue scaffolds with adjustable mechanical properties (Kim et al., 2017).

Natural material Zein (protein) obtained from corn

Zein can be used as enzyme immobilizers or biosensors that incorporate biological material such as enzymes, micro-organisms, antibodies, proteins, etc., integrated into normally non-biodegradable carriers such as synthetic polymers used to measure biological or chemical parameters. They are mainly used to measure glucose and cholesterol levels in blood or nitrates, phosphates and ammonium in water and chlorpyrifos is one of the most widely used pesticides in the United States.

For example, a relevant investigation was carried out by Dong et al. (Dong et al., 2004) they demonstrated how the films obtained from zein could be used as biopolymeric structural units for the growth and proliferation of the human liver, based on investigations related to the fibroblasts of some rodents. Furthermore, several studies demonstrated the use of zein films as drug delivery matrices for bioactive molecules, such as antioxidants and antibiotics (Zhang et al., 2015), for either food packaging (Boyacı et al., 2019; Güçbilmez et al., 2007) or biomedical applications (El-Rashidy et al., 2018; Singh et al., 2010; Shukla and Cheryan, 2001). Although pure zein does not have congenital antimicrobial activity, it can achieve this characteristic by loading the structural unit with antibiotic molecules or specific nanoparticles (Babitha and Korrapati, 2017). On the other hand, Singh and colleagues (Singh et al., 2009) proposed a solution of zein combined with iodine, formulating a drug synthesized from a film applied to mitigate the action of pathogenic microorganisms when an infection occurs.

3.3.2 Natural polymers (Polysaccharides)

Collagen

Collagen plays an essential role in biological integrity as it provides vital support in the human body. Recently, attempts have been made to replace the natural collagen-based extracellular matrix by developing biomaterials that mimic its architecture and function as a cellular scaffold. Recombinant and animal-derived collagens are the most important biomaterials available for tissue engineering, drug delivery and cosmetic surgery. The main animal collagens used for biomedical applications include bovine or porcine skin and bovine or equine achilles tendons. Collagen materials used for skin repair have been used to treat ulcers, as skin substitutes from cell-seeded collagen, and have also been used to engineer patellar tendons in rabbits. Collagen has been used as a drug delivery vehicle for proteins, genes, plasmids and antibiotics (Song et al., 2018).

Some ceramics, such as hydroxyapatite (HA), calcium sulfate ($CaSO_4$), and calcium carbonate ($CaCO_3$), have been used in combination with collagen to make substrates and scaffolds for bone repair. The collagen matrix facilitates cell interaction and tissue formation while the ceramics provide mechanical strength and support bone regeneration (Guo et al., 2021). Antoniac et al. in 2021 fabricated a scaffold based on collagen with hydroxyapatite doped with magnesium (Mg), which obtained a 25% lower degradation rate concerning the composite without Mg, which is necessary *in vivo* since bone degradation is slow to give opportunity for the healing process and bone regeneration, so this collagen composite with ceramic is promising for bone regeneration application (Antoniac et al., 2021). On the other hand, Senra et al. (2020), made a similar composite of collagen with hydroxyapatite and carbonated hydroxyapatite, stating that both composites had the potential to be used as bone tissue grafts since they presented the formation of calcium phosphate.

Alginate

Alginate is commonly used to heal and regenerate human tissue. It has biodegradable properties; the dissolution rate of alginate can be controlled by oxidation and reduction of the molecular weight of alginate. It is used for tissue engineering and wound healing due to its ability to form a gel (Ahmad Raus et al., 2020). It is also used as a stabilizer, thickener, gelling agent, emulsifier and has excellent application as a controlled drug system. However, it has the disadvantage of being a porous matrix, very permeable and degradable, making it difficult to control the release of different substances, and it is not very resistant in acidic environments. Under these conditions, the alginate matrix will shrink. The use of other biopolymers in conjunction with alginate, among which hydrocolloids, proteins and starches, has been proposed to solve the limitations and improve porous and permeable matrixes (Ramdhan et al., 2020). Alginate is commonly used to create gels together with chitosan or other polymers. They interact ionically, and this allows them to maintain their stability, as the solubility of alginate at alkaline pH is obstructed by chitosan, which is unable to dissolve at pH above 6, and vice versa, the dissolution of chitosan at acidic pH is prevented by alginate which is insoluble at acidic pH (Ramdhan et al., 2020). Zhang et al. (2021) fabricated a sodium alginate/chitosan/zinc oxide composite hydrogel to corroborate its effectiveness when used as a wound-healing agent. The hydrogel showed good biocompatibility properties against blood cells and antimicrobial activity against *Escherichia coli*, *Staphylococcus aureus*, *Candida albicans* and *Bacillus subtilis*. These characteristics make it a promising composite for wound healing.

Chitosan

Chitin is an ordinary natural polymer that originates mainly from fungi and insect shells; conversely, chitosan belongs to a linear polysaccharide secreted from chitin (Sultankulov et al., 2019). The deacetylation of chitin through chemical or enzymatic hydrolysis also results in the development of chitosan (Ghormade et al., 2017). The properties of chitosan can be adjusted by altering the degree of deacetylation. The structure of chitosan allows the transport of nutrients for cell growth and regeneration of primary tissue cells. It is worth highlighting the immense capacity of these biopolymers for their use in nanomedicine and tissue engineering applications (Islam et al., 2020). However, the mechanical characteristics of chitosan nanostructures are incomparable to original human bones. Therefore, it is unable to bear the load when used in load-bearing conditions in implantable bone. Chitosan by itself does not have a property that makes it an osteoconductive material, (material that promotes the growth of damaged bone (osteoblasts), penetrating the cavity or surface of the inert material). Therefore, it is combined with biopolymers or numerous bioactive nano-ceramics such as silicon, zirconium and titanium oxides (SiO_2, ZrO_2, TiO_2), etc., that improve mechanical potency and achieve better structural reliability of chitosan-based products compounded with hydroxyapatite for bone tissue engineering applications (Arun Kumar et al., 2015; Chandra et al., 2017; Cheow and Hadinoto, 2011; Deepthi et al., 2016; Saravanan et al., 2016).

The combination of polymers and inorganic minerals has been used to develop new composites with properties like nacre, bone and tooth. Composites based on chitosan and calcium phosphate have been reported to achieve a high interaction between the bioactive calcium phosphate phase and chitosan, resulting in a rigid material. These composites have been used for bone regeneration, bioactive coatings and composites, controlled drug release and bioactive filler for occlusion of exposed dentinal tubules (Salama, 2021).

It is possible to appreciate the use of foams of a synthetic polymer in the presence of a compound extracted from a natural source. Such is the case of polyurethane foams with certain biocompatibility provided by chitosan. Polyurethanes are among the most crucial specialty polymer class.

They are produced by the polycondensation reaction of a diisocyanate compound in conjunction with a material containing hydroxyl groups such as a polyol (Chu et al., 1992; Javaid et al., 2019). This pre-polymer formulation commonly used to produce polyurethanes is used to manufacture small molds and elastomer synthesis (Hostettler et al., 2000).

These biomaterials are used in a variety of applications. Concerning the earlier case, biopolymers are the primary type of biomaterials. They can be used for implants, such as bone substitutes or bone fixation materials, to develop some dental fillings that can perform their function for a long time, etc. (Philip et al., 2007). In-depth research has been carried out related to natural polymers since they are widely used as matrices to manufacture nanofibrous films to design bandages or gauze for skin regeneration, etc. (Kossyvaki et al., 2020; Suarato et al., 2018). However, most of them are soluble in hydrophilic media, such as water; this leads to a distinctive feature that could limit prolonged administration in the supply of pharmaceuticals, etc. (Kimna et al., 2019).

Pectin

Pectin is another material with biocompatibility since, like the rest of those mentioned, it is a polymer derived from plant sources, qualified as a substance for food use according to the regulations registered by the United States FDA (FDA). It is the most abundant polysaccharide found in the cell wall of plants and is widely used as a gelling agent in the production of jams, sugary jellies, etc. (Mishra et al., 2012). Most biocomposites are obtained from renewable resources such as biomass; they have been investigated and suggested as possible substitutes for fossil materials (Jung et al., 2013; May, 1990).

Dextran

It is a polysaccharide that has been used in the biomedical area for imaging, including fluorescent dextran derivatives used as macrophage markers for fluorescence microscopy (Soni and Rodell, 2021). Other applications due to their biocompatibility, biodegradability and cost-effectiveness properties are in drug delivery. Dextran electrospun fibers have been used for cell growth as scaffolds in tissue engineering (Yang et al., 2019). Yang et al., 2019, fabricated a gelatin/dextran-malic anhydride composite with improved physical and biological properties, where tensile strength improved and degradation could be increased up to 40%; biocompatibility tests favored adhesion and natural proliferation, so the composite has potential for tissue engineering applications. On the other hand, Zheng et al., 2019 obtained polyvinyl alcohol (PVA)/dextran composite hydrogel to be applied in wound healing. The hydrogel structure was highly porous and had high tensile strength, with the ability to absorb fluids up to six times its weight; it reached almost complete healing in 10 days.

Hyaluronic acid

Hyaluronic acid is an essential polysaccharide that forms part of the extracellular matrix. Hyaluronic acid is widely used in the biomedical field due to its adaptability to chemical modification and numerous biological functions, including aiding dendritic cell maturation and T-cell proliferation, and is considered an inflammatory regulator in tissue repair (Soni and Rodell, 2021). Hyaluronic acid has been used in addition to compounds made of chitosan and collagen hydrogels, and the result was beneficial for wound repair, granulation tissues, cell migration and skin healing. Chuysinuan et al. in 2020 developed a hydrogel of chitosan and collagen and incorporated hyaluronic acid to evaluate its effect when added to the hydrogel and found that the hydrogel released the active substance faster due to the high solubility of hyaluronic acid. Meng et al. (2021) prepared a chitosan/alginate/hyaluronic acid-based wound dressing, which showed promise as a wound repair agent, as it accelerated blood clotting and significantly facilitated wound closure.

Agarose

Agarose is a biocompatible and biodegradable polysaccharide widely used for DNA separation, cell encapsulation and tissue regeneration (Shan et al., 2018). The disadvantage is that when used in scaffolds, there is a lack of cell recognition. Agarose composites have been formed with other natural polymers such as chitosan, collagen and gelatin that have cell recognition to overcome such a limitation. Among the agarose-based composites, agarose-chitosan composites have gained attention as scaffolds in tissue engineering due to their improved stability along with the ability for thermosensitive gelling and cell attachment (Siyashankari and Prabaharan, 2020).

3.3.3 Synthetic polymers

Polylactic acid (PLA)

Polylactic acid is also known as polylactide (Yang et al., 2021). A polyester with slow biodegradation as it is not as susceptible to microbial attack (Sharma et al., 2021), it has been used for wound treatment, orthopedic devices, dental applications, drug delivery and tissue engineering. Ceramics have been incorporated into polymeric matrices to increase the biocompatibility and bioactivity of materials and/or implants. One example is the use of silica, PLA/SiO_2 nanocomposites that have been reported to enhance PLA biodegradation (Dziadek et al., 2017). It can also be combined with carbonated calcium phosphates because they can neutralize acidity and maintain physiological pH, and for this reason, its primary use is in musculoskeletal tissue, where it is used for the fabrication of rigid orthoses (Pugliese et al., 2021).

Polyglycolic acid (PGA)

A biodegradable and non-toxic polymer is not favorable for cell binding and proliferation because it lacks specific signals for cell recognition. Therefore, it is necessary to add bioactive materials to improve the biological performance of this material (Cao and Kuboyama, 2010).

Poly (lactic acid co-glycolic acid)

PLGA is a polymer used in different biomedical applications such as sutures and anti-cancer drug delivery systems. Implants that are formed with PLGA achieve faster healing rates and increased bone growth (Alizadeh-Osgouei et al., 2019). It is typically used in conjunction with other materials such as ceramics or bioactive glass and is modified to enhance bone regeneration (Song et al., 2018). One of the key reasons for using bioactive glass is the possibility of enhancing a wide range of biological and chemical properties. The structure and chemistry of glasses can be adapted at a molecular level by varying the composition. Bioactive silicate glasses are the most investigated for biomedical applications, such as fillers for bone tissue engineering composites and have used PLGA as a modifier.

Polyethylene glycol (PEG)

PEG is a polymer with a highly hydrated structure. It has been used in controlled drug release to extend circulation time and improve drug efficacy under safety regimens (Soni and Rodell, 2021), as tissue engineering scaffolds (Shan et al., 2018).

Poly-caprolactone (PCL)

PCL has been used in conjunction with collagen, demonstrating that this composite improves the physical characteristics of the scaffold compared to pure collagen scaffolds. The combination of PCL with chitosan provides the biological affinity of chitosan, such as cell adhesion and proliferation, providing cell recognition sites and a porous structure while providing the physicochemical characteristics of PCL, such as improved mechanical properties (Abbasian et al., 2019).

3.3.4 Applications

Table 9.2 shows the main applications according to the type of polymeric composite.

Table 9.2. Some biodegradable polymeric composites and their final application.

	Polymers	**Alloy with ceramic, metal, or polymer**	**Applications**
Natural	Collagen	Hydroxyapatite Calcium sulfate Calcium carbonate	Bone regeneration (Guo et al., 2021) Scaffolds in tissue engineering
	Fibrin	Collagen	Sealant (Kim et al., 2017)
	Chitosan	Calcium phosphate Hydroxyapatite	Bone regeneration (Salama, 2021) Wound healing (Zhang et al., 2021)
	Alginate	Chitosan	Wound healing (Zheng et al., 2019)
	Dextran	Polyvinyl alcohol (PVA)	Drug release (Chuysinuan et al., 2020)
	Hyaluronic acid	Chitosan Alginate Collagen	Wound healing (Meng et al., 2021)
	Agarose	Chitosan	Scaffolds in tissue engineering (Siyashankari and Prabaharan, 2020)
Synthetics	Polylactic acid (PLA)	Calcium phosphates	Manufacture of rigid orthoses (Pugliese et al., 2021)
	Polyglycolic acid (PGA)	Bioactive glass	Cell recognition (Cao and Kuboyama, 2010)
	Poly (lactic-co-glycolic acid (PLGA)	Bioactive glass	Bone regeneration (Song et al., 2018)
	Polycaprolactone (PCL)	Chitosan	Cell proliferation (Abbasian et al., 2019)
	Poly (trimethylene terephthalate) (PTT)	Corn starch	Textiles (Ammala et al., 2011)
	Polybutylene adipate terephthalate (PBAT)	Sugarcane bagasse	Automotive parts (Ammala et al., 2011)
	Polybutylene succinate (PBS)	Coconut, sugarcane bagasse, sisal	Automotive parts (Song et al., 2018)

3.3.5 Possible combinations of materials to obtain biocomposites

One of the main focuses of polymeric biocomposites is their use in medicine or other scientific fields.

Polymer/natural fiber and mineral

Natural fibers are considered an alternative renewable source to synthetic fiber with specific comparable properties (May Pat et al., 2013). According to their origin, these fibers can be classified into plants, animals and minerals (Pecas et al., 2019). Plant-based ones consist primarily of cellulose, hemicellulose, lignin, pectin, waxes and water-soluble components (Mohanty et al., 2001). In contrast, what could be got from plant-based ones from various sources, including agricultural wastes, currently imposing a growing problem on the farming industry worldwide (Sabiiti, 2011). Examples of such agricultural waste sources can include fresh and dehydrated fruits, rice straw, pineapple waste, sugarcane bagasse, rice husk and coffee (Dungani et al., 2016). Although agricultural residues are the most abundant form, only 10% of these come from agriculture. In short, waste is used as an alternative raw material for applications such as composites in the automotive and biomedical industries (Dorée, 1947). Composite materials are formed or structured by certain substances that act as reinforcement of the matrix, whether based on polymer, ceramic or metal; These reinforcements can be generalized as fibers, particles or nanoparticles; to seek an improvement in the physical-mechanical properties of the material (Shubhra et al., 2013). Composites can be made of synthetic or natural materials (Koronis and Silva, 2018). Biobased materials, such as natural fibers rather than synthetic fibers, can provide a more sustainable option for these composite materials (Nirmal et al., 2015). Besides, they are widely used to produce composite materials due to their lower density (El-Shekeil et al., 2012) and low cost, and higher flexural strength (Petchwattana and Covavisaruch, 2014).

Polymer/polymer

Recently, concern has increased about the use of polymeric implants with specific bio-absorbable properties in tissue engineering. This has drawn attention to numerous interests in research on materials with absorption capacity (Yang et al., 2006). Compared with metal matrix implants, those made from bioabsorbable polymers have advantages in tissue healing since they produce good presence in biocompatibility and bio-absorption in the human body. During the healing stage, these implants can undergo a degradation process due to the passage of time (Santos et al., 2019). Among the variety of existing polymers, synthetic polymers are superior to natural polymers due to their well-controlled degradation rate and excellent mechanical properties. The everyday use of synthetic polymers for fixation devices under degradability conditions is polylactic acid (PLA) and polyglycolide acid (PGA), in addition to all the numerous copolymer blends (Gentile et al., 2014; Lin et al., 2002; Liu et al., 2006). However, there is a precedent for the degradation products of these poly-lactone-type polymers, as they contain acidic matter that causes inflammation and damage to bone health and prevents the growth of new bone material (Liu et al., 2006).

To reduce the acidity caused by the degradation of PLA/PGA, it has been reported in various investigations to include a filler of mineral origin of the tricalcium phosphate type in the compound, to act as an alkaline component that neutralizes the PLA/PGA degradation products (Chen et al., 2019; Niemela, 2005; Park et al., 2017; Yang et al., 2006).

For example, blends of PLA with soft, biodegradable polyesters such as Poly (Butylene Adipate-co-Terephthalate) (PBAT) have been explored to extend their mechanical properties and make materials more flexible (Arruda et al., 2015; Lascano et al., 2019; Muthuraj et al., 2019; Xiang et al., 2020).

Ceramic biocomposites

One of the first constituents as a biocomposite is ceramics, combined with other agents and/or additives that produce better biocompatibility in these materials. Such is the case of hydroxyapatite. This material is considered the primary constituent of bone, which can act and appear as an ideal substance to substitute for a bone graft. In recent research, it has been used in dental and orthopedic surgery, and various ceramic mixtures have been used to fill bone defects and coat metal surfaces such as implants to improve their integration with bone. However, its applications in clinical tissue engineering have been limited due to its fragility, difficulty in modeling the implant and the new bone shape, as the porous network of hydroxyapatite does not provide support with the mechanical load necessary for the remodeling of the implant and host assembly (Lim et al., 2019; Wang et al., 2016).

Undoubtedly, hydroxyapatite has a high potential and has been widely used as a bone graft due to its inherent biological properties, including biocompatibility, bioaffinity, bioactivity, osteoconduction and osseointegration. There is no implantation toxicity as hydroxyapatite contains calcium (Ca^{+2}), phosphate (PO_4^{3-}), and hydroxyl ions (OH^-), which are naturally occurring elements in the human body (Jayakumar and Di Silvio, 2010; Lim et al., 2019; Liu et al., 2016).

The attachment of bone directly to an implant material is activated by forming the apatite layer, which is added before. An amorphous or hydroxyl ion-loaded calcium phosphate salt (hydroxylated apatite) is formed when the implant surface is immersed in Simulated Body Fluid (SBF). This mechanism is associated with developing the apatite layer due to partial dissolution of the hydroxyapatite and ionic exchanges between the SBF solution and the surface. Finally, applications of this material are found in numerous antibiotics, anticancer drugs and responses for genomic use (Ott, 2010; Parisi et al., 2020). In general, ceramic materials have been matched with sources of polymeric origin. Next how these mixtures have influenced the synthesis of ceramic-polymer matrix biocomposites are described.

Ceramic/polymer

In ceramic/polymer biocomposites, the copolymer additive has a good capacity for osteogenesis, osseointegration and regeneration of the bone tissue. The most common bone void filler used is hydroxyapatite, as it is the mineral composition of bone and has good biodegradability and osteoconduction, thus improving hydroxyapatite/PLA-PGA implants (Le Geros, 2002; Walton and Cotton, 2007). In these biocomposites, the study of the mechanical properties is an essential point to evaluate since the performance of these biomaterials allows a promising development in the whole field of biomedicine and a complete study of the biodegradation rate. These materials, according to the authors of the book, come in the form of screws, which are inserted into the bone tunnel or cavity, where interference between the bone tunnel and the screw can result in high shear and bending stresses acting on the core of the implants, which can lead to failure of the biomaterial. Factors affecting screw failure strength include the materials used and screw design. One of the most definitive studies was developed by Wendell et al. (Heard et al., 2013). They evaluated the mechanical properties of four commercially available bolts of different designs but found no significant differences in yield strength, stiffness or failure at maximum load. On the other hand, Weiler et al. (1998) tested different screws in six biodegradable products, measuring pull-out force and torque at failure. The results indicated that the stiffness of the fastener attachment had a relationship between the design and the torque at failure, all based on the design of the unit.

Polymeric/ceramic

Biocomposites developed from a polymeric matrix and ceramic filler are designed using polymerization techniques with specific aggregates in the form of nanoparticles (reinforcement), such as zinc oxide (ZnO). Biodegradable polymers, especially polyesters (PES), have become sustainable alternatives to standard petroleum-based polymers in a progressive number of demanding applications. A wide variety of products are already in the market today. PLA is probably one of the most environmentally friendly biodegradable polymers and has attracted the most research interest. However, some of the commercial applications of PLA are limited due to specific properties such as brittleness, slow crystallization rate and low viscosity. The alternatives for compound preparation and blending with other polymers have expanded so that these drawbacks have been mostly overcome (Arrieta et al., 2018; Echeverría et al., 2019; Peponi et al., 2018; Raquez et al., 2013; Skrlová et al., 2019; Sonseca et al., 2020).

Such biocomposites have been extensively studied in recent years. For example, the use of zinc oxide fillers within PLA matrix has gained significant interest from researchers (Jayaramudu et al., 2014) due to their non-toxicity, high availability, low cost and perfect chemical stability, and, the properties they can impart to the materials such as high ultraviolet absorption capacity (Reinosa et al., 2016; Sambandan and Ratner, 2011) and antimicrobial activity (Marra et al., 2016; Shankar et al., 2018). In addition, inorganic ZnO filler and other Zn-based compounds also have a catalytic effect on the degradation of polyester matrices through a chemical process of depolymerization and intermolecular transesterification, especially at high temperatures (Abe et al., 2004; Murariu et al., 2011). However, despite the improved degradation and other properties due to ZnO within these materials, which according to literature is considered potentially non-toxic, European regulations prevent its use in products intended to contact food or cosmetics (De Lucas et al., 2018).

4. Conclusions

The need to obtain composites with unique properties and according to the customer's needs for any type of application in all fields of science and mainly in the biomedical area, together with the imperative need for these types of materials to be biodegradable and not pollute the environment, has led researchers to develop more and more composites and biocomposites with extraordinary properties and tailored to the specific application sought, and which are also environmentally benign, because they are biodegradable. It was found that the area that has seen the most development in the use of these biomaterial composites is the biomedical area.

Therefore, biocomposites have undoubtedly acquired a prominent place among the new materials for modern medicine. Their considerable mechanical, chemical and physical possibilities give rise to a wide variety of applications to facilitate treatments for the desired therapeutic indications. These materials continue to advance thanks to the great versatility of the presentations they offer. However, one should not forget the need to control certain aspects when choosing the materials that will make up the biocomposites, such as biocompatibility, biodegradability, good insulation, adhesion to tissues and mechanical properties, among others.

This chapter provides a current overview of the materials used in biocomposites and the considerations that must be taken into account when choosing the components that will make up the material for the medical applications sought.

Acknowledgment

The National Council of Science and Technology (CONACYT) of Mexico is acknowledged for the grants (618041 and 486946) given to MSc. Claudia Gabriela Cuéllar Gaona and MSc. Ricardo Reyna Martínez and the Autonomous University of Coahuila for his doctoral studies, Science and Technology of Materials. The authors thank the research and postgraduate direction of the Autonomous University of Coahuila for supporting the project "Use of plasma technology in obtaining intelligent hydrogel materials (chitosan/citric acid/natural extract)".

References

Abbasian, M., Massoumi, B., Mohammad-Rezaei, R. and Samadian, H. (2019). Scaffolding polymeric biomaterials: Are naturally occurring biological macromolecules more appropriate for tissue engineering. *International Journal of Biological Macromolecules,* 134: 673–694.

Abe, H., Takahashi, N., Kim, K. J., Mochizuki, M. and Doi, Y. (2004). Thermal degradation processes of end-capped poly (L-lactide) s in the presence and absence of residual zinc catalyst. *Biomacromolecules,* 5(4): 1606–1614.

Ahmad Raus, R., Wan Nawawi, W. M. F. and Rahman Nasaruddin, R. (2020). Alginate and alginate composites for biomedical applications. *Asian Journal of Pharmaceutical Sciences.* https://doi.org/10.1016/j.ajps.2020.10.001.

Alizadeh-Osgouei, M., Li, Y. and Wen, C. (2019). A comprehensive review of biodegradable synthetic polymer-ceramic composites and their manufacture for biomedical applications. *Bioactive Materials* 4: 22–36.

Ammala, A., Bateman, S., Dean, K., Petinakis, E., Sangwan, P., Wong, S., Yuan, Q., Yu, L., Patrick, C. and Leong, K. H. (2011). An overview of degradable and biodegradable polyolefins. *Progress in Polymer Science (Oxford),* 36(8).

Anju, P. V., Khandelwal, M., Subahan, M. P., Kalle, A. M. and Mathaparthi, S. (2020). *In situ* synthesized hydro-lipophilic nano and micro fibrous bacterial cellulose: polystyrene composites for tissue scaffolds. *Journal of Materials Science,* 55(12): 5247–5256.

Antoniac, I. V., Antoniac, A., Vasile, E., Tecu, C., Fosca, M., Yankova, V. and Rau, J. (2021). *In vitro* characterization of novel nanostructured collagen-hydroxyapatite composite scaffolds doped with magnesium with improved biodegradation rate for hard tissue regeneration. *Bioactive Materials,* 6(10): 3383–3395.

Arrieta, M. P., Peponi, L., López, D. and Fernández-García, M. (2018). Recovery of yerba mate (Ilex paraguariensis) residue for the development of PLA-based bionanocomposite films. *Industrial Crops and Products,* 111: 317–328.

Arruda, L. C., Magaton, M., Bretas, R. E. S. and Ueki, M. M. (2015). Influence of chain extender on mechanical, thermal, and morphological properties of blown films of PLA/PBAT blends. *Polymer Testing,* 43: 27–37.

Arun Kumar, R., Sivashanmugam, A., Deepthi, S., Iseki, S., Chennazhi, K. P., Nair, S. V. and Jayakumar, R. (2015). Injectable chitin-poly (ε-caprolactone)/nanohydroxyapatite composite microgels prepared by simple regeneration technique for bone tissue engineering. *ACS Applied Materials & Interfaces,* 7(18): 9399–9409.

Babitha, S. and Korrapati, P. S. (2017). Biodegradable zein–polydopamine polymeric scaffold impregnated with TiO_2 nanoparticles for skin tissue engineering. *Biomedical Materials,* 12(5): 055008.

Boyacı, D., Iorio, G., Sozbilen, G. S., Alkan, D., Trabattoni, S., Pucillo, F. and Yemenicioğlu, A. (2019). Development of flexible antimicrobial zein coatings with essential oils for the inhibition of critical pathogens on the surface of whole fruits: Test of coatings on inoculated melons. *Food Packaging and Shelf Life,* 20: 100316.

Brien, F. J. O. (2011). Biomaterials & scaffolds Every day thousands of surgical procedures are performed to replace. *Mater. Today,* 14(3): 88–95.

Cao, H. and Kuboyama, N. (2010). A biodegradable porous composite scaffold of PGA/β-TCP for bone tissue engineering. *Bone,* 46(2): 386–395.

Chandra, M. V. L., Karthikeyan, S., Selvasekarapandian, S., Premalatha, M. and Monisha, S. (2017). Study of PVAc-PMMA-LiCl polymer blend electrolyte and the effect of plasticizer ethylene carbonate and nanofiller titania on PVAc-PMMA-LiCl polymer blend electrolyte. *Journal of Polymer Engineering,* 37(6): 617–631.

Chen, G., Chen, N. and Wang, Q. (2019). Fabrication and properties of poly (vinyl alcohol)/β-tricalcium phosphate composite scaffolds via fused deposition modeling for bone tissue engineering. *Composites Science and Technology,* 172: 17–28.

Cheow, W. S. and Hadinoto, K. (2011). Factors affecting drug encapsulation and stability of lipid–polymer hybrid nanoparticles. *Colloids and Surfaces B: Biointerfaces,* 85(2): 214–220.

Chopparapu, R. T., Bala Chennaiah, M., Srivalli, G., Kartik Raju, S., Dileep Kumar, E., Sycam, V. and Dasari, R. (2020). Biodegradable polymer filter made from fiber composites for addition of minerals and salts to water. *Materials Today: Proceedings,* 33: 5607–5611.

Chu, B., Gao, T., Li, Y., Wang, J., Desper, C. R. and Byrne, C. A. (1992). Microphase separation kinetics in segmented polyurethanes: effects of soft segment length and structure. *Macromolecules,* 25(21): 5724–5729.

Chuysinuan, P., Thanyacharoen, T., Tanga, K., Techasakul, S. and Ummariotina, S. (2020). Preparation of chitosan/hydrolyzed collagen/hyaluronic acid based hydrogel composite with caffeic acid addition. *International Journal of Biological Macromolecules,* 162: 1937–1943.

Cojocaru, C., Dorneanu, P. P., Airinei, A., Olaru, N., Samoila, P. and Rotaru, A. (2017). Design and evaluation of electrospun polysulfone fibers and polysulfone/$NiFe_2O_4$ nanostructured composite as sorbents for oil spill cleanup. *Journal of the Taiwan Institute of Chemical Engineers,* 70: 267–281.

Coreño-Alonso, J. and Méndez-Bautista, M. T. (2010). Relationship between structure and properties of polymers. *Educ. Quim.,* 21(4): 291–299, DOI: 10.1016/s0187-893x(18)30098-3.

Dattola, E., Parrotta, E. I., Scalise, S., Perozziello, G., Limongi, T., Candeloro, P. and Cuda, G. (2019). Development of 3D PVA scaffolds for cardiac tissue engineering and cell screening applications. *RSC Advances,* 9(8): 4246–4257.

De Lucas-Gil, E., Leret, P., Monte-Serrano, M., Reinosa, J. J., Enríquez, E., Del Campo, A. and Rubio-Marcos, F. (2018). ZnO nanoporous spheres with broad-spectrum antimicrobial activity by physicochemical interactions. *ACS Applied Nano Materials,* 1(7): 3214–3225.

Deepthi, S., Venkatesan, J., Kim, S. K., Bumgardner, J. D. and Jayakumar, R. (2016). An overview of chitin or chitosan/nano ceramic composite scaffolds for bone tissue engineering. *International Journal of Biological Macromolecules,* 93: 1338–1353.

Dong, J., Sun, Q. and Wang, J. Y. (2004). Basic study of corn protein, zein, as a biomaterial in tissue engineering, surface morphology, and biocompatibility. *Biomaterials,* 25(19): 4691–4697.

Dorée, C. (1947). *The Methods of Cellulose Chemistry Including Methods for the Investigation of Substances Associated with Cellulose in Plant Tissues* (Edn 2 (revised)).

Dungani, R., Karina, M., Sulaeman, A., Hermawan, D. and Hadiyane, A. (2016). Agricultural waste fibers towards sustainability and advanced utilization: a review. *Asian Journal of Plant Sciences,* 15(1/2): 42–55.

Dziadek, M., Stodolak-Zych, E. and Cholewa-Kowalska, K. (2017). Biodegradable ceramic-polymer composites for biomedical applications: A review. *Materials Science and Engineering C,* 71: 1175–1191.

Echeverría, C., Muñoz-Bonilla, A., Cuervo-Rodríguez, R., López, D. and Fernández-García, M. (2019). Antibacterial PLA fibers containing thiazolium groups as wound dressing materials. *ACS Applied Bio Materials,* 2(11): 4714–4719.

El-Rashidy, A. A., Waly, G., Gad, A., Roether, J. A., Hum, J., Yang, Y. and Boccaccini, A. R. (2018). Antibacterial activity and biocompatibility of zein scaffolds containing silver-doped bioactive glass. *Biomedical Materials,* 13(6): 065006.

El-Shekeil, Y. A., Sapuan, S. M., Abdan, K. and Zainudin, E. S. (2012). Influence of fiber content on the mechanical and thermal properties of Kenaf fiber reinforced thermoplastic polyurethane composites. *Materials & Design,* 40: 299–303.

FDA, Food and drugs, CFR - Code Fed. Regul. 3 (21CFR184.1588) (2020) 1–2.

Ferrandez-Montero, A., Lieblich, M., Benavente, R., González-Carrasco, J. L. and Ferrari, B. (2020). New approach to improve Polymer-Mg interface in biodegradable PLA/Mg composites through particle surface modification. *Surface and Coatings Technology,* 383(2020): 125285.

Gentile, P., Chiono, V., Carmagnola, I. and Hatton, P. V. (2014). An overview of poly (lactic-co-glycolic) acid (PLGA)-based biomaterials for bone tissue engineering. *International Journal of Molecular Sciences,* 15(3): 3640–3659.

Ghormade, V., Pathan, E. K. and Deshpande, M. V. (2017). Can fungi compete with marine sources for chitosan production? *International Journal of Biological Macromolecules,* 104: 1415–1421.

Giannuzzi, L. A., Kempshall, B. W., Schwarz, S. M., Lomness, J. K., Prenitzer, B. I. and Stevie, F. A. (2005). FIB lift-out specimen preparation techniques. In Introduction to focused ion beams. *Introducción to Focused Ion Beams* 201–228.

Güçbilmez, Ç. M., Yemenicioğlu, A. and Arslanoğlu, A. (2007). Antimicrobial and antioxidant activity of edible zein films incorporated with lysozyme, albumin proteins, and disodium EDTA. *Food Research International,* 40(1): 80–91.

Guo, Z., Poot, A. and Grijpma, D. (2021). Advanced polymer-based composites and structures for biomedical applications. *European Polymer Journal,* 149: 110388.

Heard, W. M., Paller, D. J., Christino, M. A., Behrens, S. B., Biercevicz, A., Fadale, P. D. and Monchik, K. O. (2013). Effect of insertion of a single interference screw on the mechanical properties of porcine anterior cruciate ligament reconstruction grafts. *Am. J. Orthop.* 42(4): 168–172.

Hostettler, F., Rhum, D., Forman, M. R., Helmus, M. N. and Ding, N. (2000). U.S. Patent No. 6,120,904. Washington, DC: U.S. Patent and Trademark Office.

Islam, M. M., Shahruzzaman, M., Biswas, S., Sakib, M. N. and Rashid, T. U. (2020). Chitosan based bioactive materials in tissue engineering applications—A review. *Bioactive Materials,* 5(1): 164–183.

Javaid, M. A., Younas, M., Zafar, I., Khera, R. A., Zia, K. M. and Jabeen, S. (2019). Mathematical modeling and experimental study of mechanical properties of chitosan based polyurethanes: Effect of diisocyanate nature by mixture design approach. *International Journal of Biological Macromolecules,* 124: 321–330.

Javaid, M. A., Zia, K. M., Ilyas, H. N., Yaqub, N., Bhatti, I. A., Rehan, M., Shoaib, M. and Bahadur, A. (2019). Influence of chitosan/1,4-butanediol blends on the thermal and surface behavior of polycaprolactone diol-based polyurethanes. *International Journal of Biological Macromolecules,* 141: 1022–1034.

Jayakumar, P. and Di Silvio, L. (2010). Osteoblasts in bone tissue engineering. Proceedings of the Institution of Mechanical Engineers, Part H: *Journal of Engineering in Medicine,* 224(12): 1415–1440.

Jayaramudu, J., Das, K., Sonakshi, M., Reddy, G. S. M., Aderibigbe, B., Sadiku, R. and Ray, S. S. (2014). Structure and properties of highly toughened biodegradable polylactide/ZnO biocomposite films. *International Journal of Biological Macromolecules,* 64: 428–434.

Jung, J., Arnold, R. D. and Wicker, L. (2013). Pectin and charge modified pectin hydrogel beads as a colon-targeted drug delivery carrier. *Colloids and Surfaces B: Biointerfaces,* 104: 116–121.

Khatti, T., Naderi-Manesh, H. and Kalantar, S. M. (2019). Application of ANN and RSM techniques for modeling electrospinning process of polycaprolactone. *Neural Computing and Applications,* 31(1): 239–248.

Kim, O. V., Litvinov, R.I., Chen, J., Chen, D. Z, Weisel, J. W. and Alber, M. S. (2017). Compression-induced structural and mechanical changes of fibrin-collagen composites. *Matrix Biology,* 60-61: 141–156.

Kimna, C., Tamburaci, S. and Tihminlioglu, F. (2019). Novel zein-based multilayer wound dressing membranes with controlled release of gentamicin. *Journal of Biomedical Materials Research Part B: Applied Biomaterials,* 107(6): 2057–2070.

Koronis, G. and Silva, A. (2018). *Green Composites for Automotive Applications.* Woodhead Publishing.

Kossyvaki, D., Suarato, G., Summa, M., Gennari, A., Francini, N., Gounaki, I., Venieri, D., Tirelli, N., Bertorelli, R., Athanassiou, A. and Papadopoulou, E. L. (2020). Keratin–cinnamon essential oil biocomposite fibrous patches for skin burn care. *Materials Advances,* 1(6): 1805–1816.

Kraus, D. and Trappe, V. (2021). Transverse damage in glass fiber reinforced polymer under thermo-mechanical loading. *Composites Part C: Open Access,* 5: 100147.

Langer, R. (2000). Biomaterials in drug delivery and tissue engineering: one laboratory's experience. *Accounts of Chemical Research,* 33(2): 94–101.

Lascano, D., Quiles-Carrillo, L., Balart, R., Boronat, T. and Montanes, N. (2019). Toughened poly (lactic acid)—PLA formulations by binary blends with poly (butylene succinate-co-adipate)—PBSA and their shape memory behaviour. *Materials,* 12(4): 622.

LeGeros, R. Z. (2002). Properties of osteoconductive biomaterials: calcium phosphates. *Clinical Orthopaedics and Related Research®,* 395: 81–98.

Leong, M. F., Lu, H. F., Lim, T. C., Du, C., Ma, N. K. and Wan, A. C. (2016). Electrospun polystyrene scaffolds as a synthetic substrate for xeno-free expansion and differentiation of human induced pluripotent stem cells. *Acta Biomaterialia,* 46: 266–277.

Lett, J. A., Sundareswari, M. and Ravichandran, K. (2016). Porous hydroxyapatite scaffolds for orthopedic and dental applications-the role of binders. *Materials Today: Proceedings,* 3(6): 1672–1677.

Lett, J. A., Sagadevan, S., Paiman, S., Mohammad, F., Schirhagl, R., Léonard, E. and Oh, W. C. (2020). Exploring the thumbprints of Ag-hydroxyapatite composite as a surface coating bone material for the implants. *Journal of Materials Research and Technology,* 9(6): 12824–12833.

Lett, J. A., Sagadevan, S., Fatimah, I., Hoque, M. E., Lokanathan, Y., Léonard, E. and Oh, W. C. (2021). Recent advances in natural polymer-based hydroxyapatite scaffolds: Properties and applications. *European Polymer Journal,* 148: 110360.

Lim, K. T., Patel, D. K., Choung, H. W., Seonwoo, H., Kim, J. and Chung, J. H. (2019). Evaluation of bone regeneration potential of long-term soaked natural hydroxyapatite. *ACS Applied Bio Materials,* 2(12): 5535–5543.

Lin, H. R., Kuo, C. J., Yang, C. Y., Shaw, S. Y. and Wu, Y. J. (2002). Preparation of macroporous biodegradable PLGA scaffolds for cell attachment with the use of mixed salts as porogen additives. *Journal of Biomedical Materials Research,* 63(3): 271–279.

Lin, J., Li, C., Zhao, Y., Hu, J. and Zhang, L. M. (2012). Co-electrospun nanofibrous membranes of collagen and zein for wound healing. *ACS Applied Materials & Interfaces,* 4(2): 1050–1057.

Liu, B., Chen, L., Shao, C., Zhang, F., Zhou, K., Cao, J. and Zhang, D. (2016). Improved osteoblasts growth on osteomimetic hydroxyapatite/$BaTiO_3$ composites with aligned lamellar porous structure. *Materials Science and Engineering: C,* 61: 8–14.

Liu, H. L., Liu, S. J., Xiao, Z. L., Chen, Q. Y. and Yang, D. W. (2006). Excess molar enthalpies of binary mixtures for (tributyl phosphate+methanol/ethanol) at 298.15 K. *Journal of Thermal Analysis and Calorimetry,* 85(3): 541–544.

Liu, H., Guan, Y., Wei, D., Gao, C., Yang, H. and Yang, L. (2016). Reinforcement of injectable calcium phosphate cement by gelatinized starches. *Journal of Biomedical Materials Research Part B: Applied Biomaterials,* 104(3): 615–625.

Mano, J. F., Sousa, R. A., Boesel, L. F., Neves, N. M. and Reis, R. L. (2004). Bioinert, biodegradable and injectable polymeric matrix composites for hard tissue replacement: State of the art and recent developments. *Composites* 64(6): 789–817.

Marra, A., Silvestre, C., Duraccio, D. and Cimmino, S. (2016). Polylactic acid/zinc oxide biocomposite films for food packaging application. *International Journal of Biological Macromolecules,* 88: 254–262.

May, C. D. (1990). Industrial pectins: sources, production, and applications. *Carbohydrate Polymers,* 12(1): 79–99.

May-Pat, A., Valadez-González, A. and Herrera-Franco, P. J. (2013). Effect of fiber surface treatments on the essential work of fracture of HDPE-continuous henequen fiber-reinforced composites. *Polymer Testing,* 32(6): 1114–1122.

Meng, X., Lu, Y., Gao, Y., Cheng, S., Tian, F., Xiao, Y. and Li, V. (2021). Chitosan/alginate/hyaluronic acid polyelectrolyte composite sponges crosslinked with genipin for wound dressing application. *International Journal of Biological Macromolecules,* 182: 512–523.

Mishra, R. K., Banthia, A. K. and Majeed, A. B. A. (2012). Pectin based formulations for biomedical applications: a review. Asian *Journal of Pharmaceutical and Clinical Research,* 5(4): 1–7.

Mohanty, A. K., Misra, M. and Drzal, L. T. (2001). Surface modifications of natural fibers and performance of the resulting biocomposites: an overview. *Composite Interfaces,* 8(5): 313–343.

Murariu, M., Doumbia, A., Bonnaud, L., Dechief, A. L., Paint, Y., Ferreira, M. and Dubois, P. (2011). High-performance polylactide/ZnO nanocomposites designed for films and fibers with special end-use properties. *Biomacromolecules,* 12(5): 1762–1771.

Muthuraj, R., Lacoste, C., Lacroix, P. and Bergeret, A. (2019). Sustainable thermal insulation biocomposites from rice husk, wheat husk, wood fibers, and textile waste fibers: Elaboration and performances evaluation. *Industrial Crops and Products,* 135: 238–245.

Naghieh, S., Badrossamay, M., Foroozmehr, E. and Kharaziha, M. (2017). Combination of PLA micro-fibers and PCL-gelatin nano-fibers for development of bone tissue engineering scaffolds. *Int. J. Swarm Intell. Evol. Comput,* 6(1): 1–4.

Nasouri, K., Shoushtari, A. M. and Mojtahedi, M. R. M. (2015). Evaluation of effective electrospinning parameters controlling polyvinylpyrrolidone nanofibers surface morphology via response surface methodology. *Fibers and Polymers,* 16(9): 1941–1954.

Neves, A. A., Medcalf, N. and Brindle, K. M. (2005). Influence of stirring-induced mixing on cell proliferation and extracellular matrix deposition in meniscal cartilage constructs based on polyethylene terephthalate scaffolds. *Biomaterials,* 26(23): 4828–4836.

Niemelä, T. (2005). Effect of β-tricalcium phosphate addition on the in vitro degradation of self-reinforced poly-l, d-lactide. *Polymer Degradation and Stability,* 89(3): 492–500.

Nirmal, U., Hashim, J. and Ahmad, M. M. (2015). A review on tribological performance of natural fibre polymeric composites. *Tribology International,* 83: 77–104.

Ojeda, T. F. M., Dalmolin, E., Forte, M. M. C., Jacques, R. J. S., Bento, F. M. and Camargo, F. A. O. (2009). Abiotic and biotic degradation of oxo-biodegradable polyethylenes. *Polymer Degradation and Stability,* 94(6): 965–970.

Ott, S. M. (2010). New aspects of normal bone biology. *The Spectrum of Mineral and Bone Disorders in Chronic Kidney Disease*, 2 editions, 15–29.

Parisi, C., Salvatore, L., Veschini, L., Serra, M. P., Hobbs, C., Madaghiele, M. and Di Silvio, L. (2020). Biomimetic gradient scaffold of collagen–hydroxyapatite for osteochondral regeneration. *Journal of Tissue Engineering,* 11: 2041731419896068.

Park, J., Lee, S. J., Jo, H. H., Lee, J. H., Kim, W. D., Lee, J. Y. and Su, A. (2017). Fabrication and characterization of 3D-printed bone-like β-tricalcium phosphate/polycaprolactone scaffolds for dental tissue engineering. *Journal of Industrial and Engineering Chemistry,* 46: 175–181.

Peças, P., Ribeiro, I., Carvalho, H., Silva, A., Salman, H. M. and Henriques, E. (2019). Ramie and jute as natural fibers in a composite part—a life cycle engineering comparison with an aluminum part. *In Green Composites for Automotive Applications*, 253–284. Woodhead Publishing.

Peponi, L., Sessini, V., Arrieta, M. P., Navarro-Baena, I., Sonseca, A., Dominici, F. and Kenny, J. M. (2018). Thermally-activated shape memory effect on biodegradable nanocomposites based on PLA/PCL blend reinforced with hydroxyapatite. *Polymer Degradation and Stability,* 151: 36–51.

Petchwattana, N. and Covavisaruch, S. (2014). Mechanical and morphological properties of wood plastic biocomposites prepared from toughened poly (lactic acid) and rubber wood sawdust (Hevea brasiliensis). *Journal of Bionic Engineering,* 11(4): 630–637.

Philip, S. E., Odell, M., Keshavarz, T. and Roy, I. (2007). Polyhydroxy-alkanoates-the biodegradable polymers. *In: International Conference on Biodegradable Polymers: Their Production, Characterisation, and Application.*

Pugliese, R., Beltrami, B., Regondi, S. and Lunetta, C. (2021). Polymeric biomaterials for 3D printing in medicine: An overview. *Annals of 3D Printed Medicine* 2: 100011.

Ramdhan, T., Ching, H., Prakash, S. and Bhandari, B. (2020). Physical and mechanical properties of alginate based composite gels. *Trends in Food Science & Technology,* 106: 150–159.

Raquez, J. M., Habibi, Y., Murariu, M. and Dubois, P. (2013). Polylactide (PLA)-based nanocomposites. *Progress in Polymer Science,* 38(10-11): 1504–1542.

Reinosa, J. J., Leret, P., Álvarez-Docio, C. M., del Campo, A. and Fernández, J. F. (2016). Enhancement of UV absorption behavior in ZnO–TiO_2 composites. *boletín de la sociedad española de cerámica y vidrio,* 55(2): 55–62.

Rendón-Villalobos, R., Ortíz-Sánchez, A., Tovar-Sánchez, E. and Flores-Huicochea, E. (2016). The role of biopolymers in obtaining environmentally friendly materials. *Composites from Renewable and Sustainable Materials*. https://doi.org/10.5772/65265.

Sabiiti, E. N. (2011). Utilising agricultural waste to enhance food security and conserve the environment. *African Journal of Food, Agriculture, Nutrition and Development,* 11(6): 1–9.

Salama, A. (2021). Recent progress in preparation and applications of chitosan/calcium phosphate composite materials. *International Journal of Biological Macromolecules,* 178: 240–252.

Sambandan, D. R. and Ratner, D. (2011). Sunscreens: an overview and update. *Journal of the American Academy of Dermatology,* 64(4): 748–758.

Santos, A. E., Bracciali, A. L., Vilela, J., Foschini, C. R. and Sanchez, L. E. (2019). Poly L, DL-lactic acid, and composite poly l, DL-lactic acid/β-tricalcium phosphate-based bioabsorbable interference screw. *Polymer Composites,* 40(6): 2197–2207.

Saravanan, S., Leena, R. S. and Selvamurugan, N. (2016). Chitosan based biocomposite scaffolds for bone tissue engineering. *International Journal of Biological Macromolecules,* 93: 1354–1365.

Satyanarayana, K. G., Arizaga, G. G. C. and Wypych, F. (2009). Biodegradable composites based on lignocellulosic fibers—An overview. *Progress in Polymer Science (Oxford),* 34(9): 982–1021.

Senra, M. R., Barbosa d. Lima, R., Saboya Souza, D., Vieira Marques, M. F. and Neves Monteiro, S. (2020). Thermal characterization of hydroxyapatite or carbonated hydroxyapatite hybrid composites with distinguished collagens for bone graft. *Journal of Materials Research and Technology,* 9(4): 7190–7200.

Shan, D., Gerhard, E., Zhang, C., Tierney, J. W., Xie, D., Liu, Z. and Yang, J. (2018). Polymeric biomaterials for biophotonic applications. *Bioactive Materials,* 3(4): 434–445.

Shankar, S., Wang, L. F. and Rhim, J. W. (2018). Incorporation of zinc oxide nanoparticles improved the mechanical, water vapor barrier, UV-light barrier, and antibacterial properties of PLA-based nanocomposite films. *Materials Science and Engineering: C,* 93: 289–298.

Sharma, S., Majumdar, A. and Butola, B. S. (2021). Tailoring the biodegradability of polylactic acid (PLA) based films and ramie-PLA green composites by using selective additives. *International Journal of Biological Macromolecules,* 181: 1092–1103.

Shubhra, Q. T., Alam, A. K. M. M. and Quaiyyum, M. A. (2013). Mechanical properties of polypropylene composites: A review. *Journal of Thermoplastic Composite Materials,* 26(3): 362–391.

Shukla, R. and Cheryan, M. (2001). Zein: the industrial protein from corn. *Industrial Crops and Products,* 13(3): 171–192.

Siew Chun Low, Sivakumar V. and Murugaiyan. (2021). Thermoplastic polymers in membrane separation. *Materials Science and Materials Engineering.* https://doi.org/10.1016/B978-0-12-820352-1.00083-3.

Singh, N., Georget, D. M., Belton, P. S. and Barker, S. A. (2009). Zein−iodine complex studied by FTIR spectroscopy and dielectric and dynamic rheometry in films and precipitates. *Journal of Agricultural and Food Chemistry,* 57(10): 4334–4341.

Singh, N., Georget, D. M., Belton, P. S. and Barker, S. A. (2010). Physical properties of zein films containing salicylic acid and acetyl salicylic acid. *Journal of Cereal Science,* 52(2): 282–287.

Sivashankari, P. R. and Prabaharan, M. (2020). Three-dimensional porous scaffolds based on agarose/chitosan/ graphene oxide composite for tissue engineering. *International Journal of Biological Macromolecules,* 146: 222–231.

Škrlová, K., Malachová, K., Muñoz-Bonilla, A., Měřinská, D., Rybková, Z., Fernández-García, M. and Plachá, D. (2019). Biocompatible polymer materials with antimicrobial properties for preparation of stents. *Nanomaterials,* 9(11): 1548.

Song, R., Murphy, M., Li, C., Ting, K., Soo, C. and Zheng, Z. (2018). Current development of biodegradable polymeric materials for biomedical applications. *Drug Design, Development, and Therapy,* 12: 3117–3145.

Soni, S. S. and Rodell C. B. (2021). Polymeric materials for immune engineering: Molecular interaction to biomaterial design. *Acta Biomaterialia.* https://doi.org/10.1016/j.actbio.2021.01.016.

Sonseca, A., Madani, S., Rodríguez, G., Hevilla, V., Echeverría, C., Fernández-García, M. and López, D. (2020). Multifunctional PLA blends containing chitosan mediated silver nanoparticles: Thermal, mechanical, antibacterial, and degradation properties. *Nanomaterials,* 10(1): 22.

Springer, I. N., Fleiner, B., Jepsen, S. and Açil, Y. (2001). Culture of cells gained from temporomandibular joint cartilage on non-absorbable scaffolds. *Biomaterials,* 22(18): 2569–2577.

Suarato, G., Bertorelli, R. and Athanassiou, A. (2018). Borrowing from Nature: biopolymers and biocomposites as smart wound care materials. *Frontiers in Bioengineering and Biotechnology,* 6: 137.

Sultankulov, B., Berillo, D., Sultankulova, K., Tokay, T. and Saparov, A. (2019). Progress in the development of chitosan-based biomaterials for tissue engineering and regenerative medicine. *Biomolecules,* 9(9): 470.

Swain, S. K., Bhattacharyya, S. and Sarkar, D. (2015). Fabrication of porous hydroxyapatite scaffold via polyethylene glycol-polyvinyl alcohol hydrogel state. *Materials Research Bulletin,* 64: 257–261.

Tomba, P., Viganò, A., Ruggieri, P. and Gasbarrini, A. (2014). Gaspare Tagliacozzi, pioneer of plastic surgery and the spread of his technique throughout Europe in "De Curtorum Chirugia per Insitionem." *Eur Rev Med Pharmacol Sci,* 18(4): 445–450.

Walton, M. and Cotton, N. J. (2007). Long-term *in vivo* degradation of poly-L-lactide (PLLA) in bone. *Journal of Biomaterials Applications,* 21(4): 395–411.

Wang, C., Liu, D., Zhang, C., Sun, J., Feng, W., Liang, X. J. and Zhang, J. (2016). Defect-related luminescent hydroxyapatite-enhanced osteogenic differentiation of bone mesenchymal stem cells via an ATP-induced cAMP/PKA pathway. *ACS Applied Materials & Interfaces,* 8(18): 11262–11271.

Weiler, A., Windhagen, H. J., Raschke, M. J., Laumeyer, A. and Hoffmann, R. F. (1998). Biodegradable interference screw fixation exhibits pull-out force and stiffness similar to titanium screws. *The American Journal of Sports Medicine,* 26(1): 119–128.

Xiang, S., Feng, L., Bian, X., Li, G. and Chen, X. (2020). Evaluation of PLA content in PLA/PBAT blends using TGA. *Polymer Testing,* 81: 106211.

Xu, Z., Wang, N., Liu, P., Sun, Y., Wang, Y., Fei, F. and Han, B. (2019). Poly (dopamine) coating on 3D-printed poly-lactic-co-glycolic acid/β-tricalcium phosphate scaffolds for bone tissue engineering. *Molecules,* 24(23): 4397.

Yang, F., Cui, W., Xiong, Z., Liu, L., Bei, J. and Wang, S. (2006). Poly (l, l-lactide-co-glycolide)/tricalcium phosphate composite scaffold and its various changes during degradation *in vitro. Polymer Degradation and Stability,* 91(12): 3065–3073.

Yang, X., Yang, D., Zhu, X., Nie, J. and Ma, G. (2019). Electrospun and photocrosslinked gelatin/dextran–maleic anhydride composite fibers for tissue engineering. *European Polymer Journal,* 113: 142–147.

Yang, X., Fan, W., Ge, S., Gao, X., Wang, S., Zhang, Y., Foong, S. Y., Liwe, R. K., Lam, S. S. and Xia, C. (2021). Advanced textile technology for fabrication of ramie fiber PLA composites with enhanced mechanical properties. *Industrial Crops and Products,* 162: 113312.

Zhang, M., Qiao, X., Wenweihan, Jiang, T., Liu, F. and Zhao, X. (2021). Alginate-chitosan oligosaccharide-ZnO composite hydrogel for accelerating wound healing. *Carbohydrate Polymers* 266: 118100.

Zhang, Q., Khan, M. U., Lin, X., Yi, W. and Lei, H. (2020). Green-composites produced from waste residue in pulp and paper industry: A sustainable way to manage industrial wastes. *Journal of Cleaner Production,* 262: 121251.

Zhang, Y., Cui, L., Che, X., Zhang, H., Shi, N., Li, C., Chen, Y. and Kong, W. (2015). Zein-based films and their usage for controlled delivery: Origin, classes and current landscape. *Journal of Controlled Release,* 206: 206–219.

Zheng, C., Liu, C., Chen, H., Wang, N., Liu, X., de Guozhen, S. and Weihongqiao. (2019). Effective wound dressing based on Poly (vinyl alcohol)/Dextran-aldehyde composite hydrogel. *International Journal of Biological Macromolecules,* 132: 1098–1105.

Chapter 10

Green Materials in the Packaging

Biodegradable Foams

Sindhu Thalappan Manikkoth, Deepthi Panoth, Kunnambeth M. Thulasi, Fabeena Jahan, Anjali Paravannoor and *Baiju Kizhakkekilikoodayil Vijayan**

1. Introduction

The advancement in biodegradable materials have been gaining attention in the present era as synthetic macromolecular polymers with a non-renewable source of origin causes serious environmental issues. The limited availability of the raw materials and the non-degradable nature of such synthetic polymers restricts their use in temporary purposes such as food packaging, thermal insulation, etc. Thus, scientists around the world are focusing on the development of 'green foams' with unique features for the production of different functional materials. Aiming at this, natural fibers and composites of natural fibers with environmentally friendly polymers grabbed the wide attention of many industries.

The significant features of the compostable foams make them suitable for packaging purposes owing to their very good resilience even when applied to multiple impact loads (Mali, 2018). The low thermal conductivity and high sensitivity towards heat or temperature make it an excellent candidate for thermal insulation purposes (Avella et al., 2012). Several biodegradable polymers have been processed into scaffolds for tissue engineering and cell transplantation. Bio foams with high porosity, large surface area, excellent structural integrity and mechanical strength, etc., meet the requirements for cell attachment, function and growth (Ma and Langer, 1998).

Natural fibers are extensively used as a reinforcement material in thermosetting and thermoplastics owing to their low cost, abundance and high specific features. In addition, it is worth highlighting the positive environmental advantages and recycling benefits obtained by employing such materials. Starch-based foams are naturally occurring polymer foams that have a large application prospect owing to their large availability and low cost. The environmentally and economically viable starch foams are a suitable alternative for expanded polystyrene in the food packaging sector (Debiagi et al., 2011). The functional properties of starch foams such as elasticity, density, thermal diffusivity, thermal conductivity, heat capacity and moisture absorption mainly depend on the synthetic method adapted and the hydrophobic additives, which control the structure and morphology of the products

Department of Chemistry/Nanoscience, Kannur University, Swami Anandha Theertha Campus, Payyannur, Edat P.O., Kerala-670 327, India.

* Corresponding author: baijuvijayan@kannuruniv.ac.in

(Mariam et al., 2008). However, the hygroscopic and brittle nature of pure starch-based foams limits its application in many fields. In order to enhance the hydrophobic and mechanical properties of starch-based foams, they were blended with different bio-based materials such as cellulose and a number of polymers to retain the biodegradable nature.

The vast abundance of natural cellulose fibers, a renewable polysaccharide extracted from wood pulp together with its extraordinary thermal and mechanical properties makes it an excellent biodegradable material for foam production. Its remarkable hydrophilicity and outstanding mechanical, chemical and biological properties demonstrate the versatility of the material in various biomedical applications including tissue-engineered scaffolds, drug delivery vehicles and *in vitro* applications (Demitri et al., 2014). Polyvinyl alcohol is a compatible water-soluble synthetic polymer vulnerable to biodegradation in the presence of selective microorganisms. Its capability to build a well-defined polymer matrix and high solubility in water or other functional organic solvents have gained the wide attention of many industries (Avella et al., 2011). Other polymer materials such as poly (lactic acid), polyurethane, etc., have also become an essential part of the biodegradable foam market (Fang et al., 2019; Zhang and Sun, 2007).

This chapter includes recent developments in biodegradable foams, synthetic processes adapted for foam production such as extrusion, baking, etc., the properties and applications of various types of foams including starch and cellulose-based foams and their polymer blends.

2. Starch-based foams

Starch-based foams can be widely utilized in biodegradable and environmental friendly foam packaging and food serving applications and thus various technologies have been initiated for producing starch-based foams as polystyrene replacements. Various processing techniques of the starch-based foams, their modification using different additives, natural fibers, modified starches, other biopolymers, nanofillers, etc., and properties of the resulting starch-based composite foams are discussed here in detail.

2.1 Starch

Starch is the major form in which carbohydrates are stored in higher plants. Its advantages are easily producible from bio-resources, cost-effective, renewable and inherently biodegradable (Anglès and Dufresne, 2000). It exists as granules (small, dense, discrete packages), that are insoluble in cold water. Amylose and amylopectin are the two major starch polymers. Amylose is a linear polymer with (1–4) linked α-glucopyranosyl units, whereas (1–4) linked α-D-glucopyranosyl units in chains are joined by (1–6) linkages to form the highly branched polymer amylopectin. The smaller one, amylose has a molecular weight in the order of 10^4 to 10^5 and a Degree of Polymerization (DP) of 250–1,000. Amylopectin possesses molecular weight in the order of 10^6 to 10^8 and a DP of around 5,000 to 50,000 (Pérez et al., 2009). Amylopectin chains are primarily responsible for the crystallinity of starch, whereas amylose is present in the amorphous structure. Normal starches, such as maize, potato, rice, wheat, etc., contain 70–80% amylopectin and 20–30% amylose. Thus the starch granule has both amorphous and crystalline regions arranged in the grain.

Various physicochemical changes such as melting, gelatinization, glass transition, change of crystal structure, crystallization, volume expansion, molecular degradation and motion of water during heating complicate the thermal behavior of starches, and thus analyzing the thermal behaviour of starches is very important in food processing. Thermal behaviour is highly dependent on the water content to the starch ratio. It was firstly reported by Shogren that the melting of corn-starch with the lower water content, 11–30% occurred at a temperature of 190–200°C and when the water content was above 30%, the amorphous region started to gelatinize at about 70°C (Shogren, 1992).

2.2 Processing of starch-based foams

2.2.1 Extrusion

There is significant interest in the utilization of starch-based foams for food packaging industries in recent years and primarily utilizes starch extrusion for its processing (Zhang and Sun, 2007). In the extrusion process, voids are formed at temperatures above 100°C as the steam expands during extrusion at the high-pressure conditions of the die. The extrusion conditions such as barrel temperature, die diameter, screw speed, etc., and material compositions such as starch type, nucleation agent concentration, feed moisture, etc., decide the structure of extruded starch-foam (Aguilar-Palazuelos et al., 2007; Guan and Hanna, 2006). Screw speed and barrel temperature used are frequently in the range of 70–400 rpm and 120–170°C respectively in the starch-based foam extrusion (Robin et al., 2011).

Water serves as an effective blowing agent as well as a notable plasticizer for starch-based foams (Robin et al., 2011). To provide the largest expansion ratio of the extruded foams, the water content in the feed was kept in an optimum range of 15–18% (Pushpadass et al., 2008). Talc is commonly used in starch-based foams as an effective inorganic nucleating agent, which provides enough nucleation sites for water vaporization. The cell size became smaller and cell size distribution became narrower and the cell size became smaller with increasing the talc content (Zhang and Sun, 2007). The starch type also significantly influences the foam structure owing to their botanic source, the ratio of amylose and amylopectin, uniqueness in granule dimension, etc. (Willett and Shogren, 2002).

2.2.2 Baking/Compression

Baking a batter of starch and water in a heated closed mold for few minutes results in the formation of starch-based foams with desired shapes (Glenn et al., 2001). The evaporation of encapsulated water during the gelatinization of starch granules into a thick paste causes the paste to expand dramatically. Gradual drying of the starch-foam then occurs and finally takes up the shape of the mold (Soykeabkaew et al., 2004). The baked foam structure depicts an outer skin layer with a small, dense and closed-cell structure; while the inner layer shows a large, loose and opened cell structure. The geometry of the mold, batter composition, batter volume, baking temperature and baking time are the major factors that decides the foam shape, thickness and density. The optimum baking time and temperature are usually in the range of 125–300 sec and 180–250°C respectively.

Magnesium stearate and guar gum are added to the batter to release the mold and to prevent the settling of the starch respectively (Shogren et al., 2002). The type of starch determines the density of the baked starch foam. Glenn et al. (2001) reported that baked foam systems made of potato and tapioca starches had lower density (~ 0.12 g/cm^3) than those made of wheat and corn (~ 0.15 g/cm^3). The increase in viscosity weakens the foam expansion and thus the viscosity of the batter affects the quality of the product. Generally, batters with higher starch contents have higher viscosity. The batter's viscosity can be increased by adding fibers and fillers (Vercelheze et al., 2012).

Glenn and Orts demonstrated a compression/explosion to reduce the long baking time in starch foaming, which took only 10 seconds to complete (Glenn and Orts, 2001). Thus it reduced time and energy input and at the same time, increased the product throughput.

2.2.3 Microwave heating

Microwave energy can be utilized for baking and expanding some cereal foods such as popcorn. When it is heated above the boiling point of water in microwaves, the water will get transformed into steam, generating high pressure inside. The starch granules start to gelatinize, expands into a cellular structure by the pressure of the steam bubbles, and eventually form the solid foam systems as moisture is lost from the matrix (Moraru and Kokini, 2003; Boischot et al., 2003).

Sjöqvist and Gatenholm reported the processing of foams from three different starches; potato amylopectin, high amylose potato and native potato by using microwave heating with a frequency of 2450 MHz for 3 minutes and concluded that amylopectin starch was a better raw material for preparing foams than amylose-rich starch foams (Sjöqvist and Gatenholm, 2005). Lee et al. processed expandable extruded pellets by the extrusion of unexpanded starch pellets followed by microwave heating and optimized the expansion at around 50% gelatinization under the starch extrusion at 90°C (Lee et al., 2000).

Zhou et al. reported the preparation of starch-foam blocks by the Microwave-Assisted Molding (MAM) method (Zhou et al., 2007). Polytetrafluoroethylene (PTFE) is found to be the most appropriate choice as a mold material for the MAM process since it has a high service temperature, non-stick nature and negligible absorption of microwave energy. In this process, microwave energy was applied to the extruded pellets loaded in the PTFE mold cavity to form the starch foam blocks. More recently, Lopez-Gil et al. reported the synthesis of foam blocks from thermoformed starch sheets by microwave heating (Lopez-Gil et al., 2015). Starch-based porous scaffolds for tissue-engineering purposes can be prepared by microwave heating.

2.2.4 Freeze drying process

In 1906, Bordas and d'Arsonval introduced the freeze-drying technique at a laboratory scale where a frozen product was dried using a vacuum. Currently, the freeze-drying technique is effectively employed in a wide range of applications in diverse fields like the food industry, biomedical and pharmaceutical fields, research, etc. (Morais et al., 2016). The freeze-drying technique utilized for the production of biodegradable foams mainly consists of three steps: freezing, primary drying and secondary drying (Soykeabkaew et al., 2015).

The first step of the freeze-drying process is freezing, where the liquid samples are cooled to a sufficiently low temperature till it becomes a solid. On lowering the temperature, the solute separates from the water and it is trapped in the interstitial regions between the ice crystals. Here the size of the ice crystals is maintained as small as possible to reduce the physical damage to the samples. The formation of ice crystals includes two steps, one is nucleation and the other is the growth of the crystals. Fast freezing and higher nucleation rate result in the formation of more homogenous and smaller ice crystals (Tang and Pikal, 2004). The freeze-drying step proceeds to the next step, primary drying where the ice crystals are sublimated by the low pressure and temperature and thus form an open network of pores (Morais et al., 2016). The sublimated vapor is eliminated through the porous layers via convection or diffusion (Soykeabkaew et al., 2015). During this stage, pressure plays a vital role, as it acts as the driving force for the movement of water vapor. Thus lower chamber pressure results in faster sublimation of the ice. The final step of freeze-drying is the secondary drying in which the residual water from the solute phase is removed via desorption (Tang and Pikal, 2004). Even after the primary drying process, 5–20% of the residual water remains in the porous product. The secondary drying is carried out to minimize the residual moisture content of the porous material to an optimum level (~ less than 1%).

In 1995, Glenn and Irving developed microcellular freeze-dried starch-based foams using semi-grid auqagels of wheat, corn, tapioca and potato starch. The slabs of aquagel were placed in a freezer at –10°C overnight after removing from the mold. After the chamber pressure of freeze-drier dropped to 1.33×10^{-5} MPa, the aquagels placed inside heating trays were heated to 60°C for 3 days to obtain foams with thin and continuous cell wall having relatively higher tensile strength (Soykeabkaew et al., 2015). Torres et al., developed starch-based scaffolds by freeze-drying starch solutions of potato, sweet potato and corn with different processing routes for tissue engineering applications (Nakamatsu et al., 2006). The samples were frozen at two different temperatures, –15°C and –196°C. At higher freezing temperature (–196°C), the foamed samples displayed micropores of 2.5 μm as small ice crystals were produced at a higher freezing rate (19.2°C/min), while at lower freezing temperature (–15°C) macropores of 430.93 μm were formed with the low freezing rate (0.12°C/min) (Fig. 10.1).

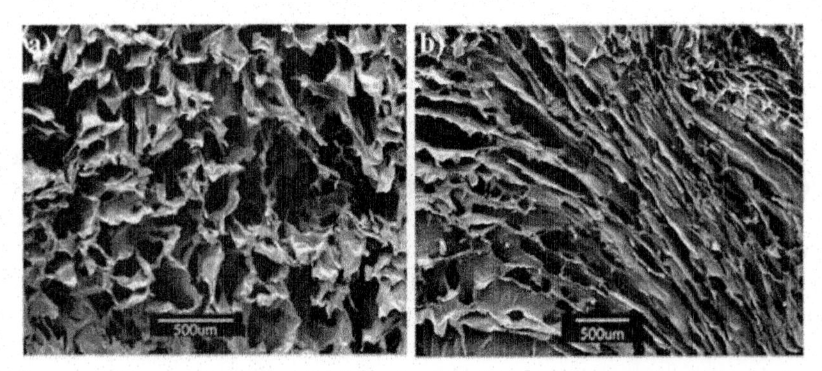

Figure 10.1. Scanning electron micrographs of chitosan scaffold showing macroporous anisotropic structures formed when (a) frozen at –15°C and (b) frozen at –196°C (Copyright: Nakamatsu et al. (2006). Processing and characterization of porous structures from chitosan and starch for tissue engineering scaffolds. *Biomacromolecules*, 7(12): 3345–3355. https://doi.org/10.1021/bm0605311).

2.2.5 Super critical fluid extraction (SCFX)

SCFX involves the introduction of super critical-CO_2 into a specially configured and modified extruder containing starch melt. At first, the melts containing SC-CO_2 flows through the nozzle, and finally, the extrudate exits from the die. The dramatic decrease in pressure grounds cell nucleation, cell growth and extrudate expansion (Manoi and Rizvi, 2010). Starch-based microcellular foams with cell size ranges of 50–200 μm were successfully prepared by SCFX. According to Alavi et al. various process and material parameters, such as nozzle temperature, melt yield stress and melt viscosity, CO_2 injection rate, post-extrusion oven temperature, etc., significantly affect the final bubble size and expansion ratio of starch extrudates (Alavi et al., 2003). Among others, SCFX is regarded as the most controllable and versatile process for the preparation of starch-based foams with desired cell size, cell density, product expansion as one can easily manipulate the processing parameters (Manoi and Rizvi, 2010).

3. Modification of starch-based foams

Starch consists of different ratios of amylose (linear) and amylopectin (branched) (Molavi et al., 2015). Earlier different starch-based foams that can be developed using various processing routes for a wide variety of applications have been described. But one of the major drawbacks of starch foams is their lack of moisture resistance and poor mechanical properties. Its hygroscopic nature is mainly due to the glucopyranosyl (repeating) unit of starch with numerous free hydroxyl groups which can easily form a hydrogen bond with moisture. Usually, starch granules are insoluble in cold water as the starch chains are held together by a strong hydrogen bond (Ribba et al., 2017). Moreover, starch foams possess dimensional instability or can collapse in the presence of water or moisture which limits their use. Extensive research has been carried out to further improve the existing mechanical properties and water resistance of porous starch foams (Xu and Hanna, 2005; Shogren et al., 1991). The feasible approaches to overcome the limitations and to improve the performance of starch-based foams will be discussed next.

3.1 Modified starch-based foams

Pure starch-based foams do not possess satisfactory functional properties mainly attributed to their hydrophilic nature and their brittleness (Fang and Hanna, 2000). In order to improve the water resistance of starch, its hydrophilic hydroxyl groups are modified using hydrophobic compounds (Xu and Hanna, 2005). Several researchers have focused on the production of foams using chemically modified starches like hydroxypropylated and acetylated starches. Starch acetate having a high

Degree of Substitution (DS) was used as a base material for extrusion with α-cellulose, corncobs, ethanol and talc at 160°C, by Guan and Hann (Guan et al., 2005).

Increased hydrophobicity of starch can be achieved with the addition of crosslinking agents to the starch material. In a humid environment, the cross-linking agents limit the direct interaction of starch with water by creating some network structures within. Narayan et al. studied the hydrophobic nature and mechanical properties of extruded hydroxypropylated high amylose corn starch foams with the addition of various functional aids like crosslinker, glyoxal and other biodegradable polymers like polyvinyl alcohol (PVA), Methylated Pectin (MP), poly-caprolactone (PCL), Cellulose Acetate (CA) and Poly (Butylene Adipate-co-Terephthalate) (PBAT) (Nabar et al., 2006). The water sensitivity of the starch foams decreased with the addition of these processing aids. The addition of these polymers and crosslinkers resulted in the formation of low-density starch foams with a higher expansion ratio. Manoi and Rizvi demonstrated the crosslinking of starch blends by phosphorylation using the supercritical fluid extrusion technique. They found that crosslinkers can improve the dimensional stability of starch-based foams along with the enhancement in water resistance. The gelatinization of starch phosphates at higher temperatures was consistent with the minimum water solubility as crosslinking restricted the movement of starch molecules with the rigid structure (Manoi and Rizvi, 2010).

3.2 Starch-polymer blend foams

One of the possible approaches to improve the mechanical properties is to blend starch with biodegradable polymers (Fang and Hanna, 2000; Simmons and Thomas, 1995). Adding various biodegradable polymers to the starch can improve their barrier and mechanical properties, processability, increase its biodegradability and reduce the hydrophilic nature of starch. In 2005, Shey et al. reported natural rubber latex incorporated into the batter of baked starch foams which enhanced the water resistance and flexibility of the resultant foams (Shey et al., 2006). The starch/latex and cassava/chitosan foams produced by blending starch with hydrophobic polymers showed a satisfactory enhancement in the water resistance of starch-blend foams (Kaisangsri, 2014).

Kaisangsri developed Cassava Starch Foam (CSF) trays blended with additive agents of natural polymers like soy protein, zein, gluten, palm oil and craft fiber and studied how the properties of biopolymers affect the various properties of the CSF (Kaisangsri et al., 2014). With increasing the concentration of kraft, gluten and zein to CSF, its structural and compressive strength of the CSF trays increased. CSF containing 15% zein and 15% protein showed low water absorption and a water solubility index. The CSF blended with 15% kraft fiber showed the highest compressive strength as there is a good interfacial interaction between fiber and starch matrix which helps the matrix to transfer stress to the reinforcing fiber effectively.

3.3 Starch/natural fibers composite foam

The incorporation of various natural fibers to the starch-based biocomposite system contributed to the satisfactory improvement in the mechanical properties of the foams. The better stress transfer capability in the composite system can be attributed to the similar chemistry of starch and natural fibers which results in good compatibility and a strong interaction between them (Soykeabkaew et al., 2015). Soykeabkaew et al. prepared Starch-based Composite Foams (SCF) by incorporating either flax or jute fibers into the starch-based batter by baking inside a hot mold (Soykeabkaew et al., 2004). They also studied the effect of moisture content, fiber type, fiber orientation, fiber content and fiber aspect ratio on the mechanical properties of SCF. The jute-reinforced SCF exhibited a greater flexural strength than the flax-reinforced SCF's as jute fibers possess higher specific surface area and stiffness than flax fibers.

The addition of cellulose, microfibrillated cellulose, crystallites displayed a major increase in water resistance and compared to vegetable cellulose fiber, composites formed with bacteria cellulose exhibited improved mechanical properties. In 2013, Silva et al. produced composite foam trays by baking cassava starch with reinforced bacterial cellulose by two different methods (Silva et al., 2013). In method 1, during the baking process, bacterial cellulose powder was directly incorporated into the starch matrix while in method 2, after production the tray surface was coated with bacterial cellulose films. The foaming ability of starch enhanced with the addition of bacterial cellulose using method 1 by producing low dense, thicker and expanded trays having low water absorption capacity. Method 2 produced composite foams with higher elongation lower water absorption capacities. Lawton et al. added aspen fibers as a reinforcing filler in the baked corn starch foam to enhance the mechanical properties (Lawton et al., 2004). The strength of the starch fiber composite foam trays got larger with the increasing fiber content until added aspen fiber content became 15% and strength gradually decreased as the fiber content went above 35%.

3.4 Starch-based nanocomposite foams

One of the popular and latest advances in the field of biodegradable foams is the reinforcement of nanoscale fillers to the starch matrix to produce nanocomposite foams to improve the functional properties. In order to overcome the drawback of the conventional starch-based or bio-polymer-based foam, the application of nanotechnology has been fruitfully used to minimize their limitations for the fabrication of nanocomposite foams for advanced multifaceted applications. In different ways, nanofillers can be presented in foams, like nanolayers, spherical or polyhedral nanoparticles and nanotubes. Some of the recent research works carried out based on nano reinforcement or nanofillers are, nano celluloses, layered silicates (clay) and hydroxyapatite was used for the production of nanocomposite foams (Soykeabkaew et al., 2015).

Lee and Hanna prepared nanocomposite foams with tapioca starch, poly(lactic acid) and cloisite NA^+ with different clay contents by melt-intercalation method and studied the effect of nanoclay content on the physical, thermal and mechanical properties of the foam. The nanocomposite foams presented more compact cells with increasing the clay content and had expanded cell density (Lee and Hanna, 2008). Qiu et al. developed PLA-based nanocomposite foams using cellulose nanocrystals (CNC) to tune the cyclic tensile properties of nanocomposite foams via a high-pressure batch foaming process (Qiu et al., 2018). In PLA/CNC based nanocomposite foams, a decrease in wall thickness and improvement in cell density was observed.

4. Cellulose based biodegradable foams

The most abundant natural biopolymer found on Earth is cellulose and is derived from plants, algae, bacteria, etc. Cellulose can also be obtained from wastes of cardboard, papers, etc. (Wang and Sánchez-Soto, 2015). It is a sturdy polymer of high molar mass which is fibrous and semi-crystalline in nature. It is a renewable, sustainable and cheap raw material that is biodegradable and thus can be broken down completely in the environment. As cellulose possesses different physical and chemical properties from that of synthetic polymers like PLA, PVA, etc., it became a key research attraction among scientists. Cellulose is a naturally occurring one, will be more eco-friendly while synthetic polymers will be harmful to environment. It also possesses excellent mechanical and thermal properties (Mohammadinejad et al., 2016). One needs to have greater knowledge about the cellulose material, i.e., about its structure and the modification that it can undergo or the reactivity of the material, to tune its properties in different ways and thus could be able to use for various purposes as cellulose is the best option to prepare materials of multifunctional properties.

Cellulosic material in their nanoscale is usually employed for many applications, as when they were incorporated into the biopolymers, will reinforce its mechanical and thermal properties. Cellulose nanostructures were categorized into two types, microfibrillated cellulose (cellulose nanofibers) and cellulose nanocrystals (Lavoine and Bergström, 2017). They differ in their processing methods and dimensions. Both these nanostructures are hydrophilic, thus their introduction into the hydrophobic petroleum-based polymers becomes difficult. To overcome this, modification of cellulose nanostructure is needed. Biodegradable polymers which are substitutes to petroleum-based foams have a low melting point, viscosity and rate of crystallization that are essential for polymer foaming. Thus cell growth during foaming results in cell coalescence and rupture. To overcome this and to obtain a high crystallization rate that can overpower the low melt strength and viscosity is obtained on binding of nanofiller material like cellulose nanostructure to the biopolymer like PLA to develop cellulose-based biopolymer foams. Cellulose here performs as a nucleating agent that increases the number of nucleation sites and crystallinity of the material which in turn can control the foam morphology and thermal and mechanical properties respectively. Thus cellulose nanostructures could be able to improve and reinforce the foam properties of a polymer.

Thus, biodegradable materials like cellulose can be easily used to improve the polymer characteristics that could be used for various applications like green packaging, lightweight construction, tissue engineering, as separation agents, sensors, in the delivery sector, for biofuel purification, supercapacitors, etc., as they show attractive properties like high biodegradability and biocompatibility with some interesting structure (Ahmadzadeh et al., 2016). They also show improved rheological and dynamic properties which indicate the effect of cellulose nanostructures in improving the foaming process. The potential applications of these cellulose-based foam are in the packaging industry itself, even though they have numerous applications across various industries. Thus cellulose based foams have emerged as a great substitute for petroleum-derived foams.

4.1 Processing of cellulose nanostructured biopolymer foam

Cellulose is not foamable, straight from the plants or any biomass. So for the preparation of cellulose foams, different processing methods are needed, depending on the properties required for the foam. For that different reinforcing methodologies are used like the preparation of composites, the inclusion of polymer and crosslinking, etc., by keeping the porosity of material at its high level. Mainly three types of processing methods were used earlier, they were batch processing, extrusion foaming and injection foaming, all with the same principle. Methods like microwave curing and freeze-drying were also employed for foaming (Motloung et al., 2019).

Batch processing can be of two types—the pressure quenched method and temperature-induced method. Both the processes involve the same steps, the difference lies in that while in the former case the cell nucleation and cell growth occurs in the autoclave itself as result of pressure drop, while in the latter case it occurs when the saturated polymer is taken out and is placed in a hot oil/ solvent bath. Dlouha et al. prepared cellulose nanofibers/PLA foams by the pressure-quench method at 60°C. They varied the foaming pressure from 12 to 20 MPa and nucleation is carried by a rapid depressurization to the atmospheric pressure. The cellulose nanofibers thus increase the nucleation, but the cell growth rate is seen to be reduced. The cell morphology of the foams also seemed to be changed on increasing the pressure. Thus with precise control of the processing conditions of pressure, one can generate cellulose nanofibers composite foams (Dlouhá et al., 2012). Qiu et al. in 2018 prepared PLA/cellulose nanocrystal foams by the temperature-induced method. Here CO_2 is used as a blowing agent and experiments were carried out for three forms of cellulose nanocrystals. In this case, the presence of cellulose nanostructures also greatly affects the cell morphology and structure, which in turn considerably improvise their mechanical properties and tensile strength. Batch processing is the cheapest among all (Qiu et al., 2018).

Injection foaming can be of two types: a low-pressure process and high-pressure process and is used to produce foams of different shapes. The addition of cellulose nanostructure results in improved cell density and cell size. Besides many benefits like less processing duration and sink mark removal, the main drawback it faces is the non-uniform cell morphology due to low-density range and nucleation. In all studies, super-critical fluids like supercritical nitrogen are used (Ries et al., 2014). Mi et al. prepared poly(ε-caprolactone)/cellulose nanocrystal foams via microcellular injection molding process and observed that there are non-uniform cells and it is due to the irregular cooling time, shear flow and mold design and thus the whole performance of the resultant foam may differ because of the size difference in the cells. The biocompatibility of the foams is also found to be good. So in order to obtain foams with uniform cell size with high cell density, it is necessary to optimize the processing parameters (Mi et al., 2014).

Extrusion foaming is usually employed where large scale industrial preparation is needed. It is a continuous process and is similar to normal extrusion. Several factors like processing temperatures, i.e., die temperature, die geometry and the pressure plays significant role to affect the cell properties like density and morphology in foam extrusion (Larsen and Neldin, 2013). In 2014 Zhao et al. prepared PVA/microfibrillated cellulose composite foam by a continuous extrusion foaming method in which supercritical CO_2 acts as a blowing agent and results in a biodegradable polymer of uniform cell structure and high density. High CO_2 content and low die temperature is found to increase the cell density (Zhao et al., 2014).

In all the techniques discussed above the processing conditions were very crucial for obtaining the desired cellular structures, and thus it is suggested to have accurate optimization of the conditions required to give a better performance at the end. For this an appropriate understanding of the material should be known, therefore it would be able to synthesis and incorporate the desired foams as required (Motloung et al., 2019).

Biodegradation of cellulose based polymer foam has been evaluated and found that several factors like nature of the material, i.e., whether it is amorphous or crystalline, molecular weight, polymer structure, hydrophobicity, etc., greatly affect the biodegradation of polymer materials. When considering the polymeric foams degradation ability also depends on the cell morphology. That is for closed cells the exposure of surface area towards the biodegrading agent will be less compared to the porous structure, hence complete biodegradation will occur effectively in the latter case within a lesser time. Various results shows that the cellulose nanostructure will increase the biodegradability of the foam material because of their great susceptibility towards microorganisms present in the environment.

5. Conclusion

The introduction of biodegradable foam was a prominent step on the road to use green materials in the packaging, biomedical and transport industries. Conventional polymers such as expanded polystyrene, expanded polypropylene, etc., with a non-renewable origin and non-biodegradable nature create severe environmental problems. The high price and reduced availability of fossil fuel-derived materials also confines the use of synthetic polymers for many reasons. This has surged a high demand for alternative options with economically and ecologically beneficial routes across industries. The environmentally sustainable natural polymer foams have changed the face of packaging, medical, automotive, logistics and other short-term applications. Starch foams are a major option as natural fiber foams owing to their good thermal insulation, excellent moisture absorption and low cost. The hydrophilicity and brittleness of the pure starch fibers trigger the development of composites or blends with other natural fibers and biopolymers with enhanced characteristics. Cellulose-based foams are also gaining much attention owing to their outstanding mechanical and thermal properties. Cellulose-based nanocomposite foams are a hot topic in both research and industries with improved features. The biodegradable or compostable market provides a fine and clean platform for sustainable materials which boosts the economic growth of industries with recent trends.

References

Aguilar-Palazuelos, E., José de Jesús Zazueta-Morales, Omar A. Jiménez-Arévalo and Fernando Martínez-Bustos. (2007). Mechanical and structural properties of expanded extrudates produced from blends of native starches and natural fibers of henequen and coconut. *Starch - Stärke,* 59(11): 533–42. https://doi.org/10.1002/star.200700608.

Ahmadzadeh, Safoura, Javad Keramat, Ali Nasirpour, Nasser Hamdami, Tayebeh Behzad, Lionel Aranda, Michel Vilasi and Stephane Desobry. (2016). Structural and mechanical properties of clay nanocomposite foams based on cellulose for the food-packaging industry. *Journal of Applied Polymer Science,* 133(2): n/a-n/a. https://doi.org/10.1002/app.42079.

Alavi, S. H., Rizvi, S. S. H. and Harriott, P. (2003). Process dynamics of starch-based microcellular foams produced by supercritical fluid extrusion. I: Model development. *Food Research International,* 36(4): 309–19. https://doi.org/10.1016/S0963-9969(02)00222-3.

Anglès, M. Neus and Alain Dufresne. (2000). Plasticized starch/tunicin whiskers nanocomposites. 1. Structural analysis. *Macromolecules,* 33(22): 8344–53. https://doi.org/10.1021/ma0008701.

Avella, M., Cocca, M., Me Errico and Gentile, G. (2011). Biodegradable PVOH-based foams for packaging applications. *Journal of Cellular Plastics,* 47(3): 271–81. https://doi.org/10.1177/0021955X11407401.

Avella, Maurizio, Mariacristina Cocca, Maria Emanuela Errico and Gennaro Gentile. (2012). Polyvinyl alcohol biodegradable foams containing cellulose fibres. *Journal of Cellular Plastics,* 48(5): 459–70. https://doi.org/10.1177/0021955X12449639.

Boischot, C., Moraru, C. I. and Kokini, J. L. (2003). Factors that influence the microwave expansion of glassy amylopectin extrudates. *Cereal Chemistry Journal,* 80(1): 56–61. https://doi.org/10.1094/CCHEM.2003.80.1.56.

Debiagi, Flávia, Suzana Mali, Maria Victória Eiras Grossmann and Fábio Yamashita. (2011). Biodegradable foams based on starch, polyvinyl alcohol, chitosan and sugarcane fibers obtained by extrusion. *Brazilian Archives of Biology and Technology,* 54(5): 1043–52. https://doi.org/10.1590/S1516-89132011000500023.

Demitri, Christian, Antonella Giuri, Maria Grazia Raucci, Daniela Giugliano, Marta Madaghiele, Alessandro Sannino and Luigi Ambrosio. (2014). Preparation and characterization of cellulose-based foams via microwave curing. *Interface Focus,* 4(1): 20130053. https://doi.org/10.1098/rsfs.2013.0053.

Dlouhá, Jana, Lisman Suryanegara and Hiroyuki Yano. (2012). The role of cellulose nanofibres in supercritical foaming of polylactic acid and their effect on the foam morphology. *Soft Matter,* 8(33): 8704. https://doi.org/10.1039/c2sm25909e.

Fang, Q. and Hanna, M. A. (2000). Mechanical properties of starch-based foams as affected by ingredient formulations and foam physical characteristics. *Transactions of the ASAE,* 43(6): 1715–23. https://doi.org/10.13031/2013.3073.

Fang, Qi and Milford A. Hanna. (2000). Functional properties of polylactic acid starch-based loose-fill packaging foams. *Cereal Chemistry Journal,* 77(6): 779–83. https://doi.org/10.1094/CCHEM.2000.77.6.779.

Fang, Zheng, Chuanhong Qiu, Dong Ji, Zhao Yang, Ning Zhu, Jingjing Meng, Xin Hu and Kai Guo. (2019). Development of high-performance biodegradable rigid polyurethane foams using full modified soy-based polyols. *Journal of Agricultural and Food Chemistry,* 67(8): 2220–26. https://doi.org/10.1021/acs.jafc.8b05342.

Glenn, G. M. and Orts, W. J. (2001). Properties of starch-based foam formed by compression/explosion processing. *Industrial Crops and Products,* 13(2): 135–43. https://doi.org/10.1016/S0926-6690(00)00060-1.

Glenn, G. M., Orts, W. J. and Nobes, G. A. R. (2001). Starch, fiber and $CaCO_3$ effects on the physical properties of foams made by a baking process. *Industrial Crops and Products,* 14(3): 201–12. https://doi.org/10.1016/S0926-6690(01)00085-1.

Guan, Junjie, Kent M. Eskridge and Milford A. Hanna. (2005). Acetylated starch-polylactic acid loose-fill packaging materials. *Industrial Crops and Products,* 22(2): 109–23. https://doi.org/10.1016/j.indcrop.2004.06.004.

Guan, Junjie and Milford A. Hanna. (2006). Selected morphological and functional properties of extruded acetylated starch–cellulose foams. *Bioresource Technology,* 97(14): 1716–26. https://doi.org/10.1016/j.biortech.2004.09.017.

Kaisangsri, Nattapon, Orapin Kerdchoechuen and Natta Laohakunjit. (2012). Biodegradable foam tray from cassava starch blended with natural fiber and chitosan. *Industrial Crops and Products,* 37(1): 542–46. https://doi.org/10.1016/j.indcrop.2011.07.034.

Kaisangsri. (2014). Characterization of cassava starch based foam blended with plant proteins, kraft fiber, and palm oil. *Carbohydrate Polymers,* 110(September): 70–77. https://doi.org/10.1016/j.carbpol.2014.03.067.

Larsen, Åge and Christoffer Neldin. (2013). Physical extruder foaming of poly(lactic acid)-processing and foam properties. *Polymer Engineering & Science,* 53(5): 941–49. https://doi.org/10.1002/pen.23341.

Lavoine, Nathalie and Lennart Bergström. (2017). Nanocellulose-based foams and aerogels: processing, properties, and applications. *Journal of Materials Chemistry A,* 5(31): 16105–17. https://doi.org/10.1039/C7TA02807E.

Lawton, J. W., Shogren, R. L. and Tiefenbacher, K. F. (2004). Aspen fiber addition improves the mechanical properties of baked cornstarch foams. *Industrial Crops and Products,* 19(1): 41–48. https://doi.org/10.1016/S0926-6690(03)00079-7.

Lee, Eun Yong, Kyung Il Lim, Jae-kag Lim and Seung-Taik Lim. (2000). Effects of gelatinization and moisture content of extruded starch pellets on morphology and physical properties of microwave-expanded products. *Cereal Chemistry Journal,* 77(6): 769–73. https://doi.org/10.1094/CCHEM.2000.77.6.769.

Lee, Siew-Yoong and Milford A. Hanna. (2008). Preparation and characterization of tapioca starch-poly(lactic acid)-cloisite Na^+ nanocomposite foams. *Journal of Applied Polymer Science,* 110(4): 2337–44. https://doi.org/10.1002/app.27730.

Lopez-Gil, A., Silva-Bellucci, F., Velasco, D., Ardanuy, M. and Rodriguez-Perez, M. A. (2015). Cellular structure and mechanical properties of starch-based foamed blocks reinforced with natural fibers and produced by microwave heating. *Industrial Crops and Products,* 66(April): 194–205. https://doi.org/10.1016/j.indcrop.2014.12.025.

Ma, Peter X. and Robert Langer. (1998). Fabrication of biodegradable polymer foams for cell transplantation and tissue engineering. *In*: *Tissue Engineering*, by Jeffrey R. Morgan and Martin L. Yarmush, 18: 47–56. New Jersey: Humana Press. https://doi.org/10.1385/0-89603-516-6:47.

Mali, Suzana. (2018). Biodegradable foams in the development of food packaging. pp. 329–45. *In*: *Polymers for Food Applications*, edited by Tomy J. Gutiérrez. Cham: Springer International Publishing. https://doi.org/10.1007/978-3-319-94625-2_12.

Manoi, Khanitta and Syed S. H. Rizvi. (2010). Physicochemical characteristics of phosphorylated cross-linked starch produced by reactive supercritical fluid extrusion. *Carbohydrate Polymers,* 81(3): 687–94. https://doi.org/10.1016/j.carbpol.2010.03.042.

Mariam, Irfana, Ki Yul Cho and Syed S. H. Rizvi. (2008). Thermal properties of starch-based biodegradable foams produced using supercritical fluid extrusion (SCFX). *International Journal of Food Properties,* 11(2): 415–26. https://doi.org/10.1080/10942910701444705.

Mi, Hao-Yang, Xin Jing, Jun Peng, Max R. Salick, Xiang-Fang Peng and Lih-Sheng Turng. (2014). Poly(ε-caprolactone) (PCL)/cellulose nano-crystal (CNC) nanocomposites and foams. *Cellulose,* 21(4): 2727–41. https://doi.org/10.1007/s10570-014-0327-y.

Mohammadinejad, Reza, Samaneh Karimi, Siavash Iravani and Rajender S. Varma. (2016). Plant-derived nanostructures: types and applications. *Green Chemistry,* 18(1): 20–52. https://doi.org/10.1039/C5GC01403D.

Molavi, Hooman, Somayyeh Behfar, Mohammad Ali Shariati, Mehdi Kaviani and Shirin Atarod. (2015). A review on biodegradable starch based film. *Journal of Microbiology, Biotechnology and Food Sciences,* 04(05): 456–61. https://doi.org/10.15414/jmbfs.2015.4.5.456-461.

Morais, Andreza Rochelle do Vale, Éverton do Nascimento Alencar, Francisco Humberto Xavier Júnior, Christian Melo de Oliveira, Henrique Rodrigues Marcelino, Gillian Barratt, Hatem Fessi, Eryvaldo Sócrates Tabosa do Egito and Abdelhamid Elaissari. (2016). Freeze-drying of emulsified systems: a review. *International Journal of Pharmaceutics,* 503(1–2): 102–14. https://doi.org/10.1016/j.ijpharm.2016.02.047.

Moraru, C. I. and Kokini, J. L. (2003). Nucleation and expansion during extrusion and microwave heating of cereal foods. *Comprehensive Reviews in Food Science and Food Safety,* 2(4): 147–65. https://doi.org/10.1111/j.1541-4337.2003.tb00020.x.

Motloung, Ojijo, Bandyopadhyay and Ray. (2019). Cellulose nanostructure-based biodegradable nanocomposite foams: a brief overview on the recent advancements and perspectives. *Polymers,* 11(8): 1270. https://doi.org/10.3390/polym11081270.

Nabar, Yogaraj, U., David Draybuck and Ramani Narayan. (2006). Physicomechanical and hydrophobic properties of starch foams extruded with different biodegradable polymers. *Journal of Applied Polymer Science,* 102(1): 58–68. https://doi.org/10.1002/app.22127.

Nakamatsu, Javier, Fernando G. Torres, Omar P. Troncoso, Yuan Min-Lin and Aldo R. Boccaccini. (2006). Processing and characterization of porous structures from chitosan and starch for tissue engineering scaffolds. *Biomacromolecules,* 7(12): 3345–55. https://doi.org/10.1021/bm0605311.

Pérez, Serge, Paul M. Baldwin and Daniel J. Gallant. (2009). Structural features of starch granules I. pp. 149–92. *In*: *Starch*. Elsevier. https://doi.org/10.1016/B978-0-12-746275-2.00005-7.

Pushpadass, Heartwin Amaladhas, Govindarajan Suresh Babu, Robert W. Weber and Milford A. Hanna. (2008). Extrusion of starch-based loose-fill packaging foams: effects of temperature, moisture and talc on physical properties. *Packaging Technology and Science,* 21(3): 171–83. https://doi.org/10.1002/pts.809.

Qiu, Yaxin, Qiaolian Lv, Defeng Wu, Wenyuan Xie, Sheng Peng, Ruyue Lan and Hui Xie. (2018). Cyclic tensile properties of the polylactide nanocomposite foams containing cellulose nanocrystals. *Cellulose,* 25(3): 1795–1807. https://doi.org/10.1007/s10570-018-1703-9.

Ribba, Laura, Nancy L. Garcia, Norma D'Accorso and Silvia Goyanes. (2017). Disadvantages of starch-based materials, feasible alternatives in order to overcome these limitations. pp. 37–76. *In*: *Starch-Based Materials in Food Packaging*. Elsevier. https://doi.org/10.1016/B978-0-12-809439-6.00003-0.

Ries, S., Spoerrer, A. and Altstaedt, V. (2014). Foam injection molding of thermoplastic elastomers: blowing agents, foaming process and characterization of structural foams. pp. 401–10. *In*: Nuremberg, Germany. https://doi.org/10.1063/1.4873809.

Robin, Frédéric, Cédric Dubois, Nicolas Pineau, Heike P. Schuchmann and Stefan Palzer. (2011). Expansion mechanism of extruded foams supplemented with wheat bran. *Journal of Food Engineering,* 107(1): 80–89. https://doi.org/10.1016/j.jfoodeng.2011.05.041.

Shey, J., Imam, S. H., Glenn, G. M. and Orts, W. J. (2006). Properties of baked starch foam with natural rubber latex. *Industrial Crops and Products,* 24(1): 34–40. https://doi.org/10.1016/j.indcrop.2005.12.001.

Shogren, R. L., Thompson, A. R., Greene, R. V., Gordon, S. H. and Cote, G. (1991). Complexes of starch polysaccharides and poly(ethylene co-acrylic acid): structural characterization in the solid state. *Journal of Applied Polymer Science,* 42(8): 2279–86. https://doi.org/10.1002/app.1991.070420819.

Shogren, R. L. (1992). Effect of moisture content on the melting and subsequent physical aging of cornstarch. *Carbohydrate Polymers,* 19(2): 83–90. https://doi.org/10.1016/0144-8617(92)90117-9.

Shogren, R. L., Lawton, J. W. and Tiefenbacher, K. F. (2002). Baked starch foams: starch modifications and additives improve process parameters, structure and properties. *Industrial Crops and Products,* 16(1): 69–79. https://doi.org/10.1016/S0926-6690(02)00010-9.

Silva, André da, Letícia Maciel Nievola, Cesar Augusto Tischer, Suzana Mali and Paula C. S. Faria-Tischer. (2013). Cassava starch-based foams reinforced with bacterial cellulose. *Journal of Applied Polymer Science,* 130(5): 3043–49. https://doi.org/10.1002/app.39526.

Simmons, Stephanie and Edwin L. Thomas. (1995). Structural characteristics of biodegradable thermoplastic starch/poly(ethylene–vinyl alcohol) blends. *Journal of Applied Polymer Science,* 58(12): 2259–85. https://doi.org/10.1002/app.1995.070581215.

Sjöqvist, Mia and Paul Gatenholm. (2005). The effect of starch composition on structure of foams prepared by microwave treatment. *Journal of Polymers and the Environment,* 13(1): 29–37. https://doi.org/10.1007/s10924-004-1213-8.

Soykeabkaew, Nattakan, Pitt Supaphol and Ratana Rujiravanit. (2004). Preparation and characterization of jute- and flax-reinforced starch-based composite foams. *Carbohydrate Polymers,* 58(1): 53–63. https://doi.org/10.1016/j.carbpol.2004.06.037.

Soykeabkaew, Nattakan, Chuleeporn Thanomsilp and Orawan Suwantong. (2015). A review: starch-based composite foams. *Composites Part A: Applied Science and Manufacturing,* 78(November): 246–63. https://doi.org/10.1016/j.compositesa.2015.08.014.

Tang, Xiaolin (Charlie) and Michael J. Pikal. (2004). Design of freeze-drying processes for pharmaceuticals: practical advice. *Pharmaceutical Research,* 21(2): 191–200. https://doi.org/10.1023/B:PHAM.0000016234.73023.75.

Vercelheze, Ana E. S., Farayde M. Fakhouri, Luiz H. Dall'Antônia, Alexandre Urbano, Elza Y. Youssef, Fábio Yamashita and Suzana Mali. (2012). Properties of baked foams based on cassava starch, sugarcane bagasse fibers and montmorillonite. *Carbohydrate Polymers,* 87(2): 1302–10. https://doi.org/10.1016/j.carbpol.2011.09.016.

Wang, Liang and Miguel Sánchez-Soto. (2015). Green bio-based aerogels prepared from recycled cellulose fiber suspensions. *RSC Advances,* 5(40): 31384–91. https://doi.org/10.1039/C5RA02981C.

Willett, J. L. and Shogren, R. L. (2002. Processing and properties of extruded starch/polymer foams. *Polymer,* 43(22): 5935–47. https://doi.org/10.1016/S0032-3861(02)00497-4.

Xu, Yixiang and Milford A. Hanna. (2005). Preparation and properties of biodegradable foams from starch acetate and poly(tetramethylene adipate-co-terephthalate). *Carbohydrate Polymers,* 59(4): 521–29. https://doi.org/10.1016/j.carbpol.2004.11.007.

Zhang, Jian-Feng and Xiuzhi Sun. (2007). Biodegradable foams of poly(lactic acid)/starch. I. Extrusion condition and cellular size distribution. *Journal of Applied Polymer Science,* 106(2): 857–62. https://doi.org/10.1002/app.26715.

Zhao, Na, Lun Howe Mark, Changwei Zhu, Chul B. Park, Qian Li, Robert Glenn and Todd Ryan Thompson. (2014). Foaming poly(vinyl alcohol)/microfibrillated cellulose composites with CO_2 and water as co-blowing agents. *Industrial & Engineering Chemistry Research,* 53(30): 11962–72. https://doi.org/10.1021/ie502018v.

Zhou, Jiang, Jim Song and Roger Parker. (2007). Microwave-assisted moulding using expandable extruded pellets from wheat flours and starch. *Carbohydrate Polymers,* 69(3): 445–54. https://doi.org/10.1016/j.carbpol.2007.01.001.

Chapter 11

Biodegradable Foams
Processes and Applications

Lorena Farías-Cepeda,[1,*] *Lucero Rosales Marines,*[1] *Anilú Rubio Rios,*[1] *Victor A. Cepeda Tovar,*[1] *Karina Y. Reyes Acosta*[1] and *Bertha T. Pérez-Martínez*[2]

1. Introduction

Currently, most of the organic compounds used in different applications are prepared by using non-renewable resources like fossil fuels and coal. The scarcity of these non-renewable resources, as well as their non-degradable nature, make petroleum-based products expensive and represent a strong impact on the environment (George et al., 2020; Singh et al., 2020). For these reasons, the industries and particularly researchers, have made multiple efforts to develop eco-friendly technologies and greener products that are petroleum-free and biobased renewable raw materials (Singh et al., 2020). In this way, biobased materials are materials produced from substances from living matter or with a biotechnological procedure and can be obtained naturally or synthetically. These biobased materials, being produced from agricultural compost, wood, vegetable oils and natural fibers, are considered greener than those produced from petroleum (which are generally not biodegradable). Some of these biobased materials are biopolymers, which means synthetic polymers chemically synthesized from renewable resources or living matter (George et al., 2020). However, not all biobased materials can be considered biodegradable materials, and synthetic biopolymers can be classified as biodegradable and non-biodegradable polymers (George et al., 2020).

There are different kinds of biopolymers, and one of them is biodegradable polymer foams. Polymer foams are porous polymer materials whose matrix contains many tiny foam holes inside. In general, these polymer foams have specific properties like low density, good heat and sound insulation, high specific strength or high corrosion resistance. These polymer foams can be classified in different ways, but one of the most common methods classifies them in three categories, hardness, density and foaming structure (Jin et al., 2019).

To produce polymer foams it is necessary to use blowing agents, which can be classified as physical blowing agents and chemical blowing agents. Physical and chemical blowing agents can be sub-classified as inorganic or organic blowing agents. The chemical blowing agents can be separated into two types, thermal decomposition and reaction blowing agents. Some of them are shown in Table 11.1 (Jin et al., 2019).

[1] Facutlad de Ciecias Químicas, Universidad Autónoma de Coahuila, Saltillo, Coahuila., México.
[2] Polymat y Departamento de Química Aplicada, Facultad de Ciencias Químicas, Univeridad del Pais Basco UPV/EHU, Donostia-San Sebastián, España.
* Corresponding author: lorenafarias@uadec.edu.mx

Table 11.1. Classification and some examples of blowing agents (Jin et al., 2019).

	Physical Blowing Agents	Chemical Blowing Agents	
		Thermal Decomposition	Reaction Blowing Agents
Inorganic	Nitrogen Carbon Dioxide Water Air	Bicarbonate Carbonate Nitrite	Sodium Bicarbonate Zinc powder Hydrogen peroxide
Organic	Pentane Hexane Dichloroethane Freon	Azo foaming agent Nitroso foaming agent Acylhydrazide foaming agent	Isocyanate Compounds

Some of the existing blowing agents include chlorofluorocarbons, hydrofluorocarbons, azodicarbonamide, azobisisobutyronitrile, butane or pentane; however, it is necessary to use the less harmful physical blowing agents like carbon dioxide or nitrogen to produce foams from biodegradable polymers like poly (lactic acid), poly (lactic-co-glycolic acid), poly (propylene carbonate), polycaprolactone, poly (ethylene oxide), poly(butylene succinate), poly(ethylene glycol), poly(vinyl alcohol). The application areas of these foams are extensive and include (1) medical devices (tissue regeneration, tissue scaffolds, drug delivery), (2) sound and heat insulation, (3) UV-absorber composites, and (4) packing (Sarver and Kiran, 2021).

However, the biodegradable polymers mentioned above are not the only ones used to produce foams; there are more polymer foams that are widely used in different applications. Some examples are summarized in Table 11.2 (Jin et al., 2019).

A particular class of materials are nanocomposites polymer foams, which are nanostructured materials that most of the time involve the use of polyurethane foaming, but emulsion templating has proven to be an effective way to prepare polymer-based foams (Sousa et al., 2017). For example, bio-based polyurethane foams can be developed from bio-polyol (castor oil-based), incorporating bentonite nanoclay into the bio-polyol mixture as nano-reinforcement, giving rise to a bio-based engineered nanocomposite foam with enhanced mechanical and thermal barrier properties (Kadam et al., 2019). Another kind of biodegradable polyurethane foam can be obtained from bioresource-based polyols like lignin, soy oil-based polyols and polymeric methyldiphenyl diisocyanate. In this case, mechanical, thermal and biodegradable properties are also improved (Luo et al., 2018a).

At the same time, there are other types of foams that can be produced from agro-industrial residues, such as starch-based foams that have been developed to reduce the use of petroleum-based packing products, particularly polystyrene, that has very good properties like resistance to water and good thermal insulation, and it has application in construction, transport or packing, but since polystyrene foams do not exhibit biodegradability, they have a negative impact on the environment. These starch-based foams are biodegradable in many environmental conditions, and that is why they are widely used for single-use packing applications (Sanhawong et al., 2017; Chaireh et al., 2020; Machado et al., 2020). One more advantage is that they can be easily processed by several processes

Table 11.2. Some polymer foams and their applications (Jin et al., 2019).

Polymer Foam	Applications
Polyurethane soft foam	Furniture, clothing, car cushions, sports equipment
Polyurethane hard foam	Refrigerators, freezers, refrigerated containers
Polypropylene foam	Daily necessities, military industries, transportation, aerospace
Phenolic foam	Architecture, automobiles, electrical and electronic applications
Polyimide foam	Aerospace, aircraft, marine

such as extrusion, baking, compression, injection molding or microwave heating. However, they exhibit some disadvantages like poor flexural properties and high-water absorption, which can be improved by the incorporation of hydrophilic reinforcing fillers like softwood, jute and flax, wheat bran, natural fibers, chitosan, nanoclays, lignin, kraft fiber, cardoon waste, grape waste and cellulose (Sanhawong et al., 2017).

The bio-based and biodegradable polymers have been considered as an important alternative for traditional petroleum-based plastics, and they are widely used in many different areas due to low-cost production, low ecological impact and improved barrier and thermal properties.

2. Biodegradable foams process

Since polymer foams can be described in terms of their density, average cell size and cell density, it is necessary to know the processing and sourcing methods of the foams to understand their characteristics and capacities. Polymer foams have a variety of classification methods, but the common classification is by hardness, density and foaming structure.

The first stage of the foaming process consists of cell formation. This stage involves adding a blowing agent to a molten polymer under certain conditions that lead to the production of a large quantity of gas. The gas formation can occur either from a series of chemical reactions or by adding a gas inside the vessel, leading to the generation of a polymer/gas solution. As the amount of gas increases, the solution gets supersaturated, and the gas escapes from the solution, resulting in the formation of the cell nucleus through nucleation (Jin et al., 2019; Álvarez et al., 2020). For rigid foams, closed cells are designed and limit the breakage of cell walls; likewise, cross-linked systems are sought to avoid contraction due to the effect of gas (Chevali and Kandare, 2016).

Once the cells have been formed, the gas inside them exerts pressure, and the smaller the size of the cell, the greater the pressure exerted by the gas. When two cells of different sizes are adjacent to each other, the gas migrates from the smallest cell to the largest one, and the cells merge. This nucleation results in an increase in the number of cells, and the diameter of the cell hole expands, allowing cell growth. The criteria for selecting blowing agents are based on their thermodynamic properties, suitability for use in the art and application. The amount of water must also be considered in the reaction mixture as this will influence the stiffness of the domains. Considering its capacity to formation and growth of cells, the foam system increases its surface area and volume, causing a thinning of the cell wall, which destabilizes the system. The stabilization of the system will then take place by cooling or by adding surfactants to assist the components mixing and benefit a uniform distribution of cells (Chevali and Kandare, 2016; Jin et al., 2019).

In the mechanical foaming method, the air is added to the polymer resins by mechanical stirring before the foaming process. This method has advantages such as that there is no need to add a foaming agent, the mechanical means being a safe, highly efficient and an economical method (Jin et al., 2019). The physical method of foaming consists of mixing a low boiling liquid with a polymer and, through a combination of pressure and heating, foaming is achieved. There are different reagents in the mix during the synthesis. Blowing agents can be used, but the foaming method generally does not produce contaminants or leave a residue at the end of the process. Another advantage of this method is the little effect it has on the properties of foamy plastics. One factor to consider is that this foaming method requires an injection molding machine, and the technical requirements of the equipment are high (Jin et al., 2019; Wu et al., 2020). One of the main contributions related to the formulation of biopolymer foams is, in particular, the physical foaming agents, like carbon dioxide or gaseous nitrogen because they are substances that do not produce harmful damage to the environment (Kiran, 2016). Due to these, it has been chosen to eliminate agents like chlorofluorocarbon compounds or CFCs (Gama et al., 2018) or single-chain aliphatic compounds such as butanes and other compounds of carbon (Song et al., 2019), because they produce a progressive degradation in the ozone layer, causing irreparable damage, in addition to their high cost and difficulty in their use. The need to synthesize environmentally-friendly blowing agents will continue to drive the exploration, research and commercialization of formation processes that use compounds like CO_2. Recent review articles cited in this paper offer insights into current challenges (Standau et al., 2019).

A trend in bio-foams is to substitute the elements of polymeric foams for derived biobased resources; some of these developed from functionalized vegetable oil with synthetic fibers, obtaining rigid bio-foams with high biodegradability, adding reinforcements and additives to increase its functionality. These bio-based boosters can also increase the resistance of the bio-foams to humidity, fire, physical aging or photo-oxidative degradation (Chevali and Kandare, 2016).

Tannin bio-based foams reinforced with cellulosic wood fiber have been reported (Wu et al., 2020) prepared from 95% of natural raw materials and through a room temperature process without the need for a pre-polymerization. This work relates to a process by adding natural materials as modifiers, fillers and fibers into phenolic foams to reinforce and enhance foam mechanical properties. The use of a blowing agent was implemented and by a mixing method in combination with the heat release in the system. The results presented allowed the capacity of these processes and formulations to substitute phenolic foams since their mechanical, thermal properties and foam cell morphology are comparable (Wu et al., 2020).

By two methods, the chemical foaming process can be undertaken. The first method is to add a foaming agent to a molten polymer, decomposition of these release gas, subsequently the polymer is foamed by pressurizing and heating or using a conventional compression molding machine. The other way to reach a chemical foaming process is by the chemical reaction between two polymers to produce inert gases (Sanhawong et al., 2017; Jin et al., 2019; Lu et al., 2021). A condition to control is the temperature of the reaction mixture since it will depend on factors such as the foam modulus, the evolution and growth of bubbles, phase cell separation and opening and curing (Chevali and Kandare, 2016).

For bio-composite foams preparation, there are different processing trends considering the raw material. Moo-Tun et al. (2020) report the preparation of starch-based foams by an initial batter preparation from starch and cellulose fibers, gelatinizing a part of the mixture and the rest mixed with a (PLA) and calcium carbonate, using guar gum as a binder additive. Magnesium stearate was used as a mold releasing agent. Finally, water was added, and the solids content adjusted. Through the final baking process, composite foams were obtained. The composites foams that they got by this formulation had improved mechanical flexural properties, dimensional stability and moisture resistance.

Among the advantages of chemical formation is that the foaming agent can be thermally decomposed into a specific temperature range and release one or more gases, so it may be appropriate for polymers that have melt viscosity in a specific temperature range. Another advantage is that foaming can be carried out by means of an ordinary injection molding machine. It is necessary to consider that for this process, high precision of mold making is required, and a second clamping pressure device is required during the high-pressure foaming process (Jin et al., 2019). Ago et al. (2016) obtained composite bio-foams from residual oil palm using a set of sulfur-free fractionation methods, they used lignocellulose nanofibrils to reinforce starch bio-foams, obtaining lightweight materials with a performance similar to that of polystyrene foams.

There are some relevant reports using renewable biomass as raw materials in the development of biodegradable foams; many of these works try to establish new development routes seeking the use of biodegradable materials and, in some cases working in combination with conventional polymers but giving priority to the use of biological bases using them in more significant proportion (Luo et al., 2018a) and also give rise to adaptations of the methods already established in order to achieve the formation of biodegradable foams with good structure properties (Ago et al., 2016; Darder et al., 2017; Sanhawong et al., 2017; de Avila Delucis et al., 2019).

2.1 Processing techniques

2.1.1 Foam extrusion molding

Among natural polymers, there are some that have thermoplastic behaviors under certain conditions, which make them attractive to be processed by extrusion. This process is carried out by means of an extruder capable of having several heating zones. Once the extruder is brought to a specific temperature, the discharge conditions and screw speed are adjusted. A polymer (polymer blend or polymer composite) is added to the hopper of the extruder. A supercritical fluid is introduced into the barrel area, which dissolves in the polymer melt, reducing its viscosity. Through reducing the foaming temperature, the cell density and resistance of the melt increase, finally obtaining microcellular foams (Jin et al., 2019). It is important to consider that in an extrusion process, there is high shear stress, which leads to degradation (breaking of the molecular bond) and increasing the gelatinization of the polymer, so it is necessary to validate that the biopolymeric material, subject in this process, it has good resistance to melting, thermal stability and limited swelling in order to support extrusion molding (Rodríguez-Castellanos et al., 2015).

2.1.2 Foam injection molding

The injection molding machine consists of various injection temperatures from the hopper to the nozzle. Injection molding parameters such as injection and mold temperature, injection speed, injection pressure, retention time and cooling time were set to fixed values according to the processing conditions for plasticizing the polymer (polymer blends or polymer compound). There are also injection molding machines that can control the mold opening distance for the prepared foamed parts. To inject a supercritical fluid is then dosed into the barrel or into the polymer melt, and the fluid melts with the polymer to form a homogeneous polymer/gas solution. The mold temperature was controlled by electric heating. The solution formed is injected into the mold cavity, and there is a rapid pressure drop. The thickness can be controlled by changing the mold opening distance, which produces a homogeneous and/or heterogeneous cell nucleation and cell growth until foam formation (Jin et al., 2019; Hou et al., 2021).

The visual appearance and structural characteristics of obtained foams are highly dependent on raw material, mold design and processing conditions; this has allowed the development of biodegradable foams to be used in many areas and has attracted the interest of numerous research groups. The processing developments have achieved structural advantages of synthetic foams and have implemented improvements in terms of sustainability.

3. Biodegradable polymers as foams

3.1 Cellulose and derivatives bio-foams

Cellulose is the most abundant polymer in nature because it can be found in plants, is a semicrystalline, highly polymerized natural homopolymer. Cellulose is a linear polymer consisting of β-d-glucose units with β-1,4-glucan bonds. Inwood, the cellulosic fibers exist as 3D matrixes, which are responsible for their crystalline form and resulting mechanical properties. In the cell wall of bacteria, yeasts and fungus, it can be found β-glucans which contain long chains of glucose molecules linked with bonds. But, β-glucans differ from cellulose in having branched chains besides the straight linear chains. β-glucans are water soluble fibers because they have branched and linear linkages that water can diffuse into the network and solubilize them, although cellulose, is water-insoluble as a consequence of its high molecular weight and strong hydrogen bonding between the molecules and crystalline structure (Shavandi et al., 2020). In recent years, cellulose has received increased attention because it has low price, biodegradability and biocompatibility. These features make cellulose an ideal candidate for the synthesis of bio-based foams (Lu et al., 2021).

Cellulose has strong inter or intra-molecular interactions, which sometimes limits its application or process. Usually, cellulose requires to be dissolved in a surfactant solution and then dried. In order to avoid this procedure Lu et al. (2021), synthesized a cellulose nanofibrils (CNF) composite, with γ-glycidoxypropyltrimethoxysilane (GPTMS) as a crosslinker and gelatin. They found that the gelatin content influences the properties of the bio-foam, for example, better mechanical properties and water stability in comparison with the CNF foams without GPTMS or gelatin. It was also found that the bio-foams have 99.1% of porosity, a density of 0.077 g/cm^3, and hydroxyl and carboxylate groups, properties that made this bio-foam a good option as an adsorbant.

Recently, Sousa et al. (2017) obtained nanocomposite bio-foams by Pickering emulsion templating. They used cyclohexane and acetylated bacterial cellulose to stabilize the Pikering water in oil emulsion and as templates for the thermosetting nanocomposite foams; they selected a monomer mixture of acrylate epoxidized soybean oil, 1,6-hexanediol diacrylate (HDD) and divinylbenzene (DVB). They found that the stabilization of the emulsion plays an important role in the morphology and porosity of the bio-foam. The incorporation of DVB or HDD also resulted in the smallest average cell diameter and highest density and Young´s modulus. The thermal stability of these bio-foams was approximately at 320°C and had a Young's Qmodulos between 3.41 and 16.39 MPa, depending on the bio-foam composition, so they claimed that tailor-made composites could be obtained by this methodology.

Tian et al. (2016) obtained a plasmonic aerogel with Bacterial Cellulose Nanofibrils (BCN); they prepared a hydrogel of BCN, and then it was freeze-dried. They cut in the desired dimension in liquid nitrogen in order to preserve the porous structure. The samples of BCN were immersed in an aqueous solution of gold nanorods (AuNRs) and washed and then freeze-dried. For this simple procedure, they created plasmonically active three-dimensional bio-foams, also demonstrating that 3D plasmonic bio-foam exhibits significantly higher sensing, photothermal and loading efficiency compared to conventional 2D counterparts. In this respect, Han et al. (2017) prepared nanocellulose-borax-polyvinyl alcohol hybrid bio-foams. They used an aqueous medium followed by a freeze casting technique. First, they mixed PVA and borax and then the nanocellulose was dispersed in the PVA-borax in order to form a 3D network; they used three different types of nanocellulose with different aspect ratios and crystallinity. They found that with increasing size and aspect ratio the bio-foam exhibited a pronounced honeycomb-like structure, smaller cell diameter and high mechanical strength; the use of a small amount of borax also promotes the properties they observed.

3.2 Polyurethane bio-foams

Polyurethane (PU) foams have moisture permeability, thermal conductivity and high performance because of strength to weight; because of that PU is widely used in all industrial fields and is present in many aspects of modern life. But PU is a petroleum-based polymer and not biodegradable, so the academic word is focused on the development of biodegradable and environment-friendly biobased PU foams. For this purpose, Luo et al. (2018b), synthetized different PU-biofoams. They used soy-based polyol (soybean phosphate ester polyol), lignin powders and methyldiphenyl diisocyanate by one-pot and the self-rising method; it is important to note that they did not use a blowing agent. They studied the lignin content and found that all the foams had analogous internal cellular morphology and microstructure. By FTIR they found that covalent bonds exist between soy-based polyurethane and lignin, because of that a 3D macromolecular structure were formed with improved mechanical, thermal and biodegradability properties. In the same way, Luo et al. (2019) obtained bio-foams nanocomposites of soy oil-based PU and hydroxyl-functionalized multiwalled carbon nanotubes (MWCNT-OH) by the one-pot method and water as blowing agent. They found good dispersion and strong interaction of MWCNT-OH with the matrix or pMDI, the bio-foams had a cell morphology, no effect of MWCTN-OH content was observed in the cell morphology. In contrast, they found that mechanical and electrical properties were improved with the increase of MWCNT-OH content. Their bio-foam had great thermal stability properties, and they proposed that it could be used as an effective EMI-shielding material.

Recently de Avila Delucis et al., 2019 studied the aesthetical features and photodegradation resistance of different PU bio-foams. They incorporated four different forest-based wastes (wood, bark, kraft lignin and paper sludge) into rigid PU foams via free rise pouring method using castor oil as polyol source and later aged by UV radiation. They found that different color patterns and photodegradation yield can be induced depending on the forest filler. The PU bio-foam were exposed to UV radiation after 20 days; by colorimetric analyses, they found that bio-foams presented similar kinetics, a decrease in L* and specular gloss, and an increase in a* and b* (CIELab method). They recommended the use of bark and kraft lignin-filled foams for applications where UV resistance is a requirement.

Ranote et al. (2019) synthesized biobased polyurethane foam using Moringa Oleifera Gum (MOG) as a polyol source for reaction with 4,4′–diphenylmethanediisocyanate via urethane linkages. By FTIR, they confirmed the urethane linkages and the absence of -NCO groups in the MOG-PU spectra; by EDX spectrum, they found peaks of N due to urethane linkages with porous morphology. They used the MOG-PUF for the adsorption of malachite green dye from an aqueous solution and tested for antibacterial activity.

3.3 Poly (lactic-acid) bio-foams

Poly (Lactic-Acid) (PLA) is classified as an aliphatic polyester due to ester bonds that connect the monomer units. PLA can degrade in situ through a hydrolysis mechanism in which water molecules could break the ester bonds of the polymer. The products obtained by PLA degradation are mainly composed of lactic acid and short oligomers, which are recognized and metabolized by the human body. Hence, PLA is considered as a biopolymer, offering good biocompatibility (Garlotta, 2001). Some applications of PLA, can be found in tissue engineering, medical application; besides, it is suitable in food packaging foams (Chevali et al., 2011; Nofar and Park, 2014; Weber et al., 2016).

The challenge of foaming PLA is based on the low melt strength of the polymer, and to enhance this fact, several modifications were investigated. For example, the molecular architecture (branched structure) using the addition of chain extender (Di et al., 2005; Marrazzo et al., 2007; Pilla et al., 2009c; Pilla et al., 2009a; Sungsanit et al., 2010; Corre et al., 2011), as well as their configuration: modifying the L/D ratio configuration of the PLA (Bigg, 2005; Dorgan et al., 2005; Mihai et al., 2009; Dorgan, 2010), besides varying the molecular weight (Dorgan et al., 1999; Dorgan et al., 2000; Dorgan et al., 2005; Dorgan, 2010). According to literature, in general three types of PLA-foams, blends, composites and nanocomposites can be found through various manufacturing methods, mainly by extrusion or injection foaming. Some efforts have been demonstrated in order to improve the properties of the PLA-foams for example, PLA blended with other polymers like starch (Preechawong et al., 2005; Mihai et al., 2007; Zhang and Sun, 2007a; Zhang and Sun, 2007b; Hao et al., 2008) or Poly(Butylene Adipate-co-Terephthalate) (PBAT) (Yuan et al., 2009; Pilla et al., 2010; Li et al., 2012), where some characteristics of PLA were improved, like resistance to water absorption, ultimate tensile strength, low melt strength and low elasticity. For PLA composites, it was found that the addition of natural fibers or powders like flax fiber (Pilla et al., 2009b), silk fibroin powder (Kang et al., 2009), wood flour (Matuana and Faruk, 2010) and Micro Fibrillated Cellulose (MFC) (Boissard et al., 2012); could improve the final mechanical properties and achieve fully degradable biocomposite foams.

Finally, for PLA nanocomposites, the main example is the use of nanoclay (Fujimoto et al., 2003; Hwang et al., 2009; Matuana and Diaz, 2010; Ameli et al., 2014) as due to their presence could promote the cell density in foamed samples; hence, they could act as heterogeneous cell nucleating agents, controlling the cell size of the PLA foamed in the range of nanoclay is increased.

3.4 Starch bio-foams

Starch is the most common carbohydrate in human diets and is contained in large amounts in staple foods like potatoes, corn or rice; this polymer is considered as biodegradable due to that can be converted by microorganism to carbon dioxide, water and mineral biomass. The polymeric structure of the starch consists of numerous glucose units joined by glycosidic bonds containing two kinds of microstructures: linear structure (amylose which has α-1,4 linked glucopyranosyl units) and branched structure (amylopectin with short α-1,4 D-glucopyranosyl chains linked by α-1,6 bonds). Regarding crystallinity of the starch, amylopectin is responsible for the double-helical crystalline structure and amylose is considered as the amorphous structure, and part of them is present as a helical complex with the lipids (Kennedy and White, 1987; Soykeabkaew et al., 2015; Jiang et al., 2020). Traditional foams such as polystyrene, polyethylene and polyurethane could be replaced by starch-based foam; however, the starch itself is somewhat brittle and water sensitive, hence in order to improve their final properties such as flexibility, the addition of plasticizer is commonly used (e.g., water and glycerol) (Neus Angles and Dufresne, 2000), besides water is used as a blowing agent. In order to improve the foam cell structure, mechanical strength and reduce processing time, the addition of some salts like NaCl (Zhou et al., 2006), $CaCl_2$ (Zhou et al., 2006), and $CaCO_3$ (Glenn et al., 2001) is used.

However, it is worth mentioning that starch has free hydroxyl groups, forming hydrogen bonds that provoke their hygroscopic nature. This fact leads to the collapse of the foams when they are in contact with water; hence to face this issue, starch foams have been modified using different strategies. For example, starch/polymer blend foams produce more flexible foams and reduction in sensitivity to water using chitosan (Shey et al., 2006); controlled degradation with polylactic acid (Ganjyal et al., 2007); and better compressive strength blending with poly (tetramethylene adipate-co-terephthalate) (EBC) (Xu and Hanna, 2005). Composites starch foams adding natural fibers were studied as well for increasing the compatibility of starch and cellulose, obtaining a significant improvement in mechanical properties (Soykeabkaew et al., 2004); adding kraft fiber (Kaisangsri et al., 2014) and wood fibers (Glenn et al., 2001), improved the flexural and compressive strength; besides the addition of cotton linter decreased in density of the foams (Bergeret and Benezet, 2011; Bénézet et al., 2012).

Finally, different starch foaming processes are widely investigated, of which, it can be mentioned: extrusion, baking/compression, microwave heating, freeze-drying/solvent exchange and supercritical fluid extrusion.

4. General applications for biopolymers foams

In a very general way, there are various applications for biopolymeric foams. Here, the main ones are listed, stating the most outstanding contributions for each one. Some of the polymers extracted from nature that can be seen more frequently in literature are polylactic acid (PLA) (Jeong et al., 2020), polyvinyl alcohol (PVA) (Yin et al., 2019), propylene polycarbonate (PCP) (Manavitehrani et al., 2019), poly-caprolactone (PCA) (Zhang et al., 2019), and polyethylene glycol (PEG) (Asikainen et al., 2019). The areas of application of the bio-foams are in medical devices, engineering materials, in the textile industry and in the field of pharmacology; being considered as alternatives for the production of degradable bioplastics, used mainly in packing. Concerns around the accumulation of plastics in the environment are a significant factor contributing to the growing interest in biodegradable polymeric foam (Milovanovic et al., 2019). Table 11.3 provides the main applications of some biopolymers widely used in the area of science and technology.

4.1 Applications in medicine

Biomedical applications have drawn considerable attention for the handling and disposal of biopolymers compared to other areas (Martín de León et al., 2019). Biodegradable compounds are being designed more frequently today to alleviate medical concerns in applications involving details such as medical implants for artificial and/or vascular grafts, especially for surgical sutures that facilitate tissue coalescence integration (Tian et al., 2012). These sutures are easily sterilizable and offer high durability, robustness and a function to regenerate tissues before they are physically removed from the body, thereby allowing them to dissolve with great ease in the body. Some

Table 11.3. General applications for bio-foams (George et al., 2020).

Polymeric Compound	Biopolymer	Applications
Polyesters (PES)	Poli-dioxanes	Medical implants and sutures for slow tissue healing.
Polyamides (PA)	Nylon; Poly-α-amides	Textile fibers and numerous plastic products. Dilation cathers from this poly-(amino-acids), variety of transfusión sets.
Polyurethane (PU)	Polyesterurhetane; Polyeterurethane	Foams and coatings absorbent materials and biomedical grafts.
Vinyl Polymers (PV)	Poly-Vinyl Alcohol; PVC	Processing of paper, adhesives and coatings, soluble of water, emulsifiers, textile pigments, plastic tubes, films, bottles, insulation for metallic wires, packaging, etc.

compounds such as Poly-Glycolic Acid (PGA), poly-lactic acid (PLA) and their copolymers are widely used as sutures as they offer reliable knot-forming robustness and exceptional flexibility in contact with human tissues (Hirvikorpi et al., 2011). Bone fixation implants from biopolymer sources are also a less-explored field, they promote the absence of disturbances during bone improvement. They do not require additional surgery for their extraction after the restoration of bone, as the entire implanted device completely solubilizes in the affected area. Biopolymer foams are used as drug delivery systems due to their biocompatibility with the environment, pharmacokinetics and immunogenicity over synthetic counterparts, facilitating viability in the removal and absorption process aforementioned (Francis et al., 2013). Some biopolymers contain ester (R-COOR) or carboxylic acid (R-COOH) functional groups, that are used as systems for the administration of pharmaceutical products. Some polyurethanes (PU) and their derivatives are identified as a group compatible with blood flow since they can suffer wear. The implementation of polyurethane foams coupled with biopolymeric substances allows flexibility and resistance properties essential to perform grafts such as a kind of syringes in artificial blood vessels (George et al., 2020; Ma et al., 2021).

4.2 Applications in the agricultural field

Biopolymer-based foams have generated attention in the farming field by developing fertilizers and containers that allow soil improvement during harvest seasons. Some products developed from these substances of natural origin have promoted plant development since the polymer films added to this raw material conserve moisture, prevent the formation of harmful weeds and maintain the optimum temperature of the soil (Raj et al., 2011).

The amount of water and soil nutrients required is also considerably and notably reduced (Avérous, 2004). As mentioned earlier, within the agricultural field, it is possible to demonstrate how the use of PVC, LDPE and PVA films is utilized as organic substitutes for various soils in numerous regions. These films are modified to undergo degradation when the cultures growth period is complete, either with the help of microbial agents (microorganisms from subsoil) or through the addition of certain particles that promote the final stage of growth. For this specific case, a comparison of these biopolymeric foams with poly-caprolactone is carried out since this is used to manufacture containers for numerous agricultural plants. These containers tend to biodegrade rapidly over a significant period due to annulus breakage during degradation, allowing the period desired for crops to sprout appropriately (Yadav et al., 2018).

4.3 Applications in the food area

Today various polymeric materials have been perfected that are used in the food area. It should be noted that the purpose of these compounds is to keep any product of animal or vegetable source in good condition for intake in humans. That is why the foams with biopolymeric structures are of high added value in this field of research. Biopolymers that serve as packaging for nutritional products have received enormous interest due to their physical characteristics that allow foreseeing alterations in the chemical composition (bacterial, oxidation, degradation) and the processing conditions to achieve the desired characteristics. These characteristics are driven by the primary source product to be packaged, its storage environment (Rohan et al., 2018). Some significance has been established about the combination of the properties of different biopolymer foams in a biodegradable film to develop a packaging material whose resulting properties are very remarkable and with a good response to conservation, this is in comparison with those observed in the materials that come from a synthetic source. However, the latter has a slight advantage over its counterparts of biological origin. Many biopolymers are based on polysaccharide-type hydrocarbon compounds such as starch, cellulose, etc., and their respective copolymers of polyesters and polyethylene are being used to develop possible packaging films (Abid et al., 2018). Currently, natural fillers applied as foams, such as starch, are mixed with polymers from a synthetic source to improve degradation properties of the slices for recovery packaging, thus achieving more excellent biodegradability and, therefore, a lower cost in comparison with those obtained from an unsustainable source. Some of the containers are not necessarily containers for liquid content. It is also possible to acknowledge them in other presentations, including grocery bags, storage containers, plates, glasses and disposable cutlery. For food and condiments, packaging made from a biofoam offers the expectation of preserving the quality of the article in the packaging from its moment of production until its final stage, consumption (Li et al., 2017).

Foams based on some biopolymeric compounds are essential components of a variety of foods and beverages, for example: edible creams, mousses, soufflés, milk ice creams, bread, cakes, sugary juices, teas and/or any other hydrocarbon drinks, etc. These new types of "bio foams" and agents capable of forming some type of nourishment and in turn forming stabilizing compounds of a polymeric base that a safe and reliable extraction of a natural source (van der VegtHendrik and Bisschop, 2012). The bubbles that occur in this type of foam for products that are for human consumption can impart specific sensory properties in which they are described: the texture, the color, visual appearance (a novelty in the market) and the reaction of the consumer (Heymans et al., 2017). The additional advantages that these materials provide are: economic, they are products during their transport, and it is not even likely to cause adverse reactions, depending on weather conditions, for example. From the chemical point of view, the shares of the foams of an aggregate from a renewable source have generated an upward impact today. Being in the first place the production of food foams, in its composition is the gaseous phase structured from Oxygen (O_2), Nitrogen (N_2), and carbon dioxide (CO_2), being the synthesis route composed of a continuous phase (higher proportion) the aqueous solution (H_2O) so that the discontinuous phase (the gas) has a (reasonable solubility) in the ongoing phase (Valadbaigi et al., 2019).

The role played by these materials in packaging systems is very important for the protection of food quality and its shelf life, especially in the supply chain. This will allow consumers to obtain products that correspond to high food quality and excellent safety expectations (Marsh and Bugusu, 2007). All packaging for commercial purposes must provide the following (Meneses et al., 2012):

1. A perfect preservation of the safety and freshness of the food.
2. A correct identification of the product.
3. A benefit during storage and distribution.

Other primary functions are to show the brand image and provide information on the composition, preparation and final handling of the packaging. For the performance of these functions, packaging generates environmental impacts that affect food supply chains, resulting in its phases within the life cycle, namely, production, transport to consumption and disposal (Bertoluci et al., 2014). Recent research has provided the use of PLA compared to containers produced or thermoformed from polyethylene terephthalate (PET) and polystyrene (PS), emphasizing different strategies in which shelf life stand out (Madival et al., 2009). Díaz et al. (2010) evaluated the effects of two packaging systems, such as vacuum bags and plastic trays, through the synthesis of foams of the PLA biopolymer reinforced with PS to keep products under refrigeration.

In this sense, food packaging was accessed by Kaisangsri et al. (2012), developing trays of a biodegradable foam from cassava starch mixed with natural polymers extracted from Kraft fiber and chitosan, obtaining results that revealed mechanical properties similar to PS foam. These comparisons made by these authors could also be extended from an environmental perspective or point of view, allowing the marketing of packaging products under certain ecological conditions (Accorsi et al., 2014). In this regard, the central point of foam applications from biopolymeric materials is centered on the synthesis and functionality of PS foam trays, so the authors believe that it could improve knowledge in the field by providing information reliable on data and dependably obtained results. Another perspective focused on food is the packaging of some oils of vegetable origin, extracted from numerous seeds and/or legumes. However, despite the functionality, the way of using these fossil-based polymeric foams is a cause for concern about the environment in general due to the use of non-renewable resources (e.g., oil, gas, etc.) and the production of waste after the useful life of these resources. Recycling is a possible solution to prevent the conventional use of these containers, which serve as containers for oils and fats of vegetable or animal origin (Davis and Song, 2006). Biodegradable green source plastics are a competent inventive alternative to those already mentioned above due to these main reasons:

1. The use of renewable and potentially more sustainable raw materials.
2. Biodegradability allows procedure through organic reuse to minimize landfill disposal in the absence of recycling alternatives.

4.4 Applications in textile fibers

Therefore, this has led the industry to create foams composed of renewable materials, especially of the polyol type from edible and non-edible vegetable oils (FAN, 2011). Recent research has also applied this area of opportunity in the extraction of groups isocyanates of some amino acids or fatty acids (Konieczny and Loos, 2019), as well as the use of natural fibers from agricultural residues such as fillers (Buggy, 2006). The use of these sustainable materials contributes effectively to reducing the emission of gases that produce a greenhouse effect, promoting sustainability and conservation, thus providing a profitable alternative to produce compounds with these functional characteristics (Mokhothu and John, 2017). Among the renewable source materials used in the industry is the use of these by adding certain fillers of inorganic or structured organic origin from fibers, which have been a novelty in science. Recent reports have confirmed that fillers produce an improvement in density in the mechanical, optical, electrical and thermal properties of these biopolymers (Bryśkiewicz et al., 2016).

In a very general way, natural fibrous materials are purchased from vegetable or animal sources (Lips and Van Dam, 2013), including totally natural cellulosic fibers such as cotton, jute, etc., and some protein-based fibers such as wool and silk. Those of vegetable origin tend to be classified mainly as woods or some derivatives of wood fibers. Wood fibers consist mainly of soft or hardwood, and of course, some recycled products. Although in some cases, it is possible to obtain some of them that are not wood, from various sources, such as a raw leaf, straw, dry grass, etc. (Shekar and Ramachandra, 2018). At present, non-wood fibers can be derived from rural or agro-industrial waste, where they have received considerable attention in the PU processing industry, especially for applications such as reinforcement (Das et al., 2019). The simplest, in common use, belong to the group of lignocellulosic, which include coconut husk, rice husk, wheat husk, jute, flax, cotton and banana fiber, etc. (Sattar et al., 2015). They are usually of low cost due to their practical and simple availability, easy accessibility, low density, they are renewable and produce a less abrasive effect during processing (Agrawal et al., 2017). In addition to this, naturally extracted fibers are widely available in various forms throughout the world. They are non-toxic and can be chemically modified to improve their compatibility with the polymeric matrix they are inserted (Rowell, 2008).

5. Conclusions

Some advances in the use of biopolymers like substitutes of synthetic polymers are made, specifically in foam applications, these efforts are focused on cleaner and environmental safer process designing. Also, some biopolymers are used as foams, like PLA, PEG, cellulose and others. For example, one application is the use of natural derived polyols in the synthesis of polyurethane or functionalized fibers to reinforce the structure. Many applications consider the use these bio-foams in food containers, environmental or medical devices, among others. The development and applications of bio-foams are growing areas of research.

6. Acknowledgements

The authors thank the National Council of Science and Technology (CONACyT) of Mexico and to Autonomous University of Coahuila, for the access to literature used for this chapter.

References

Abid, S. R., Hilo, A. N. and Daek, Y. H. (2018). Experimental tests on the underwater abrasion of engineered cementitious composites. *Constr. Build Mater.*, 171: 779–792. https://doi.org/10.1016/j.conbuildmat.2018.03.213.

Accorsi, R., Manzini, R. and Ferrari, E. (2014). A comparison of shipping containers from technical, economic and environmental perspectives. *Transp. Res. Part D Transp. Environ.*, 26: 52–59. https://doi.org/10.1016/j.trd.2013.10.009.

Ago, M., Ferrer, A. and Rojas, O. J. (2016). Starch-based biofoams reinforced with lignocellulose nanofibrils from residual palm empty fruit bunches: water sorption and mechanical strength. *ACS Sustain Chem. Eng.*, 4(10): 5546–5552. https://doi.org/10.1021/acssuschemeng.6b01279.

Agrawal, A., Kaur, R. and Walia, R. S. (2017). PU foam derived from renewable sources: Perspective on properties enhancement: An overview. *Eur. Polym. J.*, 95: 255–274. https://doi.org/10.1016/j.eurpolymj.2017.08.022.

Álvarez, I., Gutiérrez, C., Rodríguez, J. F., de Lucas, A. and García, M. T. (2020). Production of biodegradable PLGA foams processed with high pressure CO_2. *J. Supercrit. Fluids*, 164: 104886. https://doi.org/10.1016/j.supflu.2020.104886.

Ameli, A., Jahani, D., Nofar, M., Jung, P. U. and Park, C. B. (2014). Development of high void fraction polylactide composite foams using injection molding: Mechanical and thermal insulation properties. *Compos Sci. Technol.*, 90: 88–95. https://doi.org/10.1016/j.compscitech.2013.10.019.

Asikainen, S., Paakinaho, K., Kyhkynen, A. K., Hannula, M., Malin, M., Ahola, N., Kellomäki, M. and Seppälä, J. (2019). Hydrolysis and drug release from poly(ethylene glycol)-modified lactone polymers with open porosity. *Eur. Polym. J.*, 113: 165–175. https://doi.org/10.1016/j.eurpolymj.2019.01.056.

Avérous, L. (2004). Biodegradable multiphase systems based on plasticized starch: A review. *J. Macromol. Sci. - Polym. Rev.*, 44(3): 231–274. https://doi.org/10.1081/MC-200029326.

Bénézet, J. C., Stanojlovic-Davidovic, A., Bergeret, A., Ferry, L. and Crespy, A. (2012). Mechanical and physical properties of expanded starch, reinforced by natural fibres. *Ind. Crops Prod.*, 37(1): 435–440. https://doi.org/10.1016/j.indcrop.2011.07.001.

Bergeret, A. and Benezet, J. C. (2011). Natural fibre-reinforced biofoams. *Int. J. Polym. Sci.* https://doi.org/10.1155/2011/569871.

Bertoluci, G., Leroy, Y. and Olsson, A. (2014). Exploring the environmental impacts of olive packaging solutions for the European food market. *J. Clean. Prod.*, 64: 234–243. https://doi.org/10.1016/j.jclepro.2013.09.029.

Bigg, D. M. (2005). Polylactide copolymers: Effect of copolymer ratio and end capping on their properties. *Adv. Polym. Technol.*, 24(2): 69–82. https://doi.org/10.1002/adv.20032.

Boissard, C. I., Bourban, P.-E., Plummer, C. J. G., Neagu, R. C. and Månson, J.-A. E. (2012). Cellular biocomposites from polylactide and microfibrillated cellulose. *J. Cell Plast.*, 48(5): 445–458. https://doi.org/10.1177/0021955X12448190.

Bryśkiewicz, A., Zieleniewska, M., Przyjemska, K., Chojnacki, P. and Ryszkowska, J. (2016). Modification of flexible polyurethane foams by the addition of natural origin fillers. *Polym. Degrad. Stab.*, 132: 32–40. https://doi.org/10.1016/j.polymdegradstab.2016.05.002.

Buggy, M. (2006). Natural fibers, biopolymers, and biocomposites. Edited by Amar K. Mohanty, Manjusri Misra and Lawrence T. Drzal. *Polym. Int.*, 55(12): 1462–1462. https://doi.org/10.1002/pi.2084.

Chaireh, S., Ngasatool, P. and Kaewtatip, K. (2020). Novel composite foam made from starch and water hyacinth with beeswax coating for food packaging applications. *Int. J. Biol. Macromol.*, 165: 1382–1391. https://doi.org/10.1016/j.ijbiomac.2020.10.007.

Chevali, V., Fuqua, M. and Ulven, C. A. (2011). Vegetable oil based rigid foam composites. *In*: *Handbook of Bioplastics and Biocomposites Engineering Applications.*

Chevali, V. and Kandare, E. (2016). Rigid biofoam composites as eco-efficient construction materials. pp. 275–304. *In*: *Biopolymers and Biotech Admixtures for Eco-Efficient Construction Materials*. Elsevier Ltd.

Corre, Y. M., Maazouz, A., Duchet, J. and Reignier, J. (2011). Batch foaming of chain extended PLA with supercritical CO_2: Influence of the rheological properties and the process parameters on the cellular structure. *J. Supercrit. Fluids*, 58(1): 177–188. https://doi.org/10.1016/j.supflu.2011.03.006.

Darder, M., Matos, C. R. S., Aranda, P., Gouveia, R. F. and Ruiz-Hitzky, E. (2017). Bionanocomposite foams based on the assembly of starch and alginate with sepiolite fibrous clay. *Carbohydr. Polym.*, 157: 1933–1939. https://doi.org/10.1016/j.carbpol.2016.11.079.

Das, S., Kumar, S., Mohanty, S. and Nayak, S.K. (2019). Synthesis, characterization and application of bio-based polyurethane nanocomposites. *In*: Inamuddin, Thomas, S., Kumar Mishra, R. and Asiri, A. (eds.). *Sustainable Polymer Composites and Nanocomposites*. Springer, Cham. https://doi.org/10.1007/978-3-030-05399-4_39.

Davis, G. and Song, J. H. (2006). Biodegradable packaging based on raw materials from crops and their impact on waste management. *Ind. Crops Prod.*, 23(2): 147–161. https://doi.org/10.1016/j.indcrop.2005.05.004.

de Avila Delucis, R., Fischer Kerche, E., Gatto, D. A., Magalhães Esteves, W. L., Petzhold, C. L. and Campos Amico, S. (2019). Surface response and photodegradation performance of bio-based polyurethane-forest derivatives foam composites. *Polym. Test*, 80(March): 106102. https://doi.org/10.1016/j.polymertesting.2019.106102.

Di, Y., Iannace, S., Di Maio, E. and Nicolais, L. (2005). Reactively modified poly(lactic acid): properties and foam processing. *Macromol. Mater. Eng.*, 290(11): 1083–1090. https://doi.org/10.1002/mame.200500115.

Díaz, P., Garrido, M. D. and Bañón, S. (2010). The effects of packaging method (vacuum pouch vs. plastic tray) on spoilage in a cook-chill pork-based dish kept under refrigeration. *Meat Sci.*, 84(3): 538–544. https://doi.org/10.1016/j.meatsci.2009.10.009.

Dorgan, J. R., Williams, J. S. and Lewis, D. N. (1999). Melt rheology of poly(lactic acid): Entanglement and chain architecture effects. *J. Rheol.* (N Y N Y), 43(5): 1141–1155. https://doi.org/10.1122/1.551041.

Dorgan, J. R., Lehermeier, H. and Mang, M. (2000). Thermal and rheological properties of commercial-grade poly(lactic acids)s. *J. Polym. Environ.*, 8(1): 1–9. https://doi.org/10.1023/A:1010185910301.

Dorgan, J. R., Janzen, J., Clayton, M. P., Hait, S. B. and Knauss, D. M. (2005). Melt rheology of variable L-content poly(lactic acid). *J. Rheol.* (N Y N Y), 49(3): 607–619. https://doi.org/10.1122/1.1896957.

Dorgan, J. R. (2010). Rheology of poly(lactic acid). pp. 125–139. *In*: *Poly(Lactic Acid).* John Wiley & Sons, Inc., Hoboken, NJ, USA.

FAN, H. (2011). *Polyurethane Foams Made from Bio-Based Polyols.* University of Missouri.

Francis, R., Sasikumar, S. and Gopalan, G. P. (2013). Synthesis, structure, and properties of biopolymers (natural and synthetic). pp. 11–107. *In*: Thomas, S., Joseph, K., Malhotra, S. K., Goda, K. and Sreekala, M. S. (eds.). *Polymer Composites, Biocomposites*. Wiley-VCH Verlag GmbH & Co.

Fujimoto, Y., Ray, S. S., Okamoto, M., Ogami, A., Yamada, K. and Ueda, K. (2003). Well-controlled biodegradable nanocomposite foams: from microcellular to nanocellular. *Macromol. Rapid Commun.*, 24(7): 457–461. https://doi.org/10.1002/marc.200390068.

Gama, N. V., Ferreira, A. and Barros-Timmons, A. (2018). Polyurethane foams: Past, present, and future. *Materials* (Basel), 11(10). https://doi.org/10.3390/ma11101841.

Ganjyal, G. M., Weber, R. and Hanna, M. A. (2007). Laboratory composting of extruded starch acetate and poly lactic acid blended foams. *Bioresour. Technol.*, 98(16): 3176–3179. https://doi.org/10.1016/j.biortech.2006.10.030.

Garlotta, D. (2001). A literature review of poly(lactic acid). *J. Polym. Environ.*, 9(2): 63–84. https://doi.org/10.1023/A:1020200822435.

George, A., Sanjay, M. R., Srisuk, R., Parameswaranpillai, J. and Siengchin, S. (2020). A comprehensive review on chemical properties and applications of biopolymers and their composites. *Int. J. Biol. Macromol.*, 154: 329–338. https://doi.org/10.1016/j.ijbiomac.2020.03.120.

Glenn, G. M., Orts, W. J. and Nobes, G. A. R. (2001). Starch, fiber and $CaCo_3$ effects on the physical properties of foams made by a baking process. *Ind. Crops Prod.*, 14(3): 201–212. https://doi.org/10.1016/S0926-6690(01)00085-1.

Han, J., Yue, Y., Wu, Q., Huang, C., Pan, H., Zhan, X., Mei, C. and Xu, X. (2017). Effects of nanocellulose on the structure and properties of poly(vinyl alcohol)-borax hybrid foams. Cellulose, 24(10): 4433–4448. https://doi.org/10.1007/s10570-017-1409-4.

Hao, A., Geng, Y., Xu, Q., Lu, Z. and Yu, L. (2008). Study of different effects on foaming process of biodegradable PLA/starch composites in supercritical/compressed carbon dioxide. *J. Appl. Polym. Sci.*, 109(4): 2679–2686. https://doi.org/10.1002/app.27861.

Heymans, R., Tavernier, I., Dewettinck, K. and Van der Meeren, P. (2017). Crystal stabilization of edible oil foams. *Trends Food Sci. Technol.*, 69: 13–24. https://doi.org/10.1016/j.tifs.2017.08.015.

Hirvikorpi, T., Vähä-Nissi, M., Harlin, A., Salomäki, M., Areva, S., Korhonen, J. T. and Karppinen, M. (2011). Enhanced water vapor barrier properties for biopolymer films by polyelectrolyte multilayer and atomic layer deposited Al_2O_3 double-coating. *Appl. Surf. Sci.*, 257(22): 9451–9454. https://doi.org/10.1016/j.apsusc.2011.06.031.

Hou, J., Zhao, G. and Wang, G. (2021). Polypropylene/talc foams with high weight-reduction and improved surface quality fabricated by mold-opening microcellular injection molding. *J. Mater Res. Technol.*, 12: 74–86. https://doi.org/10.1016/j.jmrt.2021.02.077.

Hwang, S.-S., Hsu, P. P., Yeh, J.-M., Chang, K.-C. and Lai, Y.-Z. (2009). The mechanical/thermal properties of microcellular injection-molded poly-lactic-acid nanocomposites. *Polym. Compos.*, 30(11): 1625–1630. https://doi.org/10.1002/pc.20736.

Jeong, E. J., Park, C. K. and Kim, S. H. (2020). Fabrication of microcellular polylactide/modified silica nanocomposite foams. *J. Appl. Polym. Sci.*, 137(17): 1–10. https://doi.org/10.1002/app.48616.

Jiang, T., Duan, Q., Zhu, J., Liu, H. and Yu, L. (2020). Starch-based biodegradable materials: Challenges and opportunities. *Adv. Ind. Eng. Polym. Res.*, 3(1): 8–18. https://doi.org/10.1016/j.aiepr.2019.11.003.

Jin, F. L., Zhao, M., Park, M. and Park, S. J. (2019). Recent trends of foaming in polymer processing: A review. *Polymers* (Basel), 11(6). https://doi.org/10.3390/polym11060953.

Kadam, H., Bandyopadhyay-Ghosh, S., Malik, N. and Ghosh, S. B. (2019). Bio-based engineered nanocomposite foam with enhanced mechanical and thermal barrier properties. *J. Appl. Polym. Sci.*, 136(7): 1–7. https://doi.org/10.1002/app.47063.

Kaisangsri, N., Kerdchoechuen, O. and Laohakunjit, N. (2012). Biodegradable foam tray from cassava starch blended with natural fiber and chitosan. *Ind. Crops Prod.*, 37(1): 542–546. https://doi.org/10.1016/j.indcrop.2011.07.034.

Kaisangsri, N., Kerdchoechuen, O. and Laohakunjit, N. (2014). Characterization of cassava starch based foam blended with plant proteins, kraft fiber, and palm oil. *Carbohydr. Polym.*, 110: 70–77. https://doi.org/10.1016/j.carbpol.2014.03.067.

Kang, D. J., Xu, D., Zhang, Z. X., Pal, K., Bang, D. S. and Kim, J. K. (2009). Well-controlled microcellular biodegradable PLA/silk composite foams using supercritical CO_2. *Macromol. Mater Eng.*, 294(9): 620–624. https://doi.org/10.1002/mame.200900103.

Kennedy, J. F. and White, C. A. (1987). *Starch Chemistry and Technology.* Second. Academic Press Inc., London.

Kiran, E. (2016). Supercritical fluids and polymers - The year in review - 2014. *J. Supercrit. Fluids*, 110: 126–153. https://doi.org/10.1016/j.supflu.2015.11.011.

Konieczny, J. and Loos, K. (2019). Green polyurethanes from renewable isocyanates and biobased white dextrins. *Polymers* (Basel), 11(2). https://doi.org/10.3390/polym11020256.

Li, H., Sun, J. T., Wang, C., Liu, S., Yuan, D., Zhou, X., Tan, J., Stubbs, L. and He, C. (2017). High modulus, strength, and toughness polyurethane elastomer based on unmodified lignin. *ACS Sustain Chem. Eng.*, 5(9): 7942–7949. https://doi.org/10.1021/acssuschemeng.7b01481.

Li, K., Cui, Z., Sun, X., Turng, L.-S. and Huang, H. (2012). Effects of nanoclay on the morphology and physical properties of solid and microcellular injection molded polyactide/poly(butylenes adipate-co-terephthalate) (PLA/PBAT) nanocomposites and blends. *J. Biobased Mater Bioenergy*, 5(4): 442–451. https://doi.org/10.1166/jbmb.2011.1182.

Lips, S. J. J. and Van Dam, J. E. G. (2013). Kenaf fibre crop for bioeconomic industrial development. *In*: Monti, A. and Alexopoulou, E. (eds.). *Kenaf: A Multi-Purpose Crop for Several Industrial Applications*.

Lu, B., Lin, Q., Yin, Z., Lin, F., Chen, X. and Huang, B. (2021). Robust and lightweight biofoam based on cellulose nanofibrils for high-efficient methylene blue adsorption. *Cellulose*, 28(1): 273–288. https://doi.org/10.1007/s10570-020-03553-4.

Luo, X., Xiao, Y., Wu, Q. and Zeng, J. (2018a). Development of high-performance biodegradable rigid polyurethane foams using all bioresource-based polyols: Lignin and soy oil-derived polyols. *Int. J. Biol. Macromol.*, 115: 786–791. https://doi.org/10.1016/j.ijbiomac.2018.04.126.

Luo, X., Xiao, Y., Wu, Q. and Zeng, J. (2018b). Development of high-performance biodegradable rigid polyurethane foams using all bioresource-based polyols: Lignin and soy oil-derived polyols. Int. J. Biol. Macromol., 115: 786–791. https://doi.org/10.1016/j.ijbiomac.2018.04.126.

Luo, X., Cai, Y., Liu, L., Zhang, F., Wu, Q. and Zeng, J. (2019). Soy oil-based rigid polyurethane biofoams obtained by a facile one-pot process and reinforced with hydroxyl-functionalized multiwalled carbon nanotube. JAOCS, *J. Am. Oil Chem. Soc.*, 96(3): 319–328. https://doi.org/10.1002/aocs.12184.

Ma, X., Chen, J., Zhu, J. and Yan, N. (2021). Lignin-based polyurethane: recent advances and future perspectives. *Macromol. Rapid Commun.*, 42(3): 1–13. https://doi.org/10.1002/marc.202000492.

Machado, C. M., Benelli, P. and Tessaro, I. C. (2020). Study of interactions between cassava starch and peanut skin on biodegradable foams. *Int. J. Biol. Macromol.*, 147: 1343–1353. https://doi.org/10.1016/j.ijbiomac.2019.10.098.

Madival, S., Auras, R., Singh, S. P. and Narayan, R. (2009). Assessment of the environmental profile of PLA, PET and PS clamshell containers using LCA methodology. *J. Clean Prod.*, 17(13): 1183–1194. https://doi.org/10.1016/j.jclepro.2009.03.015.

Manavitehrani, I., Le, T. Y. L., Daly, S., Wang, Y., Maitz, P. K., Schindeler, A. and Dehghani, F. (2019). Formation of porous biodegradable scaffolds based on poly(propylene carbonate) using gas foaming technology. *Mater. Sci. Eng. C*, 96(June 2018): 824–830. https://doi.org/10.1016/j.msec.2018.11.088.

Marrazzo, C., Maio, E. Di and Iannace, S. (2007). Foaming of synthetic and natural biodegradable polymers. *J. Cell Plast.*, 43(2): 123–133. https://doi.org/10.1177/0021955X06073214.

Marsh, K. and Bugusu, B. (2007). Food packaging—Roles, materials, and environmental issues: Scientific status summary. *J. Food Sci.*, 72(3). https://doi.org/10.1111/j.1750-3841.2007.00301.x.

Martín de León, J., Bernardo, V. and Rodríguez-Pérez, M. Á. (2019). Nanocellular polymers: The challenge of creating cells in the nanoscale. *Materials* (Basel), 12(5): 1–19. https://doi.org/10.3390/MA12050797.

Matuana, L. M. and Diaz, C. A. (2010). Study of cell nucleation in microcellular poly(lactic acid) foamed with supercritical CO_2 through a continuous-extrusion process. *Ind. Eng. Chem. Res.*, 49(5): 2186–2193. https://doi.org/10.1021/ie9011694.

Matuana, L. M. and Faruk, O. (2010). Effect of gas saturation conditions on the expansion ratio of microcellular poly (lactic acid)/wood-flour composites. *Express Polym. Lett.*, 4(10): 621–631. https://doi.org/10.3144/expresspolymlett.2010.77.

Meneses, M., Pasqualino, J. and Castells, F. (2012). Environmental assessment of the milk life cycle: The effect of packaging selection and the variability of milk production data. *J. Environ. Manage*, 107: 76–83. https://doi.org/10.1016/j.jenvman.2012.04.019.

Mihai, M., Huneault, M. A., Favis, B. D. and Li, H. (2007). Extrusion foaming of semi-crystalline PLA and PLA/thermoplastic starch blends. *Macromol. Biosci.*, 7(7): 907–920. https://doi.org/10.1002/mabi.200700080.

Mihai, M., Huneault, M. A. and Favis, B. D. (2009). Crystallinity development in cellular poly(lactic acid) in the presence of supercritical carbon dioxide. *J. Appl. Polym. Sci.*, 113(5): 2920–2932. https://doi.org/10.1002/app.30338.

Milovanovic, S., Markovic, D., Mrakovic, A., Kuska, R., Zizovic, I., Frerich, S. and Ivanovic, J. (2019). Supercritical CO_2-assisted production of PLA and PLGA foams for controlled thymol release. *Mater Sci. Eng. C*, 99: 394–404. https://doi.org/10.1016/j.msec.2019.01.106.

Mokhothu, T. H. and John, M. J. (2017). Bio-based fillers for environmentally friendly composites. *Handb Compos from Renew Mater,* 1–8: 243–270. https://doi.org/10.1002/9781119441632.ch10.

Moo-Tun, N. M., Iñiguez-Covarrubias, G. and Valadez-Gonzalez, A. (2020). Assessing the effect of PLA, cellulose microfibers and $CaCO_3$ on the properties of starch-based foams using a factorial design. *Polym. Test*, 86(February). https://doi.org/10.1016/j.polymertesting.2020.106482.

Neus Angles, M. and Dufresne, A. (2000). Plasticized starch/tunicin whiskers nanocomposites. 1. Structural analysis. *Macromolecules,* 33(22): 8344–8353. https://doi.org/10.1021/ma0008701.

Nofar, M. and Park, C. B. (2014). Poly (lactic acid) foaming. *Prog. Polym. Sci.*, 39: 1721–1741.

Pilla, S., Kim, S. G., Auer, G. K., Gong, S. and Park, C. B. (2009a). Microcellular extrusion-foaming of polylactide with chain-extender. *Polym. Eng. Sci.*, 49(8): 1653–1660. https://doi.org/10.1002/pen.21385.

Pilla, S., Kramschuster, A., Lee, J., Auer, G. K., Gong, S. and Turng, L. S. (2009b). Microcellular and solid polylactide-flax fiber composites. *Compos Interfaces*, 16(7–9): 869–890. https://doi.org/10.1163/092764409X12477467990283.

Pilla, S., Kramschuster, A., Yang, L., Lee, J., Gong, S. and Turng, L. S. (2009c). Microcellular injection-molding of polylactide with chain-extender. *Mater Sci. Eng. C*, 29(4): 1258–1265. https://doi.org/10.1016/j.msec.2008.10.027.

Pilla, S., Kim, S. G., Auer, G. K., Gong, S. and Park, C. B. (2010). Microcellular extrusion foaming of poly(lactide)/poly(butylene adipate-co-terephthalate) blends. *Mater Sci. Eng. C*, 30(2): 255–262. https://doi.org/10.1016/j.msec.2009.10.010.

Preechawong, D., Peesan, M., Supaphol, P. and Rujiravanit, R. (2005). Preparation and characterization of starch/poly(l-lactic acid) hybrid foams. *Carbohydr. Polym.*, 59(3): 329–337. https://doi.org/10.1016/j.carbpol.2004.10.003.

Raj, S. N., Lavanya, S. N., Sudisha, J. and Shetty, H. S. (2011). Applications of biopolymers in agriculture with special reference to role of plant derived biopolymers in crop protection. *In*: Kalia, S. and Avérous, L. (eds.). *Biopolymers: Biomédical and Environmental Applications*. Scrivener Publishing LLC.

Ranote, S., Kumar, D., Kumari, S., Kumar, R., Chauhan, G. S. and Joshi, V. (2019). Green synthesis of Moringa oleifera gum-based bifunctional polyurethane foam braced with ash for rapid and efficient dye removal. *Chem. Eng. J.*, 361: 1586–1596. https://doi.org/10.1016/j.cej.2018.10.194.

Rodríguez-Castellanos, W., Martínez-Bustos, F., Rodrigue, D. and Trujillo-Barragán, M. (2015). Extrusion blow molding of a starch-gelatin polymer matrix reinforced with cellulose. *Eur. Polym. J.*, 73: 335–343. https://doi.org/10.1016/j.eurpolymj.2015.10.029.

Rohan, T., Tushar, B. and Mahesha, G. T. (2018). Review of natural fiber composites. *IOP Conf. Ser. Mater Sci. Eng.*, 314(1). https://doi.org/10.1088/1757-899X/314/1/012020.

Rowell, R. M. (2008). Natural fibres: Types and properties. *Prop. Perform. Nat. Compos*, 3–66. https://doi.org/10.1533/9781845694593.1.3.

Sanhawong, W., Banhalee, P., Boonsang, S. and Kaewpirom, S. (2017). Effect of concentrated natural rubber latex on the properties and degradation behavior of cotton-fiber-reinforced cassava starch biofoam. *Ind. Crops Prod.*, 108(July): 756–766. https://doi.org/10.1016/j.indcrop.2017.07.046.

Sarver, J. and Kiran, E. (2021). Foaming of polymers with carbon dioxide—the year-in-review—2019. *J. Supercrit. Fluids*, 173(September 2020): 105166. https://doi.org/10.1016/j.supflu.2021.105166.

Sattar, R., Kausar, A. and Siddiq, M. (2015). Advances in thermoplastic polyurethane composites reinforced with carbon nanotubes and carbon nanofibers: A review. *J. Plast. Film Sheeting*, 31(2): 186–224. https://doi.org/10.1177/8756087914535126.

Shavandi, A., Hosseini, S., Okoro, O. V., Nie, L., Eghbali Babadi, F. and Melchels, F. (2020). 3D bioprinting of lignocellulosic biomaterials. *Adv. Healthc. Mater.*, 9.

Shekar, H. S. S. and Ramachandra, M. (2018). Green composites: A review. *Mater Today Proc.*, 5(1): 2518–2526. https://doi.org/10.1016/j.matpr.2017.11.034.

Shey, J., Imam, S. H., Glenn, G.M. and Orts, W. J. (2006). Properties of baked starch foam with natural rubber latex. *Ind. Crops Prod.*, 24(1): 34–40. https://doi.org/10.1016/j.indcrop.2005.12.001.

Singh, I., Samal, S. K., Mohanty, S. and Nayak, S. K. (2020). Recent advancement in plant oil derived polyol-based polyurethane foam for future perspective: a review. *Eur. J. Lipid Sci. Technol.*, 122(3): 1–23. https://doi.org/10.1002/ejlt.201900225.

Song, C., Li, S., Zhang, J., Xi, Z., Lu, E., Zhao, L. and Cen, L. (2019). Controllable fabrication of porous PLGA/PCL bilayer membrane for GTR using supercritical carbon dioxide foaming. *Appl. Surf. Sci.*, 472: 82–92. https://doi.org/10.1016/j.apsusc.2018.04.059.

Sousa, A. F., Ferreira, S., Lopez, A., Borges, I., Pinto, R. J. B., Silvestre, A. J. D. and Freire, C. S. R. (2017). Thermosetting AESO-bacterial cellulose nanocomposite foams with tailored mechanical properties obtained by Pickering emulsion templating. *Polymer* (Guildf), 118: 127–134. https://doi.org/10.1016/j.polymer.2017.04.073.

Soykeabkaew, N., Supaphol, P. and Rujiravanit, R. (2004). Preparation and characterization of jute-and flax-reinforced starch-based composite foams. *Carbohydr. Polym.*, 58(1): 53–63. https://doi.org/10.1016/j.carbpol.2004.06.037.

Soykeabkaew, N., Thanomsilp, C. and Suwantong, O. (2015). A review: Starch-based composite foams. *Compos. Part A Appl. Sci. Manuf.*, 78: 246–263.

Standau, T., Zhao, C., Castellón, S. M., Bonten, C. and Altstädt, V. (2019). Chemical modification and foam processing of polylactide (PLA). *Polymers* (Basel), 11(2): 1–38. https://doi.org/10.3390/polym11020306.

Sungsanit, K., Kao, N., Bhattacharya, S. N. and Pivsaart, S. (2010). Phvsical and rheological properties of plasticized linear and branched PLA. Korean Society of Rheology/한국유변학회

Tian, H., Tang, Z., Zhuang, X., Chen, X. and Jing, X. (2012). Biodegradable synthetic polymers: Preparation, functionalization and biomedical application. *Prog. Polym. Sci.*, 37(2): 237–280. https://doi.org/10.1016/j.progpolymsci.2011.06.004.

Tian, L., Luan, J., Liu, K. K., Jiang, Q., Tadepalli, S., Gupta, M. K., Naik, R. R. and Singamaneni, S. (2016). Plasmonic biofoam: a versatile optically active material. *Nano Lett.*, 16(1): 609–616. https://doi.org/10.1021/acs.nanolett.5b04320.

Valadbaigi, P., Ettelaie, R., Kulak, A. N. and Murray, B. S. (2019). Generation of ultra-stable Pickering microbubbles via poly alkylcyanoacrylates. *J. Colloid Interface Sci.*, 536: 618–627. https://doi.org/10.1016/j.jcis.2018.10.004.

van der VegtHendrik, A. and Bisschop, J. (2012). Sweet Particulate Fat-Containing Powder, its Preparation and Use (FrieslandCampina).

Weber, H., De Grave, I., Röhrl, E. and Altstädt, V. (2016). Foamed plastics. pp. 1–54. *In*: *Ullmann's Encyclopedia of Industrial Chemistry*. Wiley-VCH Verlag GmbH & Co. KGaA, Weinheim, Germany.

Wu, X., Yan, W., Zhou, Y., Luo, L., Yu, X., Luo, L., Fan, M., Du, G. and Zhao, W. (2020). Thermal, morphological, and mechanical characteristics of sustainable tannin bio-based foams reinforced with wood cellulosic fibers. *Ind. Crops Prod.*, 158: 113029. https://doi.org/10.1016/j.indcrop.2020.113029.

Xu, Y. and Hanna, M. A. (2005). Preparation and properties of biodegradable foams from starch acetate and poly(tetramethylene adipate-co-terephthalate). *Carbohydr. Polym.*, 59(4): 521–529. https://doi.org/10.1016/j.carbpol.2004.11.007.

Yadav, A., Mangaraj, S., Singh, R., Das, K., Kumar, N. and Arora, S. (2018). Biopolymers as packaging material in food and allied industry. Int. J. Chem. Stud., 6(2): 2411–2418.

Yin, D., Xiang, A., Li, Y., Qi, H., Tian, H. and Fan, G. (2019). Effect of plasticizer on the morphology and foaming properties of poly(vinyl alcohol) foams by supercritical CO_2 foaming agents. *J. Polym. Environ.*, 27(12): 2878–2885. https://doi.org/10.1007/s10924-019-01570-4.

Yuan, H., Liu, Z. and Ren, J. (2009). Preparation, characterization, and foaming behavior of poly(lactic acid)/poly(butylene adipate-co-butylene terephthalate) blend. pp. 1004–1012. *In*: *Polymer Engineering and Science*. John Wiley & Sons, Ltd.

Zhang, J. F. and Sun, X. (2007a). Biodegradable foams of foly(lactic acid)/starch. I. Extrusion condition and cellular size distribution. *J. Appl. Polym. Sci.*, 106(2): 857–862. https://doi.org/10.1002/app.26715.

Zhang, J. F. and Sun, X. (2007b). Biodegradable foams of folydactic acid/starch. II. Cellular structure and water resistance. *J. Appl. Polym. Sci.*, 106(5): 3058–3062. https://doi.org/10.1002/app.26697.

Zhang, K., Wang, Y., Jiang, J., Wang, X., Hou, J., Sun, S. and Li, Q. (2019). Fabrication of highly interconnected porous poly(ε-caprolactone) scaffolds with supercritical CO_2 foaming and polymer leaching. *J. Mater Sci.*, 54(6): 5112–5126. https://doi.org/10.1007/s10853-018-3166-7.

Zhou, J., Song, J. and Parker, R. (2006). Structure and properties of starch-based foams prepared by microwave heating from extruded pellets. *Carbohydr. Polym.*, 63(4): 466–475. https://doi.org/10.1016/j.carbpol.2005.09.019.

Chapter 12

Thermo-Shrinkable Biodegradable Polymers

Reyes-Acosta Yadira Karina, Farias-Cepeda Lorena, Rubio-Ríos Anilú, Rosales-Marines Lucero, Reyna-Martínez Ricardo, Alonso-Montemayor Francisco Javier* and *Cepeda-Tovar Víctor Adán*

1. Introduction

Shape memory polymers are stimuli-responsive systems that can change their shape for the action of external variation on temperature, light, pH or solvent exposure (Skrzeszewska et al., 2011; Mather et al., 2009). Shape memory polymers that are "programmed" by changes in temperature are created by elongating the polymer. Simultaneously, the temperature is increased above a moderate transition, frequently crystallizing or melting temperature of the polymer. Then the polymer is cooled below the transition temperature while the polymer is still elongated, "freezing" the molecules. If the weather increases above the transition temperature again, the polymer will return to its initial shape (Kalkan-Sevinc and Strobel, 2015).

However, macroscopic changes in polymer shape are often accompanied by chain conformational changes (Zhao and Xie, 2015). So, heat-shrinkable polymers are part of the shape memory polymers that result from heat stimuli changing their shape. Heat shrinkable polymer has broad applications in various fields such as packing film or wrapping (Khankrua et al., 2019), medical devices, can also be used as self-folding robots, deformable batteries, containers for drug delivery, reconfigurable metamaterials and 3D electronic devices (Cui et al., 2017). Other applications are doll hair, intravenous needles that soften in the body, temperature-dependent moisture-permeable fabrics and rewritable digital storage media (Koerner et al., 2004).

Heat shrinkable polymers can be modeled as a two-component system. They possess crystalline and amorphous regions and are always cross-linked to enhance their warmth shrinkable (Zhao and Xie, 2015). The amorphous polymer chains represent the elastic component, and the crystalline regions represent a reversible transition component.

For example, a heat-shrinkable polymer is exposed to a crosslinking agent after the polymer has been processed (molding) to crosslink individual polymer chains. Next, the polymer is heated around its melting temperature and mechanically expanded (deformed). This expansion also generates some crystallite and amorphous chain alignment along the deformation axis. The polymer is held in its deformed state and cooled to room temperature quickly. When cooling occurs, the transition

Ingeniería y Simulación de Procesos Químicos, Facultad de Ciencias Químicas, Universidad Autónoma de Coahuila, Saltillo, Coahuila, México.
* Corresponding author: ykreyes@uadec.edu.mx

component reforms as crystallites and the elastic component (amorphous chains) are stretched in the deformed state. As a consequence of the fast cooling, the amorphous chain mobility is limited or confined. This rapid cooling locks the polymer in its deformed state and provides its capacity to change its shape. The is known as programming steps (Zhao and Xie, 2015; https://www.azom.com/article.aspx?ArticleID=18014; Rousseau, 2008).

The ensuing heating just above the melt temperature of the polymer and the subsequent heating just above the melt temperature of the polymer, the crystallites melt along with the amorphous regions. This allows the stretched and strained amorphous chains (elastic component) to relax and recover their original shape. Both areas (amorphous and crystalline) become less oriented and more compact than in the deformed state. Therefore, the recovery temperature is when the polymer starts to return to its original shape. The reheating of the expanded heat shrink is called the recovery step.

Together, the programming and recovery steps make up the shape memory cycle (Zhao and Xie, 2015; https://www.azom.com/article.aspx?ArticleID=18014; Rousseau, 2008).

The transition from a deformed shape to an original form can be seen as an energy barrier that must be overcome to affect the transition. Heat shrink materials are similar to rubbers and elastics in this way, but the energy barrier is higher to the shrinking polymer compared with rubbers. The driving force to return the deformed shape to its permanent condition comes from the stretched amorphous chains and crosslinking. This chain crosslinking provides greater elastic recovery force. The crosslinking can be chemically or physically. With the crosslinking, the polymer chains form a denser network and can store elastic energy and become more viscous. Another advantage of crosslinking is to avoid the slippage of polymer chains under strain. This restriction helps to ensure that any shape change results from an entropy change and is recoverable. The potential recovery is favored, avoiding the chain slippage. Crosslinked materials usually show improved resistance to impact and stress cracking and may even exhibit better chemical resistance (https://www.azom.com/article.aspx?ArticleID=18014; Rousseau, 2008).

This book chapter aims to show the different biopolymers with shrinkage capabilities in contact with temperature. We describe the heat shrinking biopolymers. Biopolymers are defined as a polymer that contain raw materials originating from agricultural and marine sources. There are three such categories of biopolymers (Arora, 2018; Cha and Chinnan, 2004):

1. From natural raw materials, such as starch, cellulose, protein, and marine prokaryotes were directly extracted. All these are, by nature, hydrophilic and somewhat crystalline and create problems while processing.
2. It was produced by chemical polymerization of bio-based monomers such as aliphatic, aromatic copolymers, aliphatic polyesters, poly-lactide, aliphatic copolymer (CPLA), using renewable bio-based monomers such as poly (lactic acid) and oil-based monomers like poly-caprolactones.
3. They are produced by microorganisms such as hydroxy-butyrate and hydroxy-valerate.

As an example of biopolymer for heat shrinking materials, recently, Khankrua et al. (Cha and Chinnan, 2004) studied the possible use of PLA/EVA blends as shrinkable labels and cap seals. The stretching and shrinking ratio of the reactive blend of PLA/EVA at different compositions using Joncryl and Perkadox as compatibilizers were analyzed. They found that the elongation at the break of PLA was improved when EVA is incorporated. The shrinkage of the stretched films of reactive blend PLA/EVA after two and three times was 100%, and the four tried films shrunk less than 100%. They stated it was because of excessive stretching.

Skrzeszewska et al. (2011) studied the shape-memory behavior of biodegradable and biocompatible recombinant telechelic polypeptides (collagen-like end blocks and a random coil-like middle block) and chemical cross-linking with lysine residues present in the random coil. The biopolymer could be stretched up to 200% and "pinned" in a temporary shape by lowering the temperature. They could maintain the deformed shape of the hydrogel upon heating to 50°C or higher.

Another application is testing in animals with the aim to be used in endoscopic surgery. In this case, large bulky devices could be introduced into the body in a compressed, temporary shape and then expanded to their permanent form to fit as required. For other surgery, applications of biopolymers are in the suture to apply the real force. In other words, the suture can be used loosely in its temporary shape; when the temperature is raised above its transition temperature, the suture will shrink and tighten the knot, thus applying the optimum force. Poly (Ethylene Glycol)-e-caprolactone-based polyurethane (PEG-PCL-PU) with excellent shape memory function was used as the scaffold matrix and as a drug carrier (Pattanashetti et al., 2017).

Textile wet processing use toxic chemicals thus effluents (Rani et al., 2020), studied the use of biopolymers to avoid shrinkage of wool fabrics without affecting the original properties of the fabric. They coated the wool fabric with gum arabic, chitosan and wheat starch, they found that the biopolymer treatment significantly reduced the area shrinkage (64%) of the wool fabric.

1.1 Thermo-shrinkable process

The heat-shrinkable process consists of shrinking. Thermoplastics materials soften and flow in contact with the increase in temperature, that is, they are considered non-crosslinked materials.

The production of heat shrinkable products comprises four main stages: composition, formation, crosslinking and expansion (Morshedian et al., 2003). The process consists of the transmission of elongation stresses of the cross-section of the material to be contracted and a subsequent cooling while it is still under tension. The phenomenon of heat shrink ability is explained by understanding typical thermoplastic polymers whose structure consists of crystalline and non-crystalline phases.

The way plastic films shrinks is by exposing the material to a heat source, which leads to the film being reduced in size. When the shrink films are stretched, the polymer molecules go from a random pattern to an orientation pattern. Through external stimuli, the polymer can be deformed and fixed by heating. As they cool, their shrinkage characteristics establish until heat is applied to them, causing them to shrink to their original dimensions, which is why they are also known as shrink films (Zhang et al., 2015).

The crystalline structure of the thermoplastic material weakens during heating. It is reduced by increasing the degree of crosslinking of the polymeric chains, thus modifying the elastic modulus that the material presents. Given the characteristics of thermoplastic polymers, above their melting point, the crystalline structure of the polymer disappears, compressing the molecules by crosslinking, which causes an increase in the elastic modulus of the material (Morshedian et al., 2003).

Heat shrinkability is known to be co-determined by the orientation of amorphous chains, the crystallinity and the crosslink density (Wang et al., 2016). A heat shrink film can shrink in one direction or two-way directions. The expanding process generates crystallite and amorphous. The transition section changes to crystallite as cooling takes place. The amorphous chains—elastic components are stretched in the expanded state, and the fast-cooling results in restricted amorphous chain mobility. To shrink thermoplastic sheets or films, they must be heated to an elastomeric state at a high temperature above the crystalline melting point, applying loads that allow a subsequent expansion of the plastic material to be achieved. Subsequently, the material is subjected to rapid cooling, performing the crystals domains that lead to obtaining a temporary shape. If the temperature exceeds the crystalline melting point, the polymeric material can recover from a brief form to a permanent shrunk condition (Wang et al., 2016).

Many investigations have focused on the heat shrinkage of polymers, attributing the phenomenon of heat shrinkage and the variations in the volume of these polymers to changes in the volume parameters, stating that the shrinkage capacity of these polymers appears from the relaxation of stored elastic stresses when the temperature exceeds crystalline melting (Wang et al., 2016). Heat shrinkability is known to be co-determined by the orientation of amorphous chains, the crystallinity and the crosslink density (Wang et al., 2016). A heat shrink film can shrink in one direction or two-way directions. The expanding process generates crystallite and amorphous. The transition section changes to crystallite as cooling takes place. The amorphous chains—elastic components are stretched in the expanded state and the fast-cooling results in restricted amorphous chain mobility. To shrink thermoplastic sheets or films, they must be heated to an elastomeric state at a high temperature above the crystalline melting point, applying loads that allow a subsequent expansion of the plastic material to be achieved. Subsequently, the material is subjected to rapid cooling, performing the crystals domains that lead to obtaining a temporary shape. If the temperature exceeds the crystalline melting point, the polymeric material can recover from a brief form to a permanent shrunk condition (Wang et al., 2016).

Semi-crystalline polymers are heat-shrinkable polymers that are widely used in industry. Although there are numerous studies on semi-crystalline heat shrink polymers, little systematic information on the preparation and processing parameters affects shrinkage (Morshedian et al., 2003).

It is essential to highlight that some reports of thermo-shrinkable polymeric systems refer to the importance of a crosslinked elastomeric phase to improve the heat shrinkage of the system and flame retardancy. It has also been reported that crosslinking between polymer chains improves contraction capacity 1. Thermoplastic elastomers are a material in which crosslinked networks avoid polymer chains from sliding irreversibly over each other when deforming, allowing quick and complete recovery from deformation. Elastomers comprise a diverse range of chemical structures characterized by having weak intermolecular forces. In response to an applied force, an elastomer will undergo a sudden, linear and reversible process. The chemical design and molecular architecture of elastomers are closely related to elastomeric mechanical response. High strain requires a polymer with a high molar mass. Many materials can exhibit immediate elastic response. However, only polymers can exhibit additional high stress, limiting the flexible response when the molecules are in completely extended conformations (Whelan, 2017).

As mentioned, solid-phase polymers have crystalline and amorphous regions that give them the ability to be heat-shrinkable. They also present, in most cases, a reticulated structure that benefits the heat shrinkage process.

Heat shrinkable polymers have this ability due to the presence of elastic and rigid components. After being deformed and after heat is applied to the material, memory shape materials could go back to their original shape. One can model these materials since the amorphous polymer chains act as an elastic component, and the crystalline regions give the reversible transition. During covering a product with a shrink film, the heat is just over its melting temperature. In this stage, the film tries to regain its original size and shape (memory) and mechanically expands but is prevented from doing so by the product to be wrapped. The film thus shrinks and tightly wraps the package and cools. In these processes, a hot melt adhesive, as a binder, is frequently applied to melt and seal the part shortly before shrinkage of polymeric shell occurs (Morshedian et al., 2003).

Longitudinal shrinkage can be determined in non-isothermal conditions by the Thermal-Mechanical Analysis (TMA) test using the association at a fixed stretch ratio.

$$Sh(\%) = \frac{L_{str} - L_{shr}}{L_{str}} \times 100 \quad (1)$$

where Sh (%) is the percentage of shrinkage, L_{str} the stretched length of the specimen, and L_{shr} the length of the shrinking specimen as a function of increasing temperature (Morshedian et al., 2003).

Considering that a shrinkable film is subjected to preceding the thermo-mechanical modification, it is necessary to understand its deformation and thermal shrinkage process, dilatometric measurements, width and the transitional zone of the films are essential points to consider (Kondratov, 2014).

The manufacturing process of heat shrinkable products is also based on the memory shape phenomenon. This phenomenon occurs to some extent in unprocessed thermoplastics and crosslinked thermoplastics (Morshedian et al., 2003). During the shrink process, through external stimuli, the polymer can deform and fix temporarily and then, by heating, contract in a controlled way to its original shape, which is known as a memory shape (Zhang et al., 2015), for the material to bring back its original shape, the polymer requires to have chemical or physical crosslinking such as chain entanglement and crystallization (Sliozberg et al., 2021).

The shape memory cycle occurs from programming and recovery steps. The programming step is when the film is retrieved to a new transitory shape when the polymer expands. At this stage, the crystalline phase of the polymeric structure melts together with the amorphous regions due to successive heating to just above the melting temperature, allowing the elastic component to relax and *regain its* original shape, crystalline and amorphous regions become less oriented and more compact. The temperature at which the material returns to its original shape is called the *recovery temperature,* and the reheating of the expanded heat shrink is called the *recovery* step.

When heat-shrinkable polymeric materials present crosslinking in their structure, they offer greater elastic recovery force. In turn, crosslinking generates a polymer network, which increases the potential to store elastic energy. Cross-linking also stops long-range chain slippage when under tension, helping to ensure that any alteration in shape is the result of an entropy change, which is recoverable.

Cross-linking enhances mechanical properties above room temperature. The elongation at break decreases by increasing chemical cross-links due to the shorter length of the segments available for the extension and the lower probability of chain slippage. On the other hand, a higher number of cross-links results in higher tensile stress at break. However, as the degree of cross-linking increases from the decrease in the number of crystalline domains, a point is reached where stress hardening does not occur. Stress is no longer caused by oriented segments of the main chain but by simple cross-linking, which corresponds to a decrease in tensile strength (Morshedian et al., 2003).

In a cross-linked polymer, the shrinkage temperature is affected by polymer stretching temperature. At higher stretching temperature, the more the shrinkage temperature will be. Besides, the higher the degree of cross-linking, the lower the shrinkage temperature (Morshedian et al., 2003).

There are different types of polymeric shrink films available commercially. The operating temperature for different types of shrink polymers will vary according to each polymeric material since each material, according to its structural characteristics, has its heat range, within which they can effectively shrink in. This unique property is one of the most critical factors involved in forming products for packaging application. In this way, during the heat shrink process, the change in the individual components of the material through the mobility contrast will be determined by the critical features, which will control the shape memory properties during the process (Zhang et al., 2015). Besides, during the processing of thermoplastic elastomers, which can process factors such as viscosity or rheology of the two-phase polymer and the crystalline phase's temperature, the thermal stability it presents (Whelan, 2017). Since these factors, which have a critical effect on the material, will depend on the quality and properties of final products.

1.2 Characteristics of thermo-shrinkable polymers

They are transparent polymers that have been mixed and structurally accommodated during processing to achieve the shrink effect. They are also known as shrink films. The orientation improves tensile strength, impact resistance and low-temperature flexibility (Bourg et al., 2019; Jiang et al., 2019).

With a film of these characteristics, one or several products can be packaged in such a way that, due to the effect of the increase in temperature, the film contracts, coupling itself to the shape of the product, resulting in a compact, transparent packaging, protecting from atmospheric agents and facilitating transport. There are different types and characteristics of shrink films depending on the materials with which they are made:

- *Polyethylene (PE)*: It is widely used in the multiple packaging of beverages, canning, flour and general use. One can find it in a variety of measurements and calibers. These heats shrink films are designed to provide excellent mechanical and optical characteristics such as shrinkage, sealing, strength. The material that they can use to make the films is low-density polyethylene (LDPE), high-density (HDPE) and linear low-density polyethylene (LDPE) that are mixed to improve specific properties such as rigidity and tear resistance (Charoonsuk et al., 2021; Kakar et al., 2021).
- *Polyolefin*: It is used to make multilayer films (they can be 1, 3 or 5) from polypropylene (PP) and linear low-density polyethylene (LLDPE). They are non-toxic and odor-free materials which makes this film heat shrink the safest in the market. It can be found in gauges from 0.060 "and 0.075". Due to its low specific weight, it offers superior performance to other packaging films such as PVC. It works on any machine, from a manual sealer to a high-speed automatic engine. It has a percentage greater than 60% conversion by mass of polyolefins (Chaudhry et al., 2020; Zanchin and Leone, 2021).

Although films of this material are made in smaller gauges, they have excellent resistance, significantly favoring the cost per packaging, which is why it has become the ideal substitute for corrugated cardboard.

In high-productivity equipment, it can seal quickly, has a high shrinkage capacity, has a precise coefficient of friction and manages to form the package firmly for storage and transportation. It offers better shrinkage, resistance, gloss and transparency that results in a visual appeal that is higher than expected, both for industrial products and those aimed at the end consumer.

- *PVC*: It is transparent and very shiny, suitable for cosmetics, beverages, food products, pharmaceuticals, toys, all kinds of plastic, glass and metal containers, household appliances, etc. It is widely used in applications where electrical insulation is sought, such as battery packaging. PVC shrink film has this characteristic due to the additives or plasticizers with which it is manufactured (Wypych, 2020).

1.2.1 Properties

For a film to have good heat shrink properties, the proper selection of material for its manufacture is necessary.

The same material can obtain different levels of shrinkage in the primary or Machine Direction (MD) and the Transverse Direction (TD), depending on the operating conditions. This can be achieved by modifying the orientation of the film by changing the blow ratio (BUR), the height of the cooling line (ALE) and the stretch ratio (DDR) of the film (Wang et al., 2020; Xiaonan et al., 2020; Yang et al., 2020).

1.2.2 Biodegradable characteristics

In literature, one can find different terms on the conception of the name biodegradable. The ones that attract the most attention are : biodegradable and bio-based. Bio-based does not mean that it is also biodegradable, and biodegradable does not mean that the product can be safely disposed of in nature. Some materials need optimized conditions to degrade fully (Chopparapu et al., 2020).

They can ultimately convert biodegradation material with the help of microorganisms into water, CO_2 and biomass. Therefore, biodegradable heat shrink polymers must be part of the microbial food chain and degrade (Greene, 2019).

There are two main types of biodegradable polymers: oxo-biodegradable and hydro-biodegradable. First, the degradation begins with a chemical process and in the second with a biological process, so there is a clear distinction between biodegradable plastics and oxo-degradable plastics. The latter cannot be considered bioplastics. They are conventional plastic materials with artificial additives that do not biodegrade but fragment into small pieces that remain and potentially harm the environment and jeopardize recycling and composting (Ammala et al., 2011; Ojeda et al., 2009).

Therefore, the main characteristic of a heat-shrinkable polymer should be the decomposition capacity carried out by microorganisms found universally in nature, such as bacteria, fungi and algae, and in this way be assimilated entirely, without leaving residues in the natural environment in a specific period.

1.3 Perspectives of biodegradable thermo-shrinkable polymers

Biodegradable Thermo-Shrinkable Polymers (TSP) have great potential for biomedical applications such as minimally invasive surgery (Alteheld et al., 2005). Some patents have been published for such a purpose (Marco, 2006; Shikinami, 2001). Additionally, the biodegradability of some TSP make them easily compostable wastes (Jankauskaite et al., 2009). The non-biodegradability of most polymers is responsible for many current environmental problems associated with their disposal (Mishra et al., 2011). In this manner, biodegradable polymers have currently been utilized as alternative materials to the existing commodity plastics in specific applications. As present literature review shows, polylactic acid (PLA) is one of the most market-available biodegradable and compostable TSP researched as a replacement for commodity plastics. Compared to other polymers, PLA exhibits suitable mechanical properties as well as superior optical clarity. However, PLA requires an improvement in its elongation at break, impact strength and film flexibility, while maintaining its transparency to be suitable for specific applications (Khankrua et al., 2019).

Moreover, since most TSP are synthesized using ring-opening polymerization (ROMP), it can be suggested that they can be easily applied in biomedical applications such as replacement tissue (Alteheld et al., 2005), tunable biomedical devices, thermal actuators (Ahn et al., 2010; Ahn et al., 2011), and degradable implants (Choi et al., 2006; Ji et al., 2020). Additionally, Table 12.1

Table 12.1. Potential applications of biodegradable thermo-shrinkable polymers and polymer composites.

Potentiality	Material	Reference
Bone tissue implants	Polyhydroxybutyrate/Polycaprolactone/CNFs	(Yue et al., 2021)
Catheters and vascular stents	Polycaprolactone	(Lendlein et al., 2005)
Controlled drug delivery	Polyester-urethane	(Alteheld et al., 2005)
Four-dimensional-printed devices	Polyhydroxybutyrate/Polycaprolactone/CNFs	(Yue et al., 2021)
Labels and cap seals	Polylactic acid/Ethylene vinyl acetate	(Khankrua et al., 2019)
Minimally invasive devices	Polylactic-based diol	(Choi et al., 2006)
	Polyurethane copolymers	(Gu et al., 2016)
	Polylactic acid	(Yue et al., 2021)
	Poly(glycol-glycerol-sebacate)	(Liu and Cai, 2009)
	Polylactic acid/Chitosan	(Meng et al., 2009)
Ocular tissue replacement	Polyester-urethane	(Alteheld et al., 2005)
Orthopedic and orthodontic material	Polylactic acid/Chitosan	(Meng et al., 2009)
	Polycaprolactone-urethane copolymer	(Sivakumar and Nasar, 2009)
	Polyurethane/Polycaprolactone	(Jankauskaite et al., 2009)
Packaging films	Polylactic-propylene diol	(Choi et al., 2006)
Remote controlled actuators	Polylactic acid/Graphene oxide nanoplatelets	(Lashgari et al., 2016)
	Polyhydroxybutyrate/Polycaprolactone/CNFs	(Yue et al., 2021)
Smart robotic devices	Polyhydroxybutyrate/Polycaprolactone/CNFs	(Yue et al., 2021)
Surgical sutures	Polycaprolactone copolymers	(Nagata and Sato, 2005)
Textiles	Polylactic acid	(Ji et al., 2020)

CNFs refers cellulose nanofibers.

Table 12.2. Higher values of glass transition, melting and crystalline clearing temperatures in °C (T_g, T_m, and T_c, respectively) of biodegradable thermo-shrinkable polymers and polymer composites.

Biodegradable homo- and co-polymers				
Polymer	T_g	T_m	T_c	**Reference**
Bile acids-based polymer	88	—	—	(Thérien et al., 2010)
Polycaprolactone	–39	60	17	(Zhu et al., 2003; Wu et al., 2020)
Polycaprolactone network	–60	64	60	(Lendlein et al., 2005)
Polycaprolactone-urethane	–17	58	17	(Sivakumar and Nasar, 2009)
Polyester-urethane network	66	—	—	(Alteheld et al., 2005)
Polyhydroxybutyrate	—	108	172	(Yue et al., 2021)
Poly(glycol-glycerol-sebacate)	–38	39	–35	(Liu and Cai, 2009)
Polylactic acid	65	180	116	(Khankrua et al., 2019; Ji et al., 2020; Meng et al., 2009; Lashgari et al., 2016; Ferreira and Andrade, 2021; Ray et al., 2002)
Polylactic acid-based diol (network)	–60	—	—	(Choi et al., 2006)
Polylactic acid-glycolide-caprolactone	49	158	—	(Min et al., 2005)
Polylactic acid-propylene-diol	45		—	(Choi et al., 2006)
Polyurethane copolymer	49	—	—	(Lendlein et al., 2001)
Poly(butylene adipate-co-terephthalate)	–27	126	76	(Wu et al., 2020)

Biodegradable polymer composites					
Matrixes	**Fillers***	T_g	T_m	T_c	**Reference**
Polycaprolactone	Epoxidized natural rubber	—	57	39	(Mishra et al., 2011)
	Styrene-butadiene-styrene	–33	60	36	(Abdallah et al., 2021)
	Polyurethane	75	—	—	(Jankauskaite et al., 2009)
Polyhydroxybutyrate	Polycaprolactone	65	59	107	(Yue et al., 2021)
	Polycaprolactone Cellulose nanofibers	—	60	106	(Yue et al., 2021)
Polylactic acid	Graphene oxide (reduced)	61	177	92	(Ferreira and Andrade, 2021)
	Graphene nanoplatelets	61	170	99	(Lashgari et al., 2016)
	Chitosan	61	170	—	(Meng et al., 2009)
	Ethylene vinyl acetate	67	—	—	(Khankrua et al., 2019)
	Silicon carbide	90	—	—	(Liu et al., 2016)
4,4-(adipoyldioxy) dicinnamic acid	Polycaprolactone	–64	48	–48	(Nagata Sato, 2005; Nagata and Kitazima, 2006)
	Polyethylene glycol	–62	40	–50	(Nagata and Kitazima, 2006)
	Polylactic acid	42	150	—	(Nagata and Sato, 2005)
	Polycaprolactone Polyethylene glycol	–63	45	–44	(Nagata and Kitazima, 2006)
	Polycaprolactone Polylactic acid	25	148	97	(Nagata and Sato, 2005)

* The fillers could be biodegradable or not.

summarizes possible applications of biodegradable TSP. According to literature review, biomedical applications are widely suggested. However, other exciting uses of biodegradable TSP have been proposed, such as intelligent robotic devices, remote actuators, packaging films, seals, among others.

Since the triggering shape memory effect is achieved using programming and recovery processes, which exploit one or more phase transition temperatures (*Ttrans*) such as glass transition

(*Tg*), melting (*Tm*). Liquid crystalline (*Tc*), the biomedical applications will depend on *Ttrans* around the human body temperature, for instance (Alteheld et al., 2005; Gu et al., 2016). Thus, Table 12.2 summarizes the *Ttrans* values of biodegradable TSP to facilitate access to such relevant information. The knowledge of *Ttrans* is necessary to form temporary crosslinks in TPS by vitrifying the polymer below *Ttrans*. The programming aims to create a temporary shape by applying external force and temperature above *Ttrans* to a predefined permanent form and then, under constant stress, cooling the deformed body below *Ttrans*. Finally, for the recovery step, under stress-free conditions, reheating the deformed shape above *Ttrans*. The shape memory cycle is repeated at least twice more on the same sample (Ahn et al., 2010; Ahn et al., 2011; Kalkan-Sevinc and Strobel, 2015; Lendlein et al., 2001). Generally, TSP has at least two phases, each one associated with a *Ttrans*. The phase with the higher *Ttrans* is responsible for the permanent shape, whereas the lower *Ttrans* enables the fixation of a temporary condition. By regulating the temperature above and below the lower *Ttrans*, the body can be deformed arbitrarily (Ji et al., 2020; Langer and Tirrell, 2004; Torbati et al., 2014).

For many applications, such as biomedical ones, it is preferable to have more elastic materials in their temporary shape at room temperature. For this purpose, multi-phase amorphous TSP networks with specified-mechanical properties in their temporary body below *Ttrans* have been proposed. For instance, polymer networks can be developed by crosslinking two or more non-miscible polymer blocks. Consequently, the respective chain segments should form differentiable phases with distinct *Ttrans* (Choi et al., 2006). However, crosslinked polymers are limited in their shape-memory functionality because they can undergo creeping and irreversible deformation during programming (Lendlein et al., 2005).

On the other hand, new biodegradable and biocompatible TSP, such as poly(glycol-glycerol-sebacate), polycaprolactone-polyethylene glycol (with and without incorporated polylactic acid block) copolymers and bile acid-based polymers, have been developed to potentially be used in minimally invasive devices and implants (Liu and Cai, 2009; Nagata and Sato, 2005; Thérien et al., 2010; Nagata and Kitazima, 2006). In this regard, it has been proposed that they could program implants based on biodegradable and biocompatible TSP to degrade within a time interval. Thus, removing the implant following a surgery will be unnecessary, which promises to reduce the patient's traumatism (Liu and Cai, 2009). At the same time, it has been suggested that shape memory polyurethane-based could use TSP for the manufacture of orthopedic and orthodontic materials as well as of minimally invasive coronary stents (Sivakumar and Nasar, 2009).

Conclusion

In conclusion, we can say that there are polymers that are heat shrinkable, which have certain shrinkage properties when subjected to changes in temperature. Retractable polymers have memory capacity due to thermal stimuli that modify their shape. One of its most common applications is the area of packing and packaging, in order to facilitate its transport, identification and disposal. In the packaging area, qualities such as containment, resistance to breakage, easy identification and transportation are required, making them economical, innovative and indispensable. These polymeric compounds can have the property of taking the shape of the object that contains them. In this sense, these heat shrinkable biodegradable polymers have the ability to decompose in nature, without leaving an ecological footprint on the environment.

References

Abdallah, A. B., Gamaoun, F., Kallel, A. and Tcharkhtchi, A. (2021). Molecular weight influence on shape memory effect of shape memory polymer blend (poly(caprolactone)/styrene-butadiene-styrene). *J. Appl. Polym. Sci.*, 138(5): 1–11.

Ahn, S. K., Deshmukh, S. P. and Kasi, R. M. (2010). Shape memory behavior of side-chain liquid crystalline polymer networks triggered by dual transition temperatures. *Macromolecules*, 43(17): 7330–7340.

Ahn, S. K., Deshmukh, P. and Kasi, R. M. (2011). Exploiting architecture and composition of side-chain liquid crystalline polymers for shape memory applications. *ACS Symp. Ser.*, 1066: 39–51.

Alteheld, A., Feng, Y., Kelch, S. and Lendlein, A. (2005). Biodegradable, amorphous copolyester-urethane networks having shape-memory properties. *Angew. Chemie - Int. Ed.*, 44(8): 1188–1192.

Ammala, A., Bateman, S., Dean, K., Petinakis, E., Sangwan, P., Wong, S., Yuan, Q., Yu, L., Patrick, C. and Leong, K. H. (2011). An overview of degradable and biodegradable polyolefins. *In*: *Progress in Polymer Science* (Oxford) (Vol. 36, Issue 8). Elsevier Ltd. https://doi.org/10.1016/j.progpolymsci.2010.12.002.

Applications that will Benefit from Heat Shrink Technology. Available at: https://www.azom.com/article.aspx?ArticleID=18014 (Accessed: 8th March 2021).

Arora, S. (2018). Biopolymers as packaging material in food and allied industry. *International Journal of Chemical Studies* 6.

Bourg, V., Ienny, P., Caro-Bretelle, A. S., Le Moigne, N., Guillard, V. and Bergeret, A. (2019). Modeling of internal residual stress in linear and branched polyethylene films during cast film extrusion: Towards a prediction of heat-shrinkability. *Journal of Materials Processing Technology*, 271(April): 599–608. https://doi.org/10.1016/j.jmatprotec.2019.04.0025.

Camargo, F. A. O. (2009). Abiotic and biotic degradation of oxo-biodegradable polyethylenes. *Polymer Degradation and Stability*, 94(6): 965–970. https://doi.org/10.1016/j.polymdegradstab.2009.03.011.

Cha, D. S. and Chinnan, M. S. (2004). Biopolymer-based antimicrobial packaging: A review. *Crit. Rev. Food Sci. Nutr.*, 44: 223–237.

Choi, N. Y., Kelch, S. and Lendlein, A. (2006). Synthesis, shape-memory functionality and hydrolytical degradation studies on polymer networks from poly(rac-lactide)-b-poly(propylene oxide)-b-poly(rac-lactide) dimethacrylates. *Adv. Eng. Mater.*, 8(5): 439–445.

Charoonsuk, T., Muanghlua, R., Sriphan, S., Pongampai, S. and Vittayakorn, N. (2021). Utilization of commodity thermoplastic polyethylene (PE) by enhanced sensing performance with liquid phase electrolyte for a flexible and transparent triboelectric tactile sensor. *Sustainable Materials and Technologies*, 27: e00239. https://doi.org/10.1016/j.susmat.2020.e00239.

Chaudhry, A. U., Mabrouk, A. and Abdala, A. (2020). Thermally enhanced pristine polyolefins: Fundamentals, progress and prospective. *Journal of Materials Research and Technology*, 9(5): 10796–10806. https://doi.org/10.1016/j.jmrt.2020.07.101.

Charoonsuk, T., Muanghlua, R., Sriphan, S., Pongampai, S. and Vittayakorn, N. (2021). Utilization of commodity thermoplastic polyethylene (PE) by enhanced sensing performance with liquid phase electrolyte for a flexible and transparent triboelectric tactile sensor. *Sustainable Materials and Technologies*, 27: e00239. https://doi.org/10.1016/j.susmat.2020.e00239.

Cui, J., Adams, J. G. M. and Zhu, Y. (2017). Pop-up assembly of 3D structures actuated by heat shrinkable polymers. *Smart Mater. Struct.*, 26.

Ferreira, W. H. and Andrade, C. T. (2021). The role of graphene on thermally induced shape memory properties of poly(lactic acid) extruded composites. *J. Therm. Anal. Calorim.*, no. 0123456789.

Greene, J. P. (2019). Degradation and biodegradation standards for biodegradable food packaging materials. *In*: *Reference Module in Food Science* (Issue 2014). Elsevier. https://doi.org/10.1016/b978-0-08-100596-5.22437-2.

Gu, ying S., feng Gao, X., peng Jin, S. and liang Liu, Y. (2016). Biodegradable shape memory polyurethancs with controllable trigger temperature. *Chinese J. Polym. Sci. (English Ed.)*, 34(6): 720–729.

Jankauskaite, V., Laukaitiene, A. and Mickus, K. V. (2009). Shape memory properties of poly(ε-Caprolactone) based thermoplastic polyurethane secondary blends. *Medziagotyra*, 15(2): 142–147.

Ji, F., Li, J., Weng, Y. and Ren, J. (2020). Synthesis of PLA-based thermoplastic elastomer and study on preparation and properties of PLA-based shape memory polymers. *Mater. Res. Express*, 7(1).

Jiang, Z., Wang, X., Jia, H., Zhou, Y., Ma, J., Liu, X., Jiang, L. and Chen, S. (2019). Superhydrophobic polytetrafluoroethylene/heat-shrinkable polyvinyl chloride composite film with super anti-icing property. *Polymers*, 11(5). https://doi.org/10.3390/polym11050805.

Kakar, M. R., Mikhailenko, P., Piao, Z., Bueno, M. and Poulikakos, L. (2021). Analysis of waste polyethylene (PE) and its by-products in asphalt binder. *Construction and Building Materials*, 280: 122492. https://doi.org/10.1016/j.conbuildmat.2021.122492.

Kalkan-Sevinc, Z. S. and Strobel, C. T. (2015). Material characterization of heat shrinkable film. *J. Test. Eval.*, 43: 1531–1539.

Khankrua, R. et al. (2019). Development of PLA/EVA reactive blends for heat-shrinkable film. *Polymers (Basel)*, 11: 1925.

Koerner, H., Price, G., Pearce, N. A., Alexander, M. and Vaia, R. A. (2004). Remotely actuated polymer nanocomposites—Stress-recovery of carbon-nanotube-filled thermoplastic elastomers. *Nat. Mater.*, 3: 115–120.

Kondratov, A. P. (2014). Thermo shrink films with interval macrostructure for protection of packaging from falsification. *Mod. Appl. Sci.*, 8: 204–209.

Langer, R., Lendlein, A., Schmidt, A. and Grablowitz, H. (2000). Biodegradable shape memory polymers. https://patentimages.storage.googleapis.com/c6/ec/d6/58f5a4f226e348/US6160084.pdf. ES PATENTE.

Langer, R. and Tirrell, D. A. (2004). Designing materials for biology and medicine. *Nature*, 428(6982): 487–492.

Lashgari, S., Karrabi, M., Ghasemi, I., Azizi, H., Messori, M. and Paderni, K. (2016). Shape memory nanocomposite of poly(L-lactic acid)/graphene nanoplatelets triggered by infrared light and thermal heating. *Express Polym. Lett.*, 10(4): 349–359.

Lendlein, A., Schmidt, A. M. and Langer, R. (2001). Segments showing shape-memory properties. *PNAS*, 842–847.

Lendlein, A., Schmidt, A. M., Schroeter, M. and Langer, R. (2005). Shape-memory polymer networks from oligo(ε-caprolactone)dimethacrylates. *J. Polym. Sci. Part A Polym. Chem.*, 43(7): 1369–1381.

Lendlein, A. and Langer, R. S. (2006). Self-expanding device for the gastrointestinal or urogental area. https://patentimages.storage.googleapis.com/ee/0b/c3/7862acbd64abf1/US20060142794A1.pdf.

Lendlein, A. and Langer, R. (2012). Biodegradable shape memory polymeric sutures. https://patentimages.storage.googleapis.com/33/ae/0e/0923d127669554/US8303625.pdf.

Liu, L. and Cai, W. (2009). Novel copolyester for a shape-memory biodegradable material *in vivo*. *Mater. Lett.*, 63(20): 1656–1658.

Liu, W., Wu, N. and Pochiraju, K. (2016). Shape recovery characteristics of 3D printed soft polymers and their composites. *ASME International Mechanical Engineering Congress and Exposition*, 2013: 1–7.

Marco, D. (2006). Biodegradable self-inflating intragastric implants and method of curbing appetite by the same.

Mather, P. T., Luo, X. and Rousseau, I. A. (2009). Shape memory polymer research. *Annual Review of Materials Research,* 39: 445–471.

Meng, Q., Hu, J., Ho, K., Ji, F. and Chen, S. (2009). The shape memory properties of biodegradable chitosan/poly(l-lactide) composites. *J. Polym. Environ.*, 17(3): 212–224.

Min, C., Cui, W., Bei, J. and Wang, S. (2005). Biodegradable shape-memory polymer-Polylactide-co-poly(glycolide-co-caprolactone) multiblock copolymer. *Polym. Adv. Technol.*, 16(8): 608–615.

Mishra, J. K., Chang, Y. W. and Kim, W. (2011). The effect of peroxide crosslinking on thermal, mechanical, and rheological properties of polycaprolactone/epoxidized natural rubber blends. *Polym. Bull.*, 66(5): 673–681.

Morshedian, J., Khonakdar, H. A., Mehrabzadeh, M. and Eslami, H. (2003). Preparation and properties of heat-shrinkable cross-linked low-density polyethylene. *Adv. Polym. Technol.,* 22: 112–119.

Nagata, M. and Sato, Y. (2005). Synthesis and properties of photocurable biodegradable multiblock copolymers based on poly(ε-caprolactone) and poly(L-lactide) segments. *J. Polym. Sci. Part A Polym. Chem.*, 43(11): 2426–2439.

Nagata, M. and Kitazima, I. (2006). Photocurable biodegradable poly(ε-caprolactone)/poly(ethylene glycol) multiblock copolymers showing shape-memory properties. *Colloid Polym. Sci.*, 284(4): 380–386.

Ojeda, T. F. M., Dalmolin, E., Forte, M. M. C., Jacques, R. J. S., Bento, F. M. and Camargo, F. A. O. (2009). Abiotic and biotic degradation of oxo-biodegradable polyethylenes. *Polymer Degradation and Stability*, 94(6): 965–970. https://doi.org/10.1016/j.polymdegradstab.2009.03.011.

Oyama, T. G., Kimura, A., Nagasawa, N., Oyama, K. and Taguchi, M. (2020). Development of advanced biodevices using quantum beam microfabrication technology. *Quantum Beam Sci.*, 4(1) 14.

Pattanashetti, N. A., Heggannavar, G. B. and Kariduraganavar, M. Y. (2017). Smart biopolymers and their biomedical applications. *Procedia Manuf.,* 12: 263–279.

Rani, S. et al. (2020). Wheat starch, gum arabic and chitosan biopolymer treatment of wool fabric for improved shrink resistance finishing. *Int. J. Biol. Macromol.,* 163: 1044–1052.

Ray, S. S., Yamada, K., Ogami, A., Okamoto, M. and Ueda, K. (2002). New polylactide/layered silicate nanocomposite: Nanoscale control over multiple properties. *Macromol. Rapid Commun.*, 23(16): 943–947.

Rousseau, I. A. (2008). Challenges of shape memory polymers: A review of the progress toward overcoming SMP's limitations. *Polym. Eng. Sci.,* 48: 2075–2089.

Shikinami, Y. (2001). Shape-memory biodegradable and absorbable material. ES PATENTE https://patentimages.storage.googleapis.com/71/43/40/da6d2e93f57c1a/US6281262.pdf.

Sivakumar, C. and Nasar, A. S. (2009). Poly(ε-caprolactone)-based hyperbranched polyurethanes prepared via A2 + B3 approach and its shape-memory behavior. *Eur. Polym. J.*, 45(8): 2329–2337.

Skrzeszewska, P. J., Jong, L. N., De Wolf, F. A., Cohen Stuart, M. A. and Van Der Gucht, J. (2011). Shape-memory effects in biopolymer networks with collagen-like transient nodes. *Biomacromolecules,* 12: 2285–2292.

Sliozberg, Y. R. et al. (2021). Computational design of shape memory polymer nanocomposites. *Polymer* (Guildf), 217.

Thérien-Aubin, H., Gautrot, J. E., Shao, Y., Zhang, J. and Zhu, X. X. (2010). Shape memory properties of main chain bile acids polymers. *Polymer (Guildf).*, 51(1): 22–25.

Torbati, A. H., Nejad, H. B., Ponce, M., Sutton, J. P. and Mather, P. T. (2014). Properties of triple shape memory composites prepared via polymerization-induced phase separation. *Soft Matter*, 10(17): 3112–3121.

Wang, F. et al. (2016). Molecular origin of the shape memory properties of heat-shrink crosslinked polymers as revealed by solid-state NMR. *Polymer (Guildf),* 107: 61–70.

Wang, J., Wen, H. and Muhunthan, B. (2020). Development of test methods to characterize the shrinkage properties of cementitiously stabilized materials. *Transportation Geotechnics*, 25(July): 100405. https://doi.org/10.1016/j.trgeo.2020.100405.

Whelan, D. (2017). Thermoplastic elastomers. *Brydson's Plast. Mater. Eighth Ed.,* 653–703. doi:10.1016/B978-0-323-35824-8.00024-4.

Wu, X. et al. (2020). A triple-shape memory material fabricated by in situ crosslinking of poly(butylene adipate-co-terephthalate)/ε-polycaprolactone via electron-beam irradiation and its triple-shape memory effects. *J. Appl. Polym. Sci.*, 137(16): 1–12.

Wypych, G. (2020). Pvc Properties. *PVC Formulary*, 5–45. https://doi.org/10.1016/b978-1-927885-63-5.50005-7.

Xiao, G., Kim, J., Cai, X. and Cui, T. (2019). Shrink-induced highly sensitive dopamine sensor based on self-assembly graphene on microelectrode. *2019 20th Int. Conf. Solid-State Sensors, Actuators Microsystems Eurosensors XXXIII, Transducers 2019 Eurosensors XXXIII* 1120–1123. doi:10.1109/TRANSDUCERS.2019.8808235.

Xiaonan, W., Wencui, Y., Yong, G. and Decheng, F. (2020). The influence of shrinkage-reducing agent solution properties on shrinkage of cementitious composite using grey correlation analysis. *Construction and Building Materials*, 264: 120194. https://doi.org/10.1016/j.conbuildmat.2020.120194.

Yang, J., Huang, J., He, X., Su, Y. and Oh, S. K. (2020). Shrinkage properties and microstructure of high volume ultrafine phosphorous slag blended cement mortars with superabsorbent polymer. *Journal of Building Engineering*, 29(August 2019): 101121. https://doi.org/10.1016/j.jobe.2019.101121.

Yue, C., Hua, M., Li, H., Liu, Y., Xu, M. and Song, Y. (2021). Printability, shape-memory, and mechanical properties of PHB/PCL/CNFs composites. *J. Appl. Polym. Sci.*, January: 1–14.

Zanchin, G. and Leone, G. (2021). Polyolefin thermoplastic elastomers from polymerization catalysis: Advantages, pitfalls and future challenges. *Progress in Polymer Science*, 113: 101342. https://doi.org/10.1016/j.progpolymsci.2020.101342.

Zhang, Q., Yan, D., Zhang, K. and Hu, G. (2015). Pattern transformation of heat-shrinkable polymer by three-dimensional (3D) printing technique. *Sci. Rep.*, 5: 24–27.

Zhao, Q., Qi, H. J. and Xie, T. (2015). Recent progress in shape memory polymer: New behavior, enabling materials, and mechanistic understanding. *Progress in Polymer Science,* 49–50: 79–120.

Chapter 13

Applications of Biodegradable Polymers in Food Industry

Jose Fernando Solanilla-Duque,[1,2,*] *Diego Fernando Roa-Acosta,*[1]
Luis Daniel Daza,[2] *Darwin Carranza-Saavedra,*[2] *Henry Alexander Váquiro,*[2]
Juan Pablo Quintero-Cerón,[3] *Maria Julia Spotti*[3] and *Carlos Carrara*[3]

1. Introduction

Many countries exporting fruits and vegetables, meat and dairy products are exposed to complex stabilization processes of these raw materials at post-harvest and post-production levels, and this, added to the increase in consumer demand for higher quality products and the need to reduce the consumption and disposal of disposable plastic containers, has generated great interest in the development of Edible Films (EF) and Coatings (EC) to satisfy both demands. For this reason, it is necessary to study and apply technologies to obtain antifungal, antibacterial and antioxidant benefits in fruit and vegetable products that present deterioration in their quality attributes as a consequence of their low structural resistance, high water activity, weight loss, susceptibility to attack by microorganisms and short half-life during post-harvest and post-production handling. The design and optimization of bioactive packaging is of vital importance since passive packaging only provides a barrier to the outside, and the increased bacterial activity is reflected on the surface of the food and in many cases, compounds implemented in the control of such biota do not have the efficient activity as a result of the heterogeneity of the constituents in such products. This variant is attractive for the food industry due to the increasing demand for minimally processed foods free of synthetic additives (Chang et al., 2010).

Microorganisms have the ability to generate resistance to products that have been implemented recurrently and therefore could produce a greater risk to the public health of the consumer, for this reason it is essential to make use of bioactive compounds. But their synergy with biopolymeric matrices has not been evaluated, their incidence in the mechanical properties of edible films is not known, if the activity of their compounds decreases due to interactions with hydrocolloids which generate the polymeric structure, or if on the contrary they work synergistically maintaining the

[1] Departamento de Agroindustria, Facultad de Ciencias Agrarias, Universidad del Cauca, Campus Las Guacas, 190001 Popayán, Colombia.
[2] Centro de desarrollo agroindustrial del Tolima research group (CEDAGRITOL), Universidad del Tolima, Campus Santa Helena, Cl 42 # 1 – 02, 730006299 Ibagué, Colombia.
[3] Instituto de Tecnología de Alimentos, Facultad de Ingeniería Química, Universidad Nacional del Litoral. 1° de Mayo 3250 (3000) Santa Fe-Argentina.
* Corresponding author: jsolanilla@unicauca.edu.co

bioactivity of these compounds. It is interesting to consider plant-derived compounds as substitutes for synthetic chemical products according to their bioactive properties in the development of new products for use in the food industry (nutraceuticals). It is essential to determine the biological activity of bioactive plant compounds (Hossain and Rahman, 2019).

Packaging plays a fundamental role in the preservation, distribution and marketing of fresh, semi-processed and processed products. Some of its functions are to contain the food, protect it from physical, mechanical, chemical and microbiological action. An EC or EF has the ability to work synergistically with other packaging materials, as in the case of corn starch ECs added with glycerol as a plasticizer and applied on brussels sprouts (*Brassica oleracea* L. var. Gemmifera). These were treated with this solution, stored in expanded polystyrene plates and covered with polyvinyl chloride (PVC) film, preserving quality parameters such as: commercial acceptability, weight loss, firmness, color and nutritional quality. The treatment allowed the product to remain constant during 42 days of storage at a temperature of 0°C, and its ascorbic acid content, total flavonoids and antioxidant activity remained stable (Viña et al., 2007).

The use of an EC or EF in food applications, especially in highly perishable products, is based on certain characteristics such as:

- Production cost
- Availability
- Functional, mechanical (tension and flexibility) and optical (gloss and opacity) properties
- Barrier effect against gas flow
- Structural resistance to water and microorganisms
- Sensory acceptability

These characteristics are influenced by parameters (Guilbert et al., 1996; Rojas-Graü et al., 2009) such as:

- The type of material implemented as structural matrix (conformation, molecular weight, charge distribution),
- Conditions under which the films are preformed (type of solvent, pH, concentration of components, temperature, etc.),
- The type and concentration of additives (plasticizers, crosslinking agents, antimicrobials, antioxidants, emulsifiers, etc.).

1.1 New hydrocolloids of interest in CE and FE

The use of hydrocolloids in EC and EF (Fig. 13.1) has been the subject of several studies, among which Rojas-Grau et al. (2007) can be mentioned, who demonstrated the capacity of DE based on sodium alginate and gellan gum to transport N-acetylcysteine and glutathione as anti-puddling agents. It is noteworthy that these ECs have had a positive effect when vegetable oils have been added to them. This mixture increases the resistance to water vapor in minimally processed fruits, an example of this is the process of CD of the Fuji apple variety. But the relevant aspect in this study was the use of vegetable oil with a high content of essential fatty acids (3 and 6) and the ability of the EC to keep it encapsulated within the oil.

Oleoresins from different natural sources such as rosemary (Rosmarinus officinalis), oregano (Origanum vulgare), olive (Olea europea), chili (Capsicum frutescens), garlic (Allium sativum), bulb onion (Allium cepa L.) and common cranberry (Vaccinium oxycoccus) have been encapsulated and incorporated into polymeric matrices to obtain bioactive ECs. These have been formulated from sodium caseinate, carboxymethylcellulose and chitosan. Some of them have been applied in different food matrices to inhibit and/or retard the adverse effect of microorganisms. It has been reported in literature that the combined effect of Film-Forming Solutions (FFS) (Fig. 13.1) with

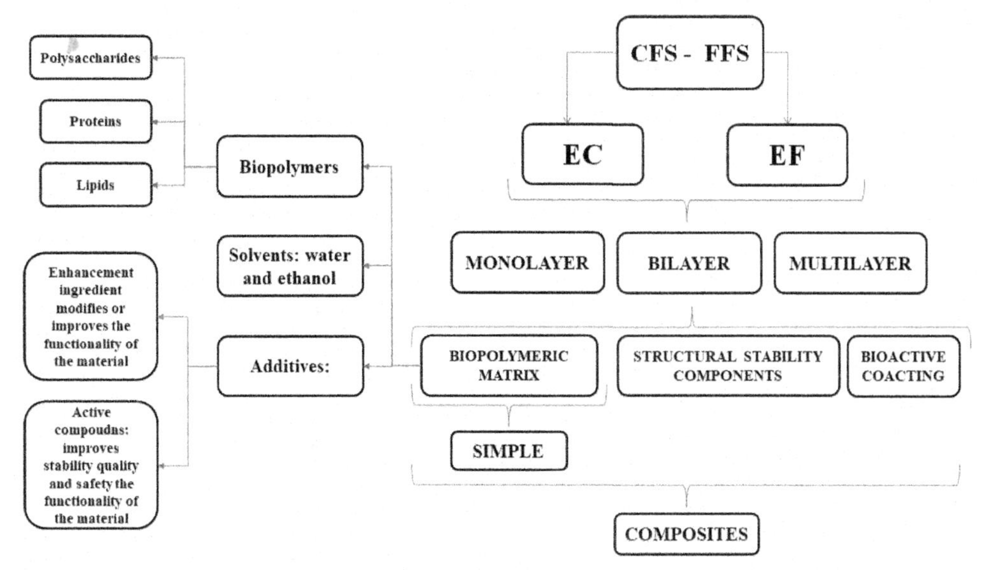

Figure 13.1. EC and EF production scheme from Coating Forming Solutions (CFS) and films (FFS) (Daza et al., 2018; Falguera et al., 2011; Galindez et al., 2019; Homez et al., 2018; Quintero-Cerón et al., 2014).

addition of oleoresins at a concentration of 1.0% w/v by agar diffusion, that the native microflora is sensitive to a chitosan-based EC enriched with olive, rosemary and chili oleoresin, as well as to CMC + rosemary solutions. While L. monocytogenes has been sensitive to CMC + rosemary-based CE and very sensitive to chitosan FFS with incorporations of rosemary extracts (Villafañe, 2017).

It is important to mention that chitosan-based EFs enriched with olive and rosemary oleoresins show clear antioxidant effects because they retard the action of peroxidase (POD) and polyphenoloxidase (PPO) enzymes during storage as a function of time (5 days). According to the above, it can also be said that a chitosan-based EC enriched with these oleoresins does not significantly affect the sensory perception of the food matrix (Ponce et al., 2008). On the other hand, the incorporation of garlic oil encapsulated in a sodium alginate-based PE as an antibacterial agent significantly improves the antibacterial activity. These antimicrobial films can be prepared with garlic oil concentrations around 0.4% v/v of the film-forming solution. This antibacterial activity has been reported to be functional against *Staphylococcus aureus* and *B. cereus*.

Another important aspect to mention is the mechanical properties (tension and elongation). These can be affected by the various intermolecular interactions between the mixture components. It can be said that when an oleoresin is added, the values of mechanical properties can decrease depending on the percentage of oil addition (Chana-Thaworn et al., 2011). In turn, the water vapor permeability can increase accordingly with the oil concentration. An example of this is when garlic oil is added, despite the hydrophobic nature of garlic oil, it appears to increase the molecular interactions of the alginate film. Water vapor permeability is one of the most important properties to quantify in the designed films, since most food products are susceptible to water vapor exchange, due to loss of turgidity or, on the contrary, to moisture gain if the latter was dried. On the other hand, the color or colorimetric measurement parameters are an indispensable factor when placing the product on the market. If the change is perceptible by the consumer, it will constitute a rejection by him. For this reason, it is important that the transparency value be the ideal in the FE so that the functional characteristics of the CE do not generate changes in the parameters (L*, a*, b*, E) or in the consumer's sensory perception Pranoto et al. (2005).

While it is important to transport bioactive components, it is also important to reduce the environmental impact. Biodegradable films based on cassava starch (Manihot esculenta Crantz) have become an alternative to this problem. They have been characterized from the mechanical point of view, as well as the effect of various components that modify their molecular structure depending on variation factors such as temperature, pressure, retrogradation, aqueous activity, thermal properties, rheological properties, among others. Many of these components have been reported, among them plasticizing agents such as glycerol and polyethylene glycol, others have crosslinking behavior such as glutaraldehyde and $CaCl_2$ that can improve water vapor transmission properties. Nowadays, studies have been guided towards the capacity of these starch films to transport natural antimicrobial ingredients (Homez-Jara et al., 2018; Quintero-Cerón et al., 2014; Vásconez et al., 2009).

When a 5.0% w/v starch-based suspension (Manihot esculenta Crantz) is formulated with the addition of ingredients with antimicrobial and antioxidant activity such as cinnamon (0.0–0.3 w/w), clove (0.0–0.3 w/w), red bell pepper (0.0–0.3 w/w) and coffee powder (0.0–0.5 w/w), propolis extract (0.0–0.7 w/w) and orange essential oil (0.0–0.2 w/w), can present a marked influence on the tensile strength of biodegradable films and on the Water Vapor Permeability (WVP) (Kechichian et al., 2010). This incorporation causes a decrease in tensile strength and percentage elongation at break, but it is possible to increase the value of water vapor permeability. On the other hand, the microbiological stability of these biodegradable films when stored in an environment with relative humidity conditions of 60.0% and temperature of 25.0°C, in contact with a food matrix, results in the presence of molds and yeasts in some cases after 7 days of storage. In this particular case, it is necessary to control the water activity (a_w) to avoid antimicrobial activity in biodegradable films.

In food, not only microbiological stability plays an indispensable role in food quality, but also factors such as sensory perception. These factors are indispensable for the successful application of technologies such as edible films and coatings. One of the relevant factors is flavor loss, which strongly affects sensory acceptability (Robles-Sánchez et al., 2013). In order to slow down flavor changes during food preservation, some authors have proposed to implement the encapsulation of aromatic compounds as a possible strategy to reduce the effect of undesired reactions such as oxidation (Hammam, 2019).

Carrageenan films result as a possible encapsulation matrix as they show affinity for polar volatile compounds. These EFs act as active packaging and may have the objective of gradually releasing aroma compounds and thus maintaining sensory characteristics such as odor and flavor for certain periods of time (Marcuzzo et al., 2010; Sason and Nussinovitch, 2021). The transport and release of various compounds (antioxidants, aromas, antimicrobial and anti-pickling compounds, vitamins, enzymes) is one of the most important aspects within the functionalities of an edible film and coating, and the use of nanotechnological solutions through the use of additive nanoparticles has been proposed to encapsulate functional and bioactive compounds that can be released from the matrices that contain them in a controlled manner (Falguera et al., 2011).

Other hydrocolloids of great interest are gum exuded from the cashew tree (Anacardium occidentale L.) and called gum policaju, galactomannans and aloe vera. First, edible films based on policaju gum have been evaluated from their mechanical properties, wettability, surface tension, opacity, tensile strength, percentage of elongation at break and water vapor permeability, with the aim of obtaining biopolymeric structures or matrices that can generate edible coatings applied to minimally processed fruits. The results reported in literature show a concentration limit of about 1.5% w/v of polycationic gum and concentrations lower than this value result in brittle films. However, the addition of Tween 80, an additive that functions as a surfactant, reduces the cohesive forces and decreases the surface tension, resulting in an increase in the wettability of the coating forming solution, improving the compatibility of the EC with the fruit surface (Carneiro-da-Cunha et al., 2009). These polymeric matrices act as an excellent barrier against mass transport by reducing weight loss as a consequence of respiration processes when evaluating the shelf life of fresh produce under refrigeration (Dhall, 2013; Li et al., 2019; Souza Almeida and Kawazoe Sato, 2019).

Galactomannans are hydrocolloids that generate interest due to their ability to structure matrices, they are stored as reserve polysaccharides and are extracted from seeds, their polymeric structure is mainly influenced by the ratio of mannose/galactose units and the distribution of galactose residues in the main chain (Cerqueira et al., 2011). Adenanthera pavonina and Caesalpinia pulcherrima two plants belonging to the legume family were used with the aim of developing coatings from new sources of galactomannans. These plants are of valuable interest because they are used for reforestation and have the ability to disperse and so far have not been commercially exploited (Lima et al., 2010; Marcuzzo et al., 2010).

Different proportions of galactomannans, collagen and glycerol have been prepared and tested in order to design possible mixtures with a high degree of wettability, which means that they have the ability to adhere and distribute homogeneously and easily in coated fruits such as mango and apple. As main conclusions, the best mixtures for mango and apple were 0.5% galactomannan from A. pavonina, 1.5% collagen and 1.5% glycerol; and 0.5% galactomannan from A. pavonina, 1.5% collagen without addition of glycerol. Lower O_2 consumption (28%) and CO_2 production (11.0%) were achieved in coated mangoes compared to control samples (uncoated). In apples, O_2 and CO_2 consumption and production were approximately 50% lower in the presence of EC. These results suggest that galactomannan-based composite coatings can reduce gas transfer and thus become useful tools to extend the shelf life of these fruits (Dubey and Dubey, 2020).

Aloe is a tropical and subtropical plant that has been used for centuries by traditional medicine for its therapeutic properties. The mucilaginous gel extracted from the Aloe barbadensis Miller plant has proven to be of interest to the food industry for its functional properties, referred to its antiradical effect due to the presence of phenolic compounds that in vitro antioxidant tests determined to be as efficient as α-tocopherol (Lee et al., 2000; Thakur et al., 2019). Hu et al. (2005), compared the radical scavenging activity of extracts obtained from aloe vera skin and gel using supercritical CO_2 and ethanol, versus the capacity of antioxidants implemented in the food industry (BHT, Trolox, α-tocopherol) the % inhibition or stabilization of radicals in descending order were as follows: Trolox (76.8%), ethanolic extract of A. vera peel (39.7%), BHT (35.9%), supercritical CO_2 extract (33.5%), α-tocopherol (25.6%), ethanolic extract of A. vera pulp (14.2%). The components belonging to the aloe vera bark or skin were the ones that generated the highest antiradical activity (ABTS, DPPH).

In the industry, this gel has been implemented in the development of new moisturizing liquids, beverages, tonics and even hair care products (Hazrati et al., 2017). But nowadays, interest has been developed in the generation of edible coatings due to the capacity to generate matrices and coatings at concentrations higher than 30% w/v of mucilaginous gel (Barzegar et al., 2020; Behbahani and Imani Fooladi, 2018).

Starch from Canna edulis, Ker, also known as achira or sago, is a perennial rhizome of the Cannacea family, native to the Andean region of South America. This plant, whose large rhizomes have been used as a traditional food for more than 4,000 years, is currently cultivated for starch production in small-scale factories in Colombia, Brazil, China, Taiwan and Vietnam (Puncha-arnon et al., 2007). New sources including roots and tubers such as achira (Canna edulis) and ulluco (Ullucus tuberosus) represent opportunities to improve starch production and utilization in the industry. Expansion in the use of these starch sources also depends on the development of new products, including edible biopolymers applied to both fresh and processed foods (Ferreira et al., 2020; Pagella et al., 2002; Thakur et al., 2019).

One of the studies of interest evaluated the effect of incorporating nanocellulose into an achira starch dispersion (3% bs) plasticized with glycerin (25% w/w dry starch). The mechanical properties, water solubility, water vapor permeability, were the experimental response parameters. As results, it was determined that the addition of cellulose nanoparticles increased the breaking strain and decreased the elongation. These changes were explained as a function of the formation of strong interactions between the hydroxyl groups of cellulose and starch. As a consequence, water vapor permeability decreased, as did water solubility values, which decreased (Choulitoudi et al., 2017; Ruan et al., 2019).

Ulluco starch is a renewable source of great interest in the production of biodegradable or edible films. The effect of glycerol-plasticized starch concentration of coating forming solutions (CFS) has been successfully reported (Fig. 13.1). The CFS formed from ulluco starch have been subjected to different drying temperatures resulting in experimental response variables such as physical properties, solubility, water vapor permeability, tensile strength, elongation at break and transparency of the CFS. To correlate all variables and determine the effect of starch concentration, glycerol concentration and temperature on the mechanical, optical and Water Affinity Properties (WAP) of the edible films, a feed-forward and cascade neural network analysis was performed. These results showed that the use of ulluco starch in the preparation of edible films has enormous potential for the replacement of non-biodegradable plastic packaging. It has been reported that the extraction of ulluco starch shows high efficiency and high purity and has a high amylose content. This is beneficial for its preparation by cooking and is a promising feature for biofilm preparation. It has been mentioned that low temperature-prepared PEs can be used as food packaging due to their favorable mechanical properties and low water vapor permeability, which would increase the shelf life of the food protected by such packaging. These PEs have low opacity; however, ulluco starch-based films are not recommended for products susceptible to degradation by exposure to light. In addition to the above, these PEs exhibit good stability against thermal degradation, which could be promising for their potential use in food transportation under various environmental conditions. In general, the application of low drying temperature allows a better reorganization of the components of ulluco starch films, which in turn influences the applicability of the material (Galindez et al., 2019).

Another study of interest in the development of CE and PE, are coatings made from chitosan, a biopolymer that has excellent characteristics for the formation of films with biodegradable and antimicrobial properties, and is also non-toxic (Palou et al., 2015; Pavinatto et al., 2020; Salazar-Sánchez et al., 2020; Xiong et al., 2020; Yan et al., 2019). These studies have reported the effect of the concentration of the polymer at different concentrations (0.5, 1.0 and 1.5%) and at three drying temperatures (2°C, 25°C and 40°C) on the physicochemical, mechanical and thermal properties of chitosan PEs. What was important in this study was the evaluation of different drying temperatures, reporting that lower temperatures had a positive effect on certain properties of the films, such as moisture content, solubility, water vapor permeability and optical properties. However, the use of higher drying temperatures (40°C), combined with a higher concentration of chitosan, showed an improvement in tensile strength, swelling power and color properties, which evidenced a decrease in their lightness. The authors of this work concluded that these chitosan-based films showed desirable characteristics, which may allow their future use as packaging for food products (Ruzaina et al., 2017; Silva et al., 2019; Won et al., 2018; Zahedi et al., 2019).

1.2 Chemistry of film-formation

The study of interactions between polysaccharides and protein is of great practical interest in several areas, such as the study of biological systems and in the development of innovative products for industry in general. Hence, the physicochemical properties and interactions among proteins and other compounds in the formulation of products determine their utility and applications as functional properties.

These functional properties have a significant influence on the processed, preparation, attributes and quality of the final product. Thus, these properties are influenced by many factors of variation (aw, °Brix, pH, Eh, I, T, P, etc.), becoming an object of study to understand the physicochemical, rheological, thermal and structural behaviors in the food. Therefore, the intermolecular interactions and the capacity of stabilizing structurally their compounds in different dispersed systems (emulsions, foams or gels) are determinate factors in the studies in food (Fig. 13.2). Moreover, in the formulations of food, there is a wide set of substances that interact in different ways at the interface (a/w or o/w) to reach specific properties. These substances compete to quickly fill the interface and even to acquire one or another configuration in this interface, thus influencing the stability in this type of dispersed system (Dickinson and McClements, 1995; Klinkesorn et al., 2006; McClements, 2009).

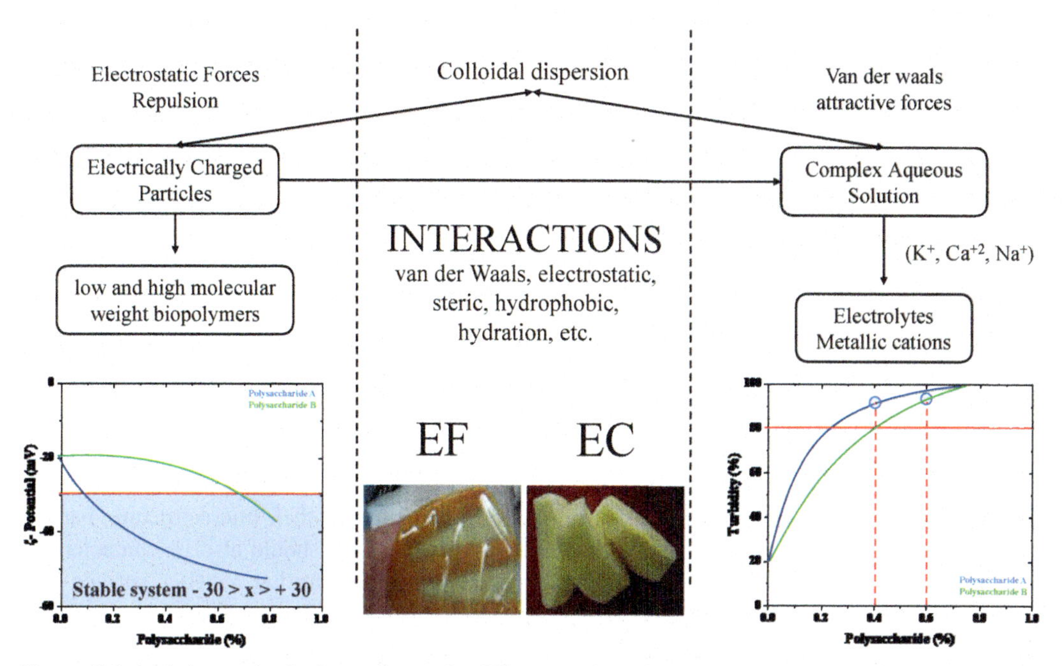

Figure 13.2. Main intermolecular interactions and stability evaluation of EC and EF forming solutions as a function of the effect of electrolyte presence, charge and molecular weight changes (Carranza-Saavedra et al., 2016a; Dickinson and McClements, 1995; McClements, 2009; Solanilla-Duque et al., 2020).

In the case of a mixture of proteins and lipids that depend on the interfacial composition, protein-lipid ration in dissolution the adsorption isotherms could show in them the grade of interaction among different components adsorbed at the air-water interface (Martinez et al., 2005; Rodríguez et al., 2007a; Rodríguez et al., 2007b; Rodríguez et al., 2003). On the whole, the activity of the adsorption system is given by the protein, due to the fact that protein saturates the interface to high concentrations. Nonetheless, the lipids in low concentrations in this solution increase the Superficial Activity (AS) of proteins. The AS of the mixed system from protein-polysaccharides is more complex, possibly because of competition and thermodynamic incompatibility between both biopolymers within the phase and at in the interface (Pérez et al., 2007). Referring to proteins and soluble lipids in water adsorption towards the air-water, it depends on molecules' chemistry and their system composition (Álvarez Gómez et al., 2008; Álvarez and Patino, 2006; Gómez and Rodríguez Patino, 2007). There are two conditions that are:

a. Critical micelle concentration in lipids
b. Adsorption efficacy of proteins

These two concentrations limit the prevalence or compatibility of these components (protein and lipid) at the interface air-water, with important repercussions on the adsorption rate. The adsorption is more complex, when the system is made up of a mixture of different protein kind, due to the competition and incompatibility at the air-water interface that occurs between proteins (Damodaran, 2004; Mackie et al., 2001; Sengupta and Damodaran, 2000, 2001).

This behavior observed in the protein induces a phase transition of the lipid monolayer (Vollhardt and Fainerman, 2000). This penetration process depends on the type of protein and surfactant, the electrostatic interactions that take place and the variation factors mentioned above. In the case of dispersions obtained from phospholipids and proteins (Kanthe et al., 2020; McClellan and Franses, 2003; Powell and Chauhan, 2016) it has been reported, at equilibrium, the film will be formed and governed mainly by phospholipid. However, when the interfacial film expands and contracts, the protein is adsorbed and governs film formation.

Regarding the protein-polysaccharide mixture, there are studies that mention the existence of intermolecular interactions between them at the air-water interface (Baeza et al., 2005; Damodaran and Razumovsky, 2003; Martinez et al., 2005, 2007), which are related with the improvement in the functionality of adsorbed proteins (Baeza et al., 2005). The dynamic behavior of biopolymer films is recognized for its great importance in the formation and stability of dispersions in foods. Besides, most polysaccharides are hydrophilic, that is, they do not have much tendency to adsorb at the air-water interface, but they can increase the stability and functionality of the proteins in the different dispersion systems, acting as a thickening agent or gelling (Dickinson, 1992). As a rule, the addition of polysaccharides produces an increasing amount of protein at the air-water interface, the polysaccharides have an important role in the formation and stabilization of dispersions (McClements, 2015), because these will be able to reach conjugating the protein to a greater or lesser degree, contributing in the same way to increase the functional properties of the protein.

1.2.1 Methods for testing films

It is important to characterize both EC and EF forming solutions. As mentioned above, methods must be established for the characterization of the FE by determining their microstructure, barrier, electromagnetic, mechanical and thermal properties. Likewise, CFS should also be characterized for their emulsifying, structural, rheological, interfacial and adsorption kinetic, surface and intermolecular interaction properties (Fig. 13.3).

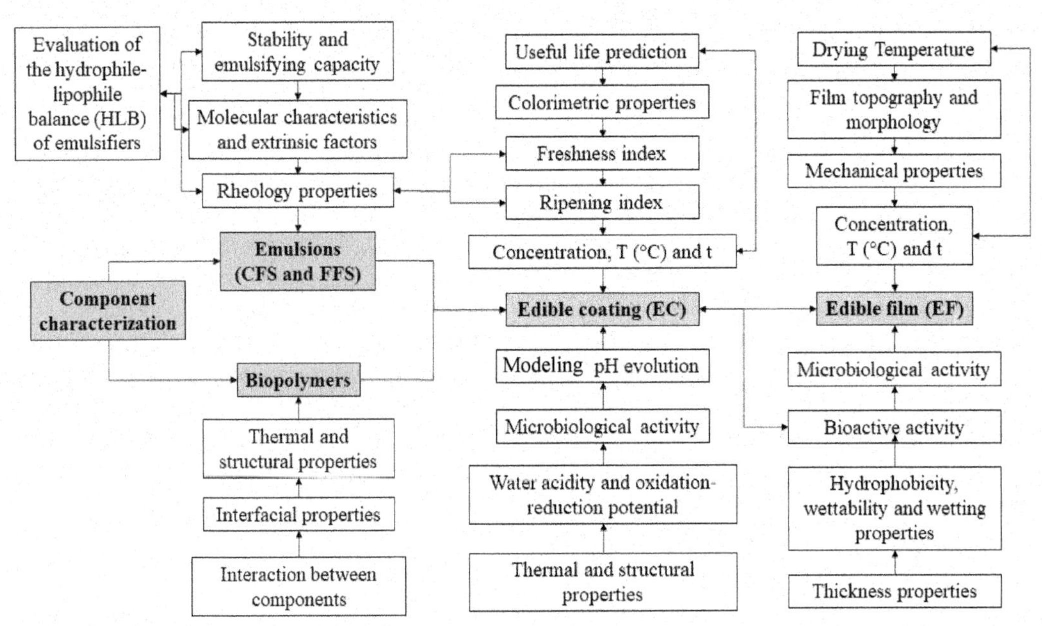

Figure 13.3. Methods for the characterization of edible coatings and films from EC and EF forming solutions (Carranza-Saavedra et al., 2016b; Daza et al., 2018; Homez-Jara et al., 2018; Homez et al., 2018; Méndez Reyes et al., 2014; Montilla-Buitrago et al., 2021; Quintero-Cerón et al., 2014; Salazar-Sánchez et al., 2019, 2022).

1.3 Probiotic incorporation in edible films/coatings

The Food and Agriculture Organization of the United Nations and the World Health Organization (FAO/WHO), define probiotics as "live microorganisms that, when administered in adequate amounts, confer a health benefit on the host". In 2013, the International Scientific Association for Probiotics and Prebiotics (ISAPP) along with FAO/WHO expert panel and other scientific experts re-examined and reinforced the aforecited concept and delivered recommendations about probiotics scope (Hill et al., 2014). Over the past 8 years, ISAPP has provided consensus statements on the definition of probiotics (Hill et al., 2014), prebiotics (Gibson et al., 2017), synbiotics (Swanson et al., 2020), postbiotics (Salminen et al., 2021) and fermented foods (Cruxen et al., 2019). The first four terms have in common the fulfillment of health benefit in the target host at functional doses.

Evidence from clinical trials in humans and animal models supports the immunomodulatory (Mahooti et al., 2019), anti-inflammatory (Din et al., 2020), anticancer (Abdolalipour et al., 2020; Rosa et al., 2020), memory improvement (Xiao et al., 2020), obesity management (Pedret et al., 2019), antioxidant and antiviral actions of probiotics. The last one has received special attention as the gut may also contribute towards the pathogenesis of COVID-19, an Emerging Infectious Disease (EID). Scientists have hypothesized that a pleiotropic mechanism could be an additional preventive/ curative approach to treat the spectrum of disease severity (Baud et al., 2020; Kurian et al., 2021). The incorporation of probiotics as supplement and probiotics containing foods to diet, are increasing with our current awareness towards the beneficial associations of gut microbiota and human health (Ávila et al., 2020; Sharma et al., 2021). An imbalance in the composition or metabolic capacity of gut microbial community is called dysbiosis. These altered microbiota increases inflammation, therefore, individual susceptibility to a broad variety of chronic diseases (Wilkins et al., 2019). In consequence, studies on new effective strategies to shape the gut microbial architecture are needed.

Among the possibilities, fermented dairy products still are the most studied and employed matrices serving as vehicle of prebiotics and probiotic strains (Simões da Silva et al., 2020). Yogurt and cheese are effective vectors to deliver probiotics in the gastrointestinal tract (GIT), additionally the dairy manufacturing technology do not need substantial modifications. These food matrices represent a source of nutritional compounds; their acidic nature, buffering capacity of proteins (cheese) and long-shelf life, protect the bacterial strains and facilitate their transference to target sites in GIT (Aljutaily et al., 2020; Zepeda-Hernández et al., 2021). However, shifts in the perception and behavior of consumers encouraged by vegetarianism, diet restrictions, high cholesterol, lactose intolerance, protein allergies, aversion to dairy derivatives and poor palatability of low-fat diets (Kumar et al., 2012), have given rise to increased demand for non-dairy probiotic foods (Pavli et al., 2018; Rivera-Espinoza and Gallardo-Navarro, 2010).

Recently, several raw materials, new vehicles (biopolymeric microcapsules), process (spray/freeze drying) (Loyeau et al., 2018; Pitigraisorn et al., 2017) and applications have been extensively investigated with a special interest in non-dairy probiotic carriers such as minimally processed fruits and vegetables, juices (da Costa et al., 2017), meats, ready-to-eat fish (Valerio et al., 2015), dried fruits (Valerio et al., 2020), cereals, desserts (Shori, 2015) and edible packaging in the form of films and coatings (Alvarez et al., 2021; Romano et al., 2014).

Edible films (EF) and coatings (EC) can be functionalized by inclusion of active compounds (phenolic compounds, oils, essential oils, organic acids, enzymes, vitamins, bacteriocins, inorganic nanoparticles) (Díaz-Montes and Castro-Muñoz, 2021; Falguera et al., 2011; Soltani et al., 2021; Xing et al., 2019) and/or beneficial living microorganisms to obtain active or bioactive packaging materials. Thus, EF and EC may be regarded as carriers or vehicles for probiotics (Pop et al., 2019). On the basis of currently available literature regarding health benefits and antagonistic activities against yeast, molds, opportunistic and foodborne pathogens (i.e., Shewanella putrefaciens, Photobacterium phosphoreum, Listeria monocytogenes), strains from Lactobacillus and Bifidobacterium genera have been frequently included in the development of probiotic and synbiotic edible packaging (Bambace et al., 2019; Hashemi and Jafarpour, 2021; Khodaei and Hamidi-Esfahani, 2019; Pop et al., 2019).

The success of this new packaging technology as a health benefit promoter and biopreservation alternative for coated or wrapped foods relies on its ability to provide a sufficient number of viable cells (about 106–109 CFU.ml^{-1}) at the end of the shelf-life (Pavli et al., 2018; Shori, 2015), its regular consumption and the material used for probiotic entrapment. The latter has proven to be a crucial factor regarding antimicrobial activity of probiotic strains (Mihalca et al., 2021; Pop et al., 2019). For instance, when incorporated in cellulose-based films, L. plantarum, L. reuteri and

L. acidophilus exhibited higher bacteriocin production but lower viability in comparison with sodium caseinate and pea protein-based films (Sánchez-González et al., 2013, 2014).

According to recent evidence probiotics viability in EF/EC can be influenced by individual or combined effects of intrinsic (type of biopolymer and plasticizer, food matrix, physical state, pH, acidity, water activity, dissolved oxygen, presence of selective prebiotics or inhibitors) (Shahrampour et al., 2020) and extrinsic factors (film/coating forming method, food processing, storage time, temperature, relative humidity, gastrointestinal stress conditions) (Akman et al., 2021; Bambace et al., 2021; Hellebois et al., 2020; Romano et al., 2014).

A variety of film/coating-forming materials (polysaccharides, lipids and composites) carrying probiotics have been fabricated and characterized in terms of cells viability, functionality, and application to foods (see Table 13.1). Sodium Alginate (SA), chitosan (Ch) (El-Sayed et al., 2021; Homez-Jara et al., 2018; Salazar Sánchez et al., 2020), methylcellulose (MC) (Romano et al., 2014), carboxymethyl cellulose (CMC) (Khodaei and Hamidi-Esfahani, 2019) hydroxypropylmethylcellulose, konjac glucomannan (Hashemi and Jafarpour, 2021), sodium caseinate, pea protein (Sánchez-González et al., 2013), delipidated egg yolk protein (Sáez-Orviz et al., 2021), have been used as a simple support matrix. In the case of composites films, alginate/pectin (Shahrampour et al., 2020), cassava starch/CMC (Li et al., 2020), pectin/SA/casein (Namratha et al., 2020) are some of the recently reported binary or ternary probiotic materials.

Structural modifications as a consequence of prebiotic and probiotic strains incorporation into simple or composite films have been reported. In this sense, microstructural, barrier, mechanical, optical and even antioxidant properties have undergone changes. Regarding microstructure, it has been visualized by SEM micrographs that diffusion and absorption of L. rhamnosus GG and L. acidophilus AA-1 in CMC and Ch/CMC films trigger the formation of micropores (Shokatayeva et al., 2020). Furthermore, heterogeneous surfaces with multiple fluctuations were described in pectin/SA/casein films carrying Enterococcus faecium Rp1 (Namratha et al., 2020). On the other hand, Sáez-Orviz et al. (2021) obtained smooth and homogenous surface microstructures in L. plantarum-loaded egg-yolk protein films.

Table 13.1. Operations involved in the viability during storage of foods packaged with biopolymeric matrices with biological activity.

Food product	Biopolymeric matrix	Probiotic strain	Operations involved	Viability during storage	Reference
Apple slices	SA	*Lactobacillus rhamnosus Bifidobacterium animalis* subsp. *lactis*	Coating/air drying (25°C × 30 min)/cold storage 5°C/Simulated GIT conditions test	> 10^9 CFU/g portion 8d/5 ± 1°C	(Alvarez et al., 2021)
Blueberries	SA	*Lacticaseibacillus casei CECT 9104 Bifidobacterium animalis* subsp. *lactis CECT 8145*	Coating/air drying (25°C × 30 min)/cold storage 5°C/Simulated GIT conditions test	> 10^6 CFU/g portion 14 d/5 ± 1°C	(Bambace et al., 2021)
Pan bread	SA SA/WPC	*Lactobacillus rhamnosus* GG	Coating/air drying (60°C × 10 min or 180°C × min)/Simulated GIT conditions test	7,57–8,98 and 6,55–6,91 log CFU/40 g portion	(Soukoulis et al., 2014)
Apple slices	Pectin	*Lactobacillus paracasei* IMPC2.1	Inclusion/coating/dehydration process/Simulated GIT conditions test	≥ 9log CFU/20 g portion 30 d/4°C	(Valerio et al., 2020)
Strawberries	CMC	*Lactobacillus plantarum*	Coating/drying at room temperature/cold storage	5,28–8,40 log CFU/g (15 d/4°C)	(Khodaei and Hamidi-Esfahani, 2019)

An increase in water vapor permeability, elongation at break and decrease in tensile and puncture strength values have been reported after prebiotics (inulin, lactobionic acid, fructo-oligosaccharides) and free or encapsulated probiotics (L. casei, L. plantarum, L. rhamnosus GG) were incorporated into films (cassava starch, egg-yolk protein, SA, whey protein isolate) (Lee et al., 2020; Orozco-Parra et al., 2020; Romano et al., 2014; Sáez-Orviz et al., 2021). Prebiotics and probiotics have shown a plasticizing effect on films and presumably can introduce discontinuities in the polymer matrix (Romano et al., 2014; Sánchez-González et al., 2013). Additionally, the hydrophilic character of wall materials of encapsulated probiotics (i.e., maltodextrine) could contribute to explain the higher mass transfer of water molecules through the films (Akman et al., 2021; Sánchez-González et al., 2014). Optical properties such a UV-visible light transmittance, opacity and color can be modified as a result of prebiotic or probiotic incorporation. Effectively, a denser structure after probiotics incorporation allowed obtaining probiotics materials with UV blocking properties. Finally, higher antioxidant capacity was measured in cassava starch/CMC materials after incorporation of Lactobacillus plantarum and Pedocococcus pentosaceus (Li et al., 2020).

Conclusions

The need to describe rarefied flows in a variety of applications that involve micro and nanoscale pores or channels has stimulated the development of a multitude of theoretical and numerical methodologies. A comparison of the predictions of the mesoscopic approaches with those of the Navier-Stokes formulation complemented by a non-slip flow condition revealed that the latter can be safely employed for very small Knudsen number values only. The DSMC method was used to determine the flow field in different types of porous media and processes.

Acknowledgements

The authors would like to thank the following institutions: Universidad del Cauca (Colombia), Universidad del Tolima (Colombia), Universidad Nacional del Litoral (Santa Fe-Argentina), Universidad del Cauca (Colombia), Universidad del Tolima (Colombia) and Universidad Nacional del Litoral (Santa Fe-Argentina).

References

Abdolalipour, E., Mahooti, M., Salehzadeh, A., Torabi, A., Mohebbi, S. R., Gorji, A. and Ghaemi, A. (2020). Evaluation of the antitumor immune responses of probiotic Bifidobacterium bifidum in human papillomavirus-induced tumor model. *Microbial Pathogenesis*, 145: 104207. https://doi.org/10.1016/j.micpath.2020.104207.

Akman, P. K., Bozkurt, F., Dogan, K., Tornuk, F. and Tamturk, F. (2021). Fabrication and characterization of probiotic Lactobacillus plantarum loaded sodium alginate edible films. *Journal of Food Measurement and Characterization*, 15(1): 84–92. https://doi.org/10.1007/s11694-020-00619-6.

Aljutaily, T., Huarte, E., Martinez-Monteagudo, S., Gonzalez-Hernandez, J. L., Rovai, M. and Sergeev, I. N. (2020). Probiotic-enriched milk and dairy products increase gut microbiota diversity: a comparative study. *Nutrition Research*, 82: 25–33. https://doi.org/10.1016/j.nutres.2020.06.017.

Álvarez Gómez, J. M. and Rodríguez Patino, J. M. (2006). Formulation engineering of food model foams containing diglycerol esters and β-lactoglobulin. *Industrial & Engineering Chemistry Research*, 45(22): 7510–7519. https://doi.org/10.1021/ie060924g.

Álvarez Gómez, J. M., Pizones Ruíz Henestrosa, V., Carrera Sánchez, C. and Rodríguez Patino, J. M. (2008). The role of static and dynamic characteristics of diglycerol esters and β-lactoglobulin mixed films foaming. 1. Dynamic phenomena at the air–water interface. *Food Hydrocolloids*, 22(6): 1105–1116. https://doi.org/10.1016/j.foodhyd.2007.06.002.

Alvarez, M. V., Bambace, M. F., Quintana, G., Gomez-Zavaglia, A. and Moreira, M. del R. (2021). Prebiotic-alginate edible coating on fresh-cut apple as a new carrier for probiotic lactobacilli and bifidobacteria. *LWT*, 137: 110483. https://doi.org/10.1016/j.lwt.2020.110483.

Ávila, B. P., da Rosa, P. P., Fernandes, T. A., Chesini, R. G., Sedrez, P. A., de Oliveira, A. P. T., Mota, G. N., Gularte, M. A. and Roll, V. F. B. (2020). Analysis of the perception and behaviour of consumers regarding probiotic dairy products. *International Dairy Journal*, 106: 104703. https://doi.org/10.1016/j.idairyj.2020.104703.

Baeza, R., Carrera Sanchez, C., Pilosof, A. M. R. and Rodríguez Patino, J. M. (2005). Interactions of polysaccharides with β-lactoglobulin adsorbed films at the air–water interface. *Food Hydrocolloids*, 19(2): 239–248. https://doi.org/10.1016/j.foodhyd.2004.06.002.

Bambace, María Florencia, Alvarez, M. V. and Moreira, M. del R. (2019). Novel functional blueberries: Fructo-oligosaccharides and probiotic lactobacilli incorporated into alginate edible coatings. *Food Research International*, 122: 653–660. https://doi.org/10.1016/j.foodres.2019.01.040.

Bambace, María F., Alvarez, M. V. and Moreira, M. R. (2021). Ready-to-eat blueberries as fruit-based alternative to deliver probiotic microorganisms and prebiotic compounds. *LWT*, 142: 111009. https://doi.org/10.1016/j.lwt.2021.111009.

Barzegar, H., Alizadeh Behbahani, B. and Mehrnia, M. A. (2020). Quality retention and shelf life extension of fresh beef using Lepidium sativum seed mucilage-based edible coating containing Heracleum lasiopetalum essential oil: an experimental and modeling study. *Food Science and Biotechnology*, 29(5): 717–728. https://doi.org/10.1007/s10068-019-00715-4.

Baud, D., Dimopoulou Agri, V., Gibson, G. R., Reid, G. and Giannoni, E. (2020). Using probiotics to flatten the curve of coronavirus disease COVID-2019 pandemic. *Frontiers in Public Health*, 8. https://doi.org/10.3389/fpubh.2020.00186.

Behbahani, B. A. and Imani Fooladi, A. A. (2018). Shirazi balangu (Lallemantia royleana) seed mucilage: Chemical composition, molecular weight, biological activity and its evaluation as edible coating on beefs. *International Journal of Biological Macromolecules*, 114: 882–889. https://doi.org/10.1016/j.ijbiomac.2018.03.177.

Carneiro-da-Cunha, M. G., Cerqueira, M. A., Souza, B. W. S., Souza, M. P., Teixeira, J. A. and Vicente, A. A. (2009). Physical properties of edible coatings and films made with a polysaccharide from Anacardium occidentale L. *Journal of Food Engineering*, 95(3): 379–385. https://doi.org/10.1016/j.jfoodeng.2009.05.020.

Carranza-Saavedra, D., Váquiro, H. A., León-Galván, M. F., Ozuna, C. and Solanilla, J. F. (2016a). Modelización de la adsorción en la interfase aire-agua de proteína de pescado mediante lógica borrosa Modeling adsorption at the air-water interface of fish protein by fuzzy logic. *Agronomía Colombiana,* 34(1Supl.)(March): S362–S366. https://doi.org/10.15446/agron.colomb.v34n1supl.58117.

Carranza-Saavedra, D., Váquiro, H., León-Galván, M., Ozuna, C. and Solanilla, J. F. (2016b). Modelización de la adsorción en la interfase aire-agua de proteína de pescado mediante lógica borrosa. *Agronomía Colombiana,* 34(1Supl): S362–S366.

Cerqueira, M. A., Bourbon, A. I., Pinheiro, A. C., Martins, J. T., Souza, B. W. S., Teixeira, J. A. and Vicente, A. A. (2011). Galactomannans use in the development of edible films/coatings for food applications. pp. 662–671. *In*: *Trends in Food Science and Technology* (Vol. 22, Issue 12). https://doi.org/10.1016/j.tifs.2011.07.002.

Chana-Thaworn, J., Chanthachum, S. and Wittaya, T. (2011). Properties and antimicrobial activity of edible films incorporated with kiam wood (Cotyleobium lanceotatum) extract. *LWT-Food Science and Technology*, 44(1): 284–292. https://doi.org/10.1016/j.lwt.2010.06.020.

Chang, L., Martino, M. and Rodríguez, J. (2010). Influencia de la incorporación de extracto de propóleos rojo cubano en las propiedades funcionales de una película base almidón. *Montero, P y Mauri, A. AGROBIOENVASES. Memorias de La II Jornadas Internacionales Sobre Avances En La Tecnología de Películas y Coberturas Funcionales En Alimentos. Buenos Aires*, 17–18.

Choulitoudi, E., Ganiari, S., Tsironi, T., Ntzimani, A., Tsimogiannis, D., Taoukis, P. and Oreopoulou, V. (2017). Edible coating enriched with rosemary extracts to enhance oxidative and microbial stability of smoked eel fillets. *Food Packaging and Shelf Life*, 12: 107–113. https://doi.org/10.1016/j.fpsl.2017.04.009.

Cruxen, C. E. dos S., Funck, G. D., Haubert, L., Dannenberg, G. da S., Marques, J. de L., Chaves, F. C., da Silva, W. P. and Fiorentini, Â. M. (2019). Selection of native bacterial starter culture in the production of fermented meat sausages: Application potential, safety aspects, and emerging technologies. pp. 371–382. *In*: *Food Research International* (Vol. 122). https://doi.org/10.1016/j.foodres.2019.04.018.

da Costa, G. M., de Carvalho Silva, J. V., Mingotti, J. D., Barão, C. E., Klososki, S. J. and Pimentel, T. C. (2017). Effect of ascorbic acid or oligofructose supplementation on L. paracasei viability, physicochemical characteristics and acceptance of probiotic orange juice. *LWT*, 75: 195–201. https://doi.org/10.1016/j.lwt.2016.08.051.

Damodaran, S. and Razumovsky, L. (2003). Competitive adsorption and thermodynamic incompatibility of mixing of β-casein and gum arabic at the air–water interface. *Food Hydrocolloids*, 17(3): 355–363. https://doi.org/10.1016/S0268-005X(02)00098-X.

Damodaran, Srinivasan. (2004). Adsorbed layers formed from mixtures of proteins. *Current Opinion in Colloid & Interface Science*, 9(5): 328–339. https://doi.org/10.1016/j.cocis.2004.09.008.

Daza, L. D., Homez-Jara, A., Solanilla, J. F. and Váquiro, H. A. (2018). Effects of temperature, starch concentration, and plasticizer concentration on the physical properties of ulluco (Ullucus tuberosus Caldas)-based edible films. *International Journal of Biological Macromolecules*, 120: 1834–1845. https://doi.org/10.1016/j.ijbiomac.2018.09.211.

Dhall, R. K. (2013). Advances in edible coatings for fresh fruits and vegetables: a review. *Critical Reviews in Food Science and Nutrition*, 53(5): 435–450. https://doi.org/10.1080/10408398.2010.541568.

Díaz-Montes, E. and Castro-Muñoz, R. (2021). Edible films and coatings as food-quality preservers: an overview. *Foods*, 10(2): 249. https://doi.org/10.3390/foods10020249.

Dickinson, E. (1992). *Introduction to Food Colloids*. Oxford University Press.

Dickinson, E. and McClements, D. J. (1995). *Advances in Food Colloids*. Springer Science & Business Media.

Din, A. U., Hassan, A., Zhu, Y., Zhang, K., Wang, Y., Li, T., Wang, Y. and Wang, G. (2020). Inhibitory effect of Bifidobacterium bifidum ATCC 29521 on colitis and its mechanism. *The Journal of Nutritional Biochemistry*, 79: 108353. https://doi.org/10.1016/j.jnutbio.2020.108353.

Dubey, N. K. and Dubey, R. (2020). Edible films and coatings. pp. 675–695. *In*: *Biopolymer-Based Formulations*. Elsevier. https://doi.org/10.1016/B978-0-12-816897-4.00027-8.

El-Sayed, H. S., El-Sayed, S. M., Mabrouk, A. M. M., Nawwar, G. A. and Youssef, A. M. (2021). Development of eco-friendly probiotic edible coatings based on chitosan, alginate and carboxymethyl cellulose for improving the shelf life of UF soft cheese. *Journal of Polymers and the Environment*, 29(6): 1941–1953. https://doi.org/10.1007/s10924-020-02003-3.

Falguera, V., Quintero, J. P., Jiménez, A., Muñoz, J. A. and Ibarz, A. (2011). Edible films and coatings: Structures, active functions and trends in their use. *Trends in Food Science and Technology*, 22(6): 292–303. https://doi.org/10.1016/j.tifs.2011.02.004.

Ferreira, D. C. M. M., Molina, G. and Pelissari, F. M. (2020). Effect of edible coating from cassava starch and babassu flour (Orbignya phalerata) on Brazilian cerrado fruits quality. *Food and Bioprocess Technology*, 13(1): 172–179. https://doi.org/10.1007/s11947-019-02366-z.

Galindez, A., Daza, L. D., Homez-Jara, A., Eim, V. S. and Váquiro, H. A. (2019). Characterization of ulluco starch and its potential for use in edible films prepared at low drying temperature. *Carbohydrate Polymers*, 215: 143–150. https://doi.org/10.1016/j.carbpol.2019.03.074.

Gibson, G. R., Hutkins, R., Sanders, M. E., Prescott, S. L., Reimer, R. A., Salminen, S. J., Scott, K., Stanton, C., Swanson, K. S., Cani, P. D., Verbeke, K. and Reid, G. (2017). Expert consensus document: The International Scientific Association for Probiotics and Prebiotics (ISAPP) consensus statement on the definition and scope of prebiotics. *Nature Reviews Gastroenterology & Hepatology*, 14(8): 491–502. https://doi.org/10.1038/nrgastro.2017.75.

Gómez, J. M. Á. and Rodríguez Patino, J. M. (2007). Interfacial properties of diglycerol esters and caseinate mixed films at the air–water interface. *The Journal of Physical Chemistry C*, 111(12): 4790–4799. https://doi.org/10.1021/jp0678610.

Guilbert, S., Gontard, N. and Gorris, L. G. M. (1996). Prolongation of the shelf-life of perishable food products using biodegradable films and coatings. pp. 10–17. *In*: *LWT - Food Science and Technology* (Vol. 29, Issues 1–2). Academic Press. https://doi.org/10.1006/fstl.1996.0002.

Hammam, A. R. A. (2019). Technological, applications, and characteristics of edible films and coatings: a review. *SN Applied Sciences*, 1(6). https://doi.org/10.1007/s42452-019-0660-8.

Hashemi, S. M. B. and Jafarpour, D. (2021). Bioactive edible film based on Konjac glucomannan and probiotic Lactobacillus plantarum strains: Physicochemical properties and shelf life of fresh-cut kiwis. *Journal of Food Science*, 86(2): 513–522. https://doi.org/10.1111/1750-3841.15568.

Hazrati, S., Beyraghdar Kashkooli, A., Habibzadeh, F., Tahmasebi-Sarvestani, Z. and Sadeghi, A. R. (2017). Evaluation of Aloe vera gel as an alternative edible coating for peach fruits during cold storage period. *Gesunde Pflanzen*, 69(3): 131–137. https://doi.org/10.1007/s10343-017-0397-5.

Hellebois, T., Tsevdou, M. and Soukoulis, C. (2020). *Functionalizing and Bio-Preserving Processed Food Products Via Probiotic and Synbiotic Edible Films and Coatings* (pp. 161–221). https://doi.org/10.1016/bs.afnr.2020.06.004.

Hill, C., Guarner, F., Reid, G., Gibson, G. R., Merenstein, D. J., Pot, B., Morelli, L., Canani, R. B., Flint, H. J., Salminen, S., Calder, P. C. and Sanders, M. E. (2014). The International Scientific Association for Probiotics and Prebiotics consensus statement on the scope and appropriate use of the term probiotic. *Nature Reviews Gastroenterology & Hepatology*, 11(8): 506–514. https://doi.org/10.1038/nrgastro.2014.66.

Homez-Jara, A., Daza, L. D., Aguirre, D. M., Muñoz, J. A., Solanilla, J. F. and Váquiro, H. A. (2018). Characterization of chitosan edible films obtained with various polymer concentrations and drying temperatures. *International Journal of Biological Macromolecules*, 113: 1233–1240. https://doi.org/10.1016/j.ijbiomac.2018.03.057.

Homez, A. K., Daza, L. D., Solanilla, J. F. and Váquiro, H. A. (2018). Effect of temperature, starch and plasticizer concentrations on color parameters of ulluco (Ullucus tuberosus Caldas) edible films. *IOP Conference Series: Materials Science and Engineering*, 437(1). https://doi.org/10.1088/1757-899X/437/1/012003.

Hossain, A. and Rahman, M. J. (2019). Safety, nutrition and functionality of the traditional foods. pp. 219–238. *In*: *Food Engineering Series*. Springer. https://doi.org/10.1007/978-3-030-24620-4_8.

Hu, Q., Hu, Y. and Xu, J. (2005). Free radical-scavenging activity of Aloe vera (Aloe barbadensis Miller) extracts by supercritical carbon dioxide extraction. *Food Chemistry*, 91(1): 85–90. https://doi.org/10.1016/j.foodchem.2004.05.052.

Kanthe, A. D., Krause, M., Zheng, S., Ilott, A., Li, J., Bu, W., Bera, M. K., Lin, B., Maldarelli, C. and Tu, R. S. (2020). Armoring the interface with surfactants to prevent the adsorption of monoclonal antibodies. *ACS Applied Materials & Interfaces*, 12(8): 9977–9988. https://doi.org/10.1021/acsami.9b21979.

Kechichian, V., Ditchfield, C., Veiga-Santos, P. and Tadini, C. C. (2010). Natural antimicrobial ingredients incorporated in biodegradable films based on cassava starch. *LWT - Food Science and Technology*, 43(7): 1088–1094. https://doi.org/10.1016/j.lwt.2010.02.014.

Khodaei, D. and Hamidi-Esfahani, Z. (2019). Influence of bioactive edible coatings loaded with Lactobacillus plantarum on physicochemical properties of fresh strawberries. *Postharvest Biology and Technology*, 156: 110944. https://doi.org/10.1016/j.postharvbio.2019.110944.

Klinkesorn, U., Sophanodora, P., Chinachoti, P., Decker, E. A. and McClements, D. J. (2006). Characterization of spray-dried tuna oil emulsified in two-layered interfacial membranes prepared using electrostatic layer-by-layer deposition. *Food Research International*, 39(4): 449–457. https://doi.org/10.1016/j.foodres.2005.09.008.

Kumar, M., Nagpal, R., Kumar, R., Hemalatha, R., Verma, V., Kumar, A., Chakraborty, C., Singh, B., Marotta, F., Jain, S. and Yadav, H. (2012). Cholesterol-lowering probiotics as potential biotherapeutics for metabolic diseases. *Experimental Diabetes Research*, 1–14. https://doi.org/10.1155/2012/902917.

Kurian, S. J., Unnikrishnan, M. K., Miraj, S. S., Bagchi, D., Banerjee, M., Reddy, B. S., Rodrigues, G. S., Manu, M. K., Saravu, K., Mukhopadhyay, C. and Rao, M. (2021). Probiotics in prevention and treatment of COVID-19: current perspective and future prospects. *Archives of Medical Research*, 52(6): 582–594. https://doi.org/10.1016/j.arcmed.2021.03.002.

Lee, K. Y., Weintraub, S. T., & and Yu, B. P. (2000). Isolation and identification of a phenolic antioxidant from Aloe barbadensis. *Free Radical Biology and Medicine*, 28(2): 261–265. https://doi.org/10.1016/S0891-5849(99)00235-X.

Lee, Y. Y., Yusof, Y. A. and Pui, L. P. (2020). Development of milk protein edible films incorporated with Lactobacillus rhamnosus GG. *BioResources*, 15(3): 6960–6973.

Li, S., Ma, Y., Ji, T., Sameen, D. E., Ahmed, S., Qin, W., Dai, J., Li, S. and Liu, Y. (2020). Cassava starch/carboxymethylcellulose edible films embedded with lactic acid bacteria to extend the shelf life of banana. *Carbohydrate Polymers*, 248: 116805. https://doi.org/10.1016/j.carbpol.2020.116805.

Li, X. yu, Du, X. long, Liu, Y., Tong, L. jing, Wang, Q. and Li, J. long. (2019). Rhubarb extract incorporated into an alginate-based edible coating for peach preservation. *Scientia Horticulturae*, 257. https://doi.org/10.1016/j.scienta.2019.108685.

Lima, Á. M., Cerqueira, M. A., Souza, B. W. S., Santos, E. C. M., Teixeira, J. A., Moreira, R. A. and Vicente, A. A. (2010). New edible coatings composed of galactomannans and collagen blends to improve the postharvest quality of fruits—Influence on fruits gas transfer rate. *Journal of Food Engineering*, 97(1): 101–109. https://doi.org/10.1016/j.jfoodeng.2009.09.021.

Loyeau, P. A., Spotti, M. J., Vanden Braber, N. L., Rossi, Y. E., Montenegro, M. A., Vinderola, G. and Carrara, C. R. (2018). Microencapsulation of Bifidobacterium animalis subsp. lactis INL1 using whey proteins and dextrans conjugates as wall materials. *Food Hydrocolloids*, 85: 129–135. https://doi.org/10.1016/j.foodhyd.2018.06.051.

Mackie, A. R., Gunning, A. P., Ridout, M. J., Wilde, P. J. and Morris, V. J. (2001). Orogenic displacement in mixed β-lactoglobulin/β-casein films at the air/water interface. *Langmuir*, 17(21): 6593–6598. https://doi.org/10.1021/la010687g.

Mahooti, M., Abdolalipour, E., Salehzadeh, A., Mohebbi, S. R., Gorji, A. and Ghaemi, A. (2019). Immunomodulatory and prophylactic effects of Bifidobacterium bifidum probiotic strain on influenza infection in mice. *World Journal of Microbiology and Biotechnology*, 35(6): 91. https://doi.org/10.1007/s11274-019-2667-0.

Marcuzzo, E., Sensidoni, A., Debeaufort, F. and Voilley, A. (2010). Encapsulation of aroma compounds in biopolymeric emulsion based edible films to control flavour release. *Carbohydrate Polymers*, 80(3): 984–988. https://doi.org/10.1016/j.carbpol.2010.01.016.

Martinez, K., Baeza, R., Millan, F. and Pilosof, A. (2005). Effect of limited hydrolysis of sunflower protein on the interactions with polysaccharides in foams. *Food Hydrocolloids*, 19(3): 361–369. https://doi.org/10.1016/j.foodhyd.2004.10.002.

Martinez, K., Sanchez, C., Ruizhenestrosa, V., Rodriguezpatino, J. and Pilosof, A. (2007). Effect of limited hydrolysis of soy protein on the interactions with polysaccharides at the air–water interface. *Food Hydrocolloids*, 21(5–6): 813–822. https://doi.org/10.1016/j.foodhyd.2006.09.008.

McClellan, S. J. and Franses, E. I. (2003). Exclusion of bovine serum albumin from the air/water interface by sodium myristate. *Colloids and Surfaces B: Biointerfaces*, 30(1–2): 1–11. https://doi.org/10.1016/S0927-7765(03)00021-3.

McClements, D. Julian. (2009). Biopolymers in food emulsions. pp. 129–166. *In*: *Modern Biopolymer Science*. Elsevier. https://doi.org/10.1016/B978-0-12-374195-0.00004-5.

McClements, D. J. (2015). *Food Emulsions: Principles, Practices, and Techniques* (David Julian McClements (ed.); third). CRC Press. https://doi.org/10.1201/b18868.

Méndez Reyes, D. A., Quintero Cerón, J. P., Vaquiro Herrera, H. A. and Solanilla Duque, J. F. (2014). Sodium alginate in the development of edible films. *Revista Venezolana De Ciencia Y Tecnología De Alimentos*, 5(2): 89–113. https://www.cabdirect.org/cabdirect/abstract/20153140847.

Mihalca, V., Kerezsi, A. D., Weber, A., Gruber-Traub, C., Schmucker, J., Vodnar, D. C., Dulf, F. V., Socaci, S. A., Fărcaş, A., Mureşan, C. I., Suharoschi, R. and Pop, O. L. (2021). Protein-based films and coatings for food industry applications. *Polymers*, 13(5): 769. https://doi.org/10.3390/polym13050769.

Montilla-Buitrago, C. E., Gómez-López, R. A., Solanilla-Duque, J. F., Serna-Cock, L. and Villada-Castillo, H. S. (2021). Effect of plasticizers on properties, retrogradation, and processing of extrusion-obtained thermoplastic starch: a review. *Starch - Stärke*, 73(9–10): 2100060. https://doi.org/10.1002/star.202100060.

Namratha, S., Sreejit, V. and Preetha, R. (2020). Fabrication and evaluation of physicochemical properties of probiotic edible film based on pectin–alginate–casein composite. *International Journal of Food Science & Technology*, 55(4): 1497–1505. https://doi.org/10.1111/ijfs.14550.

Orozco-Parra, J., Mejía, C. M. and Villa, C. C. (2020). Development of a bioactive synbiotic edible film based on cassava starch, inulin, and Lactobacillus casei. *Food Hydrocolloids*, 104: 105754. https://doi.org/10.1016/j.foodhyd.2020.105754.

Pagella, C., Spigno, G. and De Faveri, D. M. (2002). Characterization of starch based edible coatings. *Food and Bioproducts Processing*, 80(3): 193–198. https://doi.org/10.1205/096030802760309214.

Palou, L., Valencia-Chamorro, S. A. and Pérez-Gago, M. B. (2015). Antifungal edible coatings for fresh citrus fruit: A review. pp. 962–986. *In*: *Coatings* (Vol. 5, Issue 4). MDPI AG. https://doi.org/10.3390/coatings5040962.

Pavinatto, A., de Almeida Mattos, A. V., Malpass, A. C. G., Okura, M. H., Balogh, D. T. and Sanfelice, R. C. (2020). Coating with chitosan-based edible films for mechanical/biological protection of strawberries. *International Journal of Biological Macromolecules*, 151: 1004–1011. https://doi.org/10.1016/j.ijbiomac.2019.11.076.

Pavli, F., Tassou, C., Nychas, G.-J. and Chorianopoulos, N. (2018). Probiotic incorporation in edible films and coatings: bioactive solution for functional foods. *International Journal of Molecular Sciences*, 19(1): 150. https://doi.org/10.3390/ijms19010150.

Pedret, A., Valls, R. M., Calderón-Pérez, L., Llauradó, E., Companys, J., Pla-Pagà, L., Moragas, A., Martín-Luján, F., Ortega, Y., Giralt, M., Caimari, A., Chenoll, E., Genovés, S., Martorell, P., Codoñer, F. M., Ramón, D., Arola, L. and Solà, R. (2019). Effects of daily consumption of the probiotic Bifidobacterium animalis subsp. lactis CECT 8145 on anthropometric adiposity biomarkers in abdominally obese subjects: a randomized controlled trial. *International Journal of Obesity*, 43(9): 1863–1868. https://doi.org/10.1038/s41366-018-0220-0.

Pérez, O., Carrera Sánchez, C., Rodríguez Patino, J. and Pilosof, A. (2007). Adsorption dynamics and surface activity at equilibrium of whey proteins and hydroxypropyl–methyl–cellulose mixtures at the air-water interface. *Food Hydrocolloids*, 21(5–6): 794–803. https://doi.org/10.1016/j.foodhyd.2006.11.013.

Pitigraisorn, P., Srichaisupakit, K., Wongpadungkiat, N. and Wongsasulak, S. (2017). Encapsulation of Lactobacillus acidophilus in moist-heat-resistant multilayered microcapsules. *Journal of Food Engineering*, 192: 11–18. https://doi.org/10.1016/j.jfoodeng.2016.07.022.

Ponce, A. G., Roura, S. I., del Valle, C. E. and Moreira, M. R. (2008). Antimicrobial and antioxidant activities of edible coatings enriched with natural plant extracts: *in vitro* and *in vivo* studies. *Postharvest Biology and Technology*, 49(2): 294–300. https://doi.org/10.1016/j.postharvbio.2008.02.013.

Pop, O. L., Pop, C. R., Dufrechou, M., Vodnar, D. C., Socaci, S. A., Dulf, F. V., Minervini, F. and Suharoschi, R. (2019). Edible films and coatings functionalization by probiotic incorporation: a review. *Polymers*, 12(1): 12. https://doi.org/10.3390/polym12010012.

Powell, K. C. and Chauhan, A. (2016). Interfacial effects and emulsion stabilization by in situ surfactant generation through the saponification of esters. *Colloids and Surfaces A: Physicochemical and Engineering Aspects*, 504: 458–470. https://doi.org/10.1016/j.colsurfa.2016.06.002.

Pranoto, Y., Salokhe, V. M. and Rakshit, S. K. (2005). Physical and antibacte rial properties of alginate-based edible film incorporated with garlic oil. *Food Research International*, 38(3): 267–272. https://doi.org/10.1016/j.foodres.2004.04.009.

Puncha-arnon, S., Puttanlek, C., Rungsardthong, V., Pathipanawat, W. and Uttapap, D. (2007). Changes in physicochemical properties and morphology of canna starches during rhizomal development. *Carbohydrate Polymers*, 70(2): 206–217. https://doi.org/10.1016/j.carbpol.2007.03.020.

Quintero-Cerón, J. P., Váquiro, H. A., Solanilla, J. F., Murillo, E. and Méndez, J. J. (2014). *In vitro* fungistatic activity of ethanolic extract of propolis against postharvest phytopathogenic fungi: Preliminary assessment. *Acta Horticulturae*, 1016: 157–162.

Rivera-Espinoza, Y. and Gallardo-Navarro, Y. (2010). Non-dairy probiotic products. *Food Microbiology*, 27(1): 1–11. https://doi.org/10.1016/j.fm.2008.06.008

Robles-Sánchez, R. M., Rojas-Graü, M. A., Odriozola-Serrano, I., González-Aguilar, G. and Martin-Belloso, O. (2013). Influence of alginate-based edible coating as carrier of antibrowning agents on bioactive compounds and antioxidant activity in fresh-cut Kent mangoes. *LWT - Food Science and Technology*, 50(1): 240–246. https://doi.org/10.1016/j.lwt.2012.05.021.

Rodríguez Patino, J. M., Rodríguez Niño, M. R. and Sánchez, C. C. (2003). Protein–emulsifier interactions at the air–water interface. *Current Opinion in Colloid & Interface Science*, 8(4–5): 387–395. https://doi.org/10.1016/S1359-0294(03)00095-5.

Rodríguez Patino, J. M., Caro, A., Rodríguez Niño, M. R., Mackie, A., Gunning, A. and Morris, V. (2007). Some implications of nanoscience in food dispersion formulations containing phospholipids as emulsifiers. *Food Chemistry*, 102(2): 532–541. https://doi.org/10.1016/j.foodchem.2006.06.010.

Rodríguez Patino, J. M., Rodríguez Niño, M. R. and Carrera Sánchez, C. (2007). Physico-chemical properties of surfactant and protein films. *Current Opinion in Colloid & Interface Science*, 12(4–5): 187–195. https://doi.org/10.1016/j.cocis.2007.06.003.

Rojas-Graü, M. A., Soliva-Fortuny, R. and Martín-Belloso, O. (2009). Edible coatings to incorporate active ingredients to fresh-cut fruits: a review. *Trends in Food Science and Technology*, 20(10): 438–447. https://doi.org/10.1016/j.tifs.2009.05.002.

Romano, N., Tavera-Quiroz, M. J., Bertola, N., Mobili, P., Pinotti, A. and Gómez-Zavaglia, A. (2014). Edible methylcellulose-based films containing fructo-oligosaccharides as vehicles for lactic acid bacteria. *Food Research International*, 64: 560–566. https://doi.org/10.1016/j.foodres.2014.07.018.

Rosa, L. S., Santos, M. L., Abreu, J. P., Balthazar, C. F., Rocha, R. S., Silva, H. L. A., Esmerino, E. A., Duarte, M. C. K. H., Pimentel, T. C., Freitas, M. Q., Silva, M. C., Cruz, A. G. and Teodoro, A. J. (2020). Antiproliferative and apoptotic effects of probiotic whey dairy beverages in human prostate cell lines. *Food Research International*, 137: 109450. https://doi.org/10.1016/j.foodres.2020.109450.

Ruan, C., Zhang, Y., Sun, Y., Gao, X., Xiong, G. and Liang, J. (2019). Effect of sodium alginate and carboxymethyl cellulose edible coating with epigallocatechin gallate on quality and shelf life of fresh pork. *International Journal of Biological Macromolecules*, 141: 178–184. https://doi.org/10.1016/j.ijbiomac.2019.08.247.

Ruzaina, I., Zhong, F., Abd. Rashid, N., Jia, W., Li, Y., Zahrah Mohamed Som, H., Chong Seng, C., Md. Sikin, A., Ab. Wahab, N. and Zahid Abidin, M. (2017). Effect of different degree of deacetylation, molecular weight of chitosan and palm stearin and palm kernel olein concentration on chitosan as edible packaging for cherry tomato. *Journal of Food Processing and Preservation*, 41(4). https://doi.org/10.1111/jfpp.13090.

Sáez-Orviz, S., Marcet, I., Rendueles, M. and Díaz, M. (2021). Bioactive packaging based on delipidated egg yolk protein edible films with lactobionic acid and Lactobacillus plantarum CECT 9567: Characterization and use as coating in a food model. *Food Hydrocolloids*, 119: 106849. https://doi.org/10.1016/j.foodhyd.2021.106849.

Salazar-Sánchez, M. del R., Campo-Erazo, S. D., Villada-Castillo, H. S. and Solanilla-Duque, J. F. (2019). Structural changes of cassava starch and polylactic acid films submitted to biodegradation process. *International Journal of Biological Macromolecules*, 129: 442–447. https://doi.org/10.1016/j.ijbiomac.2019.01.187.

Salazar-Sánchez, M. del R., Immirzi, B., Solanilla-Duque, J. F., Zannini, D., Malinconico, M. and Santagata, G. (2022). Ulomoides dermestoides Coleopteran action on Thermoplastic Starch/Poly(lactic acid) films biodegradation: a novel, challenging and sustainable approach for a fast mineralization process. *Carbohydrate Polymers*, 279: 118989. https://doi.org/10.1016/j.carbpol.2021.118989.

Salazar-sánchez, R., Cañas-montoya, J. A. and Villada-castillo, H. S. (2020). *Biogenerated polymers: an enviromental alternative • Polímeros biogenerados : una alternativa medioambiental*. 87(214): 75–84.

Salazar Sánchez, M. D. R., Cañas Montoya, J. A., Villada Castillo, H. S., Solanilla Duque, J. F., Rodríguez Herrera, R. and Avalos Belmotes, F. (2020). Biogenerated polymers: an enviromental alternative. *DYNA*, 87(214): 75–84. https://doi.org/10.15446/dyna.v87n214.82163.

Salminen, S., Collado, M. C., Endo, A., Hill, C., Lebeer, S., Quigley, E. M. M., Sanders, M. E., Shamir, R., Swann, J. R., Szajewska, H. and Vinderola, G. (2021). The International Scientific Association of Probiotics and Prebiotics (ISAPP) consensus statement on the definition and scope of postbiotics. *Nature Reviews Gastroenterology & Hepatology*, 18(9): 649–667. https://doi.org/10.1038/s41575-021-00440-6.

Sánchez-González, L., Quintero Saavedra, J. I. and Chiralt, A. (2013). Physical properties and antilisterial activity of bioactive edible films containing Lactobacillus plantarum. *Food Hydrocolloids*, 33(1): 92–98. https://doi.org/10.1016/j.foodhyd.2013.02.011.

Sánchez-González, L., Quintero Saavedra, J. I. and Chiralt, A. (2014). Antilisterial and physical properties of biopolymer films containing lactic acid bacteria. *Food Control*, 35(1): 200–206. https://doi.org/10.1016/j.foodcont.2013.07.001.

Sason, G. and Nussinovitch, A. (2021). Hydrocolloids for edible films, coatings, and food packaging. pp. 195–235. *In*: Phillips, G. O. and Williams, P. A. (eds.). *Handbook of Hydrocolloids* (Third). Elsevier. https://doi.org/10.1016/B978-0-12-820104-6.00023-1.

Sengupta, T. and Damodaran, S. (2000). Incompatibility and phase separation in a bovine serum albumin/β-casein/water ternary film at the air–water interface. *Journal of Colloid and Interface Science*, 229(1): 21–28. https://doi.org/10.1006/jcis.2000.6992.

Sengupta, T. and Damodaran, S. (2001). Lateral phase separation in adsorbed binary protein films at the air–water interface. *Journal of Agricultural and Food Chemistry*, 49(6): 3087–3091. https://doi.org/10.1021/jf001111k.

Shahrampour, D., Khomeiri, M., Razavi, S. M. A. and Kashiri, M. (2020). Development and characterization of alginate/pectin edible films containing Lactobacillus plantarum KMC 45. *LWT*, 118: 108758. https://doi.org/10.1016/j.lwt.2019.108758.

Sharma, M., Wasan, A. and Sharma, R. K. (2021). Recent developments in probiotics: An emphasis on Bifidobacterium. *Food Bioscience*, 41: 100993. https://doi.org/10.1016/j.fbio.2021.100993.

Shokatayeva, D. H., Talipova, A., Savitskaya, I., Pogrebnjak, A., Kistaubayeva, A. and Ignatova, L. V. (2020). *Quality Parameters of Cellulose–Chitosan Based Edible Films for Probiotic Entrapment* (pp. 169–177). https://doi.org/10.1007/978-981-15-3996-1_17.

Shori, A. B. (2015). The potential applications of probiotics on dairy and non-dairy foods focusing on viability during storage. *Biocatalysis and Agricultural Biotechnology*, 4(4): 423–431. https://doi.org/10.1016/j.bcab.2015.09.010.

Silva, O. A., Pellá, M. G., Pellá, M. G., Caetano, J., Simões, M. R., Bittencourt, P. R. S. and Dragunski, D. C. (2019). Synthesis and characterization of a low solubility edible film based on native cassava starch. *International Journal of Biological Macromolecules*, 128: 290–296. https://doi.org/10.1016/j.ijbiomac.2019.01.132.

Simões da Silva, T. M., Piazentin, A. C. M., Mendonça, C. M. N., Converti, A., Bogsan, C. S. B., Mora, D. and de Souza Oliveira, R. P. (2020). Buffalo milk increases viability and resistance of probiotic bacteria in dairy beverages under in vitro simulated gastrointestinal conditions. *Journal of Dairy Science*, 103(9): 7890–7897. https://doi.org/10.3168/jds.2019-18078.

Solanilla-Duque, J. F., Roa-Acosta, D. F. and Arrazola-Paternina, G. (2020). Colloidal applications in the food industry: Prospects and trends in healthy products. *SYLWAN*, 164(11).

Soltani, S., Hammami, R., Cotter, P. D., Rebuffat, S., Said, L. Ben, Gaudreau, H., Bédard, F., Biron, E., Drider, D. and Fliss, I. (2021). Bacteriocins as a new generation of antimicrobials: toxicity aspects and regulations. *FEMS Microbiology Reviews*, 45(1). https://doi.org/10.1093/femsre/fuaa039.

Soukoulis, C., Yonekura, L., Gan, H.-H., Behboudi-Jobbehdar, S., Parmenter, C. and Fisk, I. (2014). Probiotic edible films as a new strategy for developing functional bakery products: The case of pan bread. *Food Hydrocolloids*, 39: 231–242. https://doi.org/10.1016/j.foodhyd.2014.01.023.

Souza Almeida, F. and Kawazoe Sato, A. C. (2019). Structure of gellan gum–hydrolyzed collagen particles: Effect of starch addition and coating layer. *Food Research International*, 121: 394–403. https://doi.org/10.1016/j.foodres.2019.03.057.

Swanson, K. S., Gibson, G. R., Hutkins, R., Reimer, R. A., Reid, G., Verbeke, K., Scott, K. P., Holscher, H. D., Azad, M. B., Delzenne, N. M. and Sanders, M. E. (2020). The International Scientific Association for Probiotics and Prebiotics (ISAPP) consensus statement on the definition and scope of synbiotics. *Nature Reviews Gastroenterology & Hepatology*, 17(11): 687–701. https://doi.org/10.1038/s41575-020-0344-2.

Thakur, R., Pristijono, P., Bowyer, M., Singh, S. P., Scarlett, C. J., Stathopoulos, C. E. and Vuong, Q. V. (2019). A starch edible surface coating delays banana fruit ripening. *LWT*, 100: 341–347. https://doi.org/10.1016/j.lwt.2018.10.055.

Valerio, F., Lonigro, S. L., Giribaldi, M., Di Biase, M., De Bellis, P., Cavallarin, L. and Lavermicocca, P. (2015). Probiotic Lactobacillus paracasei IMPC 2.1 strain delivered by ready-to-eat swordfish fillets colonizes the human gut after alternate-day supplementation. *Journal of Functional Foods*, 17: 468–475. https://doi.org/10.1016/j.jff.2015.05.044.

Valerio, Francesca, Volpe, M. G., Santagata, G., Boscaino, F., Barbarisi, C., Di Biase, M., Bavaro, A. R., Lonigro, S. L. and Lavermicocca, P. (2020). The viability of probiotic Lactobacillus paracasei IMPC2.1 coating on apple slices during dehydration and simulated gastro-intestinal digestion. *Food Bioscience*, 34: 100533. https://doi.org/10.1016/j.fbio.2020.100533.

Vásconez, M. B., Flores, S. K., Campos, C. A., Alvarado, J. and Gerschenson, L. N. (2009). Antimicrobial activity and physical properties of chitosan–tapioca starch based edible films and coatings. *Food Research International*, 42(7): 762–769. https://doi.org/10.1016/j.foodres.2009.02.026.

Villafañe, F. (2017). Edible coatings for carrots. pp. 84–103. *In*: *Food Reviews International* (Vol. 33, Issue 1). Taylor and Francis Inc. https://doi.org/10.1080/87559129.2016.1150291.

Viña, S. Z., Mugridge, A., García, M. A., Ferreyra, R. M., Martino, M. N., Chaves, A. R. and Zaritzky, N. E. (2007). Effects of polyvinylchloride films and edible starch coatings on quality aspects of refrigerated Brussels sprouts. *Food Chemistry*, 103(3): 701–709. https://doi.org/10.1016/j.foodchem.2006.09.010.

Vollhardt, D. and Fainerman, V.. (2000). Penetration of dissolved amphiphiles into two-dimensional aggregating lipid monolayers. *Advances in Colloid and Interface Science*, 86(1–2): 103–151. https://doi.org/10.1016/S0001-8686(00)00034-8.

Wilkins, L. J., Monga, M. and Miller, A. W. (2019). Defining dysbiosis for a cluster of chronic diseases. *Scientific Reports*, 9(1): 12918. https://doi.org/10.1038/s41598-019-49452-y.

Won, J. S., Lee, S. J., Park, H. H., Song, K. Bin and Min, S. C. (2018). Edible coating using a chitosan-based colloid incorporating grapefruit seed extract for cherry tomato safety and preservation. *Journal of Food Science*, 83(1): 138–146. https://doi.org/10.1111/1750-3841.14002.

Xiao, J., Katsumata, N., Bernier, F., Ohno, K., Yamauchi, Y., Odamaki, T., Yoshikawa, K., Ito, K. and Kaneko, T. (2020). Probiotic Bifidobacterium breve in improving cognitive functions of older adults with suspected mild cognitive impairment: a randomized, double-blind, placebo-controlled trial. *Journal of Alzheimer's Disease*, 77(1): 139–147. https://doi.org/10.3233/JAD-200488.

Xing, Y., Li, W., Wang, Q., Li, X., Xu, Q., Guo, X., Bi, X., Liu, X., Shui, Y., Lin, H. and Yang, H. (2019). Antimicrobial nanoparticles incorporated in edible coatings and films for the preservation of fruits and vegetables. *Molecules*, 24(9): 1695. https://doi.org/10.3390/molecules24091695.

Xiong, Y., Chen, M., Warner, R. D. and Fang, Z. (2020). Incorporating nisin and grape seed extract in chitosan-gelatine edible coating and its effect on cold storage of fresh pork. *Food Control*, 110. https://doi.org/10.1016/j.foodcont.2019.107018.

Yan, J., Luo, Z., Ban, Z., Lu, H., Li, D., Yang, D., Aghdam, M. S. and Li, L. (2019). The effect of the layer-by-layer (LBL) edible coating on strawberry quality and metabolites during storage. *Postharvest Biology and Technology*, 147: 29–38. https://doi.org/10.1016/j.postharvbio.2018.09.002.

Zahedi, S. M., Hosseini, M. S., Karimi, M. and Ebrahimzadeh, A. (2019). Effects of postharvest polyamine application and edible coating on maintaining quality of mango (Mangifera indica L.) cv. Langra during cold storage. *Food Science and Nutrition*, 7(2): 433–441. https://doi.org/10.1002/fsn3.802.

Zepeda-Hernández, A., Garcia-Amezquita, L. E., Requena, T. and García-Cayuela, T. (2021). Probiotics, prebiotics, and synbiotics added to dairy products: Uses and applications to manage type 2 diabetes. *Food Research International*, 142: 110208. https://doi.org/10.1016/j.foodres.2021.110208.

Chapter 14

Health Applications of Biodegradable Polymers

Sandra Cecilia Esparza González, Aide Saenz Galindo, Raúl Rodriguez Herrera, Claudia Magdalena López Badillo, Lissethe Palomo-Ligas, Isai Medina Fernandez* and *Victor de Jesús Suarez Valencia*

1. Introduction

In medicine, biodegradable polymers have been widely used, from drug transporters, tissue engineering, gene therapy, coatings for the controlled release of drugs, among others. There are important characteristics when selecting a biodegradable polymer such as: mechanical properties, the rate of degradation must be according to what is required in the treatment, that the metabolic products of degradation are not toxic, an adequate relationship between conservation and stability (Valero-Valdivieso et al., 2013).

1.1 Biodegradable polymers

Polymers are monomeric chains joined through bonds generating a molecule with long chains. Biodegradable polymers were identified in nature and have the ability to be degraded through biological activity (cellular action) through enzymatic hydrolysis of sensitive bonds in the polymer chain. In general, the carbon-carbon bonds are very stable and show resistance to degradation compared to the bonds where heteroatoms participate. Knowing this, the biodegradable polymers were synthesized in the laboratory using this principle in the modifications in the bonds. Biodegradable polymers are divided into two groups, natural and synthetic. Synthetic polymers are the most frequently used in the biomedical area since the possibility of being created in the laboratory allows adjusting the properties of the biomaterial according to the needs of its application (Sepúlveda et al., 2016).

The elimination of biodegradable polymers is important to avoid inflammation associated with the presence of a foreign material for a long time. Consequently, once the polymers have fulfilled their function they must degrade and, preferably, be excreted eliminating the polymer from the body after treatment. For this reason, polymers must have a high molecular weight between 40–60 kDa so that they biodegrade into smaller molecules and then bioreabsorption occurs (Pasut and Veronese, 2007).

The degradation process must be controlled in relation to the function, for example in tissue engineering, the degradation time will be in accordance with the healing time of the damaged tissue,

Universidad Autónoma de Coahuila, Saltillo, México.

* Corresponding author: sandraesparzagonzal@uadec.edu.mx

ranging from days (skin) or years (bone). With for the release of drugs, the degradation will be from the moment the carrier has reached the target tissue (hours, days or months). In addition to this, the biomimetic concept was adopted for most scaffold designs, in terms of physicochemical properties, as well as bioactivity for tissue regeneration. A variety of scaffolds with appropriate characteristics have been created using different materials, such as polymers, ceramics and their compounds (Kumar and Maiti, 2016).

Scaffolds must support tissue regeneration and repair processes, while providing mechanical support and subsequently degrading into non-toxic products, eventually eliminated from the body. In the case of synthetic polymers, the most widely used in the field of hard tissue engineering are polycaprolactone, polylactic acid, polyglycolic acid and polyethylene glycol (Velema and Kaplan, 2006).

1.2 Biodegradable polymers with medical applications

1.2.1 Chitosan

Chitosan is generated by the deacetylation of chitin, which allows it to dissolve in acidic solution, but not in neutral or alkaline solution. As a derivative of chitin, chitosan is biodegradable and biocompatible, ensuring its applications in food, medicine, and many other areas.

To achieve greater solubility in water, the chitosan is modified creating carboxymethyl chitosan (CMC). Compared to chitosan, CMC is more hydrophilic and dissolves easily in water over a wide pH range. For the tanto, CMC is most widely used as a biopolymer in biomedical research. Furthermore, CMC shows several biological properties such as antimicrobial activity, modulation of cell function and anticancer ability. The impact of carboxymethyl chitosan on biological functions (glucose and fat metabolism, inflammation, intestinal permeability and colon microbiota) has been evaluated in animals in vivo; finding that it has an antimicrobial capacity that alters the intestinal microbit, it was also identified as a possible factor in the development of metabolic syndrome (Liu et al., 2020).

1.2.2 Hyaluronic acid

It is a biodegradable polymer of natural origin with a high molecular weight (105–107 Da). It is a component of the fundamental substance of the extracellular matrix of connective tissue. It is a non-sulfated unbranched glycosaminoglycan (GAG) composed of repeating disaccharides (β-1,4-D-glucuronic acid (known as uronic acid) and β-1,3-N-acetyl-D-glucosamide). Hyaluronic acid is a polyanion that has a great affinity for water molecules (when it is not bound to other molecules), which gives it a viscous and rigid quality like gelatin. In nature this biopolymer works as a scaffold, binding other molecules of the extracellular matrix, including aggrecan.

It is also involved in several important biological functions, such as the regulation of cell adhesion and cell motility, the manipulation of cell differentiation and proliferation and the provision of mechanical properties to tissues.

Hyaluronic acid has been shown to interact with various cell surface receptors (CD44, RHAMM and ICAM-1) influencing cellular processes (morphogenesis, wound repair, inflammation and metastasis). The biocompatibility and hydrophilicity characteristics of hyaluronic acid have made it an excellent moisturizer in cosmetic dermatology and skin care products. Furthermore, its unique viscoelasticity and limited immunogenicity have led to its use in various biomedical applications, such as viscosupplementation in the treatment of osteoarthritis, as an aid in eye surgery and for wound regeneration (Fakhari and Berkland, 2013).

1.2.3 Polyesters

Polyesters are the most widely used biodegradable polymers in medical applications. This family includes polyglycolic, polylactic, polycaprolactone and their copolymers. However, they have been chemically modified through functionalizations to achieve properties according to their function, such as mechanical resistance, bioadesion, stability, among others (Leja and Lewandowicz, 2010).

1.2.4 Polyhydroxy acids (PHA)

They are long chains of organic carboxylic acids, made up of a hydroxide group in the alpha or beta position. The linear aliphatic nature of the structure of PHAs influences their solubility in water, giving them the hydrophilic characteristic. Due to the high molecular weight, it cannot penetrate the skin. Among the alpha hydroxy acids there are: Poly (Glycolic Acid, PGA), Poly (Lactic Acid, PLA) and a range of their copolymers, Poly (Lactic-co-Glycolic Acid, PLGA).

These synthetic polymers have been studied since the 70s, having the first suture approved by the FDA (Food Drug Administration) from the PGA. This material has low solubility characteristics and good mechanical properties (resistance and elasticity). The polymerization of PLA produces crystalline polymers in the L (PLLA) or D (PDLLA) variants; from which bone grafts have been developed that can also be released from drugs.

As PLA is more hydrophobic than PGA, its degradation rate is slower. Polylactic and polyglycolic copolymers (PLGA) have been prepared to improve the properties of biomaterials. PLGA is being investigated as a scaffold for tissue regeneration, transport, as well as drug and protein release (Leja and Lewandowicz, 2010).

1.2.5 Polylactones (PCL)

Polycaprolactone is obtained from the polymerization of caprolactone, has a semi-crystalline appearance and is soluble in common organic solvents, so it is easily processable. However, it degrades more slowly than PLA, PGA and PLGA, which is why it is mainly used in long-term implants and in drug delivery and delivery systems.

Poly-dioxanone (PDS), another polylactone, breaks down into glycoxylate, which is converted to glycine. It is a semi-crystalline polymer with slow degradation due to the non-specific breaking of the ester bond. It has been used to cover bone fixation screws and development of monofilaments as sutures (Leja and Lewandowicz, 2010).

1.2.6 Polyurethanes

These are synthetic thermoplastic polymers that have been used for the development of long-lasting implants. Its excellent biocompatibility, biological activity, mechanical properties and synthetic versatility have led to the production of numerous biodegradable polyurethanes. Conventional polyurethanes are synthesized using three monomers: a diisocyanate, a diol or a diamine. The polymerization reaction of equimolar amounts of diisocyanate and diol produces biodegradable polyurethanes whose composition is decisive in the rate of degradation. The degradation product obtained (diamine obtained after hydrolysis of the polyurethane) is what determines the choice of starting diisocyanate. That is, aliphatic diisocyanates produce less toxic diamines than aromatic diisocyanates. For example, hexamethylene diisocyanate and 1,4-butanediisocyanate are among the most widely used diisocyanates in the formulation of biodegradable polyurethanes, while diphenyl or toluene diisocyanate are used very little due to the high toxicity of their respective degradation products. The fact that the chemical and mechanical properties of polyurethanes can be adapted to needs is what explains that applications are found in various fields such as the regeneration of neurons, vasculature, cartilage and bones (Leja and Lewandowicz, 2010).

1.2.7 Polyanhydrides

Polyanhydros are biodegradable polymers built from repeating monomers linked by anhydride bonds. They are hydrophobic in nature for which they are used in the controlled release of drugs, the degradation rate being modified by changes in the polymer structure.

Aliphatic polyanhydrides are soluble in organic solvents and degrade faster than aromatic polyanhydrides, which on the other hand, are insoluble in organic solvents. They are biocompatible and degrade *in vivo* giving non-toxic diacid by-products that can be eliminated by the body. The most studied polyanhydride is poly [(carboxyphenoxy) propane-sebacic acid] (PCPP-SA), it is used as a matrix for the controlled release of carmustine, a drug used in chemotherapy against brain cancer. In bones, a copolymer based on sebacic acid and erucic acid dimer in a 1:1 ratio is used for the release of gentamicin in the treatment of osteomyelitis (Leja and Lewandowicz, 2010).

1.3 Tissue engineering

Currently, the technology of future medicine is called tissue engineering, in which, from a small fragment of tissue, the global functionality of the damaged tissue or organ can be recovered.

Alternatives have been proposed for a long time to help the regeneration of tissues once they have been damaged, however, each tissue has specific characteristics that help it perform its function, for which each tissue must be studied independently. New technologies developed based on regenerative medicine and tissue engineering have emerged as an option to address the shortage of organ donors and histocompatibility problems. The generation of organs and tissues from the patient's own cells combined with biomaterials have been shown to minimize the risks of immune rejection as well as the transmission of diseases. There are different materials that have been used for this purpose such as metals, salts and polymers of different nature.

Materials created from synthetic polymers have been widely used as scaffolds for tissue engineering, and some are currently approved for human application by the Food and Drug Administration. One of the advantages of using synthetic polymer scaffolds is that they can be produced on a large scale and thus control their physical properties such as tensile strength, degradation rate and the three-dimensional design suitable for the damage to be repaired (Velema and Kaplan, 2006).

1.3.1 Use of polymers for vascular repair

The vascular system is the set of blood vessels that form a closed system to distribute nutrients and gases. Histologically, the various types of blood vessels are distinguished from each other by the thickness of the vascular wall and the differences in the composition of each of the layers.

The walls of blood vessels (arteries, veins and capillaries) are made up of three layers called tunics.

The three layers of the vascular wall, from the lumen outward, are as follows:

- Tunica intima, is the innermost layer of the vessel wall, composed mainly of three tissue components: (1) Simple flat epithelium known as endothelium; (2) basal lamina of endoltelial cells, which is a thin layer of extracellular matrix (collagen, proteoglycans and glycoproteins) and (3) the subendothelial layer, which consists of loose connective tissue underlying the epithelium.
- The tunica media is mainly composed of concentric layers organized in layers of smooth muscle cells. In arteries, due to its function of bearing pressure, this layer is thick (1–45 concentric layers of cells) and extends from the inner elastic membrane to the outer elastic membrane.
- Tunica adventitia, is the outermost layer of dense irregular connective tissue, its main component is collagen fibers and some elastic fibers. The thickness of the tunica adventitia varies depending on the caliber of the blood vessel; it can be thin in most of the arterial system to quite thick in the venules and veins, where it is the main component of the vascular wall.

The endothelium as already described is a simple flat epithelium with autocrine, paracrine and endocrine functions. Under healthy conditions, endothelial cells produce a variety of vasoactive substances that maintain vascular homeostasis and vasomotor tone associated with blood pressure (Pawlina and Ross, 2018). When the endothelium is damaged, there is cardiovascular disease, which continues to be the main cause of morbidity and mortality throughout the world.

These conditions in the blood vessels are treated with the placement of a synthetic vascular graft (stent). The polymeric materials used for the synthesis of the stent can be divided into: non-degradable (expanded polytetrafluoroethylene, polyethylene terephthalate and polyurethane) and biodegradable (PLA, PGA). Some of the non-degradable synthetic polymers have been successful in revascularization with a diameter greater than 6 mm. However, since these polymers are not biodegradable, they present problems of acute thrombosis and stenosis caused by inflammatory processes developed from the implantation of foreign material for a long time. The permanent interaction of the non-biodegradable synthetic polymer has been shown to influence delayed arterial healing and poor re-endothelialization, both of which play an important role in the pathogenesis of late stent thrombosis as well as persistent inflammation can lead to deterioration in the endothelial function and adjacent graft sites. It has been seen that implantation of a non-biodegradable polymeric stent in an animal model generates progressive granulomatous and eosinophilic reactions, beginning around 28 days with a continuous increase up to 6 months. These findings have been associated with a hypersensitivity reaction to non-biodegradable polymers incorporated in first-generation stents synthesized based on polyethylene-vinyl acetate and poly-n-butyl methacrylate. In addition, granulomas have been reported in 10% of pigs a at 28 days and in 23.1% at 180 days after implantation (Oyamada et al., 2010).

The problem of thrombosis has been addressed by two alternatives: with the use of biodegradable polymers and the coating of the luminal surface with antithrombogenic materials, such as heparin or ethylene oxide (Cheng et al., 2014; Park et al., 2017).

The biodegradable polymer (polylactic acid) has been used to generate a stent coupled with a coating with the drug (biolimus A9) on the abluminal surface, which provides a small initial burst and sustained release of the drug and the degradation of the polymer that takes place during a period of more than 6 months. Animal studies demonstrated that stenting caused a significantly lower inflammatory response and rapid recovery of endothelial function compared to the non-biodegradable stent. This success was associated with endothelial relaxation, lower superoxide production in the vessel segments proximal and distal to the stent (Oyamada et al., 2010).

Long-term patency of small diameter vascular grafts remains a major challenge in the field of cardiovascular research. A second phenomenon occurs with the placement of a vascular graft, the proliferation of the smooth muscle cells of the tunica media that occlude the graft lumen. The grafts have been added with drugs such as paclitaxel and rampamycin, which are inhibitors of the cellular cycle. As proof of this, a stent was coated with a biodegradable polymer on the inside that releases rapamycin (140 $\mu g/mm^2$) in a gradual and sustained way (80% during 28 days) and has drastically reduced restenosis in *de novo* lesions in comparison with the conventional stent (Byrne et al., 2009). Besides, stents in biodegradable polymeric matrices have been releasing paclitaxel, which showed the highest degree of inhibition of neointimal hyperplasia, high biocompatibility and complete re-endothelialization. The findings demonstrate that a new abluminally coated bioabsorbable polymer stent, compared to one without the coating, results in comparable vasomotor function at 3 months post implantation in an iliac rabbit model (Nakamura et al., 2012).

Mechanobiological interactions between cells and scaffolds have been shown to have a crucial influence on cell behavior. Associated with this, biodegradable polymers present a loss of mechanical properties due to degradation, inducing thrombosis and hyperplasia of the intima. For this reason, it was decided to use biodegradable polymers for the controlled release of the drug coupled with durable polymers. The three-block biomaterial contains two important components: a poly (styrene-b-isobutylene-b-styrene), which is a non-biodegradable synthetic molecule and a biodegradable polymer is used to modulate optimal drug release over time and the drug paclitaxel. With which it has been possible to maintain the mechanical properties necessary to reestablish the functionality of the vessel.

Looking for new alternatives to achieve greater compatibility and bioactivity, a new generation of vascular grafts has been developed, where vascular cells are seeded on a three-dimensional biodegradable scaffold and then stimulated by conditioning in a bioreactor to promote tissue formation *in vitro* (Borhani et al., 2018).

These biodegradable scaffolds from PGA have been seeded with endothelial cells differentiated from peripheral blood mononuclear cells, the cells have improved the functionality of the scaffolds. Furthermore, cellular functionality is of great importance because obtaining venous blood is not as invasive as a puncture to obtain bone marrow (Generali et al., 2019).

1.3.2 Use of polymers for bone repair

Contrary to popular belief, the bone is a dynamic and heterogeneous tissue that is essentially composed of hydroxyapatite crystals (70%) embedded in a well-ordered network of collagen fibers. Under healthy circumstances, the bone is in continuous remodeling, which involves both anabolic (bone formation) and catabolic (bone resorption) processes (Pawlina and Ross, 2018).

Due to the nature of bone tissue, an ideal bone graft material should include four characteristics: (i) osseointegration, the ability to bind directly to the surface of the host bone; (ii) osteoconduction, the ability to serve as a scaffold to guide bone growth; (iii) osteoinduction, the ability to stimulate the differentiation of osteoprogenitor cells into osteoblasts; (iv) osteogenesis, the formation of new bone by osteoblasts presented within the graft material.

1.3.2.1 Polilactida

PLA has the highest mechanical strength of biopolymers, which is why it is widely used (tensile strength 50 to 70 MPa, tensile modulus of 3000 to 4000 MPa, flexural strength of 100 MPa and flexural modulus of 4000 to 5000 MPa). To improve the characteristics of the scaffolds, combinations have been created in the percentage of PLLA, where it has been shown that increasing the percentage of the polymer improves mechanical properties (Ko, 2020).

However, ceramic materials have been extensively studied to simulate the hardness and strength of bones. This property has been considered in the development of biopolymer/bioceramics composites, combining good biopolymer processability with excellent bio-ceramic bioactivity. For this purpose, PLLA scaffolds with Calcium Hydroxyapatite (HP) added to PGA have been created to accelerate degradation, as well as the bioactivity and osteoconductivity produced mainly by the ceramic portion. The incorporation of PGA improved the hydrophilicity of the scaffold, as well as the degradation rate, for which the embedded HP is exposed in the PLLA matrix, which favors an exchange of ions with the body fluid, facilitating the deposition of apatite similar to that of bone inducing growth and osteoblastic proliferation (Shuai et al., 2021).

PLA biomaterials with calcium carbonate have also been developed by controlling their properties (particle size and low viscosity) during selective laser synthesis. These biomaterials showed high resistance to fracture (up to 75 MPa) and compatibility with osteoblasts (Gayer et al., 2019).

1.3.2.2 Polycarbolactone

Based on PCL, scaffolds have also been developed for bone repair. Matrices of PCL combined with nanodialysis have improved the tensile force of the scaffold and in addition to induce cell differentiation and proliferation, as well as calcium precipitation at the graft site (Ahn et al., 2018).

PCL scaffolds have also been used to control drug release giving the advantage of avoiding bone infections (osteomyelitis) at the time of surgery. This antimicrobial effect has been achieved by loading rifampicin, which is to release it controlled by the use of the biodegradable polymer prolonging the drug at the site and thus inhibiting the growth of the bacteria *S. aureus* and *E. coli* (Lee et al., 2020).

With the PCL, biopolymer/bioceramic scaffolds have also been developed, even different morphologies of the calcium hydroapatite nanoparticles have been porbed to improve the properties (bioactivity, mechanical strength and biocompatibility). The nanorod form was shown to increase mechanical strength in addition to being non-toxic in *in vitro* cell experiments (Moeini et al., 2017).

Calcium oxide nanoparticles were also incorporated into the PCL, improving the differentiation and adhesion conditions of preosteoblasts (Münchow et al., 2016).

The use of grafts has also been focused on treating bone cancer, creating scaffolds comprising outer layers made of PCL and high contents of graphene oxide and inner layers of PCL, studies show that the incorporation of graphene has an effect inhibitor on malignant Saos-2 cells at the same time increases the modulus of compression and resistance to compression, thus allowing to reach values similar to those of human trabecular bone (Hou, 2020).

1.3.2.3 Poly-lactic-co-glycolic acid

PGLA is used to create surgical meshes, resorbable sutures, from which composites have been generated, such as enrichment with microfibers that mimic the structure of collagen to promote cell migration and tissue regeneration. This membrane has a two-layer structure, a dense film to prevent epithelial tissue growth and a microfiber layer to support cell colonization and bone regeneration. These membranes were placed in both rats and rabbits showing success in their application of bone regrowth (Hoornaert et al., 2016).

The synthesis of hydrogels from PLGA has been used for various investigations. An interesting proposal was the synthesis of the biodegradable thermosensitive hydrogel loaded with Bone Morphogenic Protein 2 (BMP-2) as an inducer of bone generation. In literature it is shown that this method is effective for the treatment of femoral bone defects in rabbits, placing them as a therapeutic strategy for the treatment of large bone defects (Peng et al., 2016).

Hybrid matrices have also been developed combining a gel phase with a porous biodegradable matrix and charge carrier, the combination of a hydrogel with a poly (85% lactide-co-15% glycolide) acid scaffold (PLGA 85:15), It was in order to generate a cartilage mold to subsequently achieve regeneration through the endochondral ossification mechanism. This hybrid matrix was tested *in vivo* in mice, and it was observed that mesenchymal stem cells differentiated into chondrocytes which hypertrophied, initiating endochondral ossification. This phenomenon was coupled with mineralization confirming bone regeneration (Mikael et al., 2020).

1.3.2.4 Poly (sebacic-co-ricinoleic acid)

This biodegradable polymer has been used to develop drug release scaffolds such as gentamicin, useful for the treatment of osteomyelitis caused by *S. aureus*, which can occur in a surgical procedure.

The implantation of the biomaterial in an animal model with a contaminated fracture, was able to considerably reduce the infection and promote the repair of bone calluses. However, the combination of local and systemic treatment is indicated to achieve potential therapeutic effect. These findings are encouraging regarding the future use of the scaffold in human clinical trials (Ramot et al., 2020).

1.3.3 Use of polymers for cartilaginous repair

A cartilage is considered a specialized connective tissue that, when damaged by trauma, disease or aging, shows limited capacity for very restricted regeneration because it is avascular and can evolve into osteoarthritis.

Fibrous cartilage exists between the cervical vertebrae, which over time the components are modified and the mechanical properties can cause alterations in the mechanical stresses and physiological loads applied, which subsequently result in disk degeneration and back pain (Pawlina and Ross, 2018).

The hydrated nature of cartilage has made hydrogels a good choice for developing biomaterial scaffolds for tissue engineering applications. Hydrogels provide properties such as biocompatibility, a flexible method of formation, physical characteristics, qualities similar in structure and composition to the extracellular matrix (ECM), thereby inducing cell proliferation and survival (Mikael et al., 2020).

The development of a 3D photopolymerizable, biodegradable synthetic polymer-based hydrogel using polyhydroxyl ethyl methacrylate-co-N-(3-aminopropyl) methacrylamide combined with polyamidoamine (p (HEMA-co-APMA) g PAA) has been investigated. The novelty of these studies is the ability of the polymer to be photopolymerizable in addition to being enriched with mesenchymal stem cells that promote differentiation and proliferation for cartilage regeneration.

High levels of collagen II and aggrecan, as well as chemical and hypoxic signals, are required to achieve differentiation into a chondrogenic lineage. Regarding the characterization of the mechanical properties of the hydrogels, they presented appropriate values of stiffness and modulus of the hydrogels for their application in the repair of cartilages. This injectable and photocurable hydrogel shows good potential for future in vivo applications, which can not only restore disk height with relevant mechanical properties, but also have the opportunity to administer therapeutic agents (Kumar et al., 2014).

Osteoarthritis (OA), the most common articular cartilage disorder, is recognized as a serious condition associated with increased mortality. Inflammation plays a fundamental role in the pathogenesis of OA, which is why the use of corticosteroids (CS) is widely used to relieve the pain of OA. The application of CS through prolonged release therapy is based on the use of poly (lactic-co-glycolic acid) (PLGA) microspheres designed to release the drug slowly, over a period of time. The success of the treatment is that achieving prolonged release prevents inflammation in addition to relieving pain in this condition (Bodick et al., 2018).

Physical properties such as pore size and porosity of a scaffold play a vital role in osteochondral tissue engineering. Scaffolds with a pore size of 100–200 mm in the chondral layer and a pore size of 300–450 mm in the bone layer showed the best effect in repairing the osteochondral defect.

From (PLGA) it has been possible to manufacture integrated bilayer scaffolds with adequate porosities of the two layers (pore size of 100–200 mm in the chondral layer and pore size of 300–450 mm in the bone layer) including the seeding of mesenchymal cells on them. With this modification, by controlling the pore size, it was possible to improve the repair of the endochondral layer due to the cells being able to differentiate into cartilage cells, in addition the biomechanical tests demonstrated the same order of magnitude as that of normal osteochondral tissue (Duan et al., 2019).

The efficiency between using one polymer or a different one has also been compared. Porous scaffolds have been created using PLLA and PLGA to later be seeded with chondrocytes to induce cartilage regeneration. These scaffolds were implanted in a canine animal model and PLLA-based scaffolds were reported to be suitable for transplantation of autologous chondrocytes (Asawa et al., 2012). These findings are also proven in other studies where scaffolds based on polyethylene glycol diacry (PEGDA) are developed, they are frameworks manufactured to place cells which, when in contact with TGF-β3, induces the differentiation of cells into a chondrogenic line and achieving repair, cartilage (Sun et al., 2015).

1.3.4 Use of polymers for connective tissue repair

The main components of a connective tissue are cells and extracellular matrix, produced by these cells. The components of the extracellular matrix are fundamental substances and fibers, which in relation to these the connective tissue is divided into two: loose and dense. Loose connective tissue is characterized by having an extracellular matrix abundant in fundamental substance (glycosaminoglycans, multi-adhesive glycoptroteins and proteoglycans). On the other hand, dense connective tissue has a high concentration of fibers, which can be organized in a parallel sense, giving rise to the tendon and ligaments (Pawlina and Ross, 2018).

Tendons and ligaments have little regenerative capacity with low cell density and low nutrient and oxygen requirements. Injuries to these tissues, such as the Anterior Cruciate Ligament (ACL), are frequent in athletes and in the elderly, cause joint instability accompanied by pain, disability, progression of degenerative diseases and, often, surgical interventions.

To address this problem, scaffolds enriched with collagen fibers based on PLGA compounds (50:50) have been developed which have exhibited higher tensile properties than PLGA composite scaffolds (85:15). In this regard, PLGA (50:50) -ColI-PU demonstrates the potential for use in tissue engineering that requires tensile properties similar to the ligaments found in the knee (Silva et al., 2020).

1.3.5 Use of polymers for brain repair

There are limited tissue engineering studies in the brain, due to the complexity of its structure and function. However, there are reports of the use of PLGA-based scaffolds with which axonal regeneration has been sought.

The destruction and atrophy of axons together with the formation of glial scars and cysts form a physical barrier that restricts axonal regeneration after spinal cord injury (Pawlina and Ross, 2018). To regenerate damaged axons, scaffolds have been created from PLGA, PCL and a new OPF biopolymer which was developed based on PEG with the incorporation of fumarate, forming a biocompatible and biodegradable hydrogel as an implant in the nervous system. All scaffolds with the different polymers have been found to support axonal growth of injured spinal cord neurons. Being the OPF polymer the one that propitiated axonal regeneration with a centralized pattern that resulted in a concentrated number of regenerated axons in the scaffold channels (Chen et al., 2011).

In this same context, scaffolds have been developed from methoxy poly (ethylene glycol) monoacrylate (mPEGA) with hydrophobic, semicrystalline poly (ε-caprolactone) diacrylate (PCLDA), on which the behavior of the precursor cells of the shawn's cell (SpL201). These heterogeneous (mPEGA/PCLDA) nerve channels can promote regenerative functions of nerve cells and promote axonal growth on the inside, while inhibiting cell attachment on the outside during nerve regeneration.

The use of biodegradable polymers has also served to encapsulate and release proteins that promote nerve fiber growth (GDNF) in a controlled manner. These have been tested in animal models with success promoting nerve regeneration, although regeneration did not produce any change in contraction force considered the gold standard for functional nerve repair among treated animals (Kokai et al., 2011).

1.4 Use of polymers in gene therapy

Gene therapy is an experimental technique to treat diseases by altering the genetic material of the patient. Gene therapies often involves the introduction of a healthy copy of a defective gene into the patient's cells. Biodegradable polymers have also been used in order to carry genetic sequences in order to express themselves and thus contribute to correcting this alteration that is usually associated with some pathology.

Poly (B-Amino Esters) (PBAE) are a family of hydrolytically biodegradable polymers, which can carry DNA into human stem cells with high efficiency and low toxicity. In in vitro models, polymeric carriers based on PBAE have been tested by introducing VEGF in a murine model, where angiogenesis was induced and thus limb sparing, due to the expression of the VEGF protein that induces blood vessel regeneration (Mastorakos et al., 2016).

Due to the presence of the blood-brain barrier in the central nervous system, the development of therapies is complicated, however, nowadays, different strategies have been developed that allow the administration of drugs directed and controlled to the brain. The use of biopolymers as nanoparticles (NP) represents an alternative system with enormous potential for the targeted delivery of drugs or biological macromolecules to the CNS. Using synthesized nanoparticles based on the biodegradable polymer poly (β-amino ester), it was possible to transport DNA and the expression of the transgene in the striatum of the CNS in a murine model has been observed. This tool is considered promising for the generation of new therapies targeting the CNS (Zhang et al., 2017).

2. Conclusions

The use of biodegradable polymers has wide applications in medicine. Applications are various drug delivery, scaffolds to induce tissue regeneration, DNA carriers such as gene therapies. The selection of biodegradable polymers is based on the characteristics of the tissue to be repaired. A great variety of biopolymers have been developed with successful medical applications, however there is still a lot more to be done in the research field.

References

Ahn, G. Y., Ryu, T. K., Choi, Y. R., Park, J. R., Lee, M. J. and Choi, S. W. (2018). Fabrication and optimization of Nanodiamonds-composited poly (ε-caprolactone) fibrous matrices for potential regeneration of hard tissues. *Biomaterials Research*, 22(1): 1–8.

Asawa, Y., Sakamoto, T., Komura, M., Watanabe, M., Nishizawa, S., Takazawa, Y. and Hoshi, K. (2012). Early stage foreign body reaction against biodegradable polymer scaffolds affects tissue regeneration during the autologous transplantation of tissue-engineered cartilage in the canine model. *Cell Transplantation*, 21(7): 1431–1442.

Bodick, N., Williamson, T., Strand, V., Senter, B., Kelley, S., Boyce, R. and Lightfoot-Dunn, R. (2018). Local effects following single and repeat intra-articular injections of triamcinolone acetonide extended-release: results from three nonclinical toxicity studies in dogs. *Rheumatology and Therapy*, 5(2): 475–498.

Borhani, S., Hassanajili, S., Tafti, S. H. A. and Rabbani, S. (2018). Cardiovascular stents: overview, evolution, and next generation. *Progress in Biomaterials*, 7(3): 175–205.

Byrne, R. A., Kufner, S., Tiroch, K., Massberg, S., Laugwitz, K. L., Birkmeier, A. and Intracoronary Stenting and Angiographic Restenosis–Test Efficacy of Rapamycin-Eluting STents with Different Polymer Coating Strategies (ISAR-TEST-3) Investigators. (2009). Randomised trial of three rapamycin-eluting stents with different coating strategies for the reduction of coronary restenosis: 2-year follow-up results. *Heart*, 95(18): 1489–1494.

Chen, B. K., Knight, A. M., Madigan, N. N., Gross, L., Dadsetan, M., Nesbitt, J. J. and Windebank, A. J. (2011). Comparison of polymer scaffolds in rat spinal cord: a step toward quantitative assessment of combinatorial approaches to spinal cord repair. *Biomaterials*, 32(32): 8077–8086.

Cheng, C., Sun, S. and Zhao, C. (2014). Progress in heparin and heparin-like/mimicking polymer-functionalized biomedical membranes. *Journal of Materials Chemistry B*, 2(44): 7649–7672.

Duan, P., Pan, Z., Cao, L., Gao, J., Yao, H., Liu, X. and Ding, J. (2019). Restoration of osteochondral defects by implanting bilayered poly (lactide-co-glycolide) porous scaffolds in rabbit joints for 12 and 24 weeks. *Journal of Orthopaedic Translation*, 19: 68–80.

Fakhari, A. and Berkland, C. (2013). Applications and emerging trends of hyaluronic acid in tissue engineering, as a dermal filler and in osteoarthritis treatment. *Acta Biomaterialia*, 9(7): 7081–7092.

Gayer, C., Ritter, J., Bullemer, M., Grom, S., Jauer, L., Meiners, W. and Schleifenbaum, J. H. (2019). Development of a solvent-free polylactide/calcium carbonate composite for selective laser sintering of bone tissue engineering scaffolds. *Materials Science and Engineering: C*, 101: 660–673.

Generali, M., Casanova, E. A., Kehl, D., Wanner, D., Hoerstrup, S. P., Cinelli, P. and Weber, B. (2019). Autologous endothelialized small-caliber vascular grafts engineered from blood-derived induced pluripotent stem cells. *Acta Biomaterialia*, 97: 333–343.

Hoornaert, A., d'Arros, C., Heymann, M. F. and Layrolle, P. (2016). Biocompatibility, resorption and biofunctionality of a new synthetic biodegradable membrane for guided bone regeneration. *Biomedical Materials*, 11(4): 045012.

Hou, Y., Wang, W. and Bártolo, P. (2020). Novel poly (ε-caprolactone)/graphene scaffolds for bone cancer treatment and bone regeneration. *3D Printing and Additive Manufacturing*, 7(5): 222–229.

Ko, Y. G. (2020). Formation of oriented fishbone-like pores in biodegradable polymer scaffolds using directional phase-separation processing. *Scientific Reports*, 10(1): 1–6.

Kokai, L. E., Bourbeau, D., Weber, D., McAtee, J. and Marra, K. G. (2011). Sustained growth factor delivery promotes axonal regeneration in long gap peripheral nerve repair. *Tissue Engineering Part A*, 17(9-10): 1263–1275.

Kumar, D., Gerges, I., Tamplenizza, M., Lenardi, C., Forsyth, N. R. and Liu, Y. (2014). Three-dimensional hypoxic culture of human mesenchymal stem cells encapsulated in a photocurable, biodegradable polymer hydrogel: a potential injectable cellular product for nucleus pulposus regeneration. *Acta Biomaterialia*, 10(8): 3463–3474.

Kumar, S. and Maiti, P. (2016). Controlled biodegradation of polymers using nanoparticles and its application. *RSC Advances*, 6(72): 67449–67480.

Lee, J. H., Baik, J. M., Yu, Y. S., Kim, J. H., Ahn, C. B., Son, K. H. and Lee, J. W. (2020). Development of a heat labile antibiotic eluting 3D printed scaffold for the treatment of osteomyelitis. *Scientific Reports*, 10(1): 1–8.

Leja, K. and Lewandowicz, G. (2010). Polymer biodegradation and biodegradable polymers-a review. *Polish Journal of Environmental Studies*, 19(2).

Liu, Y., Zong, S. and Li, J. (2020). Carboxymethyl chitosan perturbs inflammation profile and colonic microbiota balance in mice. *Journal of Food and Drug Analysis*, 28(1): 175–182.

Mastorakos, P., Song, E., Zhang, C., Berry, S., Park, H. W., Kim, Y. E. and Hanes, J. (2016). Biodegradable DNA nanoparticles that provide widespread gene delivery in the brain. *Small*, 12(5): 678–685.

Mikael, P. E., Golebiowska, A. A., Xin, X., Rowe, D. W. and Nukavarapu, S. P. (2020). Evaluation of an engineered hybrid matrix for bone regeneration via endochondral ossification. *Annals of Biomedical Engineering*, 48(3): 992–1005.

Moeini, S., Mohammadi, M. R. and Simchi, A. (2017). *In-situ* solvothermal processing of polycaprolactone/hydroxyapatite nanocomposites with enhanced mechanical and biological performance for bone tissue engineering. *Bioactive Materials*, 2(3): 146–155.

Münchow, E. A., Pankajakshan, D., Albuquerque, M. T., Kamocki, K., Piva, E., Gregory, R. L. and Bottino, M. C. (2016). Synthesis and characterization of CaO-loaded electrospun matrices for bone tissue engineering. *Clinical Oral Investigations*, 20(8): 1921–1933.

Nakamura, T., Winsor-Hines, D., Yin, X., Sushkova, N., Chen, J. P., III, S. B. K. and Hou, D. (2012). Vasomotor function and re-endothelialisation after implantation of biodegradable abluminal polymer coated paclitaxel-eluting stents in rabbit iliac arteries: a time-course study. *EuroIntervention*, 8(4): 493–500.

Oyamada, S., Ma, X., Wu, T., Robich, M. P., Bianchi, C., Sellke, F. W. and Laham, R. J. (2010). *In vivo* biocompatibility comparison between novel biodegradable polymer and existing non-biodegradable polymer coated stents. *Journal of the American College of Cardiology*, 55(10S): A122–E1140.

Park, S. B., Lih, E., Park, K. S., Joung, Y. K. and Han, D. K. (2017). Biopolymer-based functional composites for medical applications. *Progress in Polymer Science*, 68: 77–105.

Pasut, G. and Veronese, F. M. (2007). Polymer–drug conjugation, recent achievements and general strategies. *Progress in Polymer Science*, 32(8-9): 933–961.

Pawlina, W. and Ross, M. H. (2018). *Histology: a Text and Atlas: With Correlated Cell and Molecular Biology*. Lippincott Williams & Wilkins.

Peng, K. T., Hsieh, M. Y., Lin, C. T., Chen, C. F., Lee, M. S., Huang, Y. Y. and Chang, P. J. (2016). Treatment of critically sized femoral defects with recombinant BMP-2 delivered by a modified mPEG-PLGA biodegradable thermosensitive hydrogel. *BMC Musculoskeletal Disorders*, 17(1): 1–14.

Ramot, Y., Steiner, M., Amouyal, N., Lavie, Y., Klaiman, G., Domb, A. J. and Hagigit, T. (2020). Treatment of contaminated radial fracture in Sprague-Dawley rats by application of a degradable polymer releasing gentamicin. *Journal of Toxicologic Pathology*.

Sepúlveda, L. J. R. and Alzate, C. E. O. (2016). Aplicaciones de mezclas de biopolimeros y polimeros sinteticos: revision bibliografica/Applications of biopolymers and synthetic polymers blends: literature review/Aplicacoes de misturas de biopolimeros e polimeros sinteticoc: revisao da literatura. *Revista Científica*, (25): 252–265.

Shuai, C., Yang, W., Feng, P., Peng, S. and Pan, H. (2021). Accelerated degradation of HAP/PLLA bone scaffold by PGA blending facilitates bioactivity and osteoconductivity. *Bioactive Materials*, 6(2): 490–502.

Silva, M., Ferreira, F. N., Alves, N. M. and Paiva, M. C. (2020). Biodegradable polymer nanocomposites for ligament/tendon tissue engineering. *Journal of Nanobiotechnology*, 18(1): 23.

Sun, A. X., Lin, H., Beck, A. M., Kilroy, E. J. and Tuan, R. S. (2015). Projection stereolithographic fabrication of human adipose stem cell-incorporated biodegradable scaffolds for cartilage tissue engineering. *Frontiers in Bioengineering and Biotechnology*, 3: 115.

Valero-Valdivieso, M. F., Ortegón, Y. and Uscategui, Y. (2013). Biopolímeros: avances y perspectivas. *Dyna*, 80(181): 171–180.

Velema, J. and Kaplan, D. (2006). Biopolymer-based biomaterials as scaffolds for tissue engineering. *Tissue Engineering I*, 187–238.

Zhang, C., Mastorakos, P., Sobral, M., Berry, S., Song, E., Nance, E. and Suk, J. S. (2017). Strategies to enhance the distribution of nanotherapeutics in the brain. *Journal of Controlled Release*, 267: 232–239.

Chapter 15

Natural Polymers

Applications in the Health Field

Carneiro-da-Cunha, M. G.,[1,2,*] *Granja, R. C. B.,*[1,2] *Souza, A. A.,*[1,2] *Melo, E. C. C.,*[1] *Oliveira, W. F.*[1] and *Correia, M. T. S.*[1,*]

1. Introduction

Polymers are macromolecules formed by the combination of numerous small units (monomers) that are repeated. They are classified as homopolymers, derived from only one type of monomer or as copolymers, derived from two or more types of monomers; they can also be natural or synthetic.

Natural polymers are biodegradable and are made by sugar monomeric units (e.g., cellulose, starch and chitin), amino acids (proteins and peptides) or nucleotides (nucleic acids), through covalent bonds. They are called biopolymers according to the International Union of Pure and Applied Chemicals (IUPAC) since they are produced by living organisms (Nagel et al., 1992), while synthetic polymers, polymerized by man, many of them are not biodegradable. The most well-known synthetic polymers that are non-biodegradable comprises the plastics of fossil origin (derived from petroleum) present in different forms, widely consumed and discarded. However, according to data, recently presented at the World Economic Forum in 2020, the world produces more than 400 million tons of plastics/year, many of which are poorly managed after use, causing incalculable damage to the environment and societies.

Thus, natural and modified natural polymers, capable of undergoing hydrolytic or enzymatic degradation, have been used in the most diverse areas. In this chapter, recent scientific advances about the application of natural polymers (proteins and carbohydrates) at micro- and nanometric scale, will be discussed, which have been developed and demonstrated excellent biomedical applications, such as antipathogenic, healing, antitumoral, mitogenic agents among others, especially to develop drug delivery systems.

2. Natural polymers: applications in the health field

As a result of chemical composition of natural polymers, they can be degraded and thus converted into other natural compounds, which make them good candidates for medical applications, such as use in biomaterials, due to the absence or low toxicity and the ability to adsorb bioactive substances (Reddy et al., 2015; Park et al., 2017). These advantages are related to its

[1] Departamento de Bioquímica, Universidade Federal de Pernambuco/UFPE, Recife, Pernambuco, Brasil.
[2] Laboratório de Imunopatologia Keizo Asami/LIKA/UFPE, Recife, Brasil.
* Corresponding author: mgcc@ufpe.br; mtscorreia@gmail.com

Table 15.1. Natural biopolymers applied in the health field.

Biological application	Biopolymer	Source	Reference
Protein			
Anticoagulant	Lectin	*Bauhinia forficata seeds; Moringa oleífera* seeds; *Crataeva tapia* bark;	Silva et al., 2012; Luz et al., 2013; Araújo et al., 2011; Salu et al., 2014
	Peptide	Mouse stomach and small intestine; *Crassostrea gigas*	Tu et al., 2019; Cheng et al., 2018
	Protease	*Agrocybe aegerita* fruiting bodies; *Diopatra sugokai; Clerodendrum colebrookianum* leaves	Li et al., 2021; Kim et al., 2018; Gogoi et al., 2019
Anti-inflammatory	Lectin	*Amansia multifida red algae*; *Cicerarietinum black* seeds and *Prunus dulcis* nut raw	Mesquita et al., 2021; Krishnaveni et al., 2020
	Peptide	*Tenebrio molitor, Schistocerca gregaria and Gryllodes sigillatus; Marphysa sanguinea; Charybdis natator*	Zielińska et al., 2017; Park et al., 2020; Narayanasamy et al., 2020
	Trypsin inhibitor	*Cajanus cajan* and *Phaseolus limensis*	Shamsi et al., 2018
Wound healing	Lectin	*Eugenia malaccensis* seeds; *Bryothamnion seaforthii* *Cratylia mollis* seeds	Brustein et al., 2012; Nascimento-Neto et al., 2012; Albuquerque et al., 2017
	Peptide	*Rhopilema esculentum*	Felician et al., 2019
	Protease	*Ficus drupacea* látex	Manjuprasanna et al., 2020
Antibacterial	Lectin	*Eugenia uniflora* seeds; *Phthirusa pyrifolia* leaf; *Bothrops leucurus* venom	Oliveira et al., 2008; Costa et al., 2010; Nunes et al., 2012
	Thionin	*Nigella sativa L.* seeds	Vasilchenko et al., 2016
	Cyclotide	*Viola odorata L.* and *Viola tricolor L.* aerial parts	Slazak et al., 2018
	Protease	*Marsupenaeus japonicus* shrimp	Zhao et al., 2014
	Peptide	*Spirulina platensis*	Sun et al., 2016
	Trypsin inhibitor	*Albizia amara* seeds	Dabhade et al., 2016
Antiprotozoal	Lectin	*Bothrops leucurus* venom; *Cratylia mollis* seeds	Aranda-Souza et al., 2018; Fernandes et al., 2014
	Protease	Potato tuber	Paik et al., 2020
Antiviral	Protease	*Streptomyces chromofuscus* 34-1	Serkedjieva et al., 2012
	Peptide	*Quercus infectoria* fruits	Seetaha et al., 2021
Antitumoral	Parasporin	*Bacillus thuringiensis*	Moazamian et al., 2018
	Peptide	*Glycine max*	González-Montoya et al., 2018
	Lectin	*Arisaema tortuosum; Aspergillus niger; Praecitrullus fistulosus* fruit	Dhuna et al., 2005; Jagadeesh et al., 2021; Shivamadhu et al., 2017
	Protease	*Trichosanthes kirilowii* fruit; *Cajanus cajan; Enterolobium contortisiliquum* seeds	Song et al., 2016; Shamsi et al., 2017; Lobo et al., 2020
Polysaccharide			
Anticoagulant	Sulfated	*Phallusia nigra; Ulva conglobata; Holothuria fuscopunctata*	Abreu et al., 2019; Mao et al., 2006; Gao et al., 2020
	Glycoconjugates/ Heteropolysaccharid	*Camellia sinensis*	Cai et al., 2013

Table 15.1 contd. ...

...*Table 15.1 contd.*

Biological application	Biopolymer	Source	Reference
	Purified	*Marsypianthes chamaedrys* aerial parts; *Auricularia auricula*	Coelho et al., 2019; Yoon et al., 2003
Anti-inflammatory	Alkali-soluble extract	Purple sweet potato; *Arctium lappa* L. roots	Chen et al., 2019; Zhang et al., 2020
	Purified	*Pholiota nameko*	Li et al., 2008
	Water-soluble	*Turbinaria ornata*	Ananthi et al., 2010
	Sulfated	*Ganoderma lucidum*	Zhang et al., 2018
	Fractions	*Opuntia ficus-indica* cladodes; *Bletilla striata* root; *Gracilaria lemaneiformis* thallus	Trombetta et al., 2006; Zhang et al., 2019; Veeraperumal et al., 2020
Antibacterial	Like lentinan	Lentinus edodes	Zhu et al., 2012
	Purified	*Streptomyces virginia; Periploca laevigata root barks; Chaetomium globosum*	He et al., 2010; Hajji et al., 2018; Wang et al., 2018
	Water-soluble	*Cordyceps cicadae*	Zhang et al., 2017
	Sulfated	*Sargassum swartzii*	Vijayabaskar et al., 2012
	Chitosan	*Litopenaeus vanammei*	Santos et al., 2017
Antitumoral	Sulfated	*Monostroma nitidum*	Karnjanapratum and You, 2011
	Purified	*Trichoderma kanganensis*; *Floccularia luteovirens; Porphyra haitanensis;*	Lu et al., 2018; Liu et al., 2019; Yao et al., 2020
	Water-soluble extract	*Ganoderma applanatum* Leaves	Hanyu et al., 2020
Ophthalmological	Purified	*Lycium barbarum*	Wang et al., 2020; Yang et al., 2020

physical-chemical properties and characteristics of the backbone structure, for example, its chemical groups determine its solubility and surface charge, which influence its stability and interaction with neighboring molecules, and this association may be necessary to exert some biological effect (Wróblewska-Krepsztul et al., 2019). Over the years, there is vast literature on different applications manly of proteins and polysaccharides in health (Table 15.1).

Different pathogens involved in human infections (bacteria, fungi, protozoa and viruses) can be targets for bioactive proteins (Fig. 15.1). However, many natural polymers have been shown to be effective in combating these pathologies (Table 15.1).

2.1 *Antimicrobial activity*

According to the World Health Organization (WHO), antimicrobial resistance occurs when pathogens mutate, and the drugs used in their treatment become ineffective (WHO, 2021). To provide the need of new antibiotics, the use of natural products containing different biomolecules has been studied (Table 15.1).

Lectins, proteins used in different biological applications, can have antimicrobial activity due to their characteristics of binding specifically and reversibly to carbohydrates. Since the cell surfaces of parasites have a diversity of glycids, the lectins could recognize them, and this interaction can lead to death. For example, the lectin from *Lablab purpureus* seeds (specific for glucose-mannose) exerted an action against SARS-CoV-2, which causes the *Coronavirus Disease 2019* (COVID-19), when interacting with the Spike protein (a mannose-rich glycoprotein) of this virus (Liu et al., 2020). Some protozoa characterized as public health problems can also be combated through these biopolymers; even lectins without direct action against pathogen, can stimulate the infected organism to respond more effectively against an infection (Silva et al., 2015; Jandú et al., 2017). Although

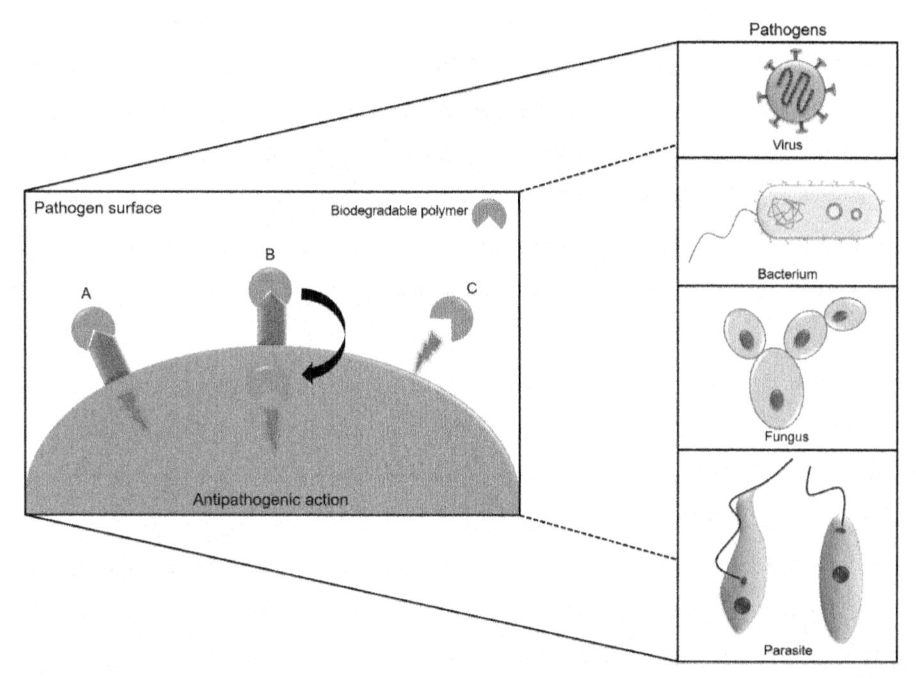

Figure 15.1. Antipathogenic action that some biodegradable polymers can present and their possible mechanisms of action: (A) interacting with molecules on the pathogen surface and inducing intracellular signals, (B) being internalized and acting in the pathogenic internal environment or (C) directly destabilizing the membrane.

proteins can have antimicrobial effects, some proteases may be considered as virulence factors of various pathogens and, therefore, considered important therapeutic targets to fight infections. Thus, protease inhibitors can block the action of proteolytic enzymes, exerting an antipathogenic effect.

Protease inhibitors can also be expressed recombinantly and continue to perform their possible antipathogenic actions. For example, a recombinant trypsin inhibitor inhibited the growth of pathogenic bacterial species (Cisneros et al., 2020), as well as a recombinant protease inhibitor showed antifungal activity (Zhang et al., 2020). It should be noted that peptides obtained by protein hydrolysis may also have an antipathogenic capacity (Lima et al., 2015); thus, antimicrobial peptides also emerge as a viable alternative to antibiotics (Thapa et al., 2020).

Polysaccharides have also been shown to have an inhibitory effect against different pathogens. Sulfated polysaccharides can present themselves as bioactive agents for medical and industrial applications, such as those of red algae *Corallina officinalis* and *Pterocladia capillacea* due to its antibacterial, antifungal, antioxidant, anti-inflammatory, anticoagulant and anti-fouling activities (Ismail and Amer, 2021). Similarly, hydrogels consisting of sulfated polysaccharides of red microalgae from different sources of water associated to chitosan promoted antimicrobial activity and, when added with zinc, shared similar release profiles with potential to function as antimicrobial dressings (Liberman et al., 2021).

2.2 Antitumoral activity

Cancer is a disease of global impact, whose number of patients increases every year, and the use of natural products extracted from plants in the treatment of this malignant neoplasm has shown to be promising.

Changes in the glycosylation pattern of normal cells may be associated with their malignant transformation and by capacity of these tumor cells to metastasize; these altered glycans in neoplastic cells can be recognized by lectins, which can induce them to cell death (Fig. 15.2A), such as the lectin extracted from *Aspergillus niger* (Jagadeesh et al., 2021) and from *Praecitrullus*

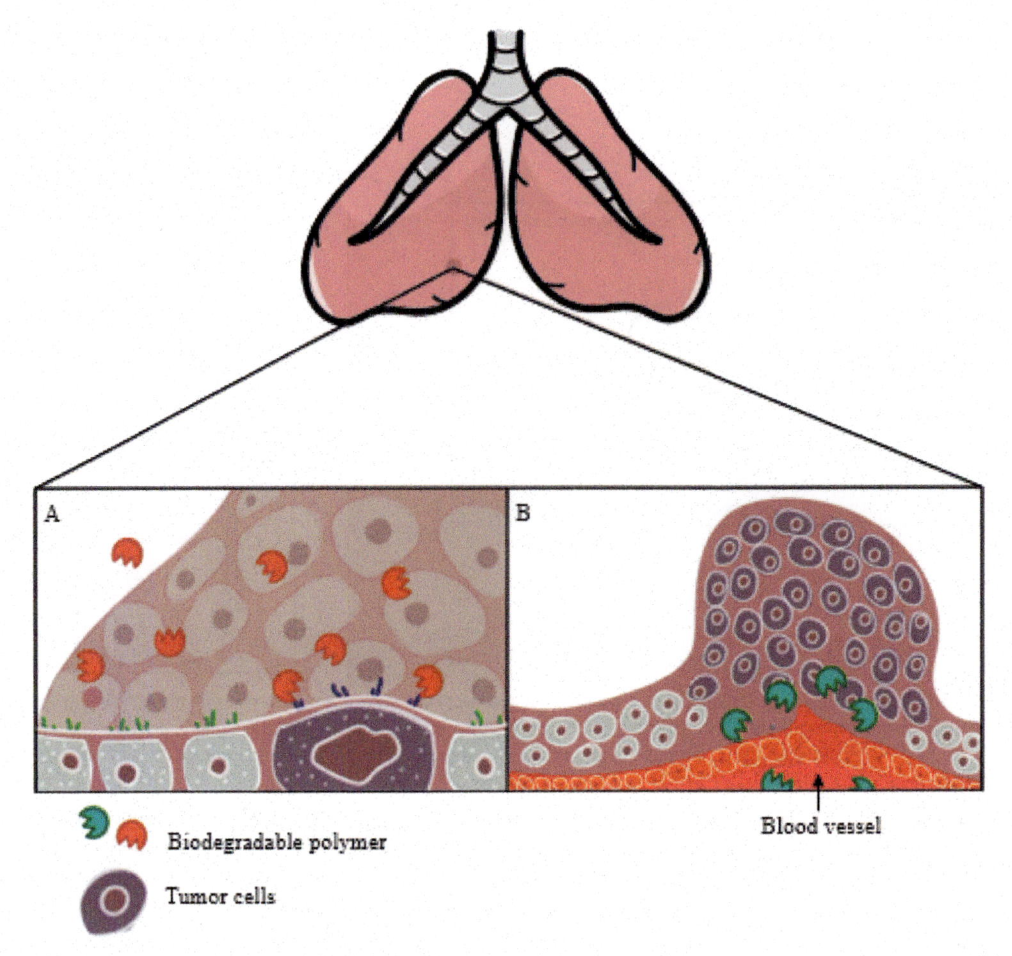

Figure 15.2. Mechanism of action of antitumoral natural polymers: (A) recognition of proteins and carbohydrates exhibited by tumor cells and (B) antiangiogenic action in tumor sites.

fistulosus fruit (Shivamadhu et al., 2017). Furthermore, since different proteases are correlated with tumor development and progression, they may be the target of their inhibitors to prevent cancer proliferation (Eatemadi et al., 2017; Lobo et al., 2020), as well as a purified bifunctional protein from *Crataeva tapia* bark, which has lectinic activity and of protease inhibitor (Bonturi et al., 2019).

Antitumor action of many polysaccharides has also been widely reported (Table 15.1). The polysaccharide of *Diospyros kaki* leaves showed an antiangiogenic effect (Fig. 15.2B) by inhibiting vascular endothelial growth factor and matrix metalloproteinases, both associated with angiogenesis, in endothelial cells of human umbilical vein (Park and Shin, 2021). Polysaccharide-based systems, such as acacia gum (Devi et al., 2020) and alginate (Ren et al., 2020), used as carriers for chemotherapeutic agents in the form of nanoparticles have been seen as effective in the treatment of cancer.

However, one of the main obstacles to cancer treatment is the development of an efficient drug delivery agent that does not induce severe adverse effects on adjacent tissues.

2.3 Drug delivery

The development of efficient technologies for drug delivery generates benefits to health. Therefore, the study of drug delivery systems is important to ensure the expected therapeutic effect, that is, drug release should occur in a specific target site, at the right time.

Accordingly, colloidal systems prepared from different modified polysaccharides formulated in nanoscale scale (nanoparticles) and loaded with a range of active models, demonstrated to be promising nanocarriers to overcome the difficulty of drugs in crossing the blood-brain barrier in the treatment of brain disorders (Bostanudin et al., 2020). Moreover, nanogels systems prepared by carbodiimide coupling between glycyl-prednisolone and anionic polysaccharides showed stronger and prolonged suppression of inflammation than prednisolone used alone as a routine anti-arthritis drug (Mizuno et al., 2020). Chitosan nanoparticles interact with the anionic surfaces of microbial cell membranes can promote cell death. Lecithin and chitosan-based nanoparticles have been described as suitable carriers for hydrophilic (Liu et al., 2016) and hydrophobic (Souza et al., 2014) molecules, as well as on a micrometric scale, hydrogel films based on cellulose and chitosan loaded with Ibuprofen presented themselves as transdermal drug delivery vehicles (Wang et al., 2020).

Protein nanoparticles, mainly of fibroin and albumin, have also shown several advantages as different material delivery carriers (genetics, anticancer drugs, peptide hormones, growth factors), since they are more stable and easier to manufacture in comparison with other coloidal carriers. However, more research should be conducted to make them an ideal material or process to be applied (Hong et al., 2020).

3. Conclusions

Natural polymers have greatly contributed to assist and/or solve health-related problems, both in macrometric and nanometric scales, whether in their isolated, conjugated, trapped forms, as encapsulating material, etc. Furthermore, it is well known that there is a permanent and growing interest in the development of new technologies and products and/or improve existing ones, in order to provide a better and equal quality of life for all humanity.

4. Acknowledgements

The authors are grateful to Conselho Nacional de Desenvolvimento Científico e Tecnológico (CNPq), to Coordenação de Aperfeiçoamento de Pessoal de Nivel Superior (CAPES/PDEE-Brazil) and Fundação de Amparo à Ciência e Tecnologia do Estado de Pernambuco (FACEPE, Brazil) for granting scholarships.

References

Abreu, W. S., Soares, P. A. G., Motta, J. M., Kozlowski, E. O., Teixeira, F. C.O. B., Soares, M. A., Borsig, L. and Mourão, P. A. S. (2019). Tunicate heparan sulfate enriched in 2-sulfated-glucuronic acid: structure, anticoagulant activity, and inhibitory effect on the binding of human colon adenocarcinoma cells to immobilized p-selectin. *Marine Drugs*, 17: 351.

Albuquerque, P. B. S., Soares, P. A. G., Aragão-Neto, A. C., Albuquerque, G. S., Silva, L. C. N., Lima-Ribeiro, M. H. M., Neto, J. C. S., Coelho, L. C. B. B., Correia, M. T. S., Teixeira, J. A. C. and Carneiro-da-Cunha, M. G. (2017). Healing activity evaluation of the galactomannan film obtained from Cassia grandis seeds with immobilized Cratylia mollis seed lectin. *International Journal of Biological Macromolecules*, 102: 749–757.

Ananthi, S., Raghavendran, H. R. B., Sunil, A. G., Gayathri, V., Ramakrishnan, G. and Vasanthi, H. R. (2010). *In vitro* antioxidant and *in vivo* anti-inflammatory potential of crude polysaccharide from Turbinaria ornata (Marine Brown Alga). *Food and Chemical Toxicology*, 48: 187–192.

Aranda-Souza, M. Â., de Lorena, V. M. B., dos Santos Correia, M. T. and de Figueiredo, R. C. B. Q. (2018). *In vitro* effect of Bothrops leucurus lectin (BLL) against Leishmania amazonensis and Leishmania braziliensis infection. *International Journal of Biological Macromolecules*, 120: 431–439.

Araújo, R. M. S., Vaz, A. F. M., Santos, M. E., Zingali, R. B., Coelho, L. C. B. B., Paiva, P. M. G., Correia, M. T. S., Oliva, M. L. V. and Ferreira, R. S. (2011). A new exogen anticoagulant with high selectivity to intrinsic pathway of coagulation. *Thrombosis Research*, 128(4): 395–397.

Bonturi, C. R., Silva, M. C. C., Motaln, H., Salu, B. R., da Silva Ferreira, R., Batista, F. P., Correia, M. T. S., Paiva, P. M. G., Turnšek, T. L. and Oliva, M. L. V. (2019). A bifunctional molecule with lectin and protease inhibitor activities isolated from crataeva tapia bark significantly affects cocultures of mesenchymal stem cells and glioblastoma cells. *Molecules*, 24(11): 2019.

Bostanudin, M. F., Lalatsab, A., Góreckib, D. C. and Barbu, E. (2020). Engineering butylglyceryl-modified polysaccharides towards nanomedicines for brain drug delivery. *Carbohydrate Polymers*, 236: 116060.

Brustein, V. P., Souza-Araújo, F. V., Vaz, A. F. M., Araújo, R. V. S., Paiva, P. M. G., Coelho, L. C. B. B. and Correia, M. T. S. (2012). A novel antimicrobial lectin from Eugenia malaccensis that stimulates cutaneous healing in mice model. *Inflammopharmacology*, 20: 315–322.

Cai, W., Xie, L., Chen, Y. and Zhang, H. (2013). Purification, characterization and anticoagulant activity of the polysaccharides from green tea. *Carbohydrate Polymers*, 92: 1086–1090.

Chen, H., Sun, J., Liu, J., Gou, Y., Zhang, X., Wu, X. and Jin, C. (2019). Structural characterization and anti-inflammatory activity of alkali-soluble polysaccharides from purple sweet potato. *International Journal of Biological Macromolecules*, 131: 484–494.

Cheng, S., Tu, M., Chen, H., Xu, Z., Wang, Z., Liu, H., Zhao, G., Zhu, B. and Du, M. (2018). Identification and inhibitory activity against α-thrombin of a novel anticoagulant peptide derived from oyster (Crassostrea gigas) protein. *Food & Function*, 9: 6391–6400.

Cisneros, J. S., Cotabarren, J., Parisi, M. G., Vasconcelos, M. W. and Obregón,W. D. (2020). Purification and characterization of a novel trypsin inhibitor from Solanum tuberosum subsp. andigenum var. overa: Study of the expression levels and preliminary evaluation of its antimicrobial activity. *International Journal of Biological Macromolecules*, 158: 1279–1287.

Coelho, M. N., Soares, P. A. G., Frattani, F. S., Camargo, L. M. M. and Tovar, A. M. F. (2019). Polysaccharide composition of an anticoagulant fraction from the aqueous extract of Marsypianthes chamaedrys (Lamiaceae). *International Journal of Biological Macromolecules*, 145: 668–681.

Costa, R. M. P. B., Vaz, A. F. M., Oliva, M. L. V., Coelho, L. C. B. B., Correia, M. T. S. and Carneiro-da-Cunha, M. G. (2010). A new mistletoe Phthirusa pyrifolia leaf lectin with antimicrobial properties. *Process Biochemistry*, 45(4): 526–533.

Dabhade, A. R., Mokashe, N. U. and Patil, U. K. (2016). Purification, characterization, and antimicrobial activity of nontoxic trypsin inhibitor from Albizia amara Boiv. *Process Biochemistry*, 51(5): 659–674.

Devi, L., Gupta, R., Jain, S. K., Singh, S. and Kesharwani, P. (2020). Synthesis, characterization and *in vitro* assessment of colloidal gold nanoparticles of Gemcitabine with natural polysaccharides for treatment of breast cancer. *Journal of Drug Delivery Science and Technology*, 56: 101565.

Dhuna, V., Bains, J. S., Kamboj, S. S., Singh, J., Kamboj, S. and Saxena, A. K. (2005). Purification and characterization of a lectin from Arisaema tortuosum Schott having *in-vitro* anticancer activity against human cancer cell lines. *Journal of Biochemistry and Molecular Biology*, 38: 526–532.

Eatemadi, A., Aiyelabegan, H. T., Negahdari, B., Mazlomi, M. A., Daraee, H., Daraee, N., Eatemadi, R. and Sadroddiny, E. (2017). Role of protease and protease inhibitors in cancer pathogenesis and treatment. *Biomedicine & Pharmacotherapy*, 86: 221–231.

Felician, F. F., Yu, R.-H., Li, M.-Z., Li, C.-J., Chen, H.-Q., Jiang, Y. and Xu, H.-M. (2019). The wound healing potential of collagen peptides derived from the jellyfish Rhopilema esculentum. *Chinese Journal of Traumatology*, 22: 12–20.

Fernandes, M. P., Leite, A. C. R., Araújo, F. F. B., Saad, S. T. O., Baratti, M. O., Correia, M. T. S., Coelho, L. C. B. B., Gadelha, F. R. and Vercesi, A. E. (2014). The Cratylia mollis seed lectin induces membrane permeability transition in isolated rat liver mitochondria and a cyclosporine a-insensitive permeability transition in Trypanosoma cruzi mitochondria. *Journal of Eukaryotic Microbiology*, 61(4): 381–388.

Gao, N., Chen, R., Mou, R., Xiang, J., Zhou, K., Li, Z. and Zhao, J. (2020). Purification, structural characterization and anticoagulant activities of four sulfated polysaccharides from sea cucumber Holothuria fuscopunctata. *International Journal of Biological Macromolecules*, 164: 3421–3428.

Gogoi, D., Ramani, S., Bhartari, S., Chattopadhyay, P. and Mukherjee, A. K. (2019). Characterization of active anticoagulant fraction and a fibrin(ogen)olytic serine protease from leaves of Clerodendrum colebrookianum, a traditional ethno-medicinal plant used to reduce hypertension. *Journal of Ethnopharmacology*, 243: 112099.

González-Montoya, M., Hernández-Ledesma, B., Silván, J. M., Mora-Escobedo, R. and Martínez-Villaluenga, C. (2018). Peptides derived from in vitro gastrointestinal digestion of germinated soybean proteins inhibit human colon cancer cells proliferation and inflammation. *Food Chemistry*, 242: 75–82.

Hajji, M., Hamdi, M., Sellimi, S., Ksouda, G., Laouer, H., Li, S. and Nasri, M. (2018). Structural characterization, antioxidant and antibacterial activities of a novel polysaccharide from Periploca laevigata root barks. *Carbohydrate Polymers*, 206: 380–388.

Hanyu, X., Lanyue, L., Miao, D., Wentao, F., Cangran, C. and Hui, S. (2020). Effect of Ganoderma applanatum polysaccharides on MAPK/ERK pathway affecting autophagy in breast cancer MCF-7 cells. *International Journal of Biological Macromolecules*, 146: 353–362.

He, F., Yang, Y., Yang, G. and Yu, L. (2010). Studies on antibacterial activity and antibacterial mechanism of a novel polysaccharide from Streptomyces virginia H03. *Food Control*, 21: 1257–1262.

Hong, S., Choi, D. W., Kim, H. N., Park, C. G., Lee, W. and Park, H. H. (2020). Protein-based nanoparticles as drug delivery systems. *Pharmaceutics*, 12: 604.

Ismail, M. M. and Amer, M. S. (2021). Characterization and biological properties of sulfated polysaccharides of Corallina officinalis and Pterocladia capillacea. *Acta Botanica Brasilica*, 34(4): 623–632.

Jagadeesh, N., Belur, S., Campbell, B. J. and Inamdar, S. R. (2021). The fucose- specific lectin ANL from *Aspergillus niger* possesses anti-cancer activity by inducing the intrinsic apoptosis pathway in hepatocellular and colon cancer cells. *Cell Biochemistry and Function*, 2021: 1–12.

Jandú, J. J., Costa, M. C., Santos, J. R. A., Andrade, F. M., Magalhães, T. F., Silva, M. V., Castro, M. C. A. B., Coelho, L. C. B. B., Gomes, A. G., Paixão, T. A., Santos, D. A. and Correia, M. T. S. (2017). Treatment with pCramoll alone and in combination with fluconazole provides therapeutic benefits in C. gattii infected mice. *Frontiers in Cellular and Infection Microbiology*, 7: 211.

Karnjanapratum, S. and You, S. (2011). Molecular characteristics of sulfated polysaccharides from Monostroma nitidum and their *in vitro* anticancer and immunomodulatory activities. *International Journal of Biological Macromolecules*, 48: 311–318.

Kim, H. J., Shim, K. H., Yeon, S. J. and Shin, H. S. (2018). A novel thrombolytic and anticoagulant serine protease from polychaeta, Diopatra sugokai. *Journal of Microbiology and Biotechnology*, 28: 275–283.

Krishnaveni, M., Kavipriya, M. and Jayasudha, J. B. (2020). Extraction and characterization of lectin from Cicerarietinum black and Prunus dulcis nut raw, its anti-inflammatory, antibacterial activity against oral pathogens. *Materials Today: Proceedings.*

Li, G., Liu, X., Cong, S., Deng, Y. and Zheng, X. (2021). A novel serine protease with anticoagulant and fibrinolytic activities from the fruiting bodies of mushroom Agrocybe aegerita. *International Journal of Biological Macromolecules*, 168: 631–639.

Li, H., Lu, X., Zhang, S., Lu, M. and Liu, H. (2008). Anti-inflammatory activity of polysaccharide from Pholiota nameko. *Biochemistry (Moscow)*, 73: 669–675.

Liberman, G. N., Ochbaum, G., Bitton, R. and Arad, S. (M.). (2021). Antimicrobial hydrogels composed of chitosan and sulfated polysaccharides of red microalgae. *Polymer*, 215: 123353.

Lima, C. A., Campos, J. F., Filho, J. L. L., Converti, A., da Cunha, M. G. C. and Porto, A. L. F. (2015). Antimicrobial and radical scavenging properties of bovine collagen hydrolysates produced by Penicillium aurantiogriseum URM 4622 collagenase. *Journal of Food Science and Technology*, 52(7): 4459–4466.

Liu, Y., Liu, L., Zhou, C. and Xia, X. (2016). Self-assembled lecithin/chitosan nanoparticles for oral insulin delivery: preparation and functional evaluation. *International Journal of Nanomedicine*, 11: 761–769.

Liu, Y. M., Shahed-Al-Mahmud, M., Chen, X., Chen, T. H., Liao, K. S., Lo, J. M., Wu, Y. M., Ho, M. C., Wu, C. Y., Wong, C. H., Jan, J. T. and Ma, C. (2020). A carbohydrate-binding protein from the edible lablab beans effectively blocks the infections of influenza viruses and SARS-CoV-2. *Cell Reports*, 32(6): 108016.

Liu, Z., Jiao, Y., Lu, H., Shu, X. and Chen, Q. (2019). Chemical characterization, antioxidant properties and anticancer activity of exopolysaccharides from Floccularia luteovirens. *Carbohydrate Polymers*, 229: 115432.

Lobo, Y. A., Bonazza, C., Batista, F. P., Castro, R. A., Bonturi, C. R., Salu, B. R., Sinigaglia, R. C., Toma, L., Vicente, C. M., Pidde, G., Tambourgi, D. V., Alvarez-Flores, M. P., Chudzinski-Tavassi, A. M. and Oliva, M. L. V. (2020). EcTI impairs survival and proliferation pathways in triple-negative breast cancer by modulating cell-glycosaminoglycans and inflammatory cytokines. *Cancer Letters*, 491: 108–120.

Lu, Y., Xu, L., Cong, Y., Song, G., Han, J., Wang, G. and Chen, K. (2018). Structural characteristics and anticancer/antioxidant activities of a novel polysaccharide from Trichoderma kanganensis. *Carbohydrate Polymers*, 205: 63–71.

Luz, L. A., Silva, M. C. C., Ferreira, R. S., Santana, L. A., Silva-Lucca, R. A., Mentele, R. and Coelho, L. C. B. B. (2013). Structural characterization of coagulant Moringa oleifera Lectin and its effect on hemostatic parameters. *International Journal of Biological Macromolecules*, 58: 31–36.

Manjuprasanna, V. N., Rudresha, G. V., Urs, A. P., Milan Gowda, M. D., Rajaiah, R. and Vishwanath, B. S. (2020). Drupin, a cysteine protease from Ficus drupacea latex accelerates excision wound healing in mice. *International Journal of Biological Macromolecules*, 165: 691–700.

Mao, W., Zang, X., Li, Y. and Zhang, H. (2006). Sulfated polysaccharides from marine green algae Ulva conglobata and their anticoagulant activity. *Journal of Applied Phycology*, 18: 9–14.

Mesquita, J. X., Brito, T. V., Fontenelle, T. P. C., Damasceno, R. A. S., Souza, M. H. L. P., Lopes, J. L. S., Beltramini, L. M., Barbosa, A. L. R. and Freitas, A. L. P. (2021). Lectin from red algae Amansia multifida Lamouroux: Extraction, characterization and anti-inflammatory activity. *International Journal of Biological Macromolecules*, 170: 532–539.

Mizuno, K., Ikeuchi-Takahashi, Y., Hattori, Y. and Onishi, H. (2020). Preparation and evaluation of conjugate nanogels of glycyl-prednisolone with natural anionic polysaccharides as anti-arthritic delivery systems, *Drug Delivery*, 28: 144–152.

Moazamian, E., Bahador, N., Azarpira, N. and Rasouli, M. (2018). Anti-cancer parasporin toxins of new Bacillus thuringiensis against human colon (HCT-116) and blood (CCRF-CEM) cancer cell lines. *Current Microbiology*, 75: 1090–1098.

Nagel, B., Dellweg, H. and Gierasch, L. M. (1992). International union of pure and applied chemistry; applied chemistry division commission on biotechnology; glossary for chemists of terms used in biotechnology. *Pure & Appl. Chem.*, 64(1): 143–168.

Narayanasamy, A., Balde, A., Raghavender, P., Shashanth, D., Abraham, J., Joshi, I. and Nazeer, R. A. (2020). Isolation of marine crab (Charybdis natator) leg muscle peptide and its anti-inflammatory effects on macrophage cells. *Biocatalysis and Agricultural Biotechnology*, 25: 101577.

Nascimento-Neto, L. G., Carneiro, R. F., da Silva, S. R., da Silva, B. R., Arruda, F. V. S., Carneiro, V. A. and Nagano, C. S. (2012). Characterization of isoforms of the lectin isolated from the red algae Bryothamnion seaforthii and its pro-healing effect. *Marine Drugs*, 10: 1936–1954.

Nunes, E. S., Souza, M. A. A., Vaz, A. F. M., Silva, T. G., Aguiar, J. S., Batista, A. M., Guerra, M. M. P., Guarnieri, M. C., Coelho, L. C. B. B. and Correia, M. T. S. (2012). Toxicon cytotoxic effect and apoptosis induction by Bothrops leucurus venom lectin on tumor cell lines. *Toxicon*, 59(7–8): 667–671.

Oliveira, M. D. L., Andrade, C. A. S., Santos-Magalhães, N. S., Coelho, L. C. B. B., Teixeira, J. A., Carneiro-da-Cunha, M. G. and Correia, M. T. S. (2008). Purification of a lectin from Eugenia uniflora L. seeds and its potential antibacterial activity. *Letters in Applied Microbiology*, 46: 371–376.

Paik, D., Pramanik, P. K. and Chakraborti, T. (2020). Curative efficacy of purified serine protease inhibitor PTF3 from potato tuber in experimental visceral leishmaniasis. *International Immunopharmacology*, 85: 106623.

Park S., Lih, E., Park, K., Ki, Y. and Keun, D. (2017). Progress in polymer science biopolymer-based functional composites for medical applications. *Progress in Polymer Science*, 68: 77–105.

Park, Y. R., Park, C-II. and Soh, Y. (2020). Antioxidant and anti-inflammatory effects of NCW peptide from clam worm (Marphysa sanguinea). *Journal of Microbiology and Biotechnology*, 30: 1387–1394.

Park, J. Y. and Shin, M.-S. (2021). Inhibitory effects of pectic polysaccharide isolated from diospyros kaki leaves on tumor cell angiogenesis via VEGF and MMP-9 regulation. *Polymers*, 13(1): 64.

Reddy, N., Reddy, R. and Jiang, Q. (2015). Crosslinking biopolymers for biomedical applications. *Trends in Biotechnology*, 33(6): 362–369.

Ren, X., Yi, Z., Sun, Z., Ma, X., Chen, G., Chen, Z. and Li, X. (2020). Natural polysaccharide-incorporated hydroxyapatite as size changeable, nuclear targeting nanocarriers for efficient cancer therapy. *Biomaterials Science*, 8(19): 5390–5401.

Salu, B. R., Ferreira, R. S., Brito, M. V, Ottaiano, T. F., C, J. W. M., Silva, M. C. C., Correia, M. T. S., Painva, P. M. G., Maffei, H. A. and Oliva, V. (2014). CrataBL, a lectin and Factor Xa inhibitor, plays a role in blood coagulation and impairs thrombus formation. *Biological Chemistry*, 395(9): 1027–1035.

Santos, F. M. S., Da Silva, A. I. M., Vieira, C. B., De Araújo, M. H., Da Silva, A. L. C., Carneiro-Da-Cunha, M. G., De Souza, B. W. S. and De Souza Bezerra, R. (2017). Use of chitosan coating in increasing the shelf life of liquid smoked Nile tilapia (Oreochromis niloticus) fillet. *Journal of Food Science and Technology*, 9: 1–8.

Seetaha, S., Hannongbua, S., Rattanasrisomporn, J. and Choowongkomon, K. (2021). Novel peptides with HIV-1 reverse transcriptase inhibitory activity derived from the fruits of Quercus infectoria. *Chemical Biology & Drug Design*, 97(1): 157–166.

Serkedjieva, J., Dalgalarrondo, M., Angelova-Duleva, L. and Ivanova, I. (2012). Antiviral potential of a proteolytic inhibitor from Streptomyces Chromofuscus 34-1. *Biotechnology & Biotechnological Equipment*, 26: 2786–2793.

Shamsi, T. N., Parveen, R., Afreen, S., Azam, M., Sen, P., Sharma, Y., Haque, Q. M. R., Fatma, T., Monzoor, N. and Fatima, S. (2018). Trypsin inhibitors from cajanus cajan and phaseolus limensis possess antioxidant, anti-inflammatory, and antibacterial activity. *Journal of Dietary Supplements*, 15(6): 939–950.

Shamsi, T. N., Parveen, R., Ahamad, S. and Fatima, S. (2017). Structural and biophysical characterization of Cajanus cajan protease inhibitor. *Journal of Natural Science, Biology and Medicine*, 8(2): 186–192.

Shivamadhu, M. C., Srinivas, B. K., Jayarama, S. and Sharada, A. C. (2017). Anti-cancer and anti-angiogenic effects of partially purified lectin from Praecitrullus fistulosus fruit on *in vitro* and *in vivo* model. *Biomedicine & Pharmacotherapy*, 96: 1299–1309.

Silva, M. C. C., Santana, L. A., Mentele, R., Ferreira, R. S., de Miranda, A., Silva-Lucca, R. A. and Oliva, M. L. V. (2012). Purification, primary structure and potential functions of a novel lectin from Bauhinia forficata seeds. *Process Biochemistry*, 47: 1049–1059.

Silva, L. C. N., Alves, N. M. P., Castro, M. C. A. B., Pereira, V. R. A., Paz, N. V. N., Coelho, L. C. B. B., Figueiredo, R. C. B. Q. and Correia, M. T. dos S. (2015). Immunomodulatory effects of pCramoll and rCramoll on peritoneal exudate cells (PECs) infected and non-infected with *Staphylococcus aureus*. *International Journal of Biological Macromolecules*, 72: 848–854.

Slazak, B., Kapusta, M., Strömstedt, A. A., Słomka, A., Krychowiak, M., Shariatgorji, M. and Göransson, U. (2018). How does the sweet violet (Viola odorata L.) fight pathogens and pests – cyclotides as a comprehensive plant host defense system. *Frontiers in Plant Science*, 9: 1296.

Song, L., Chang, J. and Li, Z. (2016). A serine protease extracted from Trichosanthes kirilowii induces apoptosis via the PI3K/AKT-mediated mitochondrial pathway in human colorectal adenocarcinoma cells. *Food and Function*, 7(2): 843–854.

Souza, M. P., Vaz, A. F. M., Costa, T. B., Cerqueira, M. A., Vicente, A.A. and Carneiro-da-Cunha, M. G. (2014). Quercetin-loaded lecithin/chitosan nanoparticles for functional food applications. *Food and Bioprocess Technology*, 7: 1149–1159.

Sun, Y., Chang, R., Li, Q. and Li, B. (2016). Isolation and characterization of an antibacterial peptide from protein hydrolysates of Spirulina platensis. *European Food Research and Technology*, 242(5): 685–692.

Thapa, R. K., Diep, D. B. and Tønnesen, H. H. (2020). Topical antimicrobial peptide formulations for wound healing: Current developments and future prospects. *Acta Biomaterialia*, 103: 52–67.

Trombetta, D., Puglia, C., Perri, D., Licata, A., Pergolizzi, S., Lauriano, E. R. and Bonina, F. P. (2006). Effect of polysaccharides from Opuntia ficus-indica (L.) cladodes on the healing of dermal wounds in the rat. *Phytomedicine*, 13: 352–358.

Tu, M., Liu, H., Cheng, S., Mao, F., Chen, H., Fan, F. and Du, M. (2019). Identification and characterization of a novel casein anticoagulant peptide derived from *in vivo* digestion. *Food & Function*, 5: 2552–2559.

Vasilchenko, A. S., Smirnov, A. N., Zavriev, S. K., Grishin, E. V., Vasilchenko, A. V. and Rogozhin, E. A. (2016). Novel thionins from black seed (Nigella sativa L.) demonstrate antimicrobial activity. *International Journal of Peptide Research and Therapeutics*, 23: 171–180.

Veeraperumal, S., Qiu, H.-M., Zeng, S.-S., Yao, W.-Z., Wang, B.-P., Liu, Y. and Cheong, K.-L. (2020). Polysaccharides from Gracilaria lemaneiformis promote the HaCaT keratinocytes wound healing by polarised and directional cell migration. *Carbohydrate Polymers*, 241.

Vijayabaskar, P., Vaseela, N. and Thirumaran, G. (2012). Potential antibacterial and antioxidant properties of a sulfated polysaccharide from the brown marine algae Sargassum swartzii. *Chinese Journal of Natural Medicines*, 10: 421–428.

Wang, S., Xu, Z., Zhao, Y. and Liu, C. (2020c). On the protective effect of lycium barbarum polysaccharide (LBP) on optic nerve tissue and retinal ganglion cells. *Journal of Biomaterials and Tissue Engineering*, 10(4): 443–448.

Wang, X.-H., Su, T., Zhao, J., Wu, Z., Wang, D., Zhang, W.-N. and Chen, Y. (2020). Fabrication of polysaccharides-based hydrogel films for transdermal sustained delivery of Ibuprofen. *Cellulose*, 27: 10277–10292.

Wang, Z., Xue, R., Cui, J., Wang, J., Fan, W., Zhang, H. and Zhan, X. (2018). Antibacterial activity of a polysaccharide produced from Chaetomium globosum CGMCC 6882. *International Journal of Biological Macromolecules*, 125: 376–382.

World Health Organization. Antimicrobial resistance [online]. 2021 [2021, February, 22]. Available from: URL: https://www.who.int/health- topics/antimicrobial-resistance.

Wróblewska-Krepsztul, J., Rydzkowski, T., Michalska-Pozoga, I. and Thakur, V. K. (2019). Biopolymers for biomedical and pharmaceutical applications: recent advances and overview of alginate electrospinning. *Nanomaterials*, 9: 404.

Yang, M., Lo, A. C. Y. and Lam, W. C. (2020). Lycium barbarum polysaccharides protected against Amyloid beta1-40 oligomers-induced adult retinal pigment epithelium 19 cell damage. *Investigative Ophthalmology & Visual Science*, 61: 4917.

Yao, W.-Z., Veeraperumal, S., Qiu, H.-M., Chen, X.-Q. and Cheong, K.-L. (2020). Anti-cancer effects of Porphyra haitanensis polysaccharides on human colon cancer cells via cell cycle arrest and apoptosis without causing adverse effects *in vitro*. *3 Biotech*, 10(9): 386.

Yoon, S.-J., Yu, M.-A., Pyun, Y.-R., Hwang, J.-K., Chu, D.-C., Juneja, L. R. and Mourão, P. A. S. (2003). The nontoxic mushroom Auricularia auricula contains a polysaccharide with anticoagulant activity mediated by antithrombin. *Thrombosis Research*, 112: 151–158.

Zhang, C., He, Y., Chen, Z., Shi, J., Qu, Y. and Zhang, J. (2019). Effect of polysaccharides from bletilla striata on the healing of dermal wounds in mice. *Evidence-Based Complementary and Alternative Medicine*, 2019: 1–9.

Zhang, K., Liu, Y., Zhao, X., Tang, Q., Dernedde, J., Zhang, J. and Fan, H. (2018). Anti-inflammatory properties of GLPss58, a sulfated polysaccharide from Ganoderma lucidum. *International Journal of Biological Macromolecules*, 107: 486–493.

Zhang, X., Guo, K., Dong, Z., Chen, Z. and Zhu, H. (2020). Kunitz-type protease inhibitor BmSPI51 plays an antifungal role in the silkworm cocoon. *Insect Biochemistry and Molecular Biology*, 116: 103258.

Zhang, X., Zhang, N., Kan, J., Sun, R., Tang, S., Wang, Z. and Jin, C. (2020). Anti-inflammatory activity of alkali-soluble polysaccharides from Arctium lappa L. and its effect on gut microbiota of mice with inflammation. *International Journal of Biological Macromolecules*, 154: 773–787.

Zhang, Y., Wu, Y.-T., Zheng, W., Han, X.-X., Jiang, Y.-H., Hu, P.-L. and Shi, L.-E. (2017). The antibacterial activity and antibacterial mechanism of a polysaccharide from Cordyceps cicadae. *Journal of Functional Foods*, 38: 273–279.

Zhao, Y. R., Xu, Y. H., Jiang, H. S., Xu, S., Zhao, X. F. and Wang, J. X. (2014). Antibacterial activity of serine protease inhibitor 1 from kuruma shrimp Marsupenaeus japonicus. *Developmental and Comparative Immunology*, 44(2): 261–269.

Zhu, H., Sheng, K., Yan, E., Qiao, J. and Lv, F. (2012). Extraction, purification and antibacterial activities of a polysaccharide from spent mushroom substrate. *International Journal of Biological Macromolecules*, 50: 840–843.

Zielińska, E., Baraniak, B. and Karas, M. (2017). Antioxidant and anti-inflammatory activities of hydrolysates and peptide fractions obtained by enzymatic hydrolysis of selected heat-treated edible insects. *Nutrients*, 9(9): 970.

Chapter 16

Bioplastics

Challenges and Opportunities

Lily Marcela Palacios,[1,*] *Germán Antonio Arboleda Muñoz,*[1] *Héctor Samuel Villada Castillo*[2] and *Hugo Portela Guarín*[3]

1. Introduction

Plastics are materials with wide use around the world. Their applications are wide and diverse, ranging from the automotive, construction and biomedical industries to agricultural films for farmers, disposable packaging to ensure food quality (Price et al., 2020). However, their extensive use has resulted in a serious environmental problem. The accumulation of millions of tons of non-degradable plastics is a reality and shows a growing trend. Its negative environmental impact is enhanced by inadequate disposal systems. Added to the fact that they come from petroleum resources that are non-renewable (Pathak et al., 2014). Plastic is one of the main sources of current pollution so new alternatives must be developed (Shivam, 2016).

Consequently, new social demands have boosted the transition of many industries and researches towards the search for materials that serve as alternatives to traditional petroleum-derived plastics such as polyethylene (PE), polystyrene (PES) and polyethylene terephthalate (PET). For example, bio-based plastics have a growing interest around the world (Storz and Vorlop, 2013). The so-called "bioplastics" have emerged as a sustainable response to reduced dependence on oil and more sustainable plastic disposal practices. In this sense, the central focus of bioplastics is on the use of abundant, accessible, renewable, economical and degradable materials. Although the market still presents a number of significant challenges, there are also a number of demands around their functional properties such as durability, strength and low price required for use in large-scale consumer products (Pathak et al., 2014). However, there is an initiative and interest in different parts of the world to drive economic empowerment around the bioplastics industry for incorporation into society (Shivam, 2016).

In its definition "Bioplastics (bio-based plastics, biodegradable or both) have the same or similar properties as conventional plastics, but offer additional benefits, such as a reduced carbon footprint, improved functionalities or additional waste management options, such as organic recycling" (European Bioplastics, 2020). It is worth mentioning that there are three groups of bioplastics each

[1] Research Group on Science and Technology of Biomolecules of Agroindustrial Interest. Faculty of Agricultural Sciences. University of Cauca. Popayán, Cauca.

[2] Vice Rector's Office of Research. University of Cauca. Popayán, Cauca.

[3] Antropos Research Group. Faculty of Human and Social Sciences. University of Cauca. Popayán, Cauca.

* Corresponding author: lilymarcelap@unicauca.edu.co

with particular characteristics: (1) Bio-based (or partially bio-based), non-biodegradable plastics, such as bio-based polyethylene (PE), polypropylene (PP), polyethylene terephthalate (PET); (2) Bio-based and biodegradable plastics, such as polylactic acid (PLA), polyhydroxyalkanoates (PHA), polybutylene succinate (PBS) and starch blends and (3) Plastics based on fossil resources and are biodegradable, such as polybutylene adipate terephthalate (PBAT), but which may well be produced, at least in part, bio-based in the future (European Bioplastics, 2020). Within these various types of bioplastics, for example, plastics such as bio-based PET or PE are essentially identical to their petroleum-derived counterparts and are referred to as "drop-in" bio-based plastics for this reason (Alaerts et al., 2018). Bioplastics that are degradable can be obtained from a variety of biomass sources from various natural and renewable feedstocks, such as sugarcane, starch, wood, waste paper, vegetable oils and fats, bacteria, algae, etc. (Sidek et al., 2019).

One of the facilities for the transition from traditional plastics to bioplastics has to do with the suitability for traditional processing infrastructure. Bioplastics can be processed into a wide range of products using conventional plastic processing technologies. In these cases, the process parameters for the transformation of each type of bioplastic need to be adjusted (European Bioplastics, 2020).

Taking into account the relevance and relevance of bioplastics, it is convenient to analyze the trends around the research processes, in order to establish and identify the challenges that are outlined for strengthening this industry, which although, has a wide support, also faces multiple demands to consolidate as elements of everyday use as traditional plastics have done for many decades.

2. The bioplastics market

The bioplastics industry due to its innovative character, presents a fast-growing profile and has the possibility to effective economic benefit and positive environmental impact (European Bioplastics, 2020). In this sense the global bioplastics production capacity reached 2.11 million tons in 2019. The main material that was produced was starch blends with 21.3% of the total, followed by Poly (Butylene Adipate-co-Terephthalate) (PBAT) with 13.4%. Regarding bio-based bioplastics, polyethylene (11.8%), polyamides (PA) (11.6%) and PET (9.8%) represented the materials with the highest production in 2019. Taking into account the degradability character, biodegradable plastics reached 55.5% and non-biodegradable bio-based plastics 44.5% of the global production in 2019.

Similarly, projections for 2024 show a growth in global production capacity, where it is estimated to reach slightly more than 2.4 million tons, where it is projected that starch blends will maintain current production levels and increase the levels of polyhydroxyalkanoates (PHA) in the case of biodegradable bioplastics. For bio-based bioplastics, production is expected to reach 1 million tons by 2024, and for biodegradable bioplastics, around 1.3 million tons. Biobased bioplastics derived from PE together with PA show a similar prominence to that currently presented (European Bioplastics, 2020).

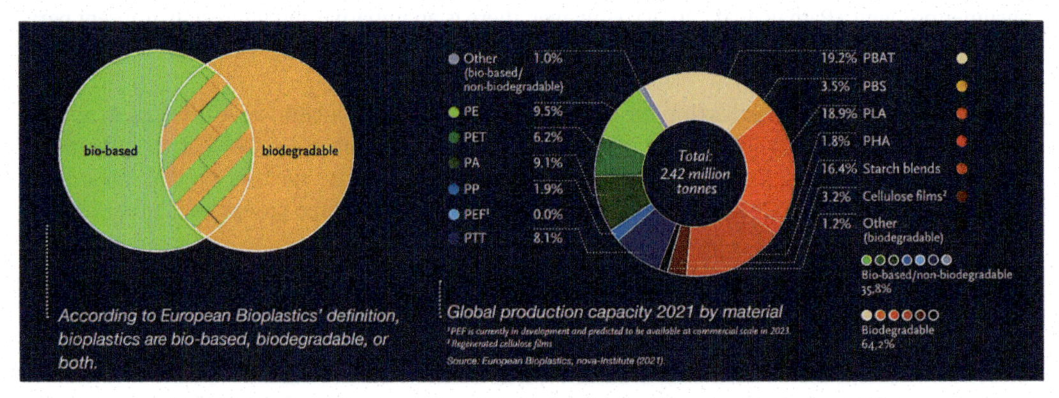

Figure 16.1. Production capacities of bioplastics. Source: European Bioplastics, 2020.

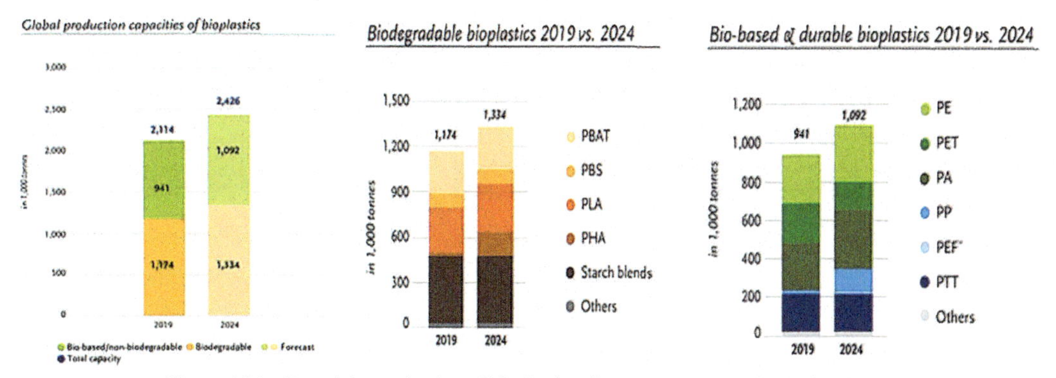

Figure 16.2. Growth in production of bioplastics. Source: European Bioplastics, 2020.

3. Types of bioplastics

As mentioned above, non-biodegradable biobased bioplastics and biodegradable bioplastics can be identified. Some of them will be explained below.

3.1 Bio-based bioplastics

Biopolyethylene (Bio-PE): PE is the most important conventional plastic in terms of annual production volume. It is produced by radical polymerization of petrochemically produced monomer ethylene. Alternatively, bio-based ethylene can be obtained from ethanol. For this, existing polymerization recuperators are used. Feedstock costs represent the most important segment of production costs, so the availability of ethanol feedstock represents a critical factor in the assembly of bio-PE plants (Storz and Vorlop, 2013).

Bio-PET: PET is a thermoplastic polyester produced by polycondensation of Ethylene Glycol (EG) and terephthalic acid (TPA). It is one of the most widely used plastics and is predominantly used in fiber (textiles) and packaging (bottles and food containers) applications (Storz and Vorlop, 2013).

3.2 Biodegradable bioplastics

Within biodegradable bioplastics, a wide variety of alternatives can be found derived from various raw material sources, but according to Shivam (2016), they can be obtained through four main ways: from biomass such as, polysaccharides and proteins, derived from microorganisms, from biotechnological processes or derived from petroleum.

Carbohydrates include the most widely known polymers such as cellulose and starch and their derivatives, as well as other less commonly used polymers such as chitosan and pectin (Nakajima et al., 2017).

3.2.1 Starch blends

Starch is a polysaccharide produced by plants primarily intended for energy storage (Storz and Vorlop, 2013). Starch is stored in granules containing linear amylose and branched amylopectin. Both feature repeating d-glucose units linked by α-1,4 and 1,6 linkages (Nakajima et al., 2017). However, due to their semi-crystalline structure, thermoplastic processing (TPS) of native starch is not possible. To correct this behavior, mixtures are made with plasticizers such as water and glycerol, which on heating processes derive in destructuring of the chain (Storz and Vorlop, 2013). These plasticizers contribute to the stability of TPS in order to reduce the possibility of its return to its natural state by retrogradation (Nakajima et al., 2017). TPS is also blended with hydrophobic polymers to reduce direct water absorption and increase water resistance (Storz and Vorlop, 2013). Some of these can be PLA, PBAT, PBS, PCL and PHA, where in addition to barrier properties mechanical properties and processing are also enhanced (Gadhave et al., 2018).

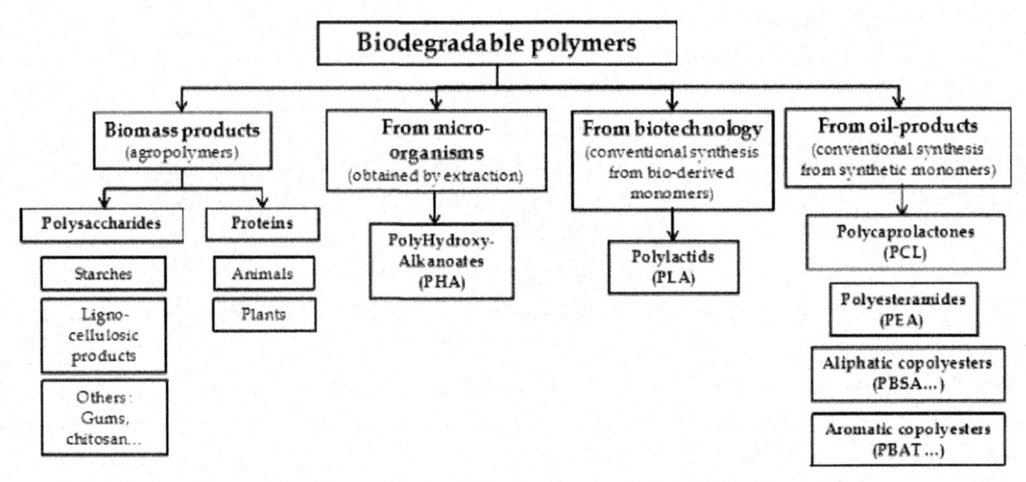

Figure 16.3. Types of biodegradable bioplastics. Source: Shivam, 2016.

3.2.2 Cellulose

The cellulose molecule has a linear conformation with compounds linked by β 1-4 glycosidic bonds. The number of n-chain repeats varies according to the source and provide rigidity to the cellulose, providing good mechanical properties and thermal stability (Rosseto et al., 2020). As with starch, cellulose degrades before melting, although it has a higher degree of crystallinity, being more stable than starch, and common plasticizers are not used for its processing. It is generally used in the form of pulp, which must be chemically modified (Storz and Vorlop, 2013). The modification of cellulose has been of great interest in order to evaluate it as a substitute raw material for synthetic polymers, fibers, films and membranes, hydrogels and aerogels, bioplastics, beads and microspheres (Rosseto et al., 2020). At the same time, it is also used as additives in the form of fibers (e.g., wood, hemp and flax) to polymeric matrices where it has been found to increase the mechanical properties of the final product. This research has also revolved around the development of nanoparticles to obtain cellulose nanofibers (Nakajima et al., 2017).

On the other hand, there are polyesters such as PLA, PHA and succinate which have bio-based polymers that have been successfully applied in the biodegradable plastic industry (Nakajima et al., 2017).

3.2.3 Polylactic acid (PLA)

They correspond to a family of amorphous or semi-crystalline polyesters, obtained from renewable Lactic Acid (LA) fermentation. Among its advantages are its degradability and possibility of processing in traditional equipment (Storz and Vorlop, 2013). Specific benefits of PLA in packaging applications are its transparency, gloss, stiffness, printability, processability and excellent aroma barrier (Gadhave et al., 2018).

3.2.4 Polyhydroxyalkanoates (PHA)

These polymers exist as pure polymer granules in bacteria, employed as an energy storage medium (similar to fat for animals and starch for plants). Commercially, they are obtained by using energy-rich feedstock transformed into fatty acids on which bacteria feed (Nakajima et al., 2017). Several types of processes have been proposed for their production, grouped into three main classes: (i) microbiological, (ii) enzymatic and (iii) chemical processes (Garcia et al., 2020). However, the market for PHA biopolymers is in early stages, so it has become a research topic of interest to scientists and industrialists worldwide (Yadav et al., 2020).

4. Research dynamics

In order to have a perspective of the dynamics regarding publications on bioplastics, the Scopus database was used using the TITLE-ABS-KEY (bioplastic) search equation (24/07/2020), where a total of 1,501 results were obtained, distributed as follows: Articles (1,034), Conferences (228); Review (102); Book chapters (76); Short survey (18); Notes (13); Books (9); Review of conferences (9); Data paper (5) and Erratum (2). Taking into account this compendium of scientific publications, its growth over the last 20 years was analyzed. The change has been noticeable, from five publications in 2000 to 214 in 2019. In 2020, 145 have been found up to July, which shows a year-on-year growth and reflects the interest generated by bioplastics in the scientific area.

At the same time, taking into account the search equation: TITLE-ABS-KEY (bioplastic) AND (LIMIT-TO (PUBYEAR, 2020) OR LIMIT-TO (PUBYEAR, 2019) OR LIMIT-TO (PUBYEAR, 2018) OR LIMIT-TO (PUBYEAR, 2017) OR LIMIT-TO (PUBYEAR, 2016) OR LIMIT-TO (PUBYEAR, 2015) 914 results were obtained, from the publications of the last 5 years, from which they were extracted according to the number of citations, in order to identify those researches that showed greater interest by the scientific community in recent times.

According to what was obtained, polylactic acid has presented a growing interest in recent years (Nagarajan et al., 2016; Murariu and Dubois, 2016; Karamanlioglu et al., 2017; Vink and Davies, 2015), as well as polyhydroxyalkanoates (Raza et al., 2018), polymer sourcing from bacteria (Angermayr et al., 2015) and microalgae (Gouveia et al., 2016), nanocrystalline cellulose (Ilyas et al., 2018). Added to the raw materials and types of bioplastics developed, the analysis of biodegradation conditions has also shown an increasing-citation dynamics in recent years (Emadian et al., 2017).

In the earlier search equations, the key terms in title, abstract or keywords were considered. In order to analyze a more specific dynamic, the search was also advanced with the equation: TITLE (bioplastic) where 460 results were found, where the publications with the highest number of citations are shown in Table 16.2. Similar to the previous exercise, it was also possible to identify a growing interest around the production of PHA (Dietrich et al., 2017; Bhatia et al., 2019), polylactic acid (Zhao et al., 2016), cellulose (Chen et al., 2018), polybutylene succinate (Muthuraj et al., 2015) and starch polymers (Elfehri et al., 2015). Combined to obtaining polymers from food waste (Tsang et al., 2019).

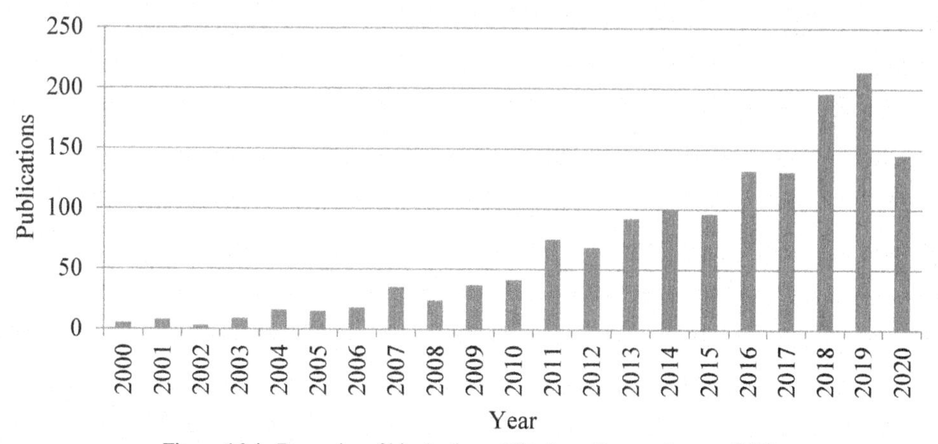

Figure 16.4. Dynamics of bioplastics publications. Source: Scopus, 2020.

Table 16.1. Most cited articles in recent years.

Title	Journal	Year	Citations	Source
Perspective on Polylactic Acid (PLA) based Sustainable Materials for Durable Applications: Focus on Toughness and Heat Resistance	ACS Sustainable Chemistry and Engineering	2016	254	(Nagarajan et al., 2016)
PLA composites: From production to properties	Advanced Drug Delivery Reviews	2016	225	(Murariu and Dubois, 2016)
Metabolic engineering of cyanobacteria for the synthesis of commodity products	Trends in Biotechnology	2015	141	(Angermayr et al., 2015)
Biodegradation of bioplastics in natural environments	Waste Management	2017	132	(Emadian et al., 2017).
Polyhydroxyalkanoates: Characteristics, production, recent developments and applications	International Biodeterioration and Biodegradation	2018	84	(Raza et al., 2018)
Abiotic and biotic environmental degradation of the bioplastic polymer poly (lactic acid): A review	Polymer Degradation and Stability	2017	83	(Karamanlioglu Robson et al., 2017)
Nanocrystalline cellulose as reinforcement for polymeric matrix nanocomposites and its potential applications: A review	Current Analytical Chemistry	2018	74	(Ilyas et al., 2018)
Life Cycle Inventory and Impact Assessment Data for 2014 Ingeo® Polylactide Production	Industrial Biotechnology	2015	74	(Vink and Davies, 2015)
A high heat-resistance bioplastic foam with efficient electromagnetic interference shielding	Chemical Engineering Journal	2017	71	(Cui et al., 2017)
Microalgae biomass production using wastewater: Treatment and costs. Scale-up considerations.	Algal Research	2016	69	(Gouveia et al., 2016)

Source: Based on Scopus, 2020.

The above shows coincidences between two different search equations, but where it was possible to establish that the citation dynamics have been oriented to topics related to the development of polyhydroxyalkanoates, analysis of polylactic acid processing, exploration of routes for obtaining polymers from microalgae and cyanobacteria and the use of alternative raw materials such as food waste. Exploration of routes to obtain polymers from microalgae and cyanobacteria and the use of alternative raw materials such as food waste.

At the same time, Zartha et al. (2015) developed a Delphi methodology, within the framework of a prospective exercise to prioritize technological innovation, new products and technology in the production of biodegradable packaging in Colombia towards the year 2032. The results showed that in terms of technological innovation, the experts agreed on natural coatings, food containers, active packaging, shrink films, foams, intelligent packaging and bioactive films. Regarding raw materials for obtaining biopolymers, corn, cassava, potato, wood fibers, agro-industrial wastes, collagen, fatty acids and monoglycerides emerged as materials with high potential for the development of bioplastics in the future. As shown above, by the year 2032, it is also projected that issues associated with PHA, microbial cellulose, polylactic acid and algae work will be relevant in the development of bioplastics in the future.

Although it seems that the development of bioplastics, for example, developed from starch, has a long way to go, it is not yet possible to declare of a process of technological or market maturity in regions such as Latin America, for example. Factors such as economic development, investment in science and technology activities or legislation, can be a determinant for the development of this

Table 16.2. Publications with most citations TITLE (bioplastic).

Title	Journal	Year	Citations	Source
Abiotic and biotic environmental degradation of the bioplastic polymer poly (lactic acid): A review	Polymer Degradation and Stability	2017	83	(Karamanlioglu et al., 2017)
A high heat-resistance bioplastic foam with efficient electromagnetic interference shielding	Chemical Engineering Journal	2017	71	(Cui et al., 2017)
Producing PHAs in the bioeconomy—Towards a sustainable bioplastic	Sustainable Production and Consumption	2017	65	(Dietrich et al., 2017)
New Superefficiently Flame-Retardant Bioplastic Poly (lactic acid): Flammability, Thermal Decomposition Behavior and Tensile Properties	ACS Sustainable Chemistry and Engineering	2016	59	(Zhao et al., 2016)
Cellulose/graphene bioplastic for thermal management: Enhanced isotropic thermally conductive property by three-dimensional interconnected graphene aerogel	Composites Part A: Applied Science and Manufacturing	2018	40	(Chen et al., 2018)
Injection Molded Sustainable Biocomposites from Poly (butylene succinate) Bioplastic and Perennial Grass	ACS Sustainable Chemistry and Engineering	2018	39	(Muthuraj et al., 2015)
Production of bioplastic through food waste valorization	Environment International	2019	32	(Tsang et al., 2019)
Bioconversion of plant biomass hydrolysate into bioplastic (polyhydroxyalkanoates) using Ralstonia eutropha 5119	Bioresource Technology	2019	30	(Bhatia et al., 2019)
Crosslinking of agarose bioplastic using citric acid	Carbohydrate Polymers	2016	29	(Awadhiya et al., 2016)
Biocomposites of Alfa fibers dispersed in the Mater-Bi® type bioplastic: Morphology, mechanical and thermal properties	Composites Part A: Applied Science and Manufacturing	2015	29	(Elfehri et al., 2015)

Source: Based on Scopus, 2020.

industry (Arboleda and Villada, 2017). In this sense, bioplastics still face an important series of challenges in terms of technological development, transfer, legislative provisions, market and social dynamics, which will also be determinants for the advancement of this industry.

5. Challenges in the development of bioplastics

A number of papers were considered, which presented challenges around bioplastics, in order to identify trends and perform a meta-analysis derived from reviews by other authors. The findings were organized in the Table 16.3.

5.1 The challenge of costs

According to the above, one of the main challenges linked to the development of bioplastics is the selection of raw materials. According to Storz and Vorlop (2013), this is considered as the most important challenge taking into account three reasons: (1) substrate costs represent 40–60% of the total costs of bio-based products (Storz and Vorlop, 2013); (2) the overall yield of production of bio-based monomers significantly lower versus those obtained from petroleum. (3) Competition with raw materials for the production of bio-based food and chemical products. In this sense, there is a tendency to explore different raw material alternatives such as those derived from the use of unused biomass such as that derived from food waste.

Table 16.3. Identification of challenges in the bioplastics industry.

Source	Issues addressed	Conclusions
Storz and Vorlop, 2013	Production of plastics of biological origin (Plastics from natural polymers. Plastics from starch. Cellulose plastics). Bio-based polyesters (Polylactic acid). Polyhydroxyalkanoate. Poly (butylene succinate). Conventional bio-based plastics (Bio-based polyethylene). Bio-poly (ethylene terephthalate).	Main challenges Reduction of production costs. Selection of raw materials for production
Pathak et al., 2014	Advantages of bioplastics (Environmentally friendly, degradation, toxicity, low energy consumption, environmental protection)	Main challenges Misconceptions Environmental impact Cost Future perspectives in the field of bioplastics Use of marine algae
Rosseto et al., 2020	Source opportunities for biodegradable polymers: cellulose, chitosan, starch, proteins, collagen, soy, casein.	Challenges: Adjustment of different formulations to obtain biodegradable plastics from cellulose, starch and proteins.
Nakajima et al., 2017	Biodegradable bio-based polymers (Polylactic acid, polyhydroxyalkanoate, polysaccharides, succinate polymers). Bio-based polymers analogous to conventional petroleum-derived polymers (Bio-based polyethylene (Bio-PE). Bio-based poly (ethylene terephthalate) (PET) and poly (trimethylene terephthalate) (PTT)). Bio-based polyamides Newly developed bio-based polymers (Poly (2,5-furandicarboxylate ethylene) (PEF), high performance modified lactide PLA, Terpen-derived bio-based polymers and other notable bio-based polymers).	Challenges Stability in production and processability Chemical modification
Sidek et al., 2019	Classification of bioplastics Advantages and disadvantages of bioplastics Processing Applications Challenges	Challenges identified Competition with potential food sources. Developments from natural fiber and green compounds.

Source: Scopus, 2020.

Taking into account the first point raised by Storz and Vorlop (2013), regarding costs, other researchers agree that this represents a key element for the development of the industry, taking into account that the costs of bioplastics are generally higher than those of conventional plastics. If one analyzes the problems associated with traditional plastics, one finds that there are two ways to address the environmental problem. On the one hand, the use of plastics can be banned or controlled, and on the other hand, an alternative can be developed. The drawback with the second arises in achieving a competitive price compared to traditional plastics (Chbib et al., 2019).

Obtaining low-cost bioplastics represents a long-term challenge. This includes not only production costs, but also those associated with externality costs, including the cost of recycling, environmental degradation and health-related costs. Which are in need of better regulation that when enacted can represent an opportunity for the inclusion of bioplastics in the dynamics of the circular bioeconomy (Karan et al., 2019).

According to the above, for example, in the case of the United Kingdom, one of the most important barriers to the development of a bioplastics industry is directly related to the cost of bioplastic resin in relation to conventional plastics. Therefore, efforts to reduce the price differential between bioplastic products and conventional plastic products will increase the ability to substitute bioplastics for traditional plastics. For which, the route that seems most appropriate, has to do with the development of economies of scale and production efficiencies, coupled with initiatives around research and innovation and the identification of more efficient production methods (Centre for Economics and Business Research, 2015).

6. New routes to obtain bioplastics

Within this dynamic to explore other alternatives for obtaining bioplastics, there has been an increase in initiatives to work and research with new and diverse processes. At present, raw materials based on agricultural crops (carbohydrates and vegetable materials) represent the main source for obtaining bioplastics. However, the use of microalgae for this purpose is emerging as the next generation of the industry, due to the possibility it offers to be located in non-arable lands (Karan et al., 2019). Added to its high biomass, the ability to grow in a wide range of environments, cost-effectiveness, non-reliance on chemicals and reduced effect on the food chain (Pathak et al., 2014).

Likewise, microalgae can contribute to expand the global photosynthetic capacity, where the ability to transform CO_2 into raw materials for bioplastics is increased. At the same time, saline solutions and/or wastewater can be employed and enable effective recycling of nutrients (e.g., nitrogen and phosphorus) in contained systems, thus reducing eutrophication and dependence on energy-intensive chemical fertilizers (Karan et al., 2019).

Faced with bioplastic production processes from microalgae, two main approaches can be considered. On the one hand, bioplastics can be obtained by blending microalgae biomass, polymers and additives transformed by thermomechanical methods such as compression molding. The other approach focuses on the cultivation of biopolymers such as polyhydroxybutyrate (PHB) within microalgal cells, which are extracted and processed for the production of bioplastics (Onen et al., 2020).

Chlorella and Spirulina species are the most widely used for the production of biopolymers and plastic blends. But according to the literature review conducted by Onen et al. (2020), there is still a need to encourage this type of development, which contributes to overcome the barriers to the economic viability of the industry. For this, the authors propose the concept of biorefinery where bioplastic is produced from by-products of high value chemical production from microalgae. Waste sources that can be used to produce PHA are domestic wastewater, food waste, molasses, olive oil mill effluent, palm oil mill effluent, lingo-cellulosic biomass, cannery waste, biodiesel industry waste, waste cooking oil, paper mill wastewater and sludge. The use of algae as feedstock, coffee waste and cheese whey (Yadav et al., 2020). Thus, working with algae to obtain bioplastics shows a growing and interesting trend for research and innovation processes.

However, PHB production using cyanobacteria is also emerging as an important path for the consolidation of this industry, where the two key factors to achieve profitability are higher PHB productivity and cheaper cyanobacteria cultivation equipment. In this regard, initiatives linked to screening, genetic modification, wastewater cultivation, downstream processing and growth optimization show significant opportunities to improve the viability of PHB obtained from cyanobacteria (Price et al., 2020).

However, questions associated with the development and consolidation of HAPs within the bioplastics industry still seem a distant reality associated, as with much of it, with the associated cost which, even for the most convenient case of the microbiological route, is still too high to compete with that of traditional plastics of fossil origin. Therefore, as mentioned above, the various efforts of academia and industry continue to focus on the identification of suitable raw materials as well as sources of waste from other industrial or civil activities. This requires efforts to identify suitable microorganisms and process engineering (Garcia et al., 2020).

Linked to the concern and interest in reducing costs, exploring new biotechnological routes and reducing the environmental impact associated with land use, possibilities have also been explored to take advantage of residues from different processes. As bioplastics are promoted as a sustainable alternative where their production should not disturb potential food sources (Sidek et al., 2019). Thus, in addition to contributing to the reduction of landfill disposal or landfill burning, from the principle of reuse also contributes positively to the economy (Rosseto et al., 2020).

7. The advancement of the bioplastics industry

The bioplastics industry has undergone a phased development, briefly summarized in Table 16.4.

Like any industry, there are also a number of internal and external factors that influence the growth of bioplastics. Regarding internal drivers, industries have developed initiatives to achieve advances in technical properties and functionality of bioplastics, which have been derived from policy support to encourage research, development and innovation, and it is expected that as their attractiveness and market size increase, progress will also be made in the consolidation of economies of scale to reduce costs. On the other hand, within the external factors, the dynamics of consumer preferences towards environmentally sustainable materials, the improved performance of bioplastics and the introduction of basic plastics produced from biological sources have been key to boost market demand (Centre for Economics and Business Research, 2015).

Similarly, the value chain of the bioplastics industry can also be expressed with the action of various actors involved from the production of raw materials, manufacture of primary and final product and in its distribution and sale to the end user. The first link in the chain is occupied by agricultural and forestry producers of starch, vegetable oil, pulp, sugars and wood cellulose, and the biochemical companies that produce the bioplastic polymers in primary form. These primary products, usually in the form of resins, are sold to companies that convert these resins into the final product. At this stage, the converter may sell the product directly to end users or to wholesale traders who may sell the product to food service or retail companies (Centre for Economics and Business Research, 2015).

However, in this scenario, it is also necessary to consider how the actors associated with the final disposal of bioplastics will be effective, once they reach the end of their useful life. In this sense, a major challenge has to do with guiding both their introduction along with developments in the recycling landscape, with a clear perspective on their incompatibilities, such as process conditions or combinations with other plastics (Alaerts et al., 2018).

Table 16.4. Stages in the development of the bioplastics industry.

Phase	Description
Phase 1	Substitution of petrochemical feedstocks with plant-based monomers and polymer inputs. Development of bioplastics produced from starch and cellulose, blended with other polymers for commercial production.
Phase 2	Parallel development of new bioplastics production routes based on a much wider range of chemistry. New plastic blends with superior properties such as UV resistance, and market-ready biocompatibility. Procurement of high-value films and materials for biomedical applications and specialized bio-based feedstocks synthesized to provide improved polymers for 3D printing.
Phase 3	Next generation microalgae and cyanobacterial systems more compatible with a progressive circular bioeconomy. Contribution to CO_2 capture (in long-term infrastructure) and recycling (through biodegradable plastics). Possibility of locating microalgae and cyanobacteria systems on non-arable land (expanding photosynthetic capacity), and use waste and salt water (conserving fresh water).

Source: Prepared from Karan et al., 2019.

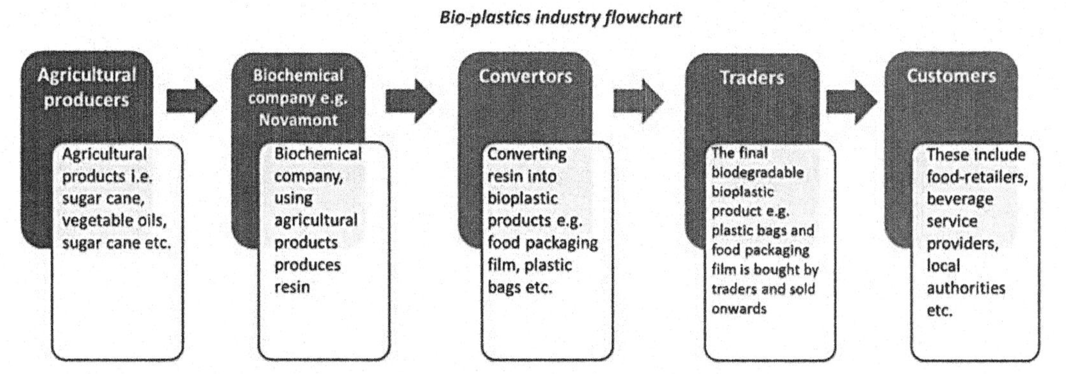

Figure 16.5. Bioplastics value chain. Source: Centre for Economics and Business Research, 2015.

8. Co-creation for the social appropriation of bioplastics

Within the framework of this bioplastics value chain, it should not be forgotten that the flows between links include materials, money and information in which human beings have a complete influence. In an industry such as this one, where the aim is to generate paradigms of change in the face of traditionally accepted practices such as the use of petroleum-derived plastics, it will be essential to promote activities to generate awareness processes among both customers and businessmen. Although a large part of the technological challenge of the industry is based on the search for viable solutions for obtaining polymers, it is relevant that through these developments inclusive research is promoted where the knowledge of society, business and academia have the same value, with the interest of generating interactions for the development of solutions to mitigate the environmental problems generated by packaging made from petrochemical sources.

In the world, approximately 90% of solid waste is only disposed of, but not used, and the generation and accumulation of disposable or plastic waste is increasing the problem, especially when 50% of these materials are single-use. In Latin America, 145,000 tons/day of solid waste are destined to landfills or open burning, where approximately only 10% are destined for recovery or use, which is reflected in "the cost of doing nothing", since the costs for society are 5–10 times higher than the financial costs per capita that would represent an adequate waste management (Pon, 2019). Consequently, there is a growing need to transform the way in which waste is managed, even when its cycle ends in landfills, since its very operation constitutes a critical factor for its environmental sustainability (Noguera-Oviedo and Olivero-Verbel, 2010). In response, waste management is part of the 2030 agenda and the Sustainable Development Goals (SDGs), which state that by 2030 it is expected to achieve the environmentally sound management of chemicals and all their wastes, significantly reduce waste generation through prevention, reduction, recycling and reuse; activities that are correlated with the efficient production and consumption of global resources (Pon, 2019).

Accordingly, venturing on the development of bioplastics is appropriate given the negative impact described and the ongoing national and international regulations that seek to reduce the use of conventional plastics. Considering this and the growing interest as a use of biopolymers, it is imperative to contemplate how the transfer of these emerging technologies should be carried out, between academia and industry, attending to the needs of the market, users, environments and regulations. In view of the situations described above, it is relevant that the conventional models, practices and strategies that have prevailed to mitigate or solve these problems also evolve, given that the way in which the transfer and appropriation of technologies has traditionally been carried out has become obsolete and has generated deep dissatisfaction in a society that is constantly undergoing changes in its environment, actors, social and economic dynamics, among others. The above has led to new opportunities for innovation, through methods that can be replicable and sustainable, and that advocate for equality and social inclusion (Camargo et al., 2017), given that it is not enough to simply design or generate new developments, but their sustainability must be guaranteed through their appropriation. Accordingly, the direct linkage with society before massifying biopolymers will provide the support to promote their viability in the short, medium and long term, therefore, social appropriation should be incorporated as a transversal articulation strategy, because this way it can be guaranteed to go beyond the mere usability of biopolymers, since it would be possible to speak of their sustainability over time, since their integration with everyday practices would be achieved.

This comes in the need to promote processes of understanding and intervention of the relationships between technoscience and society, built from the participation and symbiosis between knowledge generators and users or beneficiaries, therefore, co-design, which is directly related to co-creation, is an interesting strategy for the appropriation of biopolymers, since it allows that the knowledge produced is not implicit or integrated only in the design, but also becomes explicit, debatable, transferable and combinable (Manzini, 2015; Aguirre, 2017); although later adjustments are likely to appear, given that developments are subject to continuous improvement, therefore, it is the key that users identify themselves in the solutions generated to facilitate their adoption.

In this sense, collaboration between bioplastic packaging producers and product manufacturers should improve product functionality and innovation in packaging technologies. Accordingly, positive results can be found in the form of product innovation, increased company innovative capabilities and corporate performance (Liliani et al., 2020). Through co-design, this capability can be enhanced by learning about customer needs, improved operations, stronger customer relationships and the creation of innovative products (Liliani et al., 2020). Finally, by early adopters validating developments, errors can be prevented from reaching the industry, since a continuous and iterative process of discovery and validation, through co-design, allows for the elimination of uncertainties and efficient use of resources.

Conclusion

The current and future situation of bioplastics poses multiple challenges to be addressed, if these materials are to become increasingly sustainable and contribute to mitigate the negative effects derived from plastic waste, which affect terrestrial and aquatic environments, and represent a major concern for the environmental stability of the planet. Technological development around bioplastics is an alternative to boost sustainable industry, where there is an opportunity to promote the articulation between different actors such as academia, industry and society, consequently, for technological and social growth it is vital to implement processes of understanding and intervention between technoscience and society (social innovation processes; co-design), so that the developments and knowledge generated around bioplastics, do not remain implicit or integrated into a product, but become explicit, discussable, transferable and combinable (Manzini, 2015).

References

Aguirre, J. (2017). *Diseño social: Análisis de caso de dos plataformas implementadas para fortalecer la sostenibilidad de colectivos culturales de la ciudad de Cali, Colombia.* Doctoral thesis. Universidad de Caldas. Manizales, Colombia.

Alaerts, L., Augustinus, M. and Van Acker, K. (2018). Impact of bio-based plastics on current recycling of plastics. *Sustainability,* 10: 1487. doi: 10.3390/su10051487.

Arboleda, G. and Villada, H. (2017). Análisis de curvas en S para artículos y patentes de empaques semirrígidos biodegradables. *Revista Espacios,* 38(22): 20.

Angermayr, S., Gorchs, A. and Hellingwerf, K. (2015). Metabolic engineering of cyanobacteria for the synthesis of commodity products. *Trends in Biotechnology,* 33(6): 352–361. doi:10.1016/j.tibtech.2015.03.009.

Awadhiya, A., Kumar, D. and Verma, V. (2016). Crosslinking of agarose bioplastic using citric acid. *Carbohydrate Polymers,* 151: 60–67. doi:10.1016/j.carbpol.2016.05.040.

Bhatia, S. K., Gurav, R., Choi, T., Jung, H., Yang, S., Moon, Y., Song, H., Jeon, J., Choi, K. and Yang, Y. (2019). Bioconversion of plant biomass hydrolysate into bioplastic (polyhydroxyalkanoates) using ralstonia eutropha 5119. *Bioresource Technology,* 271: 306–315. doi:10.1016/j.biortech.2018.09.122.

Camargo, J. E. P., Contreras, F. G. and Jiménez, Y. Y. R. (2017). Estado del arte de la innovación social: una mirada a la perspectiva de Europa y Latinoamérica. *Opción: Revista de Ciencias Humanas y Sociales,* (82): 563–587.

Centre for Economics and Business Research. (2015). The future potential economic impacts of a bio-plastics industry in the UK. Available from: https://bbia.org.uk/wp-content/uploads/2015/11/BBIA-CEBR-Report.compressed.pdf.

CEPAL. (2019). Perspectivas económicas de América Latina 2012: transformación del estado para el desarrollo. OCDE.

Chbib, H., Faisal, M., El Husseiny, A., Fahim, I. and Everitt, N. (2019). The future of biodegradable plastics from an environmental and business perspective. *Modern Approaches on Material Science,* 1(2): 43–48.

Chen, L., Hou, X., Song, N., Shi, L. and Ding, P. (2018). Cellulose/graphene bioplastic for thermal management: Enhanced isotropic thermally conductive property by three-dimensional interconnected graphene aerogel. *Composites Part A: Applied Science and Manufacturing,* 107: 189–196. doi:10.1016/j.compositesa.2017.12.014.

Cui, C., Yan, D., Pang, H., Jia, L., Xu, X., Yang, S., Xu, J. and Li, Z. (2017). A high heat-resistance bioplastic foam with efficient electromagnetic interference shielding. *Chemical Engineering Journal,* 323: 29–36. doi:10.1016/j.cej.2017.04.050.

Dietrich, K., Dumont, M., Del Rio, L. F. and Orsat, V. (2017). Producing PHAs in the bioeconomy—towards a sustainable bioplastic. *Sustainable Production and Consumption,* 9: 58–70. doi:10.1016/j.spc.2016.09.001.

Elfehri, K., Carrot, C. and Jaziri, M. (2015). Biocomposites of alfa fibers dispersed in the mater-bi® type bioplastic: Morphology, mechanical and thermal properties. *Composites Part A: Applied Science and Manufacturing,* 78: 371–379. doi:10.1016/j.compositesa.2015.08.023.

Emadian, S., Onay, T. and Demirel, B. (2017). Biodegradation of bioplastics in natural environments. *Waste Management,* 59: 526–536. doi:10.1016/j.wasman.2016.10.006.

European Bioplastics. Bioplastics. Facts and figures. (2020). Available from: https://docs.european-bioplastics.org/publications/EUBP_Facts_and_figures.pdf.

Gadhave, R., Das, A., Mahanwar, P. and Gadekar, P. (2018). Starch based bio-plastics: the future of sustainable packaging. *Open Journal of Polymer Chemistry,* 8: 21–33. doi: 10.4236/ojpchem.2018.82003.

García, J., Distante, F., Storti, G., Moscatelli, D., Morbidelli, M. and Sponchioni, M. (2020). Current trends in the production of biodegradable bioplastics: The case of polyhydroxyalkanoates. *Biotechnology Advances,* 42: 107582. doi: 10.1016/j.biotechadv.2020.107582.

Gouveia, L., Graça, S., Sousa, C., Ambrosano, L., Ribeiro, B., Botrel, E., Neto, P., Ferreira, A. and Silva, C. (2016). Microalgae biomass production using wastewater: Treatment and costs. scale-up considerations. *Algal Research,* 16: 167–176. doi:10.1016/j.algal.2016.03.010.

Ilyas, R., Sapuan, S., Sanyang, M., Ishak, M. and Zainudin, E. (2018). Nanocrystalline cellulose as reinforcement for polymeric matrix nanocomposites and its potential applications: A review. *Current Analytical Chemistry,* 14(3): 203–225. doi:10.2174/1573411013666171003155624.

Kakadellis, S. and Harris, Z. (2020). Don't scrap the waste: The need for broader system boundaries in bioplastic food packaging life-cycle assessment: A critical review. *Journal of Cleaner Production,* 274: 122831. doi: 10.1016/j.jclepro.2020.122831.

Karamanlioglu, M., Preziosi, R. and Robson, G. (2017). Abiotic and biotic environmental degradation of the bioplastic polymer poly (lactic acid): A review. *Polymer Degradation and Stability,* 137: 122–130. doi:10.1016/j.polymdegradstab.2017.01.009.

Karan, H., Funk, C., Grabert, M., Oey, M. and Hankamer, B. (2019). Green bioplastics as part of a circular bioeconomy. *Trends in Plant Science,* 24(3): 237–249. doi: 10.1016/j.tplants.2018.11.010.

Liliani, L., Tjahjono, B. and Cao, D. (2020). Advancing bioplastic packaging products through co-innovation: A conceptual framework for supplier-customer collaboration. *Journal of Cleaner Production,* 252: 119861. https://doi.org/10.1016/j.jclepro.2019.119861.

Mazini, E. (2015). *Cuando todos diseñan. Una introducción al diseño para la innovación social.* Graficas Muriel. ISBN: 978-84-944817-0-3.

Muthuraj, R., Misra, M. and Mohanty, A. K. (2015). Injection molded sustainable biocomposites from poly (butylene succinate) bioplastic and perennial grass. *ACS Sustainable Chemistry and Engineering,* 3(11): 2767–2776. doi:10.1021/acssuschemeng.5b00646.

Murariu, M. and Dubois, P. (2016). PLA composites: From production to properties. *Advanced Drug Delivery Reviews,* 107: 17–46. doi: 10.1016/j.addr.2016.04.003.

Nagarajan, V., Mohanty, A. and Misra, M. (2016). Perspective on Polylactic Acid (PLA) based Sustainable Materials for Durable Applications: Focus on Toughness and Heat Resistance. *ACS Sustainable Chemistry and Engineering,* 4(6): 2899–2916. doi: 10.1021/acssuschemeng.6b00321.

Nakajima, H., Dijkstra, P. and Loos, K. (2017). The recent developments in biobased polymers toward general and engineering applications: polymers that are upgraded from biodegradable polymers, analogous to petroleum-derived polymers, and newly developed. *Polymers,* 9(10): 523. doi: 10.3390/polym9100523.

Noguera-Oviedo, Katia and Olivero-Verbel, Jesus. (2010). Los rellenos sanitarios en latinoamérica: Caso colombiano. *Revista de la Academia Colombiana de Ciencias Exactas, Físicas y Naturales*, 34: 347–356.

Onen, S., Kai, Z., Kucuker, M., Wieczorek, N., Cengiz, U. and Kuchta, K. (2020). Bioplastic production from microalgae: a review. *International Journal of Enviromental Research and Public Health,* 17: 3842; doi:10.3390/ijerph17113842.

Pathak, S., Sneha, C. and Mathew, B. (2014). Bioplastics: Its timeline based scenario & challenges. *Journal of Polymer and Biopolymer Physics Chemistry,* 2(4): 84–90. doi: 10.12691/jpbpc-2-4-5.

Price, S., Kuzhiumparambil, U., Pemice, M. and Ralph, P. (2020). Cyanobacterial polyhydroxybutyrate for sustainable bioplastic production: Critical review and perspectives. *Journal of Environmental Chemical Engineering,* 8: 104007. doi: 10.1016/j.jece.2020.104007.

Pon, Jordi. (2019). Taller Regional: Instrumentos para la implementación efectiva y coherente de la dimensión ambiental de la agenda de desarrollo. ONU Medioambiente.

Raza, Z., Abid, S. and Banat, I. (2018). Polyhydroxyalkanoates: Characteristics, production, recent developments and applications. *International Biodeterioration and Biodegradation,* 126: 45–56. doi:10.1016/j.ibiod.2017.10.001.

Rosseto, M., Rigueto, C., Krein, D., Balbé, N., Massuda, L. and Dettmer, A. (2020). Biodegradable Polymers: Opportunities and challenges. *In*: Sand, A. and Zaki, E. (eds.). Organic Polymers. InTechOpen.

Shivam, P. (2016). Recent developments on biodegradable polymers and their future trends. *International Research Journal of Science and Engineering,* 4(1): 17–26.

Sidek, I., Syed, S., Sheikh, S. and Anuar, N. (2019). Current development on bioplastics and its future prospects: an introductory review. *I Tech Mag,* 1: 3–8. doi: 10.26480/itechmag.01.2019.03.08.

Storz and Vorlop. (2013). Bio-based plastics: status, challenges and trends. *Landbauforschung,* 63(4): 321–332. doi: 10.3220/LBF_2013_321-332.

Tsang, Y. F., Kumar, V., Samadar, P., Yang, Y., Lee, J., Ok, Y. Song, H., Kim, K., Kwon, E. and Jeon, Y. (2019). Production of bioplastic through food waste valorization. *Environment International,* 127: 625–644. doi:10.1016/j.envint.2019.03.076.

Vink, E. and Davies, S. (2015). Life cycle inventory and impact assessment data for 2014 ingeo® polylactide production. *Industrial Biotechnology,* 11(3): 167–180. doi:10.1089/ind.2015.0003.

Yadav, B., Pandey, A., Kumar, L. and Tyagi, R. (2020). Bioconversion of waste (water)/residues to bioplastics- A circular bioeconomy approach. *Bioresource Technology,* 298: 122584. doi: 10.1016/j.biortech.2019.122584.

Zartha, J., Villada, H., Hernández, R., Fernández, A., Arango, B., Orozco, G., Bermúdez, R., Joaqui, D., Cerón, A. and Moreno, J. (2015). Aplication of Delphi Method in a foresigth study on biodegradable packaging up to 2032. *Espacios,* 36(15): 3.

Zhao, X., Guerrero, F., Llorca, J. and Wang, D. (2016). New superefficiently flame-retardant bioplastic poly (lactic acid): Flammability, thermal decomposition behavior, and tensile properties. *ACS Sustainable Chemistry and Engineering,* 4(1): 202–209. doi:10.1021/acssuschemeng.5b00980.

Chapter 17

Tendencies and Applications in Biodegradable Polymers

Lucía F. Cano Salazar,[1,*] *Denis A. Cabrera Munguía,*[1] *Tirso E. Flores Guía,*[1] *Jesús A. Claudio Rizo,*[1] *Martín Caldera Villalobos*[1] and *Nayvi Y. Nava Cruz*[2]

1. Introduction

It is well known that polymers are constituted of a great number of single units of structures that are joined in a regular manner. In another way polymers are enormous molecules of high molecular weight; this is the result of the linkage of monomers (small molecules) to form macromolecules. Polymers can have different chemical structures, thermal characteristics, physical properties, mechanical behavior, etc., and based on these properties, polymers can be classified in different ways: origin, thermal response, crystallinity, physical properties and applications, among others.

Usual types of polymers are plastics, that are mainly synthetic polymers, and have certain characteristics that make them indispensable for modern life (Hong and Chen, 2019). The industries are now manufacturing a broad range of items that includes sheets, molded planks, bottles, panels, piping, structural profiles (Singh Oberoi et al., 2021). Global production of plastics had increased to around 359 million metric tons in 2018, from 245 million metric tons in 2008, and it is expected to be tripled by the year of 2050 (Chia et al., 2020).

The mass production of plastic-based items greatly exceeds their recycling rate. The decomposition periods of these commodities can be preserved in the environment for centuries (Chen and Yan, 2020). Plastic debris is actually known as white pollution, is a critical environmental issue as these kinds of materials cannot be absorbed easily by nature, and also because the petroleum sources are finite (Shivam, 2016).

For this reason, is imperative to find a new generation of polymers that disrupt this toxic circle of pollution. With this concern, the scientific community had been working hard to give the world a solution for plastic contamination (Haider et al., 2019).

For the reason mentioned above, bio-based polymers are attracting increased attention due to environmental concerns. According to the definition of the American Society for Testing and Materials (ASTM), biodegradable polymers are polymers that degrade or decompose under chemical, physical and biological interactions with microorganisms from the environment, such as bacteria, fungus and algae (Zhang, 2021). The fact that polymers are biodegradable will obviously

[1] Advanced Materials Department, Autonomous University of Coahuila, Saltillo, Coahuila, Mexico.
[2] Institute of Plant Biology and Biotechnology, Westfälische Wilhelms University Muenster, Muenster, Germany.
* Corresponding author: lucia.cano@uadec.edu.mx

help as a solution to contamination by plastics mainly from petroleum, which tend to have a short life before being discarded. The use of biodegradable materials appears to be promising, however, it is important to consider not only biodegradability but also different characteristics such as environmental sustainability, degradability time, acceptance, physical properties, economic viability and applications, among others (Agarwal, 2020; Pellis et al., 2021; Shivam, 2016).

Nowadays, several polymers that are biodegradable have been studied, in addition, their properties and the applications in which they can be used have also been tested. Among the polymers that have been the target of systematic tests are: Polycaprolactone (PCL) and Poly (Lactide Acid) (PLA), amide-containing polymer, polyurethane, polyhydroxybutyrate (PHB), Poly (Glycolide Acid) (PGA), polybutylenesuccinate (PBS), polyhydrixyakanoates (PHAs), thermal polyspartate (TPA), polypeptides and natural macromolecules such as cellulose, chitin, chitosan, amylose and lignin (Yin and Yang, 2020).

This chapter focuses on summarizing some of the biodegradable polymers that are being used today, the trends in their applications, and the advantages as substitutes for conventional plastics noting that development of biodegradable polymers is part of multidisciplinary researchers.

2. Importance of biodegradable polymers

The stability of polymers has harmed society and environment due to the accumulation of a large amount of plastic waste and its distribution around the world. Biodegradable polymers could be one of the solutions to the problem of plastic waste, helping to generate alternatives for the solution of health and environmental risks. Biodegradation is the degradation of macromolecular chains by the action of microorganisms. This process takes place in two steps (Fig. 17.1).

The first step is fragmentation, which is the cleavage of high molar mass macromolecular chains to yield monomers and oligomers with polar functional end groups. Polymers lose their specific polymer properties during fragmentation. During the second step, mineralization, both yield oligomers and monomers are transformed by the action of microorganisms forming carbon dioxide, methane, water and biomass. Specific metabolic processes that involve enzymes with high selectivity are responsible for this transformation. The generation of degradation products in the gaseous state can help prevent the bioaccumulation of polymers, avoiding environmental damage; however, they can contribute to increasing the effects of global warming and the greenhouse effect (Agarwal, 2020).

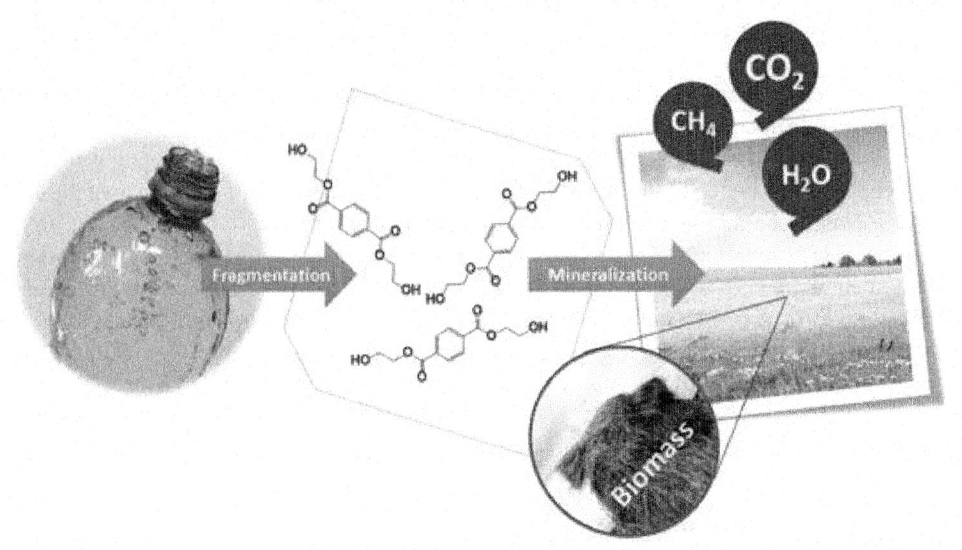

Figure 17.1. Biodegradation of polymers: Representing the biodegradation of Poly(Ethylene Terephthalate) (PET). Own authorship.

Currently, biodegradable polymers have a relevant use for packaging, agriculture, gastronomy, consumer electronics and automotives. However, biodegradable polymers represent about 1% of 300 million tons of plastic produced annually. Biodegradable plastics are used mainly in short service life applications where biodegradability is a key advantageous feature. They include compostable waste bags, biodegradable mulch films, catering products, film packaging for foods and rigid packaging (Rujnić-Sokele and Pilipović, 2017).

A current gap in the circular economy concerning plastics is the role of biodegradable plastics in it. Firstly, biodegradable plastics such as poly (lactic acid) (PLA), polyesters (PCL, PHA, PBAT and PBS) and starch can be mechanically recycled or converted by biological processes where carbon can be returned to nature safely and sustainably, e.g., composting. Further, by combining pyrolysis and microbiology, non-degradable plastics can be converted to biodegradable plastics. The above offers an unconventional route for the processing of non-degradable plastics. The improvement of enzyme-based technologies could lead to innovative methods for processing plastics in a circular economy. Depolymerization promoted by enzymes can be used to convert plastic waste to monomers. Then, the generated monomers could be used to make more biodegradable plastics, which would represent a completely biological recycling of plastics in a circular economy. The integration of hydrolytic enzymes into a microbial chassis would result in a custom microbial platform that is capable of converting plastic into biodegradable counterparts in a single cell (Narancic and O'Connor, 2019).

In summary, the current importance of biodegradable polymers depends on the problems that are expected to be alleviated or solved. They include the increasing pressure on landfills, the littering of difficult-to-recycle products and achieving of a more sustainable society by conserving the non-renewable resources. Thus, it is hoped that biodegradable plastics reduce part of the bulky plastic wastes from landfills. Also a reduction of environmental and visual pollution is expected. And finally, implementing strategies like composting for the final disposition of plastic waste will be a suitable alternative when recycling was not possible.

3. Polybutylene adipate terephtalate (PBAT)

PBAT is a synthetic polymer from fossil resources but 100% biodegradable with a Young modulus of 20–35 MPa, 32–36 MPa of tensile strength, and higher elongation at break, properties that are comparable to those of low-density polyethylene (Ferreira et al., 2017). It has been observed that the Young's modulus of PBAT increases with the content in terephthalate units, while elongation at break decreases. The molecular weight of PBAT can be modulated by reaction pressure and temperature to achieve the equilibrium conversion (Jian et al., 2020).

Typically, PBAT is synthesized by polycondensation reaction of 1,4-butanediol with both adipic and terephthalic acids (or butylene adipate), using rare earth metal compounds or zinc acetate as catalysts. The reaction conditions include long reaction time, high vacuum and temperature higher than 190°C to remove the produced water (Ferreira et al., 2017; Jian et al., 2020).

The degradation of PBAT is perfomed by (i) enzymatic action of bacteria, fungi and algae, and (ii) combined depolymerization process: chemical hydrolysis, thermal degradation. It is reported that the temperature, acid conditions, a low crystallization degree and a high concentration of terephthalic acid favors the PBAT biodegradability (Ferreira et al., 2017).

The addition of a monomer or a natural filler (e.g., cellulose nanocrystals, montmorillonites, red mud, coffee grounds) can enhance the thermomechanical properties of PBAT and reduce the production cost to obtain an accessible PBAT. PBAT composites are fabricated by three main methods: *in-situ* polymerization, melt mixing and solvent casting. However, some challenges are the filler dispersion, interaction between filler/matrix, and reduction of the filler content (Ferreira et al., 2017). The compatibility between the filler with the low polar PBAT can be overcome by the surface modification of the filler. Hence, PBAT composites based on modified bentonite have shown the best thermal and mechanical performance at room temperature. PBAT and its composites

Table 17.1. PBAT, PBS and PCL composites and their applications.

Matrix	Copolymer/Filler	Application	Reference
PBAT	Torrified coffee grounds	Food Packaging	(Moustafa et al., 2017)
PBAT	PLA	Packaging Films	(Qiu et al., 2021)
PBAT/PLA Blend	$CaCO_3$	Agriculture Films	(Belchior Rocha et al., 2018)
PBAT/PLA Blend	Babassu vegetable filler	Agriculture Films	(França et al., 2018)
PBAT/PMMA (PolyMethyl MethAcrylate)	Tricalcium phosphate ($Ca_3(PO_4)_2$)	Fabrication of Medical Devices	(Bheemaneni and Kandaswamy, 2019)
PLA/PBAT	Nanohydroxyapatite	Bone Tissue Engineering	(Yan et al., 2020)
PLA/PBS	Functionalized chitosan/dicumyl peroxide	Nanobiocomposite in the Packaging of UV-Vis Sensitive Materials	(Akhilesh Kumar Pal et al., 2018)
PBS	Zinc Oxide	Antimicrobial Food Packaging	(Petchwattana et al., 2016)
PBS	Modified Tapioca Starch	Agricultural Mulch Films	(Ayu et al., 2020)
PBS	Urea/Formaldehyde	Slow Release Application in Agriculture	(Zhang et al., 2020)
PBS	Lignin	Antioxidant and Antibacterial Properties for Biomedical Applications	(Domínguez-Robles et al., 2020)
PLA/PBS	Cellulose Nanofibril	Vascular Tissue Engineering	(Abudula et al., 2019)
PCL	Nanoclay (montmorillonite) Nanocomposites	Packaging Applications	(Yahiaoui et al., 2015)
PCL	Nanoclay	Antimicrobial Packaging Films	(Hadj-Hamou et al., 2017)
PCL	Methyl Cellulose	Antimicrobial Agricultural Mulch Films	(Boumail et al., 2013)
PCL	Coffee Husk	Agricultural Mulch Films	(Borghesi et al., 2016)
PCL	Starch	Wound Dressing Application	(Komur et al., 2017)
PCL	Bismuth Ferrite ($BiFeO_3$)	Advanced Diagnostic Imaging	(Ulag et al., 2020)

can be applied (Table 17.1) as packaging materials (trash bags, food containers, film wrapping), hygiene products (diaper back sheets, cotton swabs), biomedical field, industrial composting and agriculture (mulch films) (Ferreira et al., 2017; Jian et al., 2020).

4. Polybutylene succinate (PBS)

Polybutylene succinate (PBS) is a biodegradable polymer and compostable aliphatic polyester. It is produced by esterification at 160–190°C with succinic acid with 1, 4-butanediol to form PBS oligomers and water, which is removed by vacuum. Then, at vacuum, the oligomers are transesterified at 220–240°C to form a high molar mass PBS using titanium butoxide, zirconium, tin or germanium derivatives as catalysts. The major by-product is tetrahydrofuran that forms through dehydration of butanediol (added in excess) and water (Alias and Marsilla, 2020; Xu and Guo, 2010).

PBS is a white crystalline resin with a melting point around 100–130°C and density of 1.25 g/cm^3 which is useful to process through extrusion, injection molding, compression molding, blow molding and 3D printing. The processability of PBS depends on its Molecular Weight (MW), then, extrusion requires a MW lower than 100,000 while blow molding needs a MW around 180,000. A lower MW produces a brittle PBS, whereas a high MW gets a flexible PBS. PBS has a

Young modulus in the range of 300–500 MPa and an elongation at break higher than polypropylene, polyethylene terephthalate and polycarbonate (Alias and Marsilla, 2020; Xu and Guo, 2010).

Some techniques to modulate the thermomechanical properties of PBS are grafting with monomers into its polymer backbone or blending PBS with other polymers. The main idea is to alter the crystallization behavior of PBS to improve the thermomechanical properties of PBS. The flexibility and thermal stability of PBS can be enhanced via its copolymerization (e.g., adipic, terephthalic and methyl succinic acid) or blending with other polymers (e.g., starch, polylactide, polyhydroxybutyrate) to reduce the degree of crytallinity, and thus a better PBS biodegradability (Alias and Marsilla, 2020; Xu and Guo, 2010).

Enzymatic degradation of PBS and its copolymers is faster than hydrolysis at neutral pH without enzyme, being biodegradable in lipase solution, soil burial, water, activated sludge and compost (Gigli et al., 2016; Xu and Guo, 2010). PBS applications involves (Table 17.1) packaging, agricultural (mulch films and pots, coating materials of fertilizers), medical (drug release, tissue and bone regeneration), automotive and textile (wiping cloth and baby diapers) applications (Alias and Marsilla, 2020; Gigli et al., 2016).

5. Polycaprolactone (PCL)

Polycaprolactone (PCL) is synthetic, hydrophobic, semi-crystalline and biodegradable polymer with a good resistance to water, oil, solvent and chlorine. The MW of PCL varies from 3,000–90,000, whose crystallinity decreases as the MW increases which favors its biodegradability (Mohamed and Yusoh, 2016).

The routes to synthesize PCL are (i) ring-opening polymerization of cyclic monomer ε-caprolactone using a variety of anionic, cationic and coordination catalysts (aluminum and tin-based complexes, organ catalysts, ionic initiators) at high temperature (> 120°C) or (ii) via free radical ring-opening polymerization of 2-methylene-1-3-dioxepane (Bartnikowski et al., 2019; Guarino et al., 2017; Mandal and Shunmugam, 2020; Mohamed and Yusoh, 2016).

Due to its low melting point (59–64°C) and exceptional blend-compatibility, PCL has been applied in packaging, sutures, scaffolds for bone and cartilage regeneration, controlled drug release systems (Table 17.1). Its slow rate of degradation compared to others polylactides makes PCL a candidate for long term implantable devices for low load-bearing bone tissue engineering, and also, in dentistry (Guarino et al., 2017; Mandal and Shunmugam, 2020; Mohamed and Yusoh, 2016).

The degradation time of PCL depends on its molecular weight, pore size, degree of crystallinity and morphology. PCL degradation starts with amorphous phases followed by the crystalline domains (Guarino et al., 2017; Mandal and Shunmugam, 2020; Mohamed and Yusoh, 2016). The degradation of PCL involves chemical ester hydrolysis catalyzed in acid or base media, where the carboxylic end groups of PCL are more susceptible to hydrolysis than hydroxyl end groups (Mohamed and Yusoh, 2016). Another route is thermal degradation of PCL at 420°C through decomposition of hydroxyl end groups which is important to take into account in PCL manufacturing (Mohamed and Yusoh, 2016). The enzymatic degradation of PCL is possible by lipases (e.g., estearase) (Bartnikowski et al., 2019), and also, by bacteria and fungi microorganisms (Guarino et al., 2017).

6. Polyhydroxyalkanoates (PHA)

With the growing demand for polymeric materials and the limited availability of non-renewable sources from the petrochemical industry, numerous alternative sources have been studied to obtain various bio-plastics capable of covering a wide range of applications. Among the main advantages of bioplastics compared to traditional plastics, are the reduction of environmental impact, low costs and degradation, among others. One type of biopolymer that has attracted attention, due to its similarity in mechanical properties compared with synthetic polymers, are polyhydroxyalkanoates. The (PHA) are biogenic polyesters that present good biocompatibility, non-toxicity and can be processed to obtain a great variety of products, in addition to being environmentally friendly.

Polyhydroxyalkanoates are classified into two different groups, according to the number of carbon atoms in the monomers. The first group is considered short chain length (PHA scl) and consists of polymers containing monomers of 3 to 5 carbon atoms, while the other group considers medium chain length polymers as they contain monomers with 6–14 carbon atoms, carbon (PHA mcl) (Akaraonye et al., 2010; Akinmulewo and Nwinyi, 2019; Anjum et al., 2016). Another possible classification is depending on homogeneity/heterogeneity, this means if its base is a homopolymer (the best known being the polyhydroxy butyrate (PHB)) or a copolymer (poly(3-hydroxybutyrate-co-3-hydroxy valerate) (PHBV), poly(3-hydroxybutyrate-co-4-hydroxybutyrate) (P3HB-4HB) or poly(3-hydroxybutyrate-co-3-hydroxyhexanoate) (PHBH)) (Giubilini et al., 2021).

There are various production methods to obtain PHA and derivatives such as enzymatic catalysis, the use of genetically modified substrates and bacterial fermentation, this last one is the most widely used due to its efficiency (Sosa-Hernández et al., 2020). The production using bacterial fermentation can be divided into two steps, bacterial growth and PHA generation. This process begins when a limited supply of nutrients (oxygen, nitrogen, phosphorus, sulfur) exists in the presence of excess carbon, microorganisms can assimilate the carbon source and store it as hydroxyalkanoates to later polymerize it and accumulate them intracellularly as polymeric granules, which are considered secondary metabolites (Akinmulewo and Nwinyi, 2019; Kourmentza et al., 2017).

The final product presents interesting properties for processing, these properties depend directly on the chemical structure of the monomers that compose it. Besides, these structures are influenced by the carbon source, the microbial strain used and by the addition of precursors for subsequent modification (Kovalcik et al., 2019).

PHAs can be used in the medical and pharmaceutical fields due to their biodegradation and biocompatibility (Chen and Zhang, 2018; Voinova et al., 2019), besides, they can be used to replace some conventional petrochemical products in applications that include molded products (Chen, 2009), packaging (Bucci et al., 2005), geotextiles (Daria et al., 2020), additives (Latos-Brozio and Masek, 2020), etc. In Table 17.2, some applications of these bioplastics are listed, the relevance of the synthesized material and the source of PHA.

Among the characteristics that make the polymers of this family attractive is their biodegradability, which can occur in aerobic and anaerobic environments, without the production of toxic sub-products. The degradation of PHA occurs mainly under aerobic conditions. It is carried out if the phosphate, salts, temperature and humidity promote the growth of bacteria and fungi, which modify the properties of the polymer, deteriorating it. These microorganisms subsequently excrete depolymerase that degrades these bioplastics into oligomers and monomers. Finally, these low-molecular-weight molecules are metabolized as carbon and energy sources (Giubilini et al., 2021). These conditions are hardly found during normal use, which means that degradation occurs after they have been used (Anjum et al., 2016; Winnacker, 2019).

The industrial production of polyhydroxyalkanoate is not yet on a large scale due to the high cost of production compared to synthetic plastics. In addition to this, they present limited thermo-mechanical stability, instability when exposed to various environments and require sterile conditions for fermentation. For this reason, it is necessary to continue investigating synthesis routes and conditions to make these materials competitive on an industrial scale.

7. Poly-lactid acid (PLA)

Polylactic acid (PLA) is a thermoplastic biopolymer whose precursor molecule is lactic acid. The primary basic monomeric unit of PLA is lactide, which can have either D-lactide or L-lactide chirality. The biodegradability and mechanical properties of PLA depend on the enantiomer used and on whether the polymer is amorphous or semi-crystalline. The racemic mixture of enantiomers and the D form degrades faster than the L form (da Silva et al., 2018).

Among different conventional biopolymers, PLA is one of most widely used and promising environmental friendly alternatives because the monomer of PLA-lactide (LA) is extracted from

Table 17.2. PHA applications.

Application	Relevance	Source of PHA	Reference
Self-healing concrete	The material obtained can induce crack healing in concrete specimens.	Biomass	(Vermeer et al., 2021)
Biofuels	Methyl esters derived from PHA's could be used as fuel additives for other fuels such as propanol, butanol, gasoline and diesel.	Acid-catalyzed hydrolysis	(Zhang et al., 2009)
Wound dressing	The use of nonwoven membranes of degradable PHA derivatives as atraumatic wound dressings can reduce inflammation, enhance the angiogenic properties of the skin and facilitate its healing	Microbial culture	(Shishatskaya et al., 2016)
Food Packing	PHAs and their films present new properties as the decrease in melting temperature, the increase in the elongation at break and permeability rates.	Biomass	(Pérez-Arauz et al., 2019)
Tissue engineering	Surface modified PHA fibers with bacterial membrane produce non-cytotoxic and anti-adhesive fiber meshes for tissue regeneration purposes.	Microorganisms	(Piarali et al., 2020)
Bone Scaffolds	Composites obtained induce better cytocompatibility and osteoconductivity compared with controls and pure PHB.	Commercial	(Hajiali et al., 2012)
Drug delivery	A novel biocompatible, and biodegradable PHA nano-platform vehicle improves an effective and safe treatment to reduce the side effects of azathioprine.	Commercial	(Hu et al., 2020)
Biosensors	The biosensor was used for the determination of artemisinin in bulk and spiked human serum.	Microorganism	(Phukon et al., 2014)

natural sources (starch, cellulose, kitchen waste and fish waste) (Casalini et al., 2019; Jin et al., 2019). For the synthesis of Poly (Lactic Acid) PLA, Ring-Opening Polymerization (ROP) is the most used method (Dubey et al., 2017).

Furthermore, due to its ability to biodegrade, PLA has properties that compare favorably with plastics commonly used, for example, for packaging. This is an important factor because it allows PLA to replace polymers obtained from petroleum without redesigning products or investing heavily in new processing equipment (Malinconico et al., 2018).

The characteristics or properties that crystalline PLA has are high scratch resistance, compostability, high Young's modulus, notable hydrophilicity, high transparency, low impact strength and low heat resistance (Kühnert et al., 2018). In order to improve those properties and the processing of PLA is usually compounded and/or blended by the use of different additive packages and fillers. PLA can be formulated to be rigid or flexible and can be copolymerized with other materials. It can be made with various mechanical characteristics depending on the manufacturing process followed (Malinconico et al., 2018).

Biodegradability and remarkable properties of PLA make of it a great material for several applications, such as textile industry, medical and pharmaceutical industries, agriculture, packing, among others.

In the textile field, several researches had been made in fibers for clothing, carpets, sheets and towels and wall coverings. For example, in 2020 Kudzin et al., worked on obtaining antimicrobial hybrid materials consisting of poly(lactide) nonwoven fabrics and using phosphoro-organic compound (Fosfomycin) as a coating and modifying agent and was proved as antibacterial material (Kudzin and Mrozinska, 2020). In 2017, Sanatgar et al., proposed the manufacturing of polylactic acid nanocomposite 3D printer filaments for smart textile applications. Their aim was to develop an integrated or tailored production process for smart and functional textiles which avoid unnecessary use of water, energy, chemicals and minimize the waste to improve ecological footprint and productivity (Hashemi Sanatgar et al., 2017).

Another important field where PLA-based materials has been used is in automotive industry where manufacturers are tending to replace traditional materials such as metals and metal alloys for lightweight materials such as plastics and composites (Bouzouita et al., 2017; Macke et al., 2014). For example, Kumar and Das in 2017, published a work about fibrous biocomposites prepared by using nettle and poly(lactic acid) fibers and employing carding and compression-molding processes, the aim of this research was the use this composite for automotive dashboards panels. ÖZTÜRK, in 2020, published a study about the weight reduction of automotive components using hemp as Non-woven Natural Fiber (NNF) and Polylactic acid (PLA) non-woven fibers, the outcome of this research is to replace conventional reinforcement materials with renewable sources that have good properties and a cost benefit.

Another application that PLA has is in the packing field, packaging materials must extend shelf-life of the product preventing its deterioration due to physicochemical or biological factors and preserve or increase the overall quality and safety during storage and handling. In addition, after their useful life, it is desirable that the materials biodegrade in a reasonable time period to avoid environmental waste disposal problems (Armentano et al., 2013).

As mentioned before, the use of PLA or PLA-based materials is not only highly desired for being friendly to the environment or for its biodegradability and good mechanical properties, but it also has a high degree of biocompatibility which makes it a material that can be used for medical applications, such as drug delivery systems, tissue engineering, temporary and long-term implantable devices; constantly expanding to new fields (da Silva et al., 2018). In addition, the production costs are affordable, which is why PLA has become a very popular material.

Within the applications of PLA in the field of medicine one can find various investigations, in the year 2000, Tamai et al., reported the results of the use of biodegradable PLLA coronary stents in humans, the results indicated that coronary PLLA biodegradable stents are feasible, safe and effective in humans (Tamai et al., 2000). PLA based materials as copolymers or composites had been used for orthopedic implantable devices in the body. Copolymers of PLA and PGA (polyglycolic acid) are more useful than homopolymers of PLA and PGA because their rate of degradation can be adjusted according the final applications, besides their biocompatibility and nontoxicity (Rebelo et al., 2017).

In 2018, Mao et al., made the preparation of PLA/graphene oxide nanofiber membranes which displayed promising potential in the biomedical field, particularly as scaffolds for drug delivery and tissue engineering. The results showed that the addition of GO not only significantly improved the thermal stability and mechanical properties of the PLA nanofiber membranes, but also promoted the cumulative release and release rate of RhB (Rhodamine B, organic dye) from nanofiber membranes (Mao et al., 2018).

Since the Food and Drug Administration (FDA) has approved PLA as a harmless material to be implanted in the human body, research is also currently being carried out where PLA-based materials are developed for 3D printing (three-dimensional) of bone implants. The use of this technology has great advantages over implants manufactured by conventional methods, since it allows the customization of the implant for the needs of each patient, in addition to the previously mentioned advantages of the material (Chen et al., 2020)

As can be seen, there are many applications of PLA, PLA-based materials, however there is still much to explore in the use of this material, so the scientific community strives to achieve prominent achievements for the application of PLA-based products for its different applications.

8. Starch

Starch is a natural polymer produced by plants during photosynthesis as an energy reserve. It is found mainly in roots, stalks and seeds. The most important industrial starch sources are corn, wheat, potato, tapioca and rice. Starch is the principal food reserve of plants, and it is synthesized during photosynthesis. Starch is a mixture of two polysaccharides, amylose and amylopectin. Amylose is a linear polymer with α(1–4) glycosidic linkages. While amylopectin is a branched polymer with α(1–4) as well as α(1–6) glycosidic linkages. Both polymers occur naturally in small structures known as starch granules. Many of the chemical and physical properties of starch depend on its granular nature (Whistler and Daniel, 2000).

Starch acquires thermoplastic behavior by adding a plasticizer. Glycerol is the most common plasticizer. However, researchers recently found environmentally friendly solvents such as ionic liquids and deep eutectic solvents are suitable plasticizers for starch. Ionic liquids and deep eutectic solvents are highly efficient for the denaturation of native starch due to their strong ability to form hydrogen bonds. Ionic liquids and deep eutectic solvents depress the glass transition of plasticized starch and hinder starch recrystallization during storage. Thus, the plasticized starch has more stable thermomechanical properties (Colomines et al., 2016). Another way to favor starch plasticization is by adding cellulose nanoparticles. This reinforcement material increased the rigidity of starch films, as well as the thermal stability and moisture resistance, which is used for the generation of products for daily use that present biodegradation (Montero et al., 2017).

The most common uses and applications of thermoplastic starch are food packaging. Starch-based materials are suitable for replacing petroleum-derived plastics used for short-life applications. The starch properties are diverse depending on the extraction source. Zolek Trysnowska and Kaluza studied the effect of starch source on the mechanical and surface properties on the packaging performance. They prepared films from starches extracted from maize, potato, oat, rice and tapioca using 50% weight (%w) of glycerin as a plasticizer. The study revealed that starch-based films made from oat and tapioca exhibited the lowest tensile strength. While the highest values of tensile strength were observed for films made from potato starch. However, these values are 10 times lower than those for modern biodegradable packaging films derived from PBS, PCL and PBAT. Further, the oat and tapioca films exhibited the highest contact angle values which are related to lower wettability and lower hydrophilicity (Zolek-Trysnowska and Kaluza, 2021).

The differences in the mechanical properties of starches extracted from different sources are attributed to the amylose/amylopectin ratio. Montero and co-workers found that higher contents of amylose increase the rigidity of the polymer matrix and the water absorption capacity (Montero et al., 2017). This is because, the ratio of amylose and amylopectin affects the starch structure in terms of crystallinity, size of the granules and chemical nature and arrangement of polymers within the granule (Nawas et al., 2020).

Tzankova and La Mantia studied the photo-oxidation and the photo-stabilization of a commercial biodegradable polymer by measuring the mechanical properties as a function of photo-oxidation time. They showed that a commercial biodegradable polymer, made from maize starch, and a synthetic polyester shows poor resistance to UV irradiation. This implies a fast decay of the elongation at break. However, the presence of small amounts of conventional UV stabilizers such as benzophenone strongly improves the durability of this polymer (Dintcheva and La Mantia, 2007).

Studies performed by Ochi (2010) on the biodegradation mechanism of starch-based composites reinforced with manila hemp fibers showed that decomposition occurs first on the surface material. Decomposition of the matrix exposes fibers and the decomposition of the fiber-resin interphase results in the formation of interfacial gaps. Finally, substantial decomposition of the starch matrix and fibers occurs.

It is important to consider that reinforcement of the materials not only affect the mechanical, surface and barrier properties of starch. This also can affect the biodegradability of the biopolymer. Kwaśniewska et al. found a significant influence of kaolin on the strength and thermal and barrier properties of composite films with starch. Kaolin additives increased the barrier properties of water vapor in composite films when 9% w is used. However, an increment in kaolin content reduced the tensile strength, Young's modulus and Poisson's ratio. Further, the reinforcement with nano-clay reduced the thermal stability of composite films by 7% w and could accelerate the biodegradation process (Kwaśniewska et al., 2020).

Starch is a very water-sensitive material and its film properties are dependent on the moisture content. Starch films exhibit relatively low mechanical resistance. Blending with biodegradable polyesters is a common strategy to improve the film properties. Poly (lactic acid) films are very brittle and offer low resistance to oxygen permeation. However, their combination as blend or multilayer films with starch provide properties that are more adequate for packaging purposes based on their complementary characteristics (Muller et al., 2017). Starch blending with polyesters has a noticeable effect on biodegradability. Cho et al. found that poly(caprolactone)/starch blend was easily degraded with 88% biodegradability in 44 days under aerobic conditions. While the biodegradability under anaerobic conditions was 83% after 139 days. Conversely, the biodegradability of poly(butylene succinate) (PBS) was 31% in 80 days, and 2% in 100 days under aerobic and anaerobic conditions respectively (Cho et al., 2011). The biodegradation mechanism of plastic products derived from starch involves the generation of gaseous by-products and monomers based on glucose, by the action of enzymes such as α- and β-amylases, debranching enzymes, starch phosphorylase, cellulases, glucosidases and disproportionate enzyme that are present in microorganisms and plants.

9. Applications

Biodegradable polymers can be classified in natural and synthetic polymers. Among natural bioplastics are cellulose and starch derivatives, polylactic acid (PLA) and polyhydroxyalkanoates (PHA), while synthetic polymers from fossil fuels but biodegradables are PBAT, PBS and PCL. All of them are good alternatives for replacing non-biodegradable plastics such as polystyrene (PS), polyethylene (PE) and polypropylene (PP) (Rameshkumar et al., 2020).

Bioplastics are defined as polymers biodegradable by enzymatic action of microbes. The worldwide production of bioplastics is located in Asia which contributes with the 63.1%, followed by North America with a 13.5%, then Europe with a 13%, and finally, South America with a 10.4%. The global production of biodegradable polymers is led by PLA, in second place are biodegrable polyesters such as PBAT, followed by biodegradable starch blends and PHA (Mangaraj et al., 2019).

Packaging materials from petrochemicals such as PVC (polyvinyl chloride), PET (Polyethylene terephthalate), PS, PP and PA (polyamide); which are excellent for food packaging due to their low cost, excellent mechanical (tensile strength) and transmission properties (O_2 and CO_2 diffusion) that increase the shelf-life of the product. Some choices to replace them are biodegradable polymer than can be classified in (i) polymers which are directly extracted or removed from biomass: starch, cellulose, casein and gluten; (ii) polymeric materials which are synthesized by polymerization: PCL, PLA, PBS, PBAT; and (iii) Polymers which are produced by microorganisms or genetically modified bacteria: PHAs (Mangaraj et al., 2019).

Bioplastics can be processed in packaging films, laminated paper, trays, cups cutlery items by processing techniques like cast films, blow molding, co-extruded films. However, the shelf-life of any product depends on the barrier properties of the packaging material, thus, gas phase permeation of O_2, CO_2 and water steam is often measured (Mangaraj et al., 2019).

The biopolymers applied in biomedical applications are polyglycolic acid/or polyglycolide (PGA), Poly (lactic-co-glycolic acid) (PLGA), Poly(N-isopropylacrylamide) (PNIPAAm), polyvinyl alcohol (PVA), polyesters, PCL, and PLA. The latter is bioadsorbable and thermoplastic useful characterisitics for medical implants (screws, pins, rods, orthopedic device and as mesh). PCL has been used in medical implants, dental splints, targeted drug delivery and tissue engineering. PNIPAAm is a thermosensitive polymer applied as biosensors, tissue engineering and drug delivery. The natural polymers for biomedical applications are polysaccharides (e.g., chitin, chitosan and alginate) and proteins (collagen and gelatin) (Asghari et al., 2017).

Mechanical, degradation, biocompatibility and processability properties, as well as shelf-life/ stability and cost are determinant characteristics of biodegradable polymer for tissue and bone regeneration. In the case of drug delivery, the time of release is governed by the type of polymer (Doppalapudi et al., 2014).

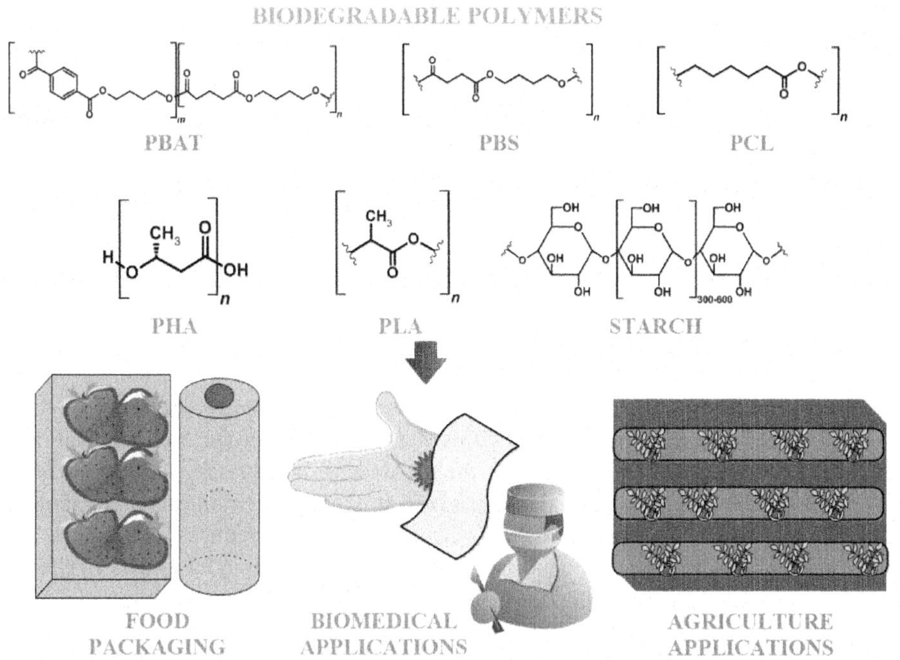

Figure 17.2. Some applications of biodegradable polymers. Own authorship.

In agriculture application, mulch films are produced from synthetic and biodegradable polymers such as PCL, PLA and PVA. The films produced should have mechanical resistance, be transparent to visible light, and opaque to infrared radiation (Puoci et al., 2008). It is also important to supply macronutrients (N, P, K, Ca, Mg and S) and micronutrients (B, Cl, Co, Cu, Fe, Mn, Mo, Ni and Zn) to the plants, another application are agrochemical delivery of fertilizers and herbicides (Fig. 17.2). There are two approaches of the application of bioplastics combined with agrochemical agents, one is the encapsulation or agrochemical agents into the bioplastic matrix, and the second one, the chemical combination of the agrochemical agent with the biodegradable polymer, where the polymer serves a carrier (Milani et al., 2017; Puoci et al., 2008). The biopolymers used for agrochemicals encapsulation are natural rubber, polysaccharides and cellulose-based materials (Milani et al., 2017).

In addition, the lack of water and the desertification of many lands, a possible solution is the use of superadsorbent polymers (SAPs). These SAPs have helped to reduce the irrigation water consumption and the death rate of plants, improve fertilizer retention in the soil, and increase plant growth rate. The main objective in the research of SAPs is their modification with copolymers to enhance their absorbency, gel strength and absorption rate. SAPs are commonly made of a non-biodegradable like polyacrylamide polymer however, it could be replaced by superadsorbent hydrogels made of the superadsorbent and biodegradable alginate (Milani et al., 2017; Puoci et al., 2008).

Finally, the uncontrolled used of agrochemical agents have caused the pollution of soil and water by heavy metals. A solution for this inconvenience is the biosorption of the heavy metals by a biopolymer, where the most common biosorbents is cellulose that serves as a metal ion sequestrant due to its high amount of hydroxyl groups than can be easily functionalized to obtain more polar or ionizable pendant groups to enhance the decontaminating water process. Nonetheless, all the possible environmental benefits for bioplastics applied in agriculture, it is still necessary for a more deeper understanding of the polymer degradation processes in soil and aqueous medium (Milani et al., 2017).

10. Conclusion

Biodegradable polymers or biopolymers are not always as biodegradable as they claim to be, their biodegradability rests with the environment, this means that it depends on environmental conditions (temperature, humidity, solar radiation, type of soil, microorganisms, etc.), that the polymer can degrade into smaller chains that can be assimilated by the ecosystem. However, as seen in this chapter, the current trend for the use of biodegradable polymers has been on the rise in recent decades. The development of materials based on biodegradable polymers must be a multidisciplinary collaboration to achieve all the desirable characteristics for these types of materials, that has been happening, which is why more and more research with more versatile applications for this type of materials are being released.

The applications in which these materials can be used include a wide range of possibilities, they can be used in textile technology, agriculture, medicine, drug delivery, bone implants (bone and teeth), tissue implants, automotive industry, packaging and storage of food, among others. And the biggest challenge continues to be finding the right materials for each application without compromising performance. For example, in the automotive industry, it is sought to lighten the weight of the car, which is why parts that were conventionally made of metal are replaced with biodegradable polymers, which in addition to lightening the vehicle, reduces costs and ensures good mechanical properties. On the other hand, for biomedical applications, it is also desired that the materials have good mechanical properties, as well as that they are biocompatible and that they will not generate problems in the organism that is hosting the material. In the area of food preservation (packaging) it is sought that the material is permeable and that it allows the preservation of the food (longer shelf life) without compromising the characteristics of the food.

As discussed in this chapter, there are materials that can be used for these purposes, some are natural biopolymers such as starch, cellulose and lignin, among others, while others can be synthetic, some of which are obtained in the same way from sources that are natural.

According to what is proposed in this work, biodegradable polymers can become the basic plastics for the future, obviously if biodegradability is something inherent for the application that will be used, as an example, in the medical area, however, it is a matter of changing the habits of consumption and use of conventional polymers.

Acknowledgements

The authors thank to Consejo Nacional de Ciencia y Tecnología (CONACyT) for the financial support (grant FORDECYT-PRONACES/6660/2020).

References

Abudula, T., Saeed, U., Memic, A., Gauthaman, K., Asif Hussain, M. and Al-Turaif, H. (2019). Electrospun cellulose Nano fibril reinforced PLA/PBS composite scaffold for vascular tissue engineering. *Journal of Polymer Research*, 26(110): 1–15. https://doi.org/https://doi.org/10.1007/s10965-019-1772-y.

Agarwal, S. (2020). Biodegradable polymers: present opportunities and challenges in providing a microplastic-free environment. *Macromolecular Chemistry and Physics*, 221(6): 1–7. https://doi.org/10.1002/macp.202000017.

Akaraonye, E., Keshavarz, T. and Roy, I. (2010). Production of polyhydroxyalkanoates: The future green materials of choice. *Journal of Chemical Technology and Biotechnology*, 85(6): 732–743. https://doi.org/10.1002/jctb.2392.

Akhilesh Kumar Pal, M., Mohan Bhasney, S., Bhagabati, P. and Katiyar, V. (2018). Effect of dicumyl peroxide on a poly(lactic acid) (PLA)/poly(butylene succinate) (PBS)/functionalized chitosan-based nanobiocomposite for packaging: a reactive extrusion study [Research-article]. *ACS Omega*, 3: 13298–13312. https://doi.org/10.1021/acsomega.8b00907.

Akinmulewo, A. B. and Nwinyi, O. C. (2019). Polyhydroxyalkanoate: A biodegradable polymer (a mini review). *Journal of Physics: Conference Series*, 1378(4): 0–12. https://doi.org/10.1088/1742-6596/1378/4/042007.

Alias, N. F. and Marsilla, K. I. K. (2020). Processes and characterization for biobased polymers from polybutylene succinate. pp. 151–170. *In*: Zhang, Y. (ed.). *Processing and Development of Polysaccharide-Based Biopolymers for Packaging Applications* (1st ed.). https://doi.org/10.1016/B978-0-12-818795-1.00006-X.

Anjum, A., Zuber, M., Zia, K. M., Noreen, A., Anjum, M. N. and Tabasum, S. (2016). Microbial production of polyhydroxyalkanoates (PHAs) and its copolymers: A review of recent advancements. *International Journal of Biological Macromolecules*, 89: 161–174. https://doi.org/10.1016/j.ijbiomac.2016.04.069.

Armentano, I., Bitinis, N., Fortunati, E., Mattioli, S., Rescignano, N., Verdejo, R., ... Kenny, J. M. (2013). Multifunctional nanostructured PLA materials for packaging and tissue engineering. *Progress in Polymer Science*, 38(10–11): 1720–1747. https://doi.org/10.1016/j.progpolymsci.2013.05.010.

Asghari, F., Samiei, M., Adibkia, K., Akbarzadeh, A. and Davaran, S. (2017). Biodegradable and biocompatible polymers for tissue engineering application: a review. *Arcifical Cells, Nanomedicine, and Biotechnology An International Journal*, 45(2): 185–192. https://doi.org/10.3109/21691401.2016.1146731.

Ayu, R. S., Abdan, K., H, A. S., Khairul, Z. and Mohd, N. (2020). Effect of empty fruit brunch reinforcement in polybutylene-succinate/modified tapioca starch blend for agricultural mulch films. *Scientific Reports*, 10(1116): 1–7. https://doi.org/10.1038/s41598-020-58278-y.

Bartnikowski, M., Dargaville, T. R. and Hutmacher, D. W. (2019). Degradation mechanisms of polycaprolactone in the context of chemistry, geometry and environment. *Progress in Polymer Science*, 96: 1–20. https://doi.org/10.1016/j.progpolymsci.2019.05.004.

Belchior Rocha, D., Souza De Carvalho, J., Aparecida de Oliveira, S. and dos Santos Rosa, D. (2018). A new approach for flexible PBAT/PLA/$CaCO_3$ films into agriculture. *Journal of Applied Polymer Science*, 46660: 1–9. https://doi.org/10.1002/app.46660.

Bheemaneni, G. and Kandaswamy, R. (2019). Melt processing and characterization of tricalcium phosphate filled polybutylene adipate-co-terephthalate/polymethyl methacrylate composites for biomedical applications. *International Journal of Polymeric Materials and Polymeric Biomaterials*, 4037. https://doi.org/10.1080/00914037.2018.1525731.

Borghesi, D. C., Gabriela, M., Campos, N. and Husk, C. (2016). Biodegradation study of a novel poly-caprolactone-coffee husk composite film. *Materials Research*, 19(4): 752–758. https://doi.org/http://dx.doi.org/10.1590/1980-5373-MR-2015-0586©.

Boumail, A., Salmieri, S., Klimas, E., Tawema, P. O., Bouchard, J. and Lacroix, M. (2013). Characterization of trilayer antimicrobial diffusion films (ADFs) based on methylcellulose–polycaprolactone composites. *Journal of Agricultural and Food Chemistry*, 61(4): 811–821. https://doi.org/dx.doi.org/10.1021/jf304439s.

Bouzouita, A., Notta-cuvier, D., Raquez, J., Lauro, F. and Dubois, P. (2017). Poly (lactic acid)-based materials for automotive applications. pp. 177–219. *In*: Di Lorenzo, M. L. and Androsch, R. (eds.). *Industrial Applications of Poly(lactic acid). Advances in Polymer Science* (1st ed.). https://doi.org/10.1007/12.

Bucci, D. Z., Tavares, L. B. B. and Sell, I. (2005). PHB packaging for the storage of food products. *Polymer Testing*, 24(5): 564–571. https://doi.org/10.1016/j.polymertesting.2005.02.008.

Casalini, T., Rossi, F., Castrovinci, A. and Perale, G. (2019). A perspective on polylactic acid-based polymers use for nanoparticles synthesis and applications. *Frontiers in Bioengineering and Biotechnology*, 7(October): 1–16. https://doi.org/10.3389/fbioe.2019.00259.

Chen, G. Q. (2009). A microbial polyhydroxyalkanoates (PHA) based bio- and materials industry. *Chemical Society Reviews*, 38(8): 2434–2446. https://doi.org/10.1039/b812677c.

Chen, G. Q. and Zhang, J. (2018). Microbial polyhydroxyalkanoates as medical implant biomaterials. *Artificial Cells, Nanomedicine and Biotechnology*, 46(1): 1–18. https://doi.org/10.1080/21691401.2017.1371185.

Chen, X. and Yan, N. (2020). A brief overview of renewable plastics. *Materials Today Sustainability*, 7–8: 100031. https://doi.org/10.1016/j.mtsust.2019.100031.

Chen, Xibao, Chen, G., Wang, G., Zhu, P. and Gao, C. (2020). Recent progress on 3D-printed polylactic acid and its applications in bone repair. Advanced Engineering Materials, 22(4). https://doi.org/10.1002/adem.201901065.

Chia, W. Y., Ying Tang, D. Y., Khoo, K. S., Kay Lup, A. N. and Chew, K. W. (2020). Nature's fight against plastic pollution: Algae for plastic biodegradation and bioplastics production. *Environmental Science and Ecotechnology*, 4: 100065. https://doi.org/10.1016/j.ese.2020.100065.

Cho, H. S., Moon, H. S., Kim, M., Nam, K. and Kim, J. Y. (2011). Biodegradability and biodegradation rate of poly(caprolactone)-starch blend and poly(butylene succinate) biodegradable polymer under aerobic and anaerobic environment. *Waste Management*, 31(3): 475–480. https://doi.org/10.1016/j.wasman.2010.10.029.

Colomines, G., Decaen, P., Lourdin, D. and Leroy, E. (2016). Biofriendly ionic liquids for starch plasticization: A screening approach. *RSC Advances*, 6(93): 90331–90337. https://doi.org/10.1039/c6ra16573g.

da Silva, D., Kaduri, M., Poley, M., Adir, O., Krinsky, N., Shainsky-Roitman, J. and Schroeder, A. (2018). Biocompatibility, biodegradation and excretion of polylactic acid (PLA) in medical implants and theranostic systems. *Chemical Engineering Journal*, 340: 9–14. https://doi.org/10.1016/j.cej.2018.01.010.

Daria, M., Krzysztof, L. and Jakub, M. (2020). Characteristics of biodegradable textiles used in environmental engineering: A comprehensive review. *Journal of Cleaner Production*, 268. https://doi.org/10.1016/j.jclepro.2020.122129.

Dintcheva, N. T. and La Mantia, F. P. (2007). Durability of a starch-based biodegradable polymer. *Polymer Degradation and Stability*, 92(4): 630–634. https://doi.org/10.1016/j.polymdegradstab.2007.01.003.

Domínguez-Robles, J., Larrañeta, E., Leon, M., Martin, N. K., Irwin, N. J., Mutjé, P., … Delgado-aguilar, M. (2020). Lignin/poly (butylene succinate) composites with antioxidant and antibacterial properties for potential biomedical applications. *International Journal of Biological Macromolecules*, 145: 92–99. https://doi.org/10.1016/j.ijbiomac.2019.12.146.

Doppalapudi, S., Jain, A., Khan, W. and Domb, A. J. (2014). Biodegradable polymers—an overview†. *Polymers Advanced Technology*, 25: 427–435. https://doi.org/10.1002/pat.3305.

Dubey, S. P., Thakur, V. K., Krishnaswamy, S., Abhyankar, H. A., Marchante, V. and Brighton, J. L. (2017). Progress in environmental-friendly polymer nanocomposite material from PLA: Synthesis, processing and applications. *Vacuum*, 146: 655–663. https://doi.org/10.1016/j.vacuum.2017.07.009.

Ferreira, F. V, Cividanes, L. S., Gouveia, R. F., Lona, L. M. F. and Cnpem, M. (2017). An overview on properties and applications of poly (butylene adipate-co-terephthalate)–PBAT based composites. *Polymer Engineering and Science*, 59: 1–9. https://doi.org/10.1002/pen.24770.

França, D. C., Almeida, T. G., Abels, G., Canedo, E. L., Carvalho, L. H., Wellen, R. M. R., … Tailoring, K. K. (2018). Tailoring PBAT/PLA/Babassu films for suitability of agriculture mulch application mulch application. *Journal of Natural Fibers*, 1–11. https://doi.org/10.1080/15440478.2018.1441092.

Gigli, M., Fabbri, M., Lotti, N., Gamberini, R., Rimini, B. and Munari, A. (2016). Poly (butylene succinate)-based polyesters for biomedical applications: a review. In Memory of our Beloved Colleague and Friend Dr. Lara Finelli. *European Polymer Journal*, (1134): 249–255. https://doi.org/10.1016/j.eurpolymj.2016.01.016.

Giubilini, A., Bondioli, F., Messori, M., Nyström, G. and Siqueira, G. (2021). Advantages of additive manufacturing for biomedical applications of polyhydroxyalkanoates. *Bioengineering*, 8(2): 1–31. https://doi.org/10.3390/bioengineering8020029.

Guarino, V., Gentile, G., Sorrentino, L. and Ambrosio, L. (2017). Polyprolactone: Synthesis, properties, and applications. pp. 1–36. *In*: *Encyclopedia of Polymer Science and Technology*. https://doi.org/10.1002/0471440264.pst658.

H. Kudzin, M. and Mrozinska, Z. (2020). Biofunctionalization of textile materials. 2. Antimicrobial modification of poly(lactide) (PLA) Nonwoven Fabricsby Fosfomycin. *Polymers*, 12(768): 1–16. https://doi.org/doi:10.3390/polym12040768.

Hadj-Hamou, A. S., Metref, F. and Yahiaoui, F. (2017). Thermal stability and decomposition kinetic studies of antimicrobial PCL/nanoclay packaging films. *Polymer Bulletin*, 74: 3833–3853. https://doi.org/10.1007/s00289-017-1929-y.

Haider, T. P., Völker, C., Kramm, J., Landfester, K. and Wurm, F. R. (2019). Plastics of the future? The impact of biodegradable polymers on the environment and on society. *Angewandte Chemie - International Edition*, 58(1): 50–62. https://doi.org/10.1002/anie.201805766.

Hajiali, H., Hosseinalipour, M., Karbasi, S. and Shokrgozar, M. A. (2012). The influence of bioglass nanoparticles on the biodegradation and biocompatibility of poly (3-hydroxybutyrate) scaffolds. *International Journal of Artificial Organs*, 35(11): 1015–1024. https://doi.org/10.5301/ijao.5000119.

Hashemi Sanatgar, R., Cayla, A., Campagne, C. and Nierstrasz, V. (2017). Manufacturing of polylactic acid nanocomposite 3D printer filaments for smart textile applications. *IOP Conference Series: Materials Science and Engineering*, 254(7): 3–7. https://doi.org/10.1088/1757-899X/254/7/072011.

Hong, M. and Chen, E. Y. X. (2019). Future directions for sustainable polymers. *Trends in Chemistry*, 1(2): 148–151. https://doi.org/https://doi.org/10.1016/j.trechm.2019.03.004.

Hu, J., Wang, M., Xiao, X., Zhang, B., Xie, Q., Xu, X., … Zhang, X. (2020). A novel long-acting azathioprine polyhydroxyalkanoate nanoparticle enhances treatment efficacy for systemic lupus erythematosus with reduced side effects. *Nanoscale*, 12(19): 10799–10808. https://doi.org/10.1039/d0nr01308k.

Jian, J., Xiangbin, Z. and Xianbo, H. (2020). An overview on synthesis, properties and applications of poly (butylene-adipate-co-terephthalate)-PBAT. *Advanced Industrial and Engineering Polymer Research*, 3(1): 19–26. https://doi.org/10.1016/j.aiepr.2020.01.001.

Jin, F., Hu, R. and Park, S. (2019). Improvement of thermal behaviors of biodegradable poly(lactic acid) polymer: A review. *Composites Part B*, 164: 287–296. https://doi.org/10.1016/j.compositesb.2018.10.078.

Komur, B., Bayrak, F., Ekren, N., Eroglu, M. S., Oktar, F. N., Sinirlioglu, Z. A., … Guler, O. (2017). Starch/PCL composite nanofibers by co-axial electrospinning technique for biomedical applications. *BioMedical Engineering OnLine*, 16(40): 1–13. https://doi.org/10.1186/s12938-017-0334-y.

Kourmentza, C., Plácido, J., Venetsaneas, N., Burniol-Figols, A., Varrone, C., Gavala, H. N. and Reis, M. A. M. (2017). Recent advances and challenges towards sustainable polyhydroxyalkanoate (PHA) production. *Bioengineering*, 4(2): 1–43. https://doi.org/10.3390/bioengineering4020055.

Kovalcik, A., Obruca, S., Fritz, I. and Marova, I. (2019). Polyhydroxyalkanoates: Their importance and future. *BioResources*, 14(2): 2468–2471. https://doi.org/10.15376/biores.14.2.2468-2471.

Kühnert, I., Spörer, Y., Brünig, H., Hoai An Tran, N. and Rudolph, N. (2018). Processing of poly (lactic acid). pp. 1–33. *In*: Di Lorenzo, M. L. and Androsch, R. (eds.). *Industrial Applications of Poly(lactic acid). Advances in Polymer Science*. https://doi.org/10.1007/12.

Kumar, N. and Das, D. (2017). Fibrous biocomposites from nettle (Girardinia diversifolia) and poly(lactic acid) fibers for automotive dashboard panel application. *Composites Part B: Engineering*, 130: 54–63. https://doi.org/10.1016/j.compositesb.2017.07.059.

Kwaśniewska, A., Chocyk, D., Gładyszewski, G., Borc, J., Świetlicki, M. and Gładyszewska, B. (2020). The influence of kaolin clay on the mechanical properties and structure of thermoplastic starch films. *Polymers*, 12(1): 1–10. https://doi.org/10.3390/polym12010073.

Latos-Brozio, M. and Masek, A. (2020). The effect of natural additives on the composting properties of aliphatic polyesters. *Polymers*, 12(9). https://doi.org/10.3390/POLYM12091856.

Macke, A., Schultz, B. F., Rohatgi, P. K. and Gupta, N. (2014). Metal matrix composites for automotive applications. *In*: Elmarakbi, A. (ed.). *Advanced Composite Materials for Automotive Applications: Structural Integrity and Crashworthiness*. https://doi.org/10.1002/9781118535288.ch13.

Malinconico, M., Vink, E. T. H. and Cain, A. (2018). Applications of poly (lactic acid) in commodities and specialties. pp. 35–50. *In*: Di Lorenzo, M. L. and Androsch, R. (eds.). *Industrial Applications of Poly(lactic acid). Advances in Polymer Science* (1st ed.). https://doi.org/10.1007/12.

Mandal, P. and Shunmugam, R. (2020). Polycaprolactone: a biodegradable polymer with its application in the field of self-assembly study. *Journal of Macromolecular Science, Part A*, 58(2): 1–19. https://doi.org/10.1080/10601325.2020.1831392.

Mangaraj, S., Yadav, A., Bal, L. M., Dash, S. K. and Mahanti, N. K. (2019). Application of biodegradable polymers in food packaging industry: a comprehensive review. *Journal of Packaging Technology and Research*, 3: 77–96. https://doi.org/10.1007/s41783-018-0049-y.

Mao, Z., Li, J., Huang, W., Jiang, H., Zimba, B. L., Chen, L., … Wu, Q. (2018). Preparation of poly(lactic acid)/graphene oxide nanofiber membranes with different structures by electrospinning for drug delivery. *RSC Advances*, 8(30): 16619–16625. https://doi.org/10.1039/c8ra01565a.

Milani, P., França, D., Balieiro, A. G. and Faez, R. (2017). Polymers and its applications in agriculture. *Polímeros*, 27(3): 256–266. https://doi.org/http://dx.doi.org/10.1590/0104-1428.09316 Polymers.

Mohamed, R. M. and Yusoh, K. (2016). A review on the recent research of polycaprolactone (PCL). *Advnced Materials Research*, 1134: 249–255. https://doi.org/10.4028/www.scientific.net/AMR.1134.249.

Montero, B., Rico, M., Rodríguez-Llamazares, S., Barral, L. and Bouza, R. (2017). Effect of nanocellulose as a filler on biodegradable thermoplastic starch films from tuber, cereal and legume. *Carbohydrate Polymers*, 157: 1094–1104. https://doi.org/10.1016/j.carbpol.2016.10.073.

Moustafa, H., Guizani, C., Dupont, C., Martin, V., Jaguirim, M. and Dufrese, A. (2017). Utilization of Torre filed coffee grounds as reinforcing agent to produce high-quality biodegradable pbat composites for food packaging applications. *Sustainable Chemistry & Engineering*, 5: 1906–1916. https://doi.org/10.1021/acssuschemeng.6b02633.

Muller, J., González-Martínez, C. and Chiralt, A. (2017). Combination of poly(lactic) acid and starch for biodegradable food packaging. *Materials*, 10(8): 1–22. https://doi.org/10.3390/ma10080952.

Narancic, T. and O'Connor, K. E. (2019). Plastic waste as a global challenge: Are biodegradable plastics the answer to the plastic waste problem? *Microbiology (United Kingdom)*, 165(2): 129–137. https://doi.org/10.1099/mic.0.000749.

Nawas, H., Waheed, R., Nawas, M. and Shahwar, D. (2020). Physical and chemical modifications in starch structure and reactivity. pp. 1–22. *In*: *Chemical Properties of Starch*.

Ochi, S. (2010). Durability of starch based biodegradable plastics reinforced with manila hemp fibers. *Materials*, 4(3): 457–468. https://doi.org/10.3390/ma4030457.

ÖZTÜRK, S. S. (2020). The Investigation of Polylactic Acid Based Natural Fiber Reinforced Biocomposites for Automotive Applications. *Kırklareli Üniversitesi Mühendislik ve Fen Bilimleri Dergisi*, 6(1): 21–31. https://doi.org/10.34186/klujes.626590.

Pellis, A., Malinconico, M., Guarneri, A. and Gardossi, L. (2021). Renewable polymers and plastics: Performance beyond the green. *New Biotechnology*, 60: 146–158. https://doi.org/10.1016/j.nbt.2020.10.003.

Pérez-Arauz, A. O., Aguilar-Rabiela, A. E., Vargas-Torres, A., Rodríguez-Hernández, A. I., Chavarría-Hernández, N., Vergara-Porras, B. and López-Cuellar, M. R. (2019). Production and characterization of biodegradable films of a novel polyhydroxyalkanoate (PHA) synthesized from peanut oil. *Food Packaging and Shelf Life*, 20(April 2018): 100297. https://doi.org/10.1016/j.fpsl.2019.01.001.

Petchwattana, N., Covavisaruch, S., Wibooranawong, S. and Naknaen, P. (2016). Antimicrobial food packaging prepared from poly (butylene succinate) and zinc oxide. *Measurement*, 93: 442–448. https://doi.org/10.1016/j.measurement.2016.07.048.

Phukon, P., Radhapyari, K., Konwar, B. K. and Khan, R. (2014). Natural polyhydroxyalkanoate-gold nanocomposite based biosensor for detection of antimalarial drug artemisinin. *Materials Science and Engineering C*, 37(1): 314–320. https://doi.org/10.1016/j.msec.2014.01.019.

Piarali, S., Marlinghaus, L., Viebahn, R., Lewis, H., Ryadnov, M. G., Groll, J., … Roy, I. (2020). Activated polyhydroxyalkanoate meshes prevent bacterial adhesion and biofilm development in regenerative medicine applications. *Frontiers in Bioengineering and Biotechnology*, 8(May): 1–14. https://doi.org/10.3389/fbioe.2020.00442.

Puoci, F., Iemma, F., Spizzirri, U. G., Cirillo, G., Farmaceutiche, S., Calabria, U. and Cs, R. (2008). Polymer in agriculture: a review. *American Journal of Agricultural and Biological Sciences*, 3(1): 299–314. https://doi.org/10.3844/ajabssp.2008.299.314.

Qiu, S., Zhou, Y., Waterhouse, I. N., Gong, R., Xie, J. and Zhang, K. (2021). Optimizing interfacial adhesion in PBAT/PLA nanocomposite for biodegradable packaging films. *Food Chemistry*, 334(June 2020): 1–9. https://doi.org/10.1016/j.foodchem.2020.127487.

Rameshkumar, S., Shaiju, P., Connor, K. E. O. and Ramesh, Babu P. (2020). Bio-based and biodegradable polymers-state-of-the-art, challenges and emerging trends. *Current Opinion in Green and Sustainable Chemistry*, 21: 75–81. https://doi.org/10.1016/j.cogsc.2019.12.005.

Rebelo, R., Fernandes, M. and Fangueiro, R. (2017). Biopolymers in medical implants: a brief review. *Procedia Engineering*, 200: 236–243. https://doi.org/10.1016/j.proeng.2017.07.034.

Rujnić-Sokele, M. and Pilipović, A. (2017). Challenges and opportunities of biodegradable plastics: A mini review. *Waste Management and Research*, 35(2): 132–140. https://doi.org/10.1177/0734242X16683272.

Shishatskaya, E. I., Nikolaeva, E. D., Vinogradova, O. N. and Volova, T. G. (2016). Experimental wound dressings of degradable PHA for skin defect repair. *Journal of Materials Science: Materials in Medicine*, 27(11): 0–1. https://doi.org/10.1007/s10856-016-5776-4.

Shivam, P. (2016). Recent developments on biodegradable polymers and their future trends. *Int. Res. J. of Science & Engineering Int. Res. Journal of Science & Engineering*, 4(41): 17–26.

Singh Oberoi, I., Rajkumar, P. and Das, S. (2021). Disposal and recycling of plastics. *Materials Today: Proceedings*. https://doi.org/https://doi.org/10.1016/j.matpr.2021.02.562.

Sosa-Hernández, J. E., Villalba-Rodríguez, A. M., Romero-Castillo, K. D., Zavala-Yoe, R., Bilal, M., Ramirez-Mendoza, R. A., … Iqbal, H. M. N. (2020). Poly-3-hydroxybutyrate-based constructs with novel characteristics for drug delivery and tissue engineering applications—A review. *Polymer Engineering and Science*, 60(8): 1760–1772. https://doi.org/10.1002/pen.25470.

Tamai, H., Igaki, K., Kyo, E., Kosuga, K., Kawashima, A., Matsui, S., … Uehata, H. (2000). Initial and 6-month results of biodegradable poly-l-lactic acid coronary stents in humans. *Circulation*, 102(4): 399–404. https://doi.org/10.1161/01.CIR.102.4.399.

Ulag, S., Kalkandelen, C., Bedir, T., Erdemir, G. and Erdem, S. (2020). Fabrication of three-dimensional PCL/$BiFeO_3$ scaffolds for biomedical applications. *Materials Science & Engineering B*, 261(114660): 1–13. https://doi.org/10.1016/j.mseb.2020.114660.

Vermeer, C. M., Rossi, E., Tamis, J., Jonkers, H. M. and Kleerebezem, R. (2021). From waste to self-healing concrete: A proof-of-concept of a new application for polyhydroxyalkanoate. *Resources, Conservation and Recycling*, 164(March 2020): 105206. https://doi.org/10.1016/j.resconrec.2020.105206.

Voinova, V., Bonartseva, G. and Bonartsev, A. (2019). Effect of poly(3-hydroxyalkanoates) as natural polymers on mesenchymal stem cells. *World Journal of Stem Cells*, 11(10): 764–786. https://doi.org/10.4252/WJSC.V11.I10.764.

Whistler, R. L. and Daniel, J. L. (2000). Satrch. *In*: *Kirk-Othmer Encyclopedia of Chemical Technology*. Wiley.

Winnacker, M. (2019). Polyhydroxyalkanoates: recent advances in their synthesis and applications. *European Journal of Lipid Science and Technology*, 121(11): 1–9. https://doi.org/10.1002/ejlt.201900101.

Xu, J. and Guo, B. (2010). Poly (butylene succinate) and its copolymers: Research, development and industrialization. *Biotechnology Journal*, 5: 1149–1163. https://doi.org/10.1002/biot.201000136.

Yahiaoui, F., Benhacine, F., Ferfera-Harrar, H., Habi, A., Hadj-Hamou, A. S. and Grohens, Y. (2015). Development of antimicrobial PCL/nanoclay nanocomposite films with enhanced mechanical and water vapor barrier properties for packaging applications. *Polymer Bulletin*, 72: 235–254. https://doi.org/10.1007/s00289-014-1269-0.

Yan, D., Wang, Z., Guo, Z., Ma, Y., Wang, C., Tan, H., & and Zhang, Y. (2020). Study on the properties of PLA/PBAT composite modified by nanohydroxyapatite. *Integrative Medicine Research*, 9(5): 11895–11904. https://doi.org/10.1016/j.jmrt.2020.08.062.

Yin, G. Z. and Yang, X. M. (2020). Biodegradable polymers: a cure for the planet, but a long way to go. *Journal of Polymer Research*, 27(2): 1–14. https://doi.org/10.1007/s10965-020-2004-1.

Zhang, Wei, Xiang, Y., Fan, H., Wang, L., Xie, Y., Zhao, G. and Liu, Y. (2020). Biodegradable urea–formaldehyde/PBS and its ternary nanocomposite prepared by a novel and scalable reactive extrusion process for slow-release applications in agriculture. *Journal of Agricultural and Food Chemistry*, 68(16): 4595–4606. https://doi.org/10.1021/acs.jafc.0c00638.

Zhang, Wenhao. (2021). Analysis on the development and application of biodegradable polymers. *IOP Conference Series: Earth and Environmental Science*, 647(1). https://doi.org/10.1088/1755-1315/647/1/012156.

Zhang, X., Luo, R., Wang, Z., Deng, Y. and Chen, G. Q. (2009). Application of (R)-3-hydroxyalkanoate methyl esters derived from microbial polyhydroxyalkanoates as novel biofuels. *Biomacromolecules*, 10(4): 707–711. https://doi.org/10.1021/bm801424e.

Zolek-Trysnowska, Zuzanna; Kaluza, A. (2021). The influence of starch origin on the properties of starch films : packaging performance. *Materials*, 14: 1146. https://doi.org/https://doi.org/10.3390/ma14051146.

Chapter 18

Tendencies in Development of Biodegradable Polymers

*Lluvia Itzel López-López**

1. Introduction

In 1833, Berzelius coined the term "Polymer", which signifies many units, emerging as a new scientific field that deals with plastics, rubbers, fibers, coatings, adhesives and packing materials (Jensen, 2008). Polymers have changed our world with low-cost products and lightweight, high-performance materials that can be found everywhere in our daily life. Since its commercial development in the 1950s, fossil-based polymers have caused serious environmental pollution. In 2018, only 25% of the 359 million metric tons of plastic production was recycled, while 22% was disposed to landfills and/or incineration, and the remaining 42% ended up directly into the environment. This situation represents a serious global concern as forecasts for 2050 estimate 1124 million tons of plastic production (Patrício Silva et al., 2020; Hong and Chen, 2017; EMF, 2016). In addition to the fact that plastic pollution and microplastics dumped into rivers and oceans represent severe consequences for aquatic life and human health. Currently, a large number of researches are focused on replacing petroleum-based polymers with bio-based polymers that have similar or better properties. Moreover, the next plastic generation must be biodegradable at the end-life, which is the principal challenge to cope with.

Biodegradable means that a material (under the appropriate conditions) can be broken down into smaller pieces, ideally naturally occurring substances without any pollution, due to the action of natural microorganisms (i.e., bacteria, fungi, or microbes). Nevertheless, polymer degradation is a complex process, where photodegradability, oxidation by chemical additives, thermal and mechanical degradation converge. The biodegradation rate depends on many factors, such as environmental conditions (temperature, humidity, pH), polymer characteristics, susceptible bonds to hydrolysis, hydrophilicity, stereochemistry, molecular weight, crystallinity, specific surface, glass transition and melting temperature, additives used, amongst others (Chandra and Rustgi, 1998). It is important to specify that bio-based polymers are not necessarily biodegradable, and not all biodegradable polymers are bio-based, e.g., polycaprolactone. The tendency of biodegradable polymer research is the use of renewable and industrial waste sources, polymer-producing microorganisms and biotechnology tools.

This chapter presents a general outlook of recent research in the field of biodegradable polymers. The first part of this chapter introduces the classification of biodegradable polymers, while the second shows the applications and tendencies in frontier research with actual reports.

Instituto de Investigación de Zonas Desérticas, Universidad Autónoma de San Luis Potosí, San Luis Potosí, México.
* Corresponding author: lluvia.lopez@uaslp.mx

2. Biodegradable polymers

Biodegradable Polymers (BPs) can be classified based on the feedstocks used for their production and their biodegradability, as shown in Fig. 18.1 Moreover, in terms of their chemical structure, BPs are divided into three groups: (1) *Polysaccharides*, e.g., starch, cellulose, chitosan and alginate; (2) *Polyesters*, such as Poly(Lactic Acid) (PLA), polyglycol (PGL), polycaprolactone (PCL), poly(lactic-co-glycolic acid) (PLGA), and poly(hydroxyalcanoates) (PHAs); and (3) *Polyvinil derivatives* as polyvinil alcohol (PVA), see Fig. 18.2 (Thomas et al., 2021). Furthermore, a third classification based on the origin and modification of BPs separate them into four main categories: (1) *Natural polymers*, such as cellulose, starch and proteins; (2) *Modified natural polymers*, such as cellulose acetate or polyalkanoates; (3) *Composite materials* that combine biodegradable particles (e.g., starch, regenerated cellulose or natural gums) with synthetic polymers (e.g., mixtures of starch

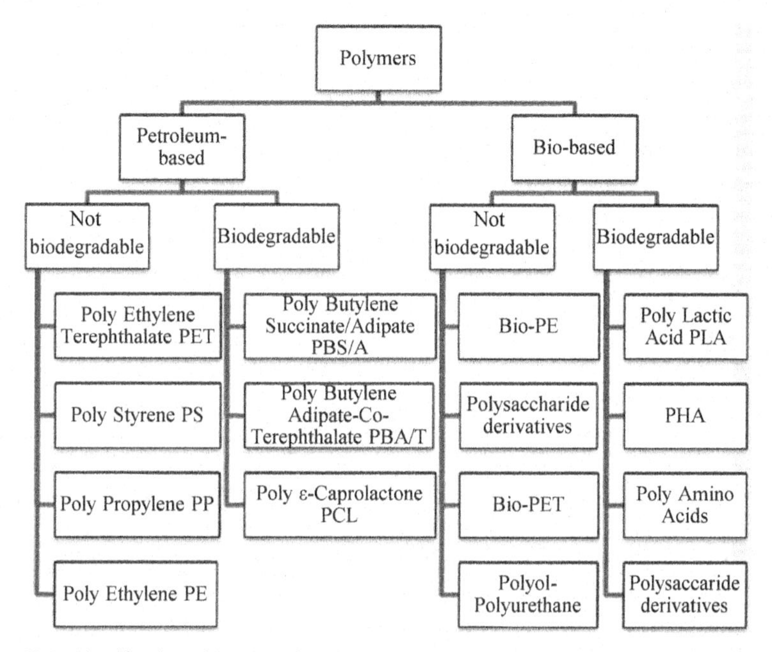

Figure 18.1. Classification of BPs based on the raw material used for production and biodegradability.

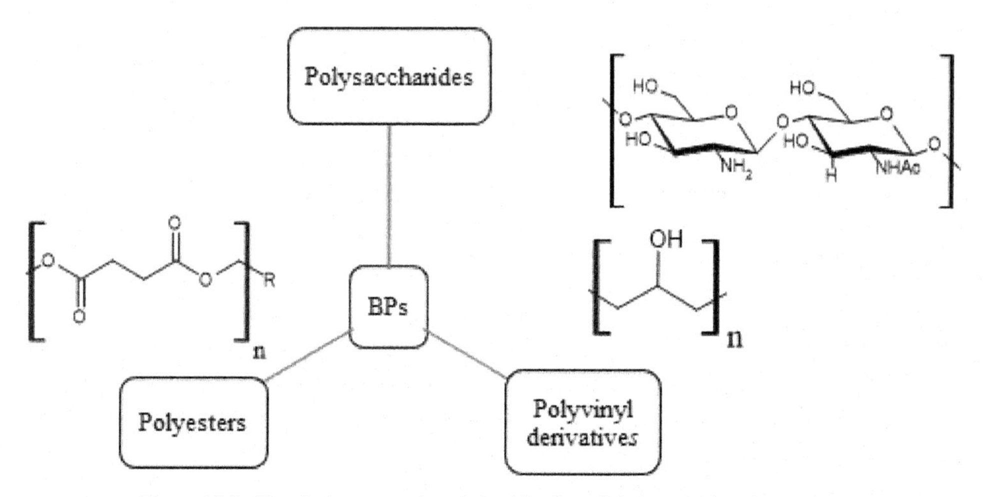

Figure 18.2. Chemical structure-based classification of biodegradable polymers BPs.

and polystyrene or starch and PCL); and (4) *Synthetic polymers*, such as polyesters, polyesteramides and polyurethanes, among others.

2.1 Natural polymers

Natural polymers have been used since ancient times such as silk, cotton, wool, cellulose fibers and proteins. Others, such as starch, alginate, gums and chitosan, etc., have been used extensively due to their biodegradable, biocompatible and eco-friendly nature (Islam et al., 2017). Nevertheless, the main use of natural polymers is to use them as a mixture to increase the chances of biodegradation in more complex matrices.

Cellulose is the most abundant natural polymer synthesized by plants as a structural material to support their weight. Cellulose is composed of D-glucose units linked by β-1,4' glycosidic bonds, this linkage makes the physical and chemical differences between cellulose and starch. In this sense, humans and mammals cannot degrade cellulose because they lack the β-glucosidase enzyme. Nevertheless, several bacteria and protozoa can hydrolyze cellulose (Wade, 2006). Due to its insolubility in water, cellulose is often used blended with plasticizers. Alternatively, a chemical modification in cellulose can turn it into a biodegradable polymer, e.g., the cellulose acetate, which is used in coatings, photographic films and in the manufacture of playing cards and cigarette filters. Moreover, nanocellulose fibers can be obtained from cellulose that can be used as nanofillers due to their high mechanical stability and optical transparency (Aliotta et al., 2019).

Starch consists of about 20% of water-soluble-amylose, and the remaining 80% is water-insoluble-amylopectin. Amylose is a linear polymer of glucose with α-1,4' glycosidic linkages, systematically named poly(1,4'-*O*-α-D-glucopyranoside). This α linkage allows a helical structure that increases hydrogen bonding water, making amylose water-soluble. Additionally, α-1,4' glucosidic linkage is easily hydrolyzed by an α-glucosidase enzyme, present in all animals. Amylopectin is primarily an α-1,4' polymer of glucose with a branch point about every 20 to 30 glucose units. These branches are connected to the principal chain by α-1,6' glucosidic linkage (Wade, 2006). Starch is the most useful biopolymer, thermoplastic, inexpensive and with a high-bioavailabilty. Usually, starch is used as an additive into synthetic polymers matrix for accelerating its degradation, via microorganisms' action mechanism.

Chitosan is the partially deacetylated form of chitin that consists of 2-acetamide-2-deoxy-β-D-glucose through the β-(1-4)-glycoside linkage. One important characteristic of chitosan is that it is soluble, unlike chitin. Moreover, chitosan is abundantly available in nature and accepted for its antifungal, antimicrobial, non-toxicity, biodegradability, biocompatibility, cytocompatibility, activity, antioxidant and anti-inflammatory actions (Hasan et al., 2020).

Alginate (AL) is an anionic polymer extracted from brown seaweed, including *Laminaria hyperborea*, *Laminaria digitata*, *Laminaria japonica*, *Ascophyllum nodosum* and *Macrocystis pyrifera*. Bacterial alginate can also be produced from *Azotobacter* and *Pseudomonas*. Alginate structure contains blocks of (1,4)-linked β-D-mannuronate (M) and α-L-guluronate (G) residues. These blocks are composed of consecutive G residues (GGGGGG), consecutive M residues (MMMMMM), and alternating M and G residues (GMGMGM) copolymers. The ratio of G to M depends on the natural source (Lee and Mooney, 2012). Alginate gels have been widely used in controlled release drug delivery systems. Alginates have also been used to encapsulate various herbicides, microorganisms and cells (Chandra and Rustgi, 1998).

Bacterial polyesters are produced as cytoplasmic reserve aggregations in various bacteria under specific conditions. PHAs as poly-3-hydroxybutyrate (PHB) and Poly-3-Hydroxy Valerate (PHV) are produced using glucose. The polyesters are characterized by the variation in their material properties from rigid brittle plastics to flexible plastics with good impact properties, depending on the structure. All polyesters that are 100% optical pure at the β-position are 100% isotactic (Chandra and Rustgi, 1998). PHB has interesting properties such as insolubility in water, resistance to hydrolytic degradation, thermal stability and total degradability in nature without any residues

(Adeleye et al., 2020). The challenge is to control the fermentation process at a large scale using several carbon sources at high yields.

Exopolisacharides (EPS) produced by lactic acid bacteria are divided into homopolysaccharides (HoPS), if the monomers unit is the same; and heteropolysaccharides (HePS), when the repeating unit is different. Usually, α-glucans (dextran, mutan, reuteran, alternan), β-glucans and fructans (levans and inulin-type polysaccharides) are the monosaccharides derivatives in HoPS. HePS have 2–8 constituting monosacharides as glucose, galactose, rhamnose, glucoronic acid, *N*-acetylglucosamine, *N*-acetylgalactosamine. EPS are particularly considered as a food additive (Korcz and Varga, 2021).

2.2 Modified natural polymers

Natural polymers modification arises from the need to match good mechanical and thermal properties of synthetic polymers, and to find alternatives to them. Natural polymers are used as BPs with or without that modification. Unfortunately, BPs bio-based have low mechanical strength and are extremely sensitive to environmental conditions, such as tensile strength, oxygen and water permeability, which are relevant parameters to determine their suitability for different applications (Cava et al., 2006).

Cellulose can be chemically modified by the reaction on its hydroxyl groups, which are present in each D-glucose monomer with specific functional groups, to modify or develop new desired characteristics. In general, modifications include the synthesis of: (1) ethers, to obtain, e.g., methylcellulose and hydroxyl-ethyl cellulose; (2) esters, to generate, e.g., cellulose acetate and cellulose xanthate; and (3) acetals, especially the cyclic acetal formed between the C2 and C3 hydroxyl groups and butyraldehyde (Chandra and Rustgi, 1998; Seddiqi et al., 2021).

Hydroxyethyl starch (HES) is a semisynthetic polysaccharide, which has been prepared by reacting amylopectin (from waxy corn or potato starch source) with ethylene oxide. HES has attracted great attention in the field of drug delivery due to its excellent physicochemical and biochemical properties. HES can be binding to small agents by different types of bonds, such as imine, ester, hydrazone and amide (Wang et al., 2021; Suwannateep et al., 2011). Li et al. (2021) reported the corn starch cross-linking/sulfonation to improve adhesion and film properties. The starch films were prepared by acid-converted starch with itaconic acid and subsequent sulfonation with $NaHSO_3$ (Li et al., 2021).

Modified chitosans have been prepared with various chemical and biological properties for a variety of applications. For example, specific reactions involving the $–NH_2$ group at the C-2 position or nonspecific reactions of –OH groups at the C-3 and C-6 positions (especially esterification and etherification). *O*-and *N*-carboxymethylchitosans (antibacterial applications, for use in cosmetics and in wound treatment), chitosan 6-*O*- and *N*-sulfate (anticoagulant agents), *N*-methylene phosphonic chitosans (complex agents) trimethylchitosan ammonium (applications to paper making) carbohydrate branched chitosans (drug targeting), chitosan-grafted copolymers (developing new biomaterials), alkylated chitosans (as amphiphilic polymers based on polysaccharides) (He et al., 2021; Rinaudo, 2006).

Graft polymerization of natural gums (xanthan gum, Arabic gum, acacia gum, etc.), is a useful technique for modification in the properties of polymers (Gandhi et al., 2019). The cross-linking significantly enhanced the mechanical strength of wheat gluten (Zhang et al., 2006).

The modification of synthetic and natural polymers with enzymes is an environmentally friendly alternative to chemical methods using harsh conditions. Specific non-destructive functionalization of polymer surfaces involves lipases, proteases, nitrilases and glycosidases. Lipases have been used for the production of optically active polyesters; oxidoreductases have been used for the cross-linking and grafting of lignaceous materials and the production of polymers from phenolics (Mgübitz and Paulo, 2003).

2.3 Composite materials

A composite material is a mixture of two or more components with different physical and chemical properties, which are specific characteristics such as flexibility, resistance to chemicals, low weight, durability, etc. The actual tendency in composite materials research is to find sustainable alternatives to non renewable petroleum, offer more versatile materials and an alternative to conventional plastics. In literature, several researchers have reported studies that combine BPs (starch, cellulose or natural gums, etc.) with synthetic polymers (mixtures of starch and PE or starch and PCL), and nanoparticles (NPs) with a specific function. Recent examples are presented below.

Ortega et al., reported nanocomposite films with antibacterial activity (against *Salmonella* spp.) composed of silver nanoparticles (AgNP) and starch-based film from application in the packing area (Ortega et al., 2021).

Nanochitosan polyurethane dispersions (NCS-PUs) were prepared and cotton fabrics were modified with antibacterial activity UV protection factor for textile application (Muzaffar et al., 2021). Annu et al. reported chitosan with PVA in active films for food packing (Annu et al., 2021).

Alginate (AL) is a BP with several applications due to its properties, biocompatible, hydrophilic, safe, non-inmunogenic, ecofriendly. Reports in literature showed the synthesis of cellulose nanofiber-based Na-AL biocomposite with antibacterial activity for potential wound dressing applications. Researchers synthesized the PVA base Na-AL blends and used them for the fabrication of a 3D scaffold and tissue engineering applications. Other research group synthesized bilayer edible conjugate films of casein-based Na-AL with silver. These films are very useful for loading antioxidants and help in sensitive product packing. It has been reported the synthesis of methylcellulose-based Na-AL hydrogel that is used to improve the strength and compatibility of cell matrixes. The synthesis of the marine collagen/agarose-based Na-AL biocomposite hydrogel and used it for 3D bioprinting has been reported (Ahmad et al., 2021).

Evon et al. reported eco-friendly Controlled-Release Fertilizers (CRF) composed of 90% Sunflower Protein Concentrate (SPC) matrix and 5–10% of a biopolymer obtained from municipal biowastes (MBW), and/or urea (U) (Evon et al., 2021).

2.4 Synthetic polymers

In general, synthetic polymers offer greater advantages over natural materials and composites because they can be designed according to the required properties. Furthermore, these can be theoretically estimated or predicted.

One of the most successful products in the commercial industry of BPs is PLA, which is a thermoplastic polyester produced in three different ways: direct condensation of lactic acid, azeotropic condensation and ring-opening polymerization after prior lactide formation (cyclic lactic acid dimers) (Nagarajan et al., 2016). On the other hand, PLC is another polyester semi-crystalline thermoplastic synthesized of cyclic form ε-caprolactone with applications in controlled drug delivery, implants for orthopedic surgery, tissue engineering and food packing. Most of the PCLs available in the market are fossil-based. Recent developments show processes for synthesizing the monomer capronic acid from corn stover, potentially rendering the polymer bio-based (Amulya et al., 2021).

PHAs are BPs that contain a fatty acid chain with an additional hydroxyl group that can be used to form ester linkages between monomers. PHAs can be synthetized by various bacteria, but their high production cost is the main drawback. While *Halomonas* spp. designed and constructed using synthetic biology not only produce low-cost intracellular PHAs, but also secrete extracellular soluble products for improved process economics (Tan et al., 2020). Polybutylene succinate (PBS) can be produced directly by fermentation using succinic acid. Engineering is used here to delete pathways for the production of unwanted by-products and obtain PBS as a single product (Amulya et al., 2021).

Recently, recombinant structural protein BP1 has been proposed as a promising alternative to conventional engineering plastics due to its good thermal and mechanical properties, its production from biomass and its potential for biodegradability. Properties like thermal degradation occur above 250°C, flexural strength and modulus are 115 ± 6 MPa and 7.38 ± 0.03 GPa, respectively. These properties are superior to those of commercially available BPs. BP1 can be spun or molded into fibers, sheets and bulk materials and used as an alternative to conventional plastics (Tachibana et al., 2021).

3. Applications

3.1 Food industry

3.1.1 Food packing

BPs have promising potential and are already applied in food packing because they provide valuable properties such as eco-friendly, biocompatible and edibility. Currently, food packing is not only about "containing the food", but satisfying other desirable functions such as sustaining freshness, increasing shelf life, taste and nutritional value of the product, and safer food supply (Videira-Quintela et al., 2021). For it, two technologies are implemented, the use of 'Active Food Packing' (AFP), and the 'Intelligent Food Packing' (IFP). AFP implies absorption or discharge of specific compounds from or into the package headspace, for example, by removal (scavenging) of gas molecules or the release of antimicrobials agents, by the interaction between food and packing. On the other hand, IFP is implemented as a material and an article that monitors the condition of packaged food or the environment surrounding the food, but they do not interact with the product (Realini and Marcos, 2014; Yildirim et al., 2018).

Several investigations on the subject have indicated the potential of AFP, especially, antimicrobial packaging systems. BPs such as starch, chitosan, soy protein, cellulose, PLA and PVA are used with the combination of synthetic or natural biopolymers (Annu et al., 2021; Chawla et al., 2021). Diverse biodegradable films are tested including *in vitro* and *in vivo* studies, see Table 18.1.

The use of nanotechnological concepts is a recent development in all science areas. The majority of the nanoparticles (NPs) explored for food packaging applications are potential antimicrobial agents, can also serve as carriers of various bioactive agents and prevent microbial contamination. The use of binary/ternary/quaternary polymer-metallic composite is designed to release antimicrobials from packing to the food surface. For example, AgNPs, CuNPs, AuNPs, SeNPs, Ag-CuNps into

Table 18.1. *In vitro* and *in vivo* studies of active food packing using BPs.

AFP BPs system	Bacteria	Food	Ref.
Starch/montmorillonite nanoclay/thyme essential oil	*Escherichia coli, Salmonella typhimurium*	Baby spinach leaves	Issa et al., 2017
Hydroxypropyl high-amylose starch/ pomegranate peel	*S. aureus, Salmonella*	-	Ali et al., 2019
Tapioca starch/chitosan nanoparticles	*Bacillus cereus, S. aureus, E. coli, S. typhimurium*	Cherry tomatoes	Shapi'I et al., 2020
Chitosan/clove essential oil/nisin	*S. aureus, S. typhimurium, E. coli, Listeria monocytogenes*	Pork patties	Venkatachalam and Lekjing, 2020
Cassava starch/chitosan/Pitanga (*Eugenia uniflora* L.) leaf extract/natamycin	*Aspergillus flavus, A. parasiticus in vitro*	-	
Fish gelatin/pomegranate (*Punicagranatum* L.) seed juice by-product	*S. aureus, Salmonella enterica in vitro*	-	Valdés et al., 2020
Na-Alg/casein/silver or Na-Alg/zataria essential oil/Agar	Antioxidant activity in bioactive food packing		Ahmad et al., 2021

the polymer matrix of PP, PE, PLA, PVA, chitosan, gelatin, starch, guar gum and others BPs (Videira-Quintela et al., 2021).

3.1.2 Food additives

Lactic Acid Bacteria (LAB)-derived exopolysaccharides (EPS) are used as prebiotics, emulsifiers, stabilizers, thickeners, gelling agents, as well as for moisture retention, in fermented dairy, bakery and meat industry areas, due to their physical properties, high viscosity in aqueous media and non-Newtonian behavior (Korcz and Varga, 2021).

3.2 Medicine

3.2.1 Drug delivery

In this science field, the actual tendency is the development of biocompatible and biodegradable "smart polymers", which can respond to the environmental stimuli (pH, temperature, chemical agents, electromagnetic radiation, electric field, etc.), and change or increase the desired properties for the controlled and targeted drug-releasing (Traitel et al., 2008; Hogan and Mikos, 2020). Biodegradable Thermoresponsive Polymers (BTPs) may be employed for *in vivo* biomedical applications, including *in situ* gelation at physiological temperatures and controlled drug delivery. BTPs such as collagen, chitosan and gelatin, as direct extracellular matrix (ECM) derivatives, offer both inherent biocompatibility and enhanced bioactivity compared to synthetic polymers. However, BTPs of natural origin have limitations such as batch variability and modulation of physicochemical properties, which are overcome by using modification or blending/copolymerization with synthetic polymers (Hogan and Mikos, 2020). On the other hand, BPs are conjugated with diverse drugs, increasing or modulating their effects (Chang et al., 2010; Tachaprutinun et al., 2014). While HES are used as drug carriers by encapsulating them to improve special features or decrease unwanted effects, see Table 18.2 (Wang et al., 2021; Suwannateep et al., 2011).

Cellulose Nanofiber-Based Na-Alginate. Cellulose has been converted into numerous nanostructures, for example, nanoparticles, films, nanofibers, hydrogels, aerogels, nanocrystals, nanowhiskers, and so on.

Table 18.2. BPs as drug delivery.

Conjugate BPs-Drug	Action
Chitosan-polyglutamic acid with amoxicillin	*H. pylori* inhibition
Ethyl cellulose-PEG-acrylate, with capsaicin and clarithromycin	*H. pylori* inhibition
Poly(vinyl sulfonate-co-vinyl alchol)-PLGA with salbutamol	Sustained bronchoprotective effect
Chitosan-PLGA-Exendin-4	Type 2 diabetes treatment
Chitosan-PLGA-Catechin hydrate	Epilepsy treatment
poly(butyl-cyanoacrylate)-polysorbate 80	With nerve growth factor, improved uptake by brain blood vessel endothelial cells
poly(2-(methacryloyloxy)ethyl-phosphorylcholine)-co-poly(2-diisopropylamino)ethylmethacrylate	DNA as intracellular delivery
HES-deferoxamine	Enhances the biosafety
HES-methotrexate	Longer half-life in plasma and consequently increased tumor accumulation
HES-5-fluorouracil-1-acetic acid	Increased half-life
HES-curcumin	Enhances the solubility and improves the cytotoxicity to HeLa cells and Caco-2 human colorectal adenocarcinoma cells
HES-dexamethasone, diclofenac sodium	Reduces dosage and frequency
HES-bovine serum albumin	Enhances immune responses

3.2.2 Tissue engineering

The term "tissue engineering" is defined as an interdisciplinary field that applies the principles of life sciences and engineering for the development of biological replacements that restore or replace the tissue function or a whole organ. Since Langer and Vacanti defined it in 1993, this research field is always a frontier and innovative due to medical applications (Langer and Vacanti, 1993). BPs are important due to their intrinsic properties of biocompatibility and bioactivity.

Thermoplastic aliphatic polyesters mimic natural bone properties, so natural polymer-based hydroxyapatite scaffolds are prepared. The great significance of this field lies in the *in vitro* growth of precise cells on porous matrices (scaffolds) to generate three-dimensional (3D) tissues that can be entrenched into the location of tissue/bone damage (Lett et al., 2021). New polymer biomaterials based on PEEK (polyether-ether-ketone) for orthopedical implantation in bone defects are documented. PEEK materials are described with several advantages as high-temperature resistance, corrosion resistance, abrasion resistance, high strength, high toughness, X-ray radiolucency, excellent sterilization performance and non-toxic. As a consequence, other binary and ternary PEEK-composite materials have been designed such as $BaSO_4$/PEEK, hydroxyapatite/PEEK, β-tricalcium phosphate (TCP)/PEEK, TCP/PLA/PEEK and carbon nanotubes/bioactive glass/PEEK (Ma et al., 2021).

Natural Hydrogels (NHs)-based fibrous protein and polysacharides contained in the human ECM are recently a research focus due to the inherent biocompatibility and bioactivity, for cardiac, neural and bone applications (Elkhoury et al., 2021). Moreover, Lee and Mooney 2012 reported the modification of amphiphilic AL with dodecylamine, microparticles as protein encapsulation, graft with poly(ε-caprolactone) (PCL), poly(butyl methacrylate) as cartilage repair and regeneration. Composites chitosan-based are prepared with cardiac tissue engineering applications, for example, nanofunctionalizaton with gold nanoparticles (AuNPs) and Graphene Oxide (GO), collagen-AuNPs, alginate-CNTs, etc. (Elkhoury et al., 2021).

Others

Conjugated polymers have special consideration as bioimaging materials using in the NIR-II (Near Infrared) imaging technology, due to intrinsical properties of fluorescence, photothermal and photodynamic. The principal controversy of these polymers is the nonbiodegradability, Wei and co-workers reported biodegradable NIR-II polymers for monitoring drugs *in vivo* and therapeutic effect feedback in real-time (Wei et al., 2021).

3.3 Industry

3.3.1 Agriculture

In this field, BPs such as cellulose, starch, chitin, alginic acid are used in Controlled Release (CR) systems of pesticides and nutrients. One advantage of CR formulations is that fewer chemicals are used with a lower impact. CR systems include microcapsules, physicals blends, dispersions in plastics, laminates, hollow fibers and membranes. On the other hand, BPs are also used as agricultural mulches to allow the growth of the plant by conserving moisture, reducing weeds and increasing soil temperatures (Chandra and Rustgi, 1998).

3.3.2 Textile

Clothing and textile goods are not only a growth media for the microorganisms, but can also serve as carriers of microbes that can be pathogenic. The surface modification of textile products has emerged as one of the interesting areas for the development of multifunctional textiles. Nanochitosan-polyurethane dispersions (NCS-Pus) can be applied as a finishing agent to enhance the antimicrobial activity along with protection from UV radiation (Muzaffar et al., 2021).

3.3.3 Electronic devices

BPs are present in materials science and nanotechnology applications as luminescent materials and conducting polymers. In this context, Nishimura reported a hybrid composite film using rGO (Graphene Oxide), Cdots (Carbon nanodots) in various weight proportions and collagen matrix with luminesce properties (Nishimura et al., 2021). The Cdots are emerging candidates for replacing metal-based semiconductors and rGO exhibits conductivity and biocompatibility. The integration of luminescence and electrical conductivity into the biodegradable hybrid film is expected to bring a paradigm shift in collagen-based applications. Much of the interest in the use of organic material is associated with the desire to design electronic components that are ecofriendly and biocompatible or even metabolizable (Han and Yoon, 2021).

3.3.4 Pollutants removal

Biomaterial-based Mixed Matrix Membrane (MMM) is a new technology in recent growth for the removal of pollutants from wastewater (Qalyoubi et al., 2021). The use of plant waste as biofiller in polyethersulfone MMM is reported, where tea waste, banana peel and shaddock peel were used with a rejection that reached up to 95% (Lin et al., 2014).

Cellulose acetate is a common filtration membrane, chitosan/cellulose acetate blend hollow fiber adsorptive membranes were prepared to improve the performance of the traditional CA hollow fibers, especially for affinity-based separation applications (Liu and Bai, 2005). Cellulose nanocrystals (CNCs) embedded in a matrix of chitosan formed a stable and nanoporous membrane structure that removed positively charges dyes (Karim et al., 2014). Chitosan nanofibrous membranes were successfully prepared as an effective adsorbent for dye removal (Li et al., 2018).

- Chitosan modification is an important method for the development of adsorbents: carboxylated chitosan (CYCS) and carboxylated nanocellulose (CNC) were used to chelate and synthesize hydrogel spheres with effective Pb(II) adsorption sites (Xu et al., 2021); He et al. reported an amidoxime-functionalized chitosan AM/AO/AEBI-CTS exhibiting porous structure, good water wettability and higher selectivity for Cu^{2+} than Ni^{2+}, with an adsorption efficiency that remained above 90% after five adsorption-desorption successive cycles (He et al., 2021).

3.4 Renewable waste

The first generation of BPs was prepared from carbohydrate rich agricultural feedstocks such as corn, potatoes and soy. However, due to competition for food and feed, the use of renewable waste from natural agricultural sources for BPs synthesis has increased, between them agricultural wastes (lignocellulosic biomass), biowastes (organic waste) and wastewaters. Consequently, all the renewable resources need processing prior to use as raw materials for the synthesis of BPs, emerging as the concept of waste biorefineries. The pre-treatment of waste/biomass to obtain monomers to produce BPs, natural reinforcements for biocomposite materials, can be integrated to waste biorefineries (Amulya et al., 2021). Banana starch, cornstarch, rice starch, natural filler as potato peel powder and sawdust are used as BPs synthesis (Shafqat et al., 2021). Yap et al. (2021) described the incorporation of 40% rice husk was able to substitute starch-based biodegradable polymer. Nano-crystals or nanofibrils can also be obtained from lignocellulosic biomass (Amulya et al., 2021).

Conclusions

This chapter has presented the recent scientific contributions of the development of biodegradable polymers. The main characteristics of natural, natural modified, composite and synthetic polymers have been described. Moreover, this chapter provided an overview of the utilization of biodegradable polymers in different areas, e.g., in medicine to improve medical implants or to enhance the release of drugs into the body, showing the advantages and areas of opportunities for future investigations.

Acknowledgements

LILL thanks IIZD/UASLP for all the facilities and Lizeth Adriana López for the English edition.

References

Adeleye, T., Odoh, C. K., Enudi, O. C., Banjoko, O. O., Osigbeminiyi, O. O., Toluwa-lope, O. E. and Louis, H. (2020). Sustainable synthesis and applications of polyhydroxyalkanoates (PHAs) from biomass. *Process Biochem*, 96: 174.

Ahmad, A., Mubarak, N. M., Jannat, F. T., Ashfaq, T., Santulli, C., Rizwan, M., Najda, A., Bin-Jumah, M., Abdel-Daim, M. M., Hussain, S. and Ali, S. (2021). Critical review on the synthesis of natural sodium alginate based composite materials: an innovative biological polymer for biomedical delivery applications. *Processes*, 9: 137.

Ali, A., Chen, Y., Liu, H., Yu, L., Baloch, Z., Khalid, S., Zhu, J. and Chen, L. (2019). Starch-based antimicrobial films functionalized by pomegranate peel. *Int. J. Biol. Macromol.*, 129: 1120.

Aliotta, L., Gigante, V., Coltelli, M. B., Cinelli, P. and Lazzeri, C. (2019). Evaluation of mechanical and interfacial properties of bio-composites based on poly(lactic acid) with natural cellulose fibres. *Int. J. Mol. Sci.*, 20: 960.

Amulya, K., Katakojwala, R., Ramakrishna, S. and Mohan, S. V. (2021). Low carbon biodegradable polymer matrices for sustainable future. *Composites Part C: Open Access*, 4: 100111.

Annu, Ali, A. and Ahmed, S. (2021). Eco-friendly natural extract loaded antioxidative chitosan/polyvinyl alcohol based active films for food packaging. *Heliyon*, 7: e06550.

Cava, D., Giménez, E., Gavara, R. and Lagaron, J. M. (2006). Comparative performance and barrier properties of biodegradable thermoplastics and nanobiocomposites versus PET for food packaging applications. *J. Plast. Film Sheeting*, 22: 265.

Chandra, R. and Rustgi, R. (1998). Biodegradable polymers. *Prog. Polym. Sci.*, 23: 1273.

Chang, C. H., Lin, Y. H., Yeh, C. L., Chen, Y. C., Chiou, S. F., Hsu, Y. M., Chen, Y. S. and Wang, C. C. (2010). Nanoparticles incorporated in pH-sensitive hydrogels as amoxicillin delivery for eradication of *Helicobacter pylori*. *Biomacromolecules*, 11: 133.

Chawla, R., Sivakumar, S. and Kaur, H. (2021). Antimicrobial edible films in food packaging: current scenario and recent nanotechnological advancements—a review. *Carbohyd. Polym. Tech App.*, 2: 100024.

EMF. (2016). The New Plastics Economy—Rethinking the Future of Plastics, Ellen MacArthur Foundation.

Elkhoury, K., Morsink, M., Sanchez-Gonzalez, L., Kahn, C., Tamayol, A. and Arab-Tehrany, E. (2021). Biofabrication of natural hydrogels for cardiac, neural, and bone Tissue engineering Applications. *Bioact. Mater.*, 6: 3904.

Evon, P., Labonne, L., Padoan, E., Vaca-Garcia, C., Montoneri, E., Boero, V. and Negre, M. (2021). A new composite biomaterial made from sunflower proteins, urea, and soluble polymers obtained from industrial and municipal biowastes to perform as slow release fertiliser. *Coatings*, 11: 43.

Gandhi, A., Verma, S., Imam, S. S. and Vyas, M. (2019). A review on techniques for grafting of natural polymers and their applications. *Plant Arch*, 19: 972.

Han, M.J. and Yoon, D. K. (2021). Advances in soft materials for sustainable electronics. *Engineering*, 7: 564.

Hasan, K. M. F., Wang, H., Mahmud, S., Jahid, Md. A., Islam, M., Jin, W. and Genyang, C. (2020). Colorful and antibacterial nylon fabric via *in-situ* biosynthesis of chitosan mediated nanosilver. *J. Mater Sci. Technol.*, 9: 16135.

He, Y., Gou, S., Zhou, L., Tang, L., Liu, T., Liu, L. and Duan, M. (2021). Amidoxime-functionalized polyacrylamide-modified chitosan containing imidazoline groups for effective removal of Cu^{2+} and Ni^{2+}. *Carbohydr. Polym.*, 252: 117160.

Hogan, K. J. and Mikos, A. G. (2020). Biodegradable thermoresponsive polymers: Applications in drug delivery and tissue engineering. *Polymer*, 211: 123063.

Hong, M. and Chen, E. Y. X. (2017). Chemically recyclable polymers: a circular economy approach to sustainability, *Green Chem.*, 19: 3692.

Islam, S., Bhuiyan, M. A. R. and Islam, M. N. (2017). Chitin and chitosan: structure, properties and applications in biomedical engineering. *J. Polym. Environ.*, 25: 854.

Issa, A., Ibrahim, S. A. and Tahergorabi, R. (2017). Impact of sweet potato starch-based nanocomposite films activated with thyme essential oil on the shelf-life of baby spinach leaves. *Foods*, 6: 43.

Jensen, W. B. (2008). The origin of the polymer concept. *J. Chem. Educ.*, 85: 624.

Karim, Z., Mathew, A. P., Grahn, M., Mouzon, J. and Oksman, K. (2014). Nanoporous membranes with cellulose nanocrystals as functional entity in chitosan: removal of dyes from water. *Carbohydr. Polym.*, 112: 668.

Korcz, E. and Varga, L. (2021). Exopolysaccharides from lactic acid bacteria: Techno-functional application in the food industry. *Trends Food Sci. Tech*, 110: 375.

Langer, R. and Vacanti, J. (1993). Tissue engineering. *Science*, 260: 920.

Lee, K. J. and Mooney D. J. (2012). Alginate: properties and biomedical applications. *Prog. Polym. Sci.*, 37: 106.

Lett, J. A., Sagadevan, S., Fatimah, I., Hoque, M. E., Lokanathan, Y., Léonard, E., Alshahateet, S. F., Schirhagl, R. and Chun Oh. W. (2021). Recent advances in natural polymer-based hydroxyapatite scaffolds: Properties and applications. *Eur. Polym. J.*, 148: 110360.

Li, C., Lou, T., Yan, X., Long, Y-Z., Cui, G. and Wang, X. (2018). Fabrication of pure chitosan nanofibrous membranes as effective absorbent for dye removal. *Int. J. Biol. Macromol.*, 106: 768.

Li, W., Zhang, Z., Wua, L., Zhu, Z., Ni, Q., Xu, Z. and Wu, J. (2021). Cross-linking/sulfonation to improve paste stability, adhesion and film properties of corn starch for warp sizing. *Int. J. Adhes Adh.*, 104: 102720.

Lin, C. H., Gung, C. H., Sun, J. J. and Suen, S. Y. (2014). Preparation of polyethersulfone/plantwaste-particles mixed matrix membranes for adsorptive removal of cationic dyes from water. *J. Membr. Sci.*, 471: 285.

Liu, C. and Bai, R. (2005). Preparation of chitosan/cellulose acetate blend hollow fibers for adsorptive performance. *J. Membr. Sci.*, 267: 68.

Ma, H., Suonan, A., Zhou, J., Yuan, Q., Liu, L., Zhao, X., Lou, X., Yang, C., Li, D. and Zhang, Y.-G. (2021). PEEK (Polyether-ether-ketone) and its composite materials in orthopedic implantation. *Arab J. Chem.*, 14: 102977.

Mgübitz, G. and Paulo, A. C. (2003). New substrates for reliable enzymes: enzymatic modification of polymers. *Curr. Opin Biotech*, 14: 577.

Muzaffar, S., Abbas, M., Siddiqua, U. H, Arshad, M., Tufail, A., Ahsan, M., Alissa, S. A, Abubshait, S. A., Abubshait, H. A. and Iqbal, M. (2021). Enhanced mechanical, UV protection and antimicrobial properties of cotton fabric employing nanochitosan and polyurethane based finishing. *J. Mater. Sci. Technol.*, 11: 946.

Nagarajan, V., Mohanty, A. K. and Misra, M. (2016). Perspective on polylactic acid (PLA) based sustainable materials for durable applications: focus on toughness and heat resistance. *ACS Sustain Chem. Eng.*, 4: 2899.

Nishimura, S., Narayan, N., Sahin, O., McWilliams, A. D. S., Miller, K. A., Salpekar, D., Wang, Z., Joyner, J., Martí, A. A., Vajtai, R., Ashokkumar, M. and Ajayan, P. M. (2021). Luminescent hybrid biocomposite films derived from animal skin waste. *Carbon Trends*, 4: 100059.

Ortega, F., Valeria, B., Arce, V. B. and Garcia, M. A. (2021). Nanocomposite starch-based films containing silver nanoparticles synthesized with lemon juice as reducing and stabilizing agent. *Carbohydr. Polym.*, 252: 117208.

Patrício Silva, A. L., Prata, J. C., Walker, T. R., Campos, D., Duarte, A. C., Soares, A. M. V. M., Barceló, D. and Rocha-Santos, T. (2020). Rethinking and optimising plastic waste management under COVID-19 pandemic: policy solutions based on redesign and reduction of single-use plastics and personal protective equipment. *Sci. Total Environ.*, 742: 140565.

Qalyoubi, L., Al-Othman, A. and Al-Asheh, S. (2021). Recent progress and challenges on adsorptive membranes for the removal of pollutants from wastewater. Part I: Fundamentals and classification of membranes. *CSCEE*, 3: 100086.

Realini, C. E. and Marcos, B. (2014). Active and intelligent packaging systems for a modern society. *Meat Sci.*, 98: 404.

Rinaudo, M. (2006). Chitin and chitosan: properties and applications. *Prog. Polym. Sci.*, 31: 603.

Seddiqi, H., Oliaei, E., Honarkar, H., Jin, J., Geonzon, L. C., Bacabac, R. G. and Klein-Nulend, J. (2021). Cellulose and its derivatives: towards biomedical applications. *Cellulose*, 28: 1893.

Shafqat, A., Al-Zaqri, N., Tahir, A. and Alsalme, A. (2021). Synthesis and characterization of starch based bioplatics using varying plant-based ingredients, plasticizers and natural fillers. *Saudi J. Bio Sci.*, 28: 1739.

Shapi'i, R. A., Othman, S. H., Nordin, N., Basha, R. K. and Naim, M. N. (2020). Antimicrobial properties of starch films incorporated with chitosan nanoparticles: *In vitro* and *in vivo* evaluation. *Carbohydr. Polym.*, 230: 115602.

Suwannateep, N., Banlunara, W., Wanichwecharungruang, S. P., Chiablaem, K., Lirdprapamongkol, K. and Svasti, J. (2011). Mucoadhesive curcumin nanospheres: biological activity, adhesion to stomach mucosa and release of curcumin into the circulation. *J. Contr. Release*, 151: 176.

Tachaprutinun, A., Pan-In, P., Samutprasert, P., Banlunara, W., Chaichanawongsaroj, N. and Wanichwecharungruang, S. (2014). Acrylate-tethering drug carrier: covalently linking carrier to biological surface and application in the treatment of *Helicobacter pylori* infection. *Biomacromolecules*, 15: 4239.

Tachibana, Y., Darbe, S., Hayashi, S., Kudasheva, A., Misawa, H., Shibata, Y. and Kasuya, K. (2021). Environmental biodegradability of recombinant structural protein. *Sci. Rep.*, 11: 242.

Tan, D., Wang, Y., Tong, Y. and Chen, G.-Q. (2021). Grand challenges for industrializing polyhydroxyalkanoates (PHAs). *Trends Biotechnol*, 39: 953. https://doi.org/10.1016/j.tibtech.2020.11.010.

Thomas, S. K., Parameswaranpillai, J., Krishnasamy, S., Begam, P. M. S., Nandi, D., Siengchin, S., George, J. J., Hameed, N., Nisa, V., Salim, N. V. and Sienkiewicz, N. (2021). A comprehensive review on cellulose, chitin, and starch as fillers in natural rubber biocomposites. *Carbohyd. Polym. Tech App.*, 2: 100095.

Traitel, T., Goldbart, R. and Kost, J. (2008). Smart polymers for responsive drug-delivery systems. *J. Biomater. Sci.*, 19: 755.

Valdés, A., Garcia-Serna, E., Martínez-Abad, A., Vilaplana, F., Jimenez, A. and Garrigós, M. C. (2020). Gelatin-based antimicrobial films incorporating pomegranate (*Punica granatum* L.) seed juice by-product. *Molecules*, 25: 166.

Venkatachalam, K. and Lekjing, S. (2020). A chitosan-based edible film with clove essential oil and nisin for improving the quality and shelf life of pork patties in cold storage. *RSC Adv.*, 10: 17777.

Videira-Quintela, D., Martin, O. and Montalvo, G. (2021). Recent advances in polymer-metallic composites for food packaging applications. *Trends Food Sci. Tech*, 109: 230.

Wade, L. G. Jr. (2006). Organic Chemistry. Sixth edition. Pearson Prentice, 1134–1136. US.

Wang, H., Hu, H., Yang, H. and Li, Z. (2021). Hydroxyethyl starch based smart nanomedicine. *RSC Adv.*, 11: 3226.

Wei, D., Yu, Y., Huang, Y., Jiang, Y., Zhao, Y., Nie, Z., Wang, F., Ma, W., Yu, Z., Huang, Y., Zhang, X.-D., Liu, Z.-Q., Zhang, X. and Xiao, H. (2021). A near-infrared-II polymer with tandem fluorophores demonstrates superior biodegradability for simultaneous drug tracking and treatment efficacy feedback. *ACS Nano*, 15: 5428.

Xu, X., Ouyang, X.-K. and Yang, L.-Y. (2021). Adsorption of Pb(II) from aqueous solutions using crosslinked carboxylated chitosan/carboxylated nanocellulose hydrogel beads. *J. Mol. Liq.*, 322: 114523.

Yap, S. Y., Sreekantan, S., Hassan, M., Sudesh, K. and Ong, M. T. 2021. Characterization and biodegradability of rice husk-filled polymer composites. *Polymers*, 13: 104.

Yildirim, S., Röcker, B., Pettersen, M. K., Nilsen-Nygaard, J., Ayhan, Z., Rutkaite, R., Radusin, T., Suminska, P., Marcos, B. and Coma, V. (2018). Active packaging applications for food. *Compr. Rev. Food Sci. Food Saf.*, 17: 166.

Zhang, X., Hoobin, P., Burgar, I. and Do, M. D. (2006). Chemical modification of wheat protein-based natural polymers: cross-linking effect on mechanical properties and phase structures. *J. Agric Food Chem.*, 54(26): 9858.

Chapter 19

Assessment of Biodegradability in Polymers

Mechanisms and Analytical Methods

Francisco J. González,[1,*] *Francisco J. Rivera-Gálvez,*[2]
Felipe Avalos Belmontes[1] and *Mario Hoyos*[3]

1. Introduction

The environmental impact caused by plastic waste around the world is one of the most significant concerns, as in 2019 its production reached almost 368 million tonnes (Plastics Europe, 2020). In addition to this, most plastic residues are usually not recycled, as the majority goes to landfills or incinerators (Ruggero et al., 2019). In an attempt to reduce plastic waste, the mass replacement of single-use and non-biodegradable polymers by biodegradable polymers in the markets can greatly reduce the global pollution impact, shortening the time polymers can be transported by the wind and decreasing their presence in terrestrial and water environments. The evident advantage of biodegradable polymers in comparison with the non-biodegradable ones is the shorter time to be completely degraded. Nevertheless, in colloquial and sometimes in technical language, the term 'biodegradable' is not adequately used. In theory, all plastics materials degrade, but degradation time can in some materials take hundreds of years. Furthermore, the term bio-based plastic/polymer and biodegradable is often confused. The group of biodegradable polymers includes both the bio-based polymers, obtained from natural sources, such as Poly (Lactic Acid) (PLA) or polyhydroxyalkanoates (PHA) and the synthetic or petroleum-based polymers such as polycaprolactone (PCL) or poly (vinyl alcohol) (PVOH) (Tokiwa et al., 2009).

Although the classification of biodegradable polymers is well defined, specific details of polymers' biodegradability are not forthright. This can be exemplified with PLA, as its complex degradation in soils (Adhikari et al., 2016) is well known. Under specific conditions, PLA can be completely degraded in 28 days; however, only 13% of degradation was determined in 60 days under other conditions (Kjeldsen et al., 2019). Some other examples of several biodegradable polymers under different conditions were reported by Emadian et al. (2017).

In order to avoid confusion and misunderstandings, the biodegradability of polymers must be specified with metrics and standards. The European Committee for Standardization (ECS) considers

[1] Faculty of Chemical Sciences, Universidad Autónoma de Coahuila, 25280 Saltillo, Coahuila, Mexico.
[2] Chemical Engineering Department, Universidad de Guadalajara, 44430 Guadalajara, Jalisco, Mexico.
[3] Instituto de Ciencia y Tecnología de Polímeros, 28002, Madrid, España.
* Corresponding author: fgonzalezgonzalez@uadec.edu.mx

a material biodegradable when it has lost 70% of its initial weight after 45 days of the test period (Capitain et al., 2020). According to The Organization for Economic Co-operation and Development (OECD) and its definition of 'ultimate biodegradability', a material is considered biodegradable when it can be degraded and used as a source of carbon for microorganisms, e.g., bacteria or fungi (Filiciotto and Rothenberg, 2021; Harrison et al., 2018). According to the European Chemicals Agency (ECHA), the biodegradability of materials can be classified into three categories: readily biodegradable, inherently biodegradable and ultimately biodegradable (Filiciotto and Rothenberg, 2021). The term biodegradable is restricted only by the action of microorganisms, and it must not be confused with the more general term 'degradable', as the latter includes other degradation factors, for example, by the action of light exposure (Jayasekara et al., 2005).

To assess the biodegradability in polymers it is necessary to understand which polymers can be biodegradable (Lucas et al., 2008). Hence, the entire biodegradation system must be considered, that is: the polymer structure (Endres and Siebert-Raths, 2011; Scott, 2002), the abiotic conditions of polymer exposure (environmental conditions) (Sivan, 2011), and the type and activity of microorganism(s) (or PSECAM system) (Alshehrei, 2017). Although the complete information in real environments is complex to be recalled (Müller, 2020; Swift, 1992), and even more in medical applications, which are not considered in this chapter (Asghari et al., 2017; Laycock et al., 2017; Ulery et al., 2011), these aspects can be used as initial information to design an assessment for each biodegradable system (Ahmed et al., 2018). In addition to this, biodegradability is highly influenced by the physicochemical properties of the polymers (Tokiwa et al., 2009), for instance the polymer composition, if it is raw or blended with recycled plastics (La Mantia et al., 2017), or its surface area (Chinaglia et al., 2017). The latter is especially interesting when evaluating the micro and nanoplastics degradation, which will be addressed next.

Environmental pollution caused by polymers requires great efforts from the scientific community to stipulate how and under what conditions polymers can be considered biodegradable (Scott, 2002). A global consensus must be efficiently disseminated to delimit the scopes and conditions of biodegradability to circumvent sharing a wrong message to the scientific community and society, that all polymers can disappear in any environment solely for obtaining the qualification of biodegradable using any norm or certification. The purpose of this chapter is to give a general overview of how to assess the biodegradation in polymers. A review of the most relevant degradation environments and mechanisms is given, along with some techniques, methods and standards used to assess biodegradability in polymers.

2. Aspects to assess the biodegradability of polymers in terrestrial, water, and artificial environments

Polymer waste can be found throughout the environment in a wide variety of shapes, sizes and compositions, even as plastic dust in locations with zero human activity due to dust storms moving a large number of plastics particles (Brahney et al., 2020). Although the assessment of biodegradability of polymers generally uses the visual disappearance of plastic fragments to determine the polymer biodegradation (see later), the production of fragments in a micro or nano scale requires a new approach to the assessment of polymers biodegradation considering the entire biodegradation system, i.e., PSECAM system, that includes the analysis of the biodegradation location. In addition, environmental specific conditions may lead to completely different degradation behavior, such as pH, temperature or microbial activity (Emadian et al., 2017).

Environments to study biodegradation can be classified into two main groups: terrestrial and water. On one side, some of the most common terrestrial environments comprise soils, landfills and compost. Each environment may present completely different conditions such as temperature and microorganisms content that affect the degradation rate. On the other hand, for water environments, degradation conditions may be classified depending on the type of water, if it is fresh or saltwater and its deep regard to the surface.

2.1 Variables for the biodegradability of polymers in terrestrial environments

In terrestrial ecosystems there exist between 4–23 times more microplastics (polymer particles in the range from a few millimeters to microns) than in marine ones (Horton et al., 2017). In terrestrial ecosystems, larger plastics allow the exchange of gases and other toxic substances that may damage environmental health and cause entanglements in species. The smallest particles can be ingested or inhaled, causing loss of appetite, blockages of the digestive tract, abrasion or irritation of the mucous membranes. Micro-fragments of polymers with a size less than 0.1 microns have biological interaction with membranes, organelles or vital molecules such as DNA. These can lead to cellular inflammation, cytotoxicity (cellular toxicity) and genotoxicity (genetic alterations) (Horton et al., 2017). On the other hand, nanoplastics, which comprise polymer particles in the sub-micrometer scale (Gigault et al., 2018), can be food for plants or fungi and later transmitted from them to herbivores such as deer, cows or rabbits (De Souza et al., 2018).

Microplastics have not been extensively studied in soils and terrestrial systems compared to those in aquatic areas because there are no well-understood routes to extract them from soils. For example, extraction from soils or minerals is usually almost artisanal (Rillig, 2012). However, there are some variables that can be considered, for instance, the abiotic degradation can increase with the exposition to the UV light, hydrolysis or the thermal-oxidative process to provide the energy leading to a statistical chain scission and subsequently the mineralization of polymer, which can be used in the final degradation of the polymer metabolizing microorganisms.

One of the most preferred methods for biodegradable polymers disposal is converting them into compost (Tosin et al., 2012). This process briefly consists of the degradation of organic matter under controlled conditions to generate carbon gas derivatives and nutrients that are added to the soil. The CO_2 produced does not have an impact on the greenhouse gas emissions as this carbon contributes to the biological carbon cycle (Folino et al., 2020; Song et al., 2009). The composting process can be carried out in an aerobic or anaerobic route. Aerobic composting consists of the degradation and the consumption of oxygen and the release of CO_2 and water. During anaerobic digestion, additional gases such as methane, carbon dioxide and traces of hydrogen and hydrogen sulfide are released (Ruggero et al., 2019).

The polymer biodegradation in landfills and composting can reduce the waste quickly due to the high content of microorganisms (Kumar et al., 2021). This fact can be enhanced via methane oxidation in semi-aerobic systems (Muenmee et al., 2016), although biodegradation occurs under anaerobic conditions (Ishigaki et al., 2004). Microorganism's content may vary according to location, environmental conditions and human activity. In agriculture fields, polymeric hydrogels are widely used to reduce water consumption; however, their high hydrophilicity and sensitivity to react with inorganic compounds may lead to premature degradation that quickly promotes material fragmentation and biodegradation (Briassoulis and Dejean, 2010; Smagin et al., 2018). Temperature is another critical issue to consider in polymer degradation rate. In simulated composts (e.g., in laboratories conditions), it is common to reach temperatures > 58°C (which is the temperature at which the hydrolysis of the material begins, and that leads to the first step of biodegradation) but in real composts, in which the temperature is not controlled, the microorganisms work very close to the ambient temperature and outside of the optimal value in compost (associated between 35°C–65°C), slowing down the rate of biodegradation.

2.2 Variables for the biodegradability of polymers in water environments

The microorganisms living in aqueous environments are immersed in a nutrient solution to maintain their biological activity. If a polymer can be degraded by abiotic factors and biological species, it can be considered as biodegradable. Nevertheless, it is important to clarify that not all polymers that are consumed by microorganisms (as the zooplankton) can be considered biodegradable, although the presence of the polymer inside the microorganism does not significantly affect their biological functions (Dussud et al., 2018; Moore et al., 2005).

2.2.1 Ocean surface

Microplastics are dangerous because the capacity of fragmentation and dispersion may release toxic additives quickly (Moore et al., 2015; Shim et al., 2018). Currents, tides, waves, whirlwinds or other types of water movement on the ocean surface can promote mechanical degradation having as a consequence, an increase in the polymer fragment production. Besides, the UV light exposition accelerates the degradation of polymers, but the salinity concentration has no major effect on degradation in hydrophobic polymers (Suhrhoff and Scholz-Böttcher, 2016). Nevertheless, it can be expected that salinity in hydrogels has a significant impact according to the gel stability with the solvated salts (Muller et al., 1979).

2.2.2 Deep sea (200 m below the surface)

In the deep sea region, if the zooplankton ingests microplastics, two options can be present: (1) the polymer biodegradation by those microorganisms or (2) transfer it to other marine creatures to enter into the food chain (Chia et al., 2020). Nevertheless, the *Bacillus paralicheniformis G1* (MN720578) isolated from 3538 meters depth sediments of the Arabian Sea can lead to the biodegradation of polystyrene (Kumar et al., 2021). Several microorganisms are able to biodegrade polymer in the deep sea (Sekiguchi et al., 2011), and probably many other types of extremophile microorganisms can be added to this list in the near future. In addition to this, submarine volcano activity can produce high temperature and acidity, having as a consequence, to trigger off the first step of biodegradation related to the abiotic degradation of the polymer, to be later used by microorganisms as an energy source or food under these conditions.

2.2.3 Freshwater environments

Unfortunately, there is not enough information to estimate the total amount of plastics in freshwater bodies. However, it is expected that in regions with fewer environmental restrictions, there will be a concentration of plastics equal or greater than reported in earlier studies (about 0.027 particles/m^3 estimated in Great Lakes of North America in 2019) (Triebskorn et al., 2019). With this, similar conditions of polymer degradation can be expected as those on the surface of the sea, i.e., UV exposure, mechanical degradation that leads to a quick fragmentation of pieces that can be later biodegraded for several microorganisms (Harrison et al., 2018; Zambrano et al., 2020).

Polymer biodegradation in water environments begins with sun irradiation (UV light), thermal degradation or chemical hydrolysis (Doble, 2014). The activity of microorganisms is higher in aqueous environments with respect to soils due to the presence of water; however, although water is a basic component of the microbial world, many of these microorganisms require or prefer the contact to a solid matrix (Doble, 2014). For example, fungi generally exhibit a better growth on solid surfaces than in agitated liquids, which can be attributed to the sensitivity of the fungal mycelium to the mechanical forces (Doble, 2014). The biodegradation in water is favored at the interface water-sediment as the activity of microorganisms for plastic degradation increases (Emadian et al., 2017). Table 19.1 summarizes the biodegradability factors that involve the biodegradability of polymers.

3. Mechanisms in polymers biodegradation

Considering only microorganisms, generally bacteria (Sivan, 2011), *eukaryotic* (especially fungi (Sánchez, 2020), and algae (Chia et al., 2020) are in the center of biodegradation while *archaea microbes* begin to have a relevant part in this field (Field, 2002). However, to assess the biodegradability of polymers it is necessary to consider both the abiotic and biotic degradation.

The biodegradation mechanisms normally initiate with an abiotic degradation followed by the corresponding macroscopic polymer changes, for instance: particle fragmentation, surface erosion, formation of cracks, color changes, variations in the refraction index, loss of the strength in stress-strain curves and a decrease in polymer viscosity. Structural polymer modifications are

Table 19.1. Variables for the biodegradability of polymers as function of the environments.

Terrestrial environments	**Beach**	**Soil**	**Landfills/Composting**
	Wide separation of polymers Hydrolysis Mechanical degradation Low content of microorganisms Variable temperature	Wide separation of polymers Minerals increase the abiotic degradation Low content of microorganisms	High concentration of polymers High content of microorganism Temperatures > 55°C can be reached Hydrolysis Chemical degradation Variable temperature
Water environments	**Surface**	**Deep sea**	**Fresh water**
	Intermediate separation of polymers Hydrolysis UV degradation Mechanical degradation Low content of microorganisms Variable temperature	Wide separation of polymers Hydrolysis High content of microorganism Stable temperature	Wide separation of polymers Hydrolysis UV degradation Mechanical degradation Low content of microorganisms
Atypical environments	**Laboratory conditions**	**Extreme environments**	
	High concentration of polymers Controlled degradation Controlled temperature Controlled content of microorganisms	Variable concentration of polymers Catalyzed degradation Thermo-oxidative degradation Temperature < 0°C or > 65°C	

generally associated with chain scission, degradation of side groups and/or structural changes in the molecular configuration. All these changes lead to mechanical degradation of polymers mostly due to exposure to the electromagnetic spectrum (UV and visible), thermal degradation, chemical degradation or hydrolysis in water presence (Gerald and Scott, 2002). Moreover, microorganisms can deteriorate polymers from a physical, chemical or enzymatic route (hydrolysis, oxidation or radical oxidation). To sum up, biodegradation essentially consists of the polymer degradation under aerobic and/or anaerobic conditions (Endres and Siebert-Raths, 2011) as is presented in Fig. 19.1.

The three main degradation mechanisms in polymers alongwith biodegradation are: depolymerization, statistics chain scission and elimination (side groups, functional groups, atoms, etc.) (Niaounakis, 2013). It is evident that the complexity to classify a polymer as biodegradable only considering their source; for example, not the entire natural or bio-based polymers are biodegradable, and not all synthetic polymers are non-biodegradable. This is because the potential biodegradability of a polymer is a function of the PSECAM system that begins with the adaptability of one or many microorganisms to the polymer structure at the environmental conditions. Considering this, several degradation mechanisms and factors can be combined synergistically in polymers' biodegradation.

The biodegradability of a polymer increases with the hydrophilicity of the polymer and decreases with hydrophobicity due to the fact that microorganisms commonly use water solutions with digestive molecules that have intermolecular interactions such as hydrogen bonds (Endres and Siebert-Raths, 2011). Another factor that increases biodegradation is the susceptibility to the thermo-oxidation in some polymers; for this reason, polymers with heteroatoms in the backbone (polymers obtained via polycondensations) are susceptible to thermal and oxidative instability (Endres and Siebert-Raths, 2011). This fact occurs especially when oxygen or nitrogen atoms are intercalated in the main chain (e.g., ester, amide or ether bonds) (Endres and Siebert-Raths, 2011) due to these chemical bonds are more sensitive to chemical dissociation by acids and other compounds released by microorganisms in comparison with C-C bonds. An increase in biodegradability can be expected in polymers that have chemical bonds with low energy. On the other hand, steric hindrances in the main chain or sides groups (e.g., phenyl groups) can drastically decrease biodegradability if these are located to the side of these chemical bonds. The increase of polarity around a chemical bond

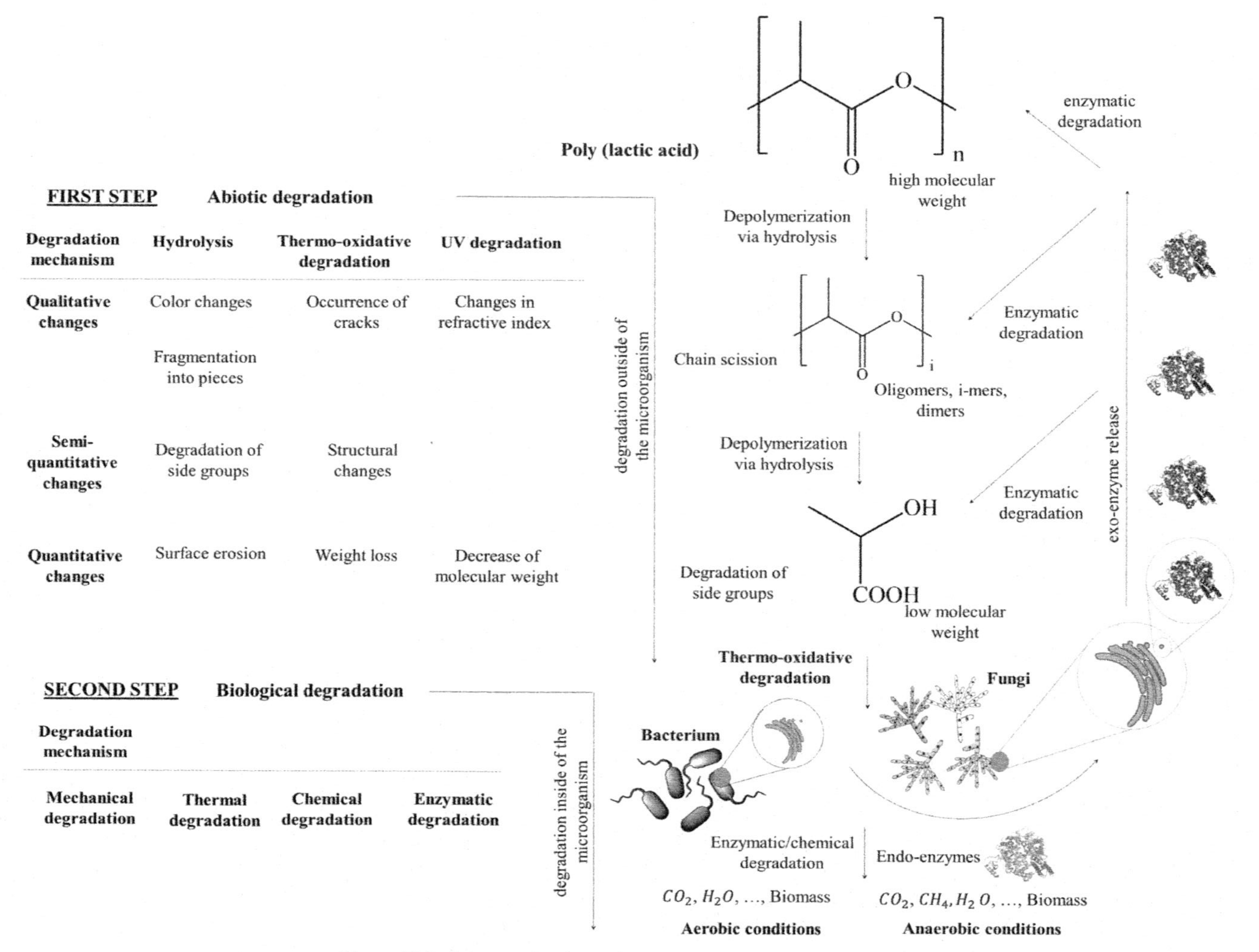

Figure 19.1. Scheme of main biodegradation mechanisms in polymers.

containing heteroatoms also tends to increase the biodegradability. These molecular properties are the essential difference in the biodegradation rate of biodegradable polyesters from other polyesters that are hardly biodegradable such as Poly (Ethylene Terephthalate) (PET). Polymers with acids, alcohols or saccharides side groups are sensitive to biodegradation (Endres and Siebert-Raths, 2011). This type of polymer degradation leads to radical degradation, producing subsequently chain scission or at least a vulnerable chain to oxidation (Lucas et al., 2008).

The biodegradation in a polymer is presented in two main phases: the first one is called primary degradation and is associated with the abiotic degradation of polymer, and in the final degradation the decomposition products are transformed in water, carbon dioxide, methane, biomass and minerals (biological degradation) (Endres and Siebert-Raths, 2011). The total biodegradation of a polymer means all by-products obtained in the decomposition can be considered as compostable. Therefore, a polymer can be classified as compostable when all its organic substances pass through the biotic and a biological degradation. It is considered as compostable from a chemical approach if the by-products produce CO_2, H_2O, CH_4 and biomass in a specific length time that change according with each norm from a few weeks to months.

3.1 Abiotic degradation

As already described, polymers obtained through condensation reactions (polycondensations) are susceptible to the abiotic degradation in the main chain through different mechanisms but especially by the hydrolysis mechanism in C-O and C-N chemical bonds. Besides, degradation can initiate in the presence of a variety of catalysts such as acids, bases, cations, nucleophiles, micellar solutions and phase transfer agents (as in a catalyzed depolymerization) that could be typically found in landfills or soils (in this case in a lower concentration) (Gerald and Scott, 2002). Important factors that affect the abiotic chemical polymer degradation and erosion include the type of chemical bond, pH, temperature, copolymer composition and water absorption (hydrophilicity).

The degradation in polymers via a homolysis or heterolysis route decreases with the increase of cohesion forces, the crystallinity degree, the number of aromatic groups and the molecular weight. The degradation can also increase with the flexibility of polymers, i.e., a lower glass transition temperature (Tg) rises biodegradability. Linear polymers are more susceptible to biodegradation in comparison with unbranched or networks. If a polymer can be swelled with polar molecules as water, biodegradability increases due to a decrease in the chains' concentration. Hence, foamed polymers present more exposed area for biodegradation in abiotic conditions. These features are illustrated in Table 19.2.

On the frontier between abiotic and biological degradation, when microorganisms release exoenzymes, the enzymatic degradation mechanisms could be carried out gradually from the surface inwards, as macromolecular enzymes cannot diffuse into the interior of the material due to their big sizes. Chemical hydrolysis of a solid material can take place throughout its cross-section, except for very hydrophobic polymers (Gerald and Scott, 2002).

3.2 Biological degradation

The biological degradation is achieved by diverse routes: (1) chemical degradation caused by products such as acids and peroxides secreted by microorganisms (bacteria, yeasts, fungi, etc.), or (2) by the action of enzymes or macroorganisms that can ingest fragments of polymers having as a consequence mechanical, chemical or enzymatic aging (Lucas et al., 2008; Müller, 2020).

There are two steps in the biological polymer degradation; a depolymerization or chain scission followed by a mineralization of by-products. Due to the insoluble nature of most polymers in polar media, the first step is carried outside the polymer, when the microorganisms release extracellular polymers as enzymes to decrease the polymer molecular weight and chain length.

When polymers are degraded to oligomers, monomers or by-products, microorganisms can be mineralized to obtain metabolic energy of this process. If enzymes are used as a catalyst in

Table 19.2. Biodegradability, abiotic and biological degradation as function of the structure and properties in polymers.

Structure/property in polymers	**Main chain with heteroatoms (specially oxygen and nitrogen atoms)**	**More than one repeating units (copolymers, terpolymers, etc.)[a]**	**High degree of crystallinity**	**High molecular weight**	**High content of aromatic groups (in the main chain or as side groups)**	**Functional groups**	**Non-linear polymers**	**High surface area**	**Polymer blends[b]**	**Polymer composites[c]**
Abiotic degradation	+	+	-	-	-	+	-	+	-	+
Biological degradation	+	-	-	-	-	-	-	+	-	+
Biodegradability	+	-	-	-	-	-	-	+	-	+

Structure/property in polymers	**Strong intermolecular interactions (e.g., hydrogen bonds)**	**High content of steric hindrance**	**High transition glass temperature**	**Hydrophobicity**	**Hydrophilicity (polarity)/ swelling**	**High thermo-oxidative stability**	**High UV-stability**			
Abiotic degradation	-	-	-	-	+	-	-			
Biological degradation	+	-	-	-	+	-	-			
Biodegradability	+	-	-	-	+	-	-			

[a] Decrease only if the second monomer added a property to the polymer that decreases their biodegradability following the rule of mixtures (statistical copolymers, alternating copolymers, periodic copolymers, etc.)

[b] Decrease if the second polymer added a property that decrease their biodegradability. In polymer blends with phase separation the effect is diminished in comparison with miscible polymer blends.

[c] Increase when the reinforcing fillers contributes to the biodegradability

biological degradation (in which the enzyme activity is related to the conformational structure and the active site), the interaction between the polymer and the enzyme at the active site leads to the chemical degradation of the polymer (Lucas et al., 2008). The optimal activity of enzymes can reach when cofactors, normally inorganic substances as metal ions or from organic origin (such as coenzyme A, ATP and vitamins), are present in biodegradation (Müller, 2020).

The enzymatic degradation of a polymer can initiate with a diffusion-controlled sequence, when the macroscopic disintegration of polymer followed by the hydrolytic process due to the enzyme are voluminous to a high rate of diffusion into the polymers (Endres and Siebert-Raths, 2011). The effects of additives in biodegradation it is not clearly understood. It is reported the acceleration of the degradation rate by the presence of additives (Kjeldsen et al., 2019); however, it depends on its composition and concentration, as a high concentration of inorganic additives can be harmful to microorganisms.

4. Methods for assessment of polymers biodegradation

The most simple (but sometimes not practical) method to identify if a polymer is biodegradable or not is to set the material on specific conditions and measure how long does the polymer take to degrade. Environmental test conditions can be either natural or artificial test conditions carried out in a laboratory. In both cases, the biodegradation test should be performed according to specific standards (Harrison et al., 2018). To have a more realistic evaluation, biodegradability test should be made on a final piece (film, container, etc.) instead of pellets of the raw starting material (Kjeldsen et al., 2019). However, as was mentioned earlier, differences in morphology have a direct impact on the biodegradation rate and the comparison of results may differ with other authors. Therefore, standardized tests suggest performing the test on polymer powders or small films (less than 5 mm × 5 mm) (Filiciotto and Rothenberg, 2021).

Simulated biodegradable environments allow tuning an optimal control of conditions such as pH, temperature, concentration, oxygen content and type of microorganisms. Therefore, it is possible to simulate atypical and accelerated biodegradation conditions. These assays are usually followed by measuring specific properties such as weight loss or tensile strength (Mergaert et al., 1993). However, as natural environments are much more complex than a controlled environment, results may not be straightforward, as other species not present in a controlled environment test may accelerate the degradation process. Moreover, laboratory conditions usually do not consider other carbon sources present in a real environment, which may decrease the polymer decomposition rate (Kjeldsen et al., 2019).

It is difficult to simulate real environmental conditions within a laboratory. However, the main drawback of performing biodegradation tests in real environments is the longer time reflecting exclusively one specific biodegradation environment (Van Der Zee, 2020). Due to the advantages and disadvantages presented, biodegradability should be evaluated in laboratory-controlled conditions and under uncontrolled natural ecosystems (Kjeldsen et al., 2019).

4.1 Biodegradation assesment techniques and analyses

Polymer biodegradability can be evaluated using different techniques and assays. These can be divided into the following groups according to Fig. 19.2.

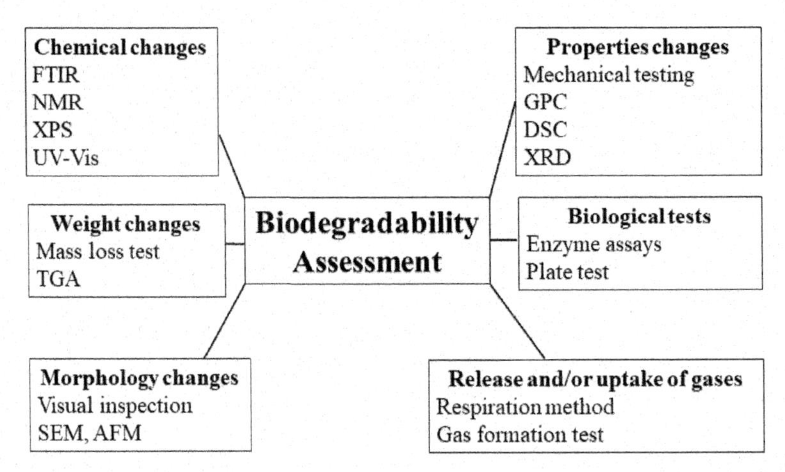

Figure 19.2. Scheme of some of the most applied techniques and analyses to assess biodegradability in polymers.

4.1.1 Morphology changes

Visual inspection is the simplest method to qualitatively evaluate the biodegradation of a material. However, this test can be subjective to the person who performs the analysis. Due to this, special softwares are needed to validate the information obtained (Van Der Zee, 2020). The standard EN 14045 details the photographic documentation criteria to evaluate their effect (Ruggero et al., 2019).

Scanning electron microscopy (SEM). This microscopy technique gives information about the morphology and surface topography of the material before and after degradation. Changes in the homogeneity, particle size and surface roughness, as well as the formation of pores and cracks, can be observed and attributed to biodegradation processes (Corrêa et al., 2008; Reis and Cunha, 1995; Ruggero et al., 2020).

In addition to SEM morphologic surface analyses, changes in roughness for pristine and biodegraded pieces can be performed with a surface roughness tester (Corrêa et al., 2008). Moreover, other microscopy techniques like Atomic Force Microscopy (AFM) can be used to complement the morphological and surface degradation analyses. For instance, AFM was used to quantify the polymer crystals' degradation (Numata et al., 2008).

4.1.2 Release and/or uptake of gases

Respiration test. This assay is based on the principle that in aerobic degradation, degradation of its initial compounds leads to the generation of its mineral constituents such as CO_2 and SO_2. Consequently, the oxygen used during this process, known as the Biological Oxygen Demand (BOD), is measured to determine the level of biodegradation (Van Der Zee, 2020). Standardized norms such as ISO 9408 or ISO 14851 can be applied to determine the oxygen demand using a respirometer. This method is an indirect way for determining biodegradation and it cannot be applied in anaerobic biodegradation.

Gas formation test. This test has the advantage that it can be performed both in aerobic and anaerobic conditions. The products obtained from aerobic and anaerobic processes can be calculated from the gaseous products obtained using Equation (1) and (2) respectively:

$$\textit{Aerobic} \quad C_T = CO_2 + C_R + C_B \tag{1}$$

$$\textit{Anaerobic} \quad C_T = CO_2 + CH_4 + C_R + C_B \tag{2}$$

where C_T is the total carbon content, C_R is the carbon content present in any residue during the degradation process and C_B is the carbon content converted to biomass (Jayasekara et al., 2005). Aerobic conditions allow obtaining CO_2 as a final product; then, the CO_2 concentration produced during composting can be measured as the degradation percentage (Balaguer et al., 2015). On the other hand, anaerobic conditions will lead to a mixture of methane and CO_2 as final products. There are standardized methods as in ISO 17556 and the modified Sturm test found in ISO 9439. A description of this last experiment can be found in Andrady (1998) and in ASTM D5338. These methods can be applied to sea, soil and water conditions (Allen et al., 1994; Filiciotto and Rothenberg, 2021).

4.1.3 Biological assays

Enzyme assay. This assay consists of using a material as a substrate in a controlled environment exposed to different enzymes. This assay is helpful to determine depolymerization kinetics or monomer release (Van Der Zee, 2020). It has the disadvantage that this method is applied for a specific type of polymer (not easy to extrapolate for modified polymers or blends).

Plate test. This method consists of placing the polymer on a petri dish within the synthetic medium agar. The formation of a clear halo demonstrates if polymers will promote the growth of the microorganisms and that they can carry out the first step of degradation detailed earlier (Niaounakis, 2013). Some standardized methods to perform this test can be found in ASTM G21-15 and ISO 846. Despite the fact that this method demonstrates that the microorganism can grow on the polymer, it does not indicate that the polymer is biodegradable (Van Der Zee, 2020).

4.1.4 Weight/mass loss changes

Mass loss. It is a fast, straightforward and widely used method, especially in the early stage of biodegradation (Arévalo-Niño et al., 1996; Jayasekara et al., 2005). The disadvantage of this method is its inaccuracy, as it can only be used with large pieces of plastics. One norm to evaluate this test is ASTM D5247-92. It is important to note that the weight loss of the material does not necessarily mean that it has been mineralized (Folino et al., 2020). For example, weight loss can also occur due to the leaking of additives or other compounds which do not reflect the polymer degradation (Laycock et al., 2017).

Thermogravimetric analysis (TGA). TGA measures the weight loss when a sample is heated at a constant rate or during isothermal conditions. Generally, a TGA curve represents the weight percentage of the sample against temperature, showing several weight loss steps, which are attributed to the decomposition of its individual components. Additionally, it is possible to determine the weight loss derivative versus time or temperature to determine the temperature in which there is maximum weight loss. Through data managing, activation energy can be calculated, where a decrease in this energy is related to hydrolysis reactions on the degraded sample (Ruggero et al., 2020). Moreover, TGA decomposition profiles can be associated to chemical degradations, as degraded samples showed different decomposition profiles than that the non-degraded materials as microorganisms accelerate the degradation rate (Matjašič et al., 2021).

4.1.5 Chemical changes

Fourier-transform infrared spectroscopy (FTIR) is an analytical technique whose operating principle is based on the interaction of matter with the infrared spectra. It is generally known as FTIR as a Fourier transform process is required to obtain the spectrum. This technique generally represents the absorption or transmittance of the sample versus wavenumber, usually between 400 and 4000 cm^{-1}. FTIR bands present in the spectrum can be associated with biodegradation processes essentially by changes in the chemical bonds of polymers. Biodegradation can be identified by the presence of characteristics bands that correspond to specific carbon bonds or amidic protein groups (Kay et al., 1993; Ruggero et al., 2020); changes in the carbonyl index (Matjašič et al., 2021; Tabasi et al., 2015); or peaks attributed to the initial break down of the polymer molecule (Tabasi et al., 2015).

Additional alternatives to follow changes in the polymer structure along biodegradation are proton Nuclear Magnetic Resonance (^1H NMR), for polymers that can be solubilized or swollen with a deuterated solvent and visible ultraviolet spectroscopy (UV-Vis) for the characterization of polymeric materials with aromatic groups, whose changes in the chemical structure are reflected by the appearance or disappearance of peaks respect to the original spectrum.

X-Ray Photoelectron Spectroscopy (XPS) is another method less used than FTIR but also important to determine the chemical composition of the surface (with a depth analysis of about ~ 5 nm from the surface inwards) of a degraded polymer. The sample is irradiated by an X-ray beam to measure its kinetic energy, which is equal to the binding energy. This technique gives information about the composition and nature of chemical bonding (Ebnesajjad, 2006). Corrêa et al. (2008) reported the composition of degraded samples of poly (3-hydroxybutyrate) (PHB) by XPS. In the case of pristine samples, they found only carbon and oxygen peaks attributed to the chemical composition of the polymer. After degradation, other peaks associated with other elements such as calcium, nitrogen and silicon present in soil appeared. An increase in the bond energy related to the C-C and/or C-H bond was due to the hydrolysis effect. Then, the decrease of the intensity of these peaks was blamed for molecule breakage, and to a subsequent decrease of molecular weight.

4.1.6 Changes in physical properties

Mechanical testing. An indirect way to determine biodegradability is through a decrease in mechanical properties. Degradation changes may be attributed from a ductile to brittle failure (Laycock et al., 2017). One norm to follow is ASTM D882-18.

Gel Permeation Chromatography (GPC). It is a type of chromatography generally used in polymer science to determine the molecular weight distribution of polymers. As mentioned earlier, biodegradation in polymers is reflected in a decrease in the polymer molecular weight. Moreover, it is known that the chain length influences biodegradation rate, as shorter chains are easier to degrade (Folino et al., 2020). GPC is a useful technique to monitor the molecular weight when performing a biodegradation test (Siotto et al., 2013). Changes in the number average molecular weight (Mn), the weight average molecular weight (Mw), and the molecular weight dispersity (also called polydispersity index, Đ = Mw/Mn) can be blamed for biodegradation, as microorganisms start the degradation of oligomers and smaller molecular chains (Matjašič et al., 2021).

Differential Scanning Calorimetry (DSC). It is a characterization technique that measures the difference in the energy/heat required to increase the temperature of a sample. It is widely used in polymer science to study the melting (Tm) and crystallization (Tc) temperatures of a polymer and thermal relaxations of long-chain segments of polymers. In that sense, it is reported that changes in the Tg, Tm and Tc can be blamed for the progressive hydrolysis of the polymer chain (Saffian et al., 2016). With the measurement of the heat of fusion, it is possible to quantify the crystallinity percentage with respect to a theoretical 100% pure polymer. As previously stated, chain flexibility and the polymer crystallinity influences the degradation rate, as the crystalline fraction of the polymer is more difficult to degrade compared to the amorphous ones (Massardier-Nageotte et al., 2006); therefore, an increase in crystallinity of degraded samples is observed (Corrêa et al., 2008). In addition to this, DSC has been probed as a tool not only for qualitative but for quantitative analyses of biodegradation (Capitain et al., 2020). In addition to DSC, other techniques like X-Ray Diffraction (XRD) can be used to complement the crystallinity analyses.

4.2 Standards and certifications

As mentioned above, assessment of biodegradability is usually carried out according to specific normalized assay. These assay are performed according to the American Society for Testing and Materials (ASTM); The International Organization for Standardization (ISO); the European Committee for Standardization (CEN); and other norms from countries. Biodegradability standards can be applied depending on conditions; for example, anaerobic conditions of biodegradability can be assessed with ASTM D5210-92, D5526-94 and ISO 14853; for aerobic conditions, the norm ASTM D5271-02 can be used. According to the environment, biodegradability of polymers in soils can be assessed with ASTM D5988-18; for composting D6002-96 or ISO 17088; and for aerobic wastewater ISO 14852.

Referring to the certifications, one European certification applied for biodegradable plastics is the TÜV AUSTRIA Certification. This certification, earlier known as Vinçotte, certificates products according to their compostability, biodegradability or biobased properties. In the case of biodegradability, a mark is indicated depending on the environment as 'OK biodegradable MARINE', 'OK biodegradable SOIL', and 'OK biodegradable WATER'. For the compostability certification, they assign the mark 'OK compost INDUSTRIAL' according to EN 13432 and 'OK compost HOME'.

Conclusions

According to Emily Woglom, Executive Vice President of the Ocean Conservancy stated: "*We all have an opportunity to do something that makes the world a better place from where we are*". This implies a re-assessment of the approach to determine when a polymer can be considered as a biodegradable (Conservancy, 2017). If the following three factors: polymer structure, the abiotic conditions of polymer exposure (conditions of the environment) and the type and activity of microorganism(s) are considered, international regulations can be specified with precision, and so the opportunities that obtain not only a polymer waste decrease but also an optimal assessment of biodegradability of polymers using a broad spectrum of microorganisms in laboratory conditions.

The aim of this chapter was to expose the complexity to determine the biodegradability of some polymers, as it depends mainly on the three factors mentioned. Although testing in real conditions is the most reliable method to evaluate its biodegradability, sometimes the duration of the assay is not practical, and it applies only to the specific conditions tested. On the other hand, the disadvantage of laboratory testing is that it is not possible to completely resemble natural ecosystems. Therefore, a whole set of different analyses, in real and simulated conditions, such as ones the mentioned in this chapter, must be performed to evaluate the biodegradability of polymers. These analyses include changes in visual appearance, physical and chemical properties, biological activity and mineralization processes. Moreover, the presence of micro and nanoplastics and the influence of other components such as additives in polymer biodegradation are a remaining challenge for biodegradability assessment. A complete characterization, preferably carried out with normalized methods and standards, will support and define with adequate metrics the environmental impact caused by biodegradable polymers.

References

Adhikari, D., Mukai, M., Kubota, K., Kai, T., Kaneko, N., Araki, K. S. and Kubo, M. (2016). Degradation of bioplastics in soil and their degradation effects on environmental microorganisms. *Journal of Agricultural Chemistry and Environment*, 5: 23–34.

Ahmed, T., Shahid, M., Azeem, F., Rasul, I., Shah, A. A., Noman, M., Hameed, A., Manzoor, N., Manzoor, I. and Muhammad, S. (2018). Biodegradation of plastics: current scenario and future prospects for environmental safety. *Environmental Science and Pollution Research,* 25: 7287–7298.

Allen, A. L., Mayer, J., Stote, R. and Kaplan, D. L. (1994). Simulated marine respirometry of biodegradable polymers. *Journal of Environmental Polymer Degradation*, 2(4): 237–244.

Alshehrei, F. (2017). Biodegradation of synthetic and natural plastic by microorganisms. *Journal of Applied & Environmental Microbiology*, 5(1): 8–19.

American Society for Testing and Materials. (1992). *Standard Test Method for Determining the Aerobic Biodegradability of Degradable Plastics by Specific Microorganisms* (ASTM D5247-92). Retrieved from www.astm.org.

American Society for Testing and Materials. (1996). *Standard Guide for Assessing the Compostability of Environmentally Degradable Plastics* (ASTM D6002-96). Retrieved from www.astm.org.

American Society for Testing and Materials. (1996). *Standard Practice for Determining Resistance of Plastics to Bacteria* (ASTM G22-76). Retrieved from www.astm.org.

American Society for Testing and Materials. (2002). *Standard Test Method for Determining the Aerobic Biodegradation of Plastic Materials in an Activated-Sludge-Wastewater-Treatment System* (ASTM D5271-02). Retrieved from www.astm.org.

American Society for Testing and Materials. (2007). *Standard Test Method for Determining the Anaerobic Biodegradation of Plastic Materials in the Presence of Municipal Sewage Sludge* (ASTM D5210-92). Retrieved from www.astm.org.

American Society for Testing and Materials. (2011). *Standard Test Method for Determining Anaerobic Biodegradation of Plastic Materials Under Accelerated Landfill Conditions* (ASTM D5526-94). Retrieved from www.astm.org.

American Society for Testing and Materials. (2018). *Standard Test Method for Determining Aerobic Biodegradation of Plastic Materials in Soil* (ASTM D5988-18). Retrieved from www.astm.org.

American Society for Testing and Materials. (2018). *Standard Test Method for Tensile Properties of Thin Plastic Sheeting* (ASTM D882-18). Retrieved from www.astm.org.

American Society for Testing and Materials. (2021). *Standard Test Method for Determining Aerobic Biodegradation of Plastic Materials Under Controlled Composting Conditions, Incorporating Thermophilic Temperatures* (ASTM D5338-15). Retrieved from www.astm.org.

American Society for Testing and Materials. (2021). *Standard Practice for Determining Resistance of Synthetic Polymeric Materials to Fungi* (ASTM G21-15). Retrieved from www.astm.org.

Andrady, A. L. (1998). Biodegradation of plastics: monitoring what happens. pp. 32–40. *In*: Pritchard, G. (ed.). *Plastics Additives*. Dordrecht: Springer.

Asghari, F., Samiei, M., Adibkia, K., Akbarzadeh, A. and Davaran, S. (2017). Biodegradable and biocompatible polymers for tissue engineering application: a review. *Artificial Cells, Nanomedicine, and Biotechnology*, 45(2): 185–192.

Arévalo-Niño, K., Sandoval, C. F., Galan, L. J., Imam, S. H., Gordon, S. H. and Greene, R. V. (1996). Starch-based extruded plastic films and evaluation of their biodegradable properties. *Biodegradation*, 7(3): 231–237.

Bahl, S., Dolma, J., Singh, J. J. and Sehgal, S. (2021). Biodegradation of plastics: A state of the art review. *Materials Today: Proceedings*, 39(1): 31–34.

Balaguer, M. P., Villanova, J., Cesar, G., Gavara, R. and Hernandez-Munoz, P. (2015). Compostable properties of antimicrobial bioplastics based on cinnamaldehyde cross-linked gliadins. *Chemical Engineering Journal*, 262: 447–455.

Brahney, J., Hallerud, M., Heim, E., Hahnenberger, M. and Sukumaran, S. (2020). Plastic rain in protected areas of the United States. *Science*, 368(6496): 1257–1260.

Briassoulis, D. and Dejean, C. (2010). Critical review of norms and standards for biodegradable agricultural plastics Part I. Biodegradation in soil. *Journal of Polymers and the Environment*, 18(3): 384–400.

Capitain, C., Ross-Jones, J., Möhring, S. and Tippkötter, N. (2020). Differential scanning calorimetry for quantification of polymer biodegradability in compost. *International Biodeterioration and Biodegradation*, 149: 104914.

Chia, W. Y., Tang, D. Y. Y., Khoo, K. S., Lup, A. N. K. and Chew, K. W. (2020). Nature's fight against plastic pollution: Algae for plastic biodegradation and bioplastics production. *Environmental Science and Ecotechnology*, 4: 100065.

Chinaglia, S., Tosin, M. and Degli-Innocenti, F. (2018). Biodegradation rate of biodegradable plastics at molecular level. *Polymer Degradation and Stability*, 147: 237–244.

Conservancy, O. (2017). Together for our Ocean-International Coastal Cleanup 2017 Report. *IC Cleanup*.

Corrêa, M. C. S., Rezende, M. L., Rosa, D. S., Agnelli, J. A. M. and Nascente, P. A. P. (2008). Surface composition and morphology of poly(3-hydroxybutyrate) exposed to biodegradation. *Polymer Testing*, 27(4): 447–452.

De Souza, Machado, A. A., Kloas, W., Zarfl, C., Hempel, S. and Rillig, M. C. (2018). Microplastics as an emerging threat to terrestrial ecosystems. *Global Change Biology*, 24(4): 1405–1416.

Doble, M. (2014). Biodegradation of polymers in marine environment. pp. 73–100. *In*: Muthukumar, T. and Doble, M. (eds.). *Polymers in a Marine Environment*. Shawbury, United Kingdom: Smithers Rapra.

Dussud, C., Hudec, C., George, M., Fabre, P., Higgs, P., Bruzaud, S., Delort, A. M., Eyheraguibel, B., Meistertzheim1, A. L., Jacquin, J., Cheng, J., Callac, N., Odobel, C., Rabouille, S. and Ghiglione, J. F. (2018). Colonization of non-biodegradable and biodegradable plastics by marine microorganisms. *Frontiers in Microbiology*, 9: 1571–1584.

Ebnesajjad, S. (2006). Surface and material characterization techniques. pp. 43–75. *In*: Ebnesajjad, S. (ed.). *Surface Treatment of Materials for Adhesion Bonding*. Walthman, MA, USA: William Andrew Publishing.

Emadian, S. M., Onay, T. T. and Demirel, B. (2017). Biodegradation of bioplastics in natural environments. *Waste Management*, 59: 526–536.

Endres, H. J. and Siebert-Raths, A. (2011). Chapter 2. State of knowledge. pp. 19–44. *In*: *Engineering biopolymers Markets, Manufacturing, Properties and Applications*. Krugzell, Germany: Hanser.

European Committee for Standardization. (2001). *Requirements for Packaging Recoverable Through Composting and Biodegradation Test Scheme and Evaluation Criteria for the Final Acceptance of Packaging* (EN 13432:2000). Retrieved from https://www.cen.eu/Pages/default.aspx.

Field, J. A. (2002). Limits of anaerobic biodegradation. *Water Science and Technology*, 45(10): 9–18.

Filiciotto, L. and Rothenberg, G. (2021). Biodegradable plastics: standards, policies, and impacts. *ChemSusChem*, 14(1): 56–72.

Folino, A., Karageorgiou, A., Calabrò, P. S. and Komilis, D. (2020). Biodegradation of wasted bioplastics in natural and industrial environments: A review. *Sustainability*, 12(15): 6030.

Gerald, S. and Scott, G. (2002). *Degradable Polymers: Principles and Applications*. Netherlands: Kluwer Academic Publishers.

Gigault, J., Halle, A. ter, Baudrimont, M., Pascal, P. Y., Gauffre, F., Phi, T. L., El Hadri, H., Grassl, B. and Reynaud, S. (2018). Current opinion: What is a nanoplastic? *Environmental Pollution*, 235: 1030–1034.

Harrison, J. P., Boardman, C., O'Callaghan, K., Delort, A. M. and Song, J. (2018). Biodegradability standards for carrier bags and plastic films in aquatic environments: A critical review. *Royal Society Open Science*, 5(5): 171792.

Harrison, J. P., Hoellein, T. J., Sapp, M., Tagg, A. S., Ju-Nam, Y. and Ojeda, J. J. (2018). Microplastic-associated biofilms: a comparison of freshwater and marine environments. pp. 181–201. *In*: Wagner, M. and Lambert, S. (eds.). *Freshwater Microplastics*. Springer, Cham.

Horton, A. A., Walton, A., Spurgeon, D. J., Lahive, E. and Svendsen, C. (2017). Microplastics in freshwater and terrestrial environments: Evaluating the current understanding to identify the knowledge gaps and future research priorities. *Science of the Total Environment*, 586: 127–141.

International Organization for Standardization. (1997). *Plastics – Evaluation of the Action of Microorganisms* (ISO 846:1997). Retrieved from https://www.iso.org/home.html.

International Organization for Standardization. (1999). *Water Quality—Evaluation of Ultimate Aerobic Biodegradability of Organic Compounds in Aqueous Medium by Determination of Oxygen Demand in a Closed Respirometer* (ISO 9408:1999). Retrieved from https://www.iso.org/home.html.

International Organization for Standardization. (1999). *Water Quality—Evaluation of Ultimate Aerobic Biodegradability of Organic Compounds in Aqueous Medium—Carbon Dioxide Evolution Test* (ISO 9439:1999). Retrieved from https://www.iso.org/home.html.

International Organization for Standardization. (2016). *Plastics—Determination of the Ultimate Anaerobic Biodegradation of Plastic Materials in an Aqueous System—Method by Measurement of Biogas Production* (ISO 14853:2016). Retrieved from from https://www.iso.org/home.html.

International Organization for Standardization. (2018). *Determination of the Ultimate Aerobic Biodegradability of Plastic Materials in an Aqueous Medium—Method by Analysis of Evolved Carbon Dioxide* (ISO 14852:2018). Retrieved from https://www.iso.org/home.html.

International Organization for Standardization. (2018). *Determination of the Ultimate Aerobic Biodegradability of Plastic Materials Under Controlled Composting Conditions—Method by Analysis of Evolved Carbon Dioxide—Part 2: Gravimetric Measurement of Carbon Dioxide Evolved in a Laboratory-Scale Test* (ISO 14855-2:2018). Retrieved from https://www.iso.org/home.html.

International Organization for Standardization. (2019). *Determination of the Ultimate Aerobic Biodegradability of Plastic Materials in an Aqueous Medium—Method by Measuring the Oxygen Demand in a Closed Respirometer* (ISO 14851:2019). Retrieved from https://www.iso.org/home.html.

International Organization for Standardization. (2019). *Determination of the Ultimate Aerobic Biodegradability of Plastic Materials in Soil by Measuring the Oxygen Demand in a Respirometer or the Amount of Carbon Dioxide Evolved* (ISO 17556:2019). Retrieved from https://www.iso.org/home.html.

International Organization for Standardization. (2021). *Plastics—Organic Recycling—Specifications for Compostable Plastics* (ISO 17088:2021). Retrieved from https://www.iso.org/home.html.

Ishigaki, T., Sugano, W., Nakanishi, A., Tateda, M., Ike, M. and Fujita, M. (2004). The degradability of biodegradable plastics in aerobic and anaerobic waste landfill model reactors. *Chemosphere*, 54(3): 225–233.

Jayasekara, R., Harding, I., Bowater, I. and Lonergan, G. (2005). Biodegradability of a selected range of polymers and polymer blends and standard methods for assessment of biodegradation. *Journal of Polymers and the Environment*, 13(3): 231–251.

Kay, M. J., McCabe, R. W. and Morton, L. H. G. (1993). Chemical and physical changes occurring in polyester polyurethane during biodegradation. *International Biodeterioration and Biodegradation*, 31(3): 209–225.

Kjeldsen, A., Price, M., Lilley, C., Guzniczak, E. and Archer, I. (2019). A Review of Standards for Biodegradable Plastics [online]. 2019. Available from: URL: https://bioplasticsnews.com/wp-content/uploads/2019/07/review-standards-for-biodegradable-plastics-IBioIC.pdf.

Kumar, A. G., Hinduja, M., Sujitha, K., Rajan, N. N. and Dharani, G. (2021). Biodegradation of polystyrene by deep-sea Bacillus paralicheniformis G1 and genome analysis. *Science of The Total Environment*, 774: 145002.

Kumar, R., Pandit, P., Kumar, D., Patel, Z., Pandya, L., Kumar, M., Joshi, C. and Joshi, M. (2021). Landfill microbiome harbour plastic degrading genes: A metagenomic study of solid waste dumping site of Gujarat, India. *Science of The Total Environment*, 779: 146184.

La Mantia, F. P., Morreale, M., Botta, L., Mistretta, M. C., Ceraulo, M. and Scaffaro, R. (2017). Degradation of polymer blends: A brief review. *Polymer Degradation and Stability*, 145: 79–92.

Laycock, B., Nikolić, M., Colwell, J. M., Gauthier, E., Halley, P., Bottle, S. and George, G. (2017). Lifetime prediction of biodegradable polymers. *Progress in Polymer Science*, 71: 144–189.

Lucas, N., Bienaime, C., Belloy, C., Queneudec, M., Silvestre, F. and Nava-Saucedo, J. E. (2008). Polymer biodegradation: Mechanisms and estimation techniques—A review. *Chemosphere*, 73(4): 429–442.

Massardier-Nageotte, V., Pestre, C., Cruard-Pradet, T. and Bayard, R. (2006). Aerobic and anaerobic biodegradability of polymer films and physico-chemical characterization. *Polymer Degradation and Stability*, 91(3): 620–627.

Matjašič, T., Simčič, T., Medvešček, N., Bajt, O., Dreo, T. and Mori, N. (2021). Critical evaluation of biodegradation studies on synthetic plastics through a systematic literature review. *Science of the Total Environment*, 752: 141959.

Mergaert, J., Webb, A., Anderson, C., Wouters, A. and Swings, J. (1993). Microbial degradation of poly (3-hydroxybutyrate) and poly (3-hydroxybutyrate-co-3-hydroxyvalerate) in soils. *Applied and Environmental Microbiology*, 59(10): 3233–3238.

Moore, C. J., Lattin, G. L. and Zellers, A. F. (2005). Density of plastic particles found in zooplankton trawls from coastal waters of California to the North Pacific Central Gyre. pp. 1–6. *In*: *The Plastic Debris Rivers to Sea Conference*. California, USA.

Moore, C. J. (2015). How much plastic is in the ocean? You tell me. *Marine Pollution Bulletine*, 92: 1–3.

Muenmee, S., Chiemchaisri, W. and Chiemchaisri, C. (2016). Enhancement of biodegradation of plastic wastes via methane oxidation in semi-aerobic landfill. *International Biodeterioration & Biodegradation*, 113: 244–255.

Müller, R. (2020). 2. Biodegradation behaviour of polymers in liquid environments. pp. 29–54. *In*: Bastioli, C. (ed.). *Handbook of Biodegradable Polymers*. Shawbury, United Kingdom: Walter de Gruyter GmbH & Co KG.

Muller, G., Laine, J. P. and Fenyo, J. C. (1979). High-molecular-weight hydrolyzed polyacrylamides. I. Characterization. Effect of salts on the conformational properties. *Journal of Polymer Science: Polymer Chemistry Edition*, 17(3): 659–672.

Niaounakis, M. (2013). Definitions and assessment of (Bio)degradation. pp. 77–94. *In*: *Biopolymers: Reuse, Recycling, and Disposal*. New York, United States of America: William Andrew.

Numata, K., Abe, H. and Doi, Y. (2008). Enzymatic processes for biodegradation of poly(hydroxyalkanoate)s crystals. *Canadian Journal of Chemistry*, 86(6): 471–483.

PlasticsEurope. Plastics - the Facts 2020. (2020). Available from: URL: https://www.plasticseurope.org/en/resources/publications/4312-plastics-facts-2020.

Reis, R. L. and Cunha, A. M. (1995). Characterization of two biodegradable polymers of potential application within the biomaterials field. *Journal of Materials Science: Materials in Medicine*, 6(12): 786–792.

Rillig, M. C. (2012). Microplastic in terrestrial ecosystems and the soil? *Enviromental Science and Technology*, 46(12): 6453–6454.

Ruggero, F., Gori, R. and Lubello, C. (2019). Methodologies to assess biodegradation of bioplastics during aerobic composting and anaerobic digestion: A review. *Waste Management and Research*, 37(10): 959–975.

Ruggero, F., Carretti, E., Gori, R., Lotti, T. and Lubello, C. (2020). Monitoring of degradation of starch-based biopolymer film under different composting conditions, using TGA, FTIR and SEM analysis. *Chemosphere*, 246: 125770.

Saffian, H. A., Abdan, K., Hassan, M. A., Ibrahim, N. A. and Jawaid, M. (2016). Characterisation and biodegradation of poly(lactic acid) blended with oil palm biomass and fertiliser for bioplastic fertiliser composites. *BioResources*, 11(1): 2055–2070.

Sánchez, C. (2020). Fungal potential for the degradation of petroleum-based polymers: An overview of macro-and microplastics biodegradation. *Biotechnology Advances*, 40: 107501.

Scott, G. (2002). Chapter 1. Why degradable polymers? pp. 1–15. *In*: *Degradable Polymers: Principles and Applications*. Dordrecht, Netherlands: Springer-Science+Bussines Media, B.V.

Sekiguchi, T., Sato, T., Enoki, M., Kanehiro, H., Uematsu, K. and Kato, C. (2011). Isolation and characterization of biodegradable plastic degrading bacteria from deep-sea environments. *JAMSTEC Report of Research and Development*, 11: 33–41.

Shim, W. J., Hong, S. H. and Eo, S. (2018). Chapter 1. Marine microplastics: abundance, distribution, and composition. pp. 1–26. *In*: *Microplastic Contamination in Aquatic Environments*. Oxford, United Kingdom: Elsevier.

Siotto, M., Zoia, L., Tosin, M., Degli Innocenti, F., Orlandi, M. and Mezzanotte, V. (2013). Monitoring biodegradation of poly(butylene sebacate) by gel permeation chromatography, ^{1}H-NMR and ^{31}P-NMR techniques. *Journal of Environmental Management*, 116: 27–35.

Sivan, A. (2011). New perspectives in plastic biodegradation. *Current Opinion in Biotechnology*, 22(3): 422–426.

Song, J. H., Murphy, R. J., Narayan, R. and Davies, G. B. H. (2009). Biodegradable and compostable alternatives to conventional plastics. *Philosophical Transactions of the Royal Society B: Biological Sciences*, 364(1526): 2127–2139.

Smagin, A. V., Sadovnikova, N. B., Vasenev, V. I. and Smagina, M. V. (2018). Biodegradation of some organic materials in soils and soil constructions: Experiments, modeling and prevention. *Materials*, 11(10): 1889.

Suhrhoff, T. J. and Scholz-Böttcher, B. M. (2016). Qualitative impact of salinity, UV radiation and turbulence on leaching of organic plastic additives from four common plastics—A lab experiment. *Marine Pollution Bulletin*, 102(1): 84–94.

Swift, G. (1992). Biodegradability of polymers in the environment: complexities and significance of definitions and measurements. *FEMS Microbiology Reviews*, 9(2-4): 339–345.

Tabasi, R. Y. and Ajji, A. (2015). Selective degradation of biodegradable blends in simulated laboratory composting. *Polymer Degradation and Stability*, 120: 435–442.

Tokiwa, Y., Calabia, B. P., Ugwu, C. U. and Aiba, S. (2009). Biodegradability of plastics. *International Journal of Molecular Sciences*, 10(9): 3722–3742.

Tosin, M., Weber, M., Siotto, M., Lott, C. and Innocenti, F. D. (2012). Laboratory test methods to determine the degradation of plastics in marine environmental conditions. *Frontiers in Microbiology*, 3: 1–9.

Triebskorn, R. Braunbeck, T., Grummt T., Hanslik, L., Huppertsberg, S., Jekel, M., Knepper, T. P., Krais, S., Müller, Y. K. Pittroff, M., Ruhl, A. S., Schmieh, H., Schür, C., Strobel, C., Wagner, M., Zumbülte, N. and Köhlera, H.-R. (2019). Relevance of nano-and microplastics for freshwater ecosystems: A critical review. *TrAC Trends in Analytical Chemistry*, 110: 375–392.

Ulery, B. D., Nair, L. S. and Laurencin, C. T. (2011). Biomedical applications of biodegradable polymers. *Journal of Polymer Science Part B: Polymer Physics*, 49(12): 832–864.

Van Der Zee, M. (2020). Methods for evaluating the biodegradability of environmentally degradable polymers. pp. 1–21. *In*: Bastioli, C. (ed.). *Handbook of Biodegradable Polymers*. Shawbury, United Kingdom: Walter de Gruyter GmbH & Co KG.

Zambrano, M. C., Pawlak, J. J., Daystar, J., Ankeny, M., Goller, C. C. and Venditti, R. A. (2020). Aerobic biodegradation in freshwater and marine environments of textile microfibers generated in clothes laundering: Effects of cellulose and polyester-based microfibers on the microbiome. *Marine Pollution Bulletin*, 151: 110826.

Chapter 20

Biodegradable Packaging

Colombian Coffee Industry

Germán Antonio Arboleda Muñoz,[1,*] *Lily Marcela Palacios,*[1]
Hugo Portela Guarín[2] and *Héctor Samuel Villada Castillo*[1]

1. Introduction

The use of plastics in agriculture has had an increasing expansion in the last 50 years, due to the increase in crop yields, related to technological requirements such as soil moisture retention, weed control, reduction of soil compaction, necessary for an increase in crop quality (Briassoulis and Giannoulis, 2018).

However, these materials are mainly composed of non-biodegradable plastic resins, which have a major focus of expansion in developing economies such as China and India. The global agricultural films market has been segmented into linear low density polyethylene (lldpe), low density polyethylene (ldpe), high density polyethylene (hdpe) and ethylene vinyl acetate/ethylene butyl acrylate (eva/eba). The other film segment includes polymerization of vinyl chloride (pvc) and ethylene vinyl alcohol (evoh). In relation to application segments, greenhouses and mulch accounted for more than 75% market share in 2017 due to its wide consumption in agriculture. In terms of value, Asia Pacific dominated the global market for that year and is expected to continue to lead this trend in future years, where this market is also expected to expand in Latin America (Transparency Market Research, 2019).

In this sense, the use of plastic materials in agriculture has been extended to a wide range of applications; however, their poor disposal has led to a complex situation of contamination in the fields. This situation began in Colombia in the 1970s and 1980s with the installation of greenhouses for the cultivation of flowers and its use has extended to tunnels, microtunnels, mulching, shading and bagging nets ss; Stavisky, 2010; Espi et al., 2006 cited by de Polania and Peña, 2013.

Faced with this reality, several efforts are being initiated in the country to generate alternatives to these types of elements to be replaced by biodegradable polymers (de Polania and Peña, 2013), which can also be adapted to other contexts where these types of problems are also involved. In this sense, the increased development of bio-based agricultural films could transform the market scenario in the near future, associated with their environmentally responsible characteristics (Transparency Market Research, 2019).

[1] Research Group on Science and Technology of Biomolecules of Agroindustrial Interest. Faculty of Agricultural Sciences. University of Cauca. Popayán, Cauca.
[2] Antropos Research Group. Faculty of Human and Social Sciences. University of Cauca. Popayán, Cauca.
* Corresponding author: garboleda@unicauca.edu.co

1.1 Experiences with biodegradable polymers in agriculture

The main application of biodegradable polymers has been the replacement of mulches or mulch-type films in crops that require them. These plastic mulches are used to prevent the growth of weeds, reduce the evaporation of irrigation water to increase moisture retention, prevent the contact of fruits with the soil (De Polanía and Peña, 2013), promote higher crop quality by reducing soil erosion and greater retention of nutrients (Virtanen et al., 2018). This application will maintain its growing demand, especially for intensive cultivation of cereals and vegetables in developing countries (Yang et al., 2015).

In this regard, China accounts for 60% of the global demand for films for agricultural purposes, leading the countries that have adopted plastic mulching (Yang et al., 2015). Likewise, crops where experiments have been conducted to test biodegradable polymers include pak-choi (Liu and Yao, 2018) and Chinese cabbage (Tan et al., 2016) in China; pumpkin (Saglam et al., 2017; Hayes et al., 2017), tomato, bell pepper (Wortman et al., 2015; Li et al., 2014) and broccoli (Cowan et al., 2013) in the United States; sweet bell pepper, in Spain (Moreno et al., 2017); vineyards in France (Touchaleaume et al., 2016); strawberries (Morra et al., 2016), melon (Iapichino et al., 2014) and Japanese privet (Sartore et al., 2013) in Italy and strawberries in Portugal (Costa et al., 2014; Andrade et al., 2014).

Different studies have been conducted around the world, with a wide diversity of materials and tested on different types of crops. However, commercial polymers such as the so-called Mater Bi® have extended their application in agriculture (Briassoulis and Giannoulis, 2018; Moreno et al., 2017; Touchaleaume et al., 2016; Barragán et al., 2016; Morra et al., 2016; Iapichino et al., 2014; Costa et al., 2014; Andrade et al., 2014).

The so-called Mater Bi® from the Italian company Novamont, made from biodegradable aliphatic/aromatic polyesters and starch, has been used in several studies. Briassoulis and Giannoulis (2018) evaluated the functionality of bio-coated films, made from commercial polymers such as Mater Bi® Reference M15 (Novamont), Ecovio Rerencia M12 (BASF) compared against low density polyethylene and found that the performance of biodegradable materials was different from that presented by traditional plastics, associated with an early reduction of maximum elongation at the breaking point and a higher rate with respect to water vapor permeability.

Similarly, Moreno et al., in addition to Mater Bi, evaluated the deterioration pattern of five other biodegradable materials under real conditions, during the useful life of the mulches and after incorporation into the soil. The materials included were Sphere 4 and Sphere 6 (thermoplastic potato starch), Bioflex (polylactic acid), Ecovio and MimGreen biodegradable paper. The main results reflected a higher deterioration of biodegradable plastics, which could generate less weed control, while they were not covered by the crop, indicating that they should be employed in low cover crops (Moreno et al., 2017).

Accordingly, Brodhagen et al. (2014) showed that within the commercial polymers employed for use in agriculture, it is important to take into account their degradation rates in soil; from polyethylene (PE) with an extremely low rate, followed by polymers such as polylactic acid (PLA) and poly(trimethylene terephthalate) (PTT) with low rates; polybutylene succinate (PBS), polybutylene succinate adipate (PBSA) and Poly(Butylene Adipate-co-Terephthalate) (PBAT) with moderately low rates; polycaprolactone (PCL), polyhydroxybutyrate (PHB) and polyhydroxyvalerate (PHV) with moderate rates; cellulose with moderately high rate and thermoplastic starch with high degradation rate in soil.

On the other hand, Innocenti (2005) described the biodegradation of polymers in the soil, where he established that polymers end their cycle in the soil by two routes, through agriculture or garbage. Regarding the end of the cycle in agriculture, there are three routes of entry: composting, mulching or other agricultural practices. In composting, the material is partially disintegrated mainly by microbial action, while in mulching the main action comes from sunlight, heating and microbial action.

Accordingly, the main factors active on the soil surface that generate a possible effect on polymer degradation are: sunlight (UV), which induces photochemical reactions that increase the fragility of the material; rainfall and irrigation, which increase water activity producing hydrolysis, separation of plasticizers and reduction of molecular weight accompanied by microbial growth; and macro-organisms, which generate fragility by physical action and an increase in the exposed area that can increase biodegradation (Innocenti, 2005).

In addition, there are also active factors in soils that can affect degradation such as: soil texture and structure; heating, associated with temperature changes that can generate hydrolysis; soil composition; water activity, which promotes hydrolysis and microbial growth; alkaline and acidic components, which induce hydrolysis and influence enzyme activity; and air, which determines O_2 and CO_2 content (Innocenti, 2005).

On the other hand, alternatives for the development of mulch-type materials have also been proposed. Commercial bioplastics used for the development of mulch-type films for agricultural applications are mostly starch-based thermoplastics developed with conventional plastic processing technologies (Virtanen et al., 2018), associated, for example, with the high rate of degradation in soil that these types of polymers have.

Among the advantages of starch, in addition to its availability, degradability and low cost, it is also associated with the possibility of being processed using the extrusion method, which is a technology used for the production of traditional plastics such as the combined application of extrusion with blown extrusion, with which it is also possible to obtain thermoplastic starch films (TPS) (Combrzyński et al., 2013). However, starch due to its hydrophilicity presents several affectations due to the influence of moisture, which negatively influence its mechanical properties.

To counteract this situation, starch is processed together with other additives such as plasticizers, so that through an extrusion process it is possible to obtain the so-called thermoplastic starch (TPS) that after being pelletized into granules is mixed with other types of hydrophobic polymers such as polylactic acid (PLA) (Arboleda et al., 2015), in order to neutralize possible deficiencies that may arise. However, although polylactic acid (PLA) and starch are biodegradable polymers derived from renewable sources, the applications of PLA may be limited by its high cost (Zuo et al., 2017).

In view of this, Fahrngruber et al., 2017 explained the possible mechanisms that may occur in a starch-polyester blend processed under blown extrusion: (1) The coupling agent replaces the TPS-glycerol interactions, but simultaneously forms less elastic bonds/impediments through its two functional carboxyl groups, which in turn affect the viscoelastic properties; (2) During blown extrusion, the hydrolytic degradation could be further accelerated, which in turn (3) affects the interfacial adhesion between polyester and dispersed TPS.

Regarding the challenges for future developments, these are framed in the degradation rate of the buried portion, non-sustained decomposition or soil assimilation after degradation and soil contamination problems that could still generate, as well as moisture resistance, deterioration of mechanical properties, added to the higher cost versus traditional plastics (Yang et al., 2015).

2. Plastics in coffee growing

After oil, aluminum, wheat and coal, coffee is one of the most important commodities traded in the international financial markets of the world economy and international trade. However, some of the adverse effects caused by coffee production include: soil erosion, intensive use of pesticides that pose health risks, large volumes of waste material in the form of pulp, bad odors, wastewater discharges into watersheds or streams (Pardo et al., 2012) and the use of plastic materials that cause strong pressure on ecological systems (Solano, 2015).

Colombian coffee growing is a sector of special relevance for the economic and social stability of the country and is emerging as an articulating axis for rural development, that is, it is an engine of development for the rural economy, which generates about 785,000 direct jobs equivalent to 26% of total agricultural employment, where the value of the harvest amounting to 5.2 billion of colombian peso (COP) is distributed among the more than 550 thousand families residing in 595 municipalities in the country (Muñoz, 2014).

The use and abuse of non-biodegradable bags from petrochemical sources for the production of coffee seedlings has generated plastic residues that have negative effects and are a risk to human health and the environment. These plastic wastes are burned, buried or dumped in open fields due to their high accumulation volume and slow natural degradation (Palacios et al., 2016).

It is important to highlight that the coffee growing area in Colombia exceeds 4,700 million trees, which are distributed in more than 911,000 hectares throughout the country (Federación Nacional de Cafeteros, 2017). Currently, coffee growers use a planting density of around 5,000 trees per hectare or more (Farfán and Sánchez, 2016), where for each of these trees that have been planted, when they were in the seedling stage it was necessary to use bags for the chapolas that were later transplanted to a definitive land, such activity has been increasing the progressive deterioration of ecosystems, a situation for which the development of packaging from renewable sources for the agricultural sector is one of the possible solutions of greatest interest.

Taking the cultivation stage as a central focus, at the beginning of this stage, bags are used for a period of 3 to 6 months, which are made from polyethylene, where a bag is filled with soil substrate, where the coffee seedling is placed to continue its growth period.

To understand this, it is worth keeping in mind the coffee cultivation process, which begins with a germinator where the seed is placed on a volume of moistened sand until it reaches a phase of growth, which allows observing a completely open chapola after about 2 months. Then it passes to the "seedling" phase, where bags filled with soil substrate are grouped together, where the seedling is sown and organized under shade to be protected from the sun's rays. Once the time estimated by the producer has elapsed, these bags with the coffee seedlings are taken to the cultivation site, where the bags are removed and the coffee seedlings are ready to be planted.

At this point, if not properly managed, large agglomerations of plastics are generated in the fields and therefore contamination due to their slow degradation, in addition to the difficulty of recycling them due to the high adherence of soil. This environmental contamination causes a progressive deterioration of ecosystems, causing a decrease in the availability of natural resources and negatively influencing people's health.

3. Application of biodegradable packaging in the Colombian coffee industry

3.1 Approach to the potential market for a biodegradable bag for coffee seedlings in the Southwest of Colombia

Taking into account this problem, the research group Science and Technology of Biomolecules of Agroindustrial Interest (Cytbia) of the Universidad del Cauca in Popayán, Colombia, has been developing alternatives for biodegradable packaging made from cassava starch. But it was in 2013, with the project financed by the Corporación Red Especializada de Centros de Investigación y Desarrollo Tecnológico del Sector Agropecuario de Colombia - CENIRED, called *"Evaluation of the potential use and identification of business development opportunities of biodegradable bags for coffee seedlings"*, where the development of biodegradable packaging with application in coffee growing was focused.

Derived from these results, it was necessary to advance in the commercial validation of this type of prototypes, so in the framework of the project "*Use of a prototype of a biodegradable bag based on starch for specialty coffee seedlings*", developed in 2015 and financed by the Administrative Department of Science, Technology and Innovation (Colciencias), the Inter-American Development Bank (IDB) and the University of Cauca, activities were carried out to study the commercial potential of a prototype of a biodegradable bag based on starch, Technology and Innovation (Colciencias), the Inter-American Development Bank (IDB) and the Universidad del Cauca, activities were carried out to study the commercial potential of a prototype of a biodegradable bag made from cassava starch for specialty coffee seedlings in the departments of Cauca, Nariño and Huila.

Within the framework of this project, more than 400 surveys were carried out with different producers of specialty coffees in the departments of Cauca, Huila and Nariño, accompanied by semi-structured interviews with leaders of associations and companies linked to the coffee sector in the region.

With respect to the characteristics of the coffee seedling bag used, it was found that the bag with specifications of 17 × 23 cm with a gusset at the bottom perforated with 6 holes (2 kg of soil substrate) was the most commonly used by coffee growers. The National Federation of Coffee Growers has established a series of technical guidelines associated with the use of the seedling bag, where it is recommended to use bags of 17.0 × 23.0 cm, with an approximate capacity of 2.0 kg of soil substrate, having as an alternative, a bag of 13.0 × 17.0 cm, with a capacity of 1.0 kg, where field planting is done after 90 days after the seedling is installed (Farfán et al., 2015); many of the producers maintain other preferences for the use of the bags.

In this sense, the definition of the size of the seedling bags is conditioned by the phenological development of the seedlings, since root growth is limited by the size of the bag, because if a root touches the bottom of the bag, it is possible that a kind of "L" folding occurs, causing problems in the support of adult plants and in the absorption of nutrients (Gaitán et al., 2011). This was also mentioned by Salazar (1996), who stated that the dynamics of acquiring seedlings in smaller bags implied a lower cost, a smaller amount of soil and a lower cost associated with filling the bags, but he found that the smaller bags had defects such as twisted stems, poor root growth and high percentages of what is known as "pig tail".

Arizaleta and Pire (2008) analyzed the response of coffee seedlings to bag size in Duaca, Lara State, Venezuela, where seedlings of *Coffea arabica* L. *'Caturra' were grown in three bag sizes (13 × 15, 15 × 19 and 18 × 23 cm).* 'Caturra' developed in three bag sizes (13 × 15, 15 × 19 and 18 × 23 cm), and found that in the smaller bags the roots had a shorter length and a greater diameter; while in the larger bags, the accumulation of biomass in the root was greater, and concluded that the 18 × 23 cm bags had advantages over other types of bags, due to the generation of vigorous plants.

Other similar results have been presented by Gil and Diaz (2016), who evaluated the root growth of coffee seedlings, cv. Castillo, planted in various types of containers of different dimensions; among these with black plastic bags of 11 × 20 cm and 17 × 23 cm. Among their conclusions, they established that in the latter a greater length of the tap root was found, associated with a greater availability of growing space and an adequate use of humidity. However, they are not recommended for long periods of time, since they also found a high percentage of tap root deformation. Likewise, root volume was lower in the 11 × 20 cm bags.

However, the dimensions of the bag, in addition to influencing root growth, have an important impact on coffee growing practices. For example, in relation to the quantity and availability of soil; for example, if bags with a capacity of 2.0 kg are used to plant 200 coffee trees, at least 400 kg of soil would be required to fill the bags, without counting the losses. This, added to the problem of each bag having a significant volume and weight, would make transplanting more complicated in inaccessible areas (Gonzalez, 2001).

Thus, the influence of cultural factors, traditional practices, soil conditions, means of transport, distance between the seedbed and the cultivation area, as well as the planning of planting, are elements that coffee growers take into account when defining the type of bags they wish to use.

On the other hand, in relation to the purchase criteria prioritized by coffee growers at the time of acquiring the seedling bag, some differences and similarities were established among the departments. In general, for Cauca, Huila and Nariño, the resistance of the material and the dimensions were the two main reasons for the purchase of nursery bags. In the case of Huila, dimensions represented the most important factor and was therefore the relevant characteristic at the time of marketing in that department. In a third line, price emerged as another important factor; where producers sought to find seedling bags at a competitive cost.

These considerations can be associated with the handling of the seedling bags in the field, where they are first opened, the substrate is deposited very quickly until they are satisfactorily filled and then they are arranged on the site provided. Once there, they are sprayed with a large amount of water, in order to sufficiently moisten the substrate and thus allow the opening of holes where the seedling will be arranged. During the seedling process, these materials are subjected to environmental conditions, irrigation management and other inputs. Then, at the time of planting, the bags are transported to the cultivation site. This represents the challenge of biodegradable materials for use as seedling bags, since they must have sufficient strength to withstand filling and handling during seedling and transport to the cultivation sites.

In this sense, Palechor et al. (2016) evaluated the structural disintegration of materials made from cassava starch and polylactic acid in the seedling stage for 120 days, finding a significant deterioration of the materials during the period of the experiment, indicating that the action of the microorganisms of the substrate, added to environmental factors such as heating, humidity, enzymatic activity could have generated ruptures in the structure of the material, where the degradation began on day 15 of the experiment. Similar results were found in another study, where factors such as temperature variations, humidity and solar radiation caused the loss of mechanical strength and increased stiffness of the material, possibly derived from photodegradation caused by exposure to ambient light and the formation of crosslinks (Bilck et al., 2014).

Accordingly, the application in seedling bags implies a condition of a kind of controlled degradation. As it is required that the material is able to withstand the period of time required for the seedbed (4 to 6 months in the case of coffee), but at the same time, when it is taken to the planting site, it must be able to degrade in the soil in the shortest possible time.

3.2 Bio-seedlings: Appropriation and use of a biodegradable packaging for seedlings obtained from cassava starch to strengthen the coffee production chain

During the development of the biodegradable packaging for coffee seedlings, it was identified that in order to achieve the transfer of technology it was necessary to generate knowledge based on the exchange of experiences, where coffee growers can be researchers that support the identification of the specific requirements that the packaging should fulfill, according to their cultural practices associated with the production of coffee in the seedling stages, with the objective of generating a dynamic and continuous process of social learning (Chaparro, 2001).

The identification of the technical characteristics desired by the coffee growers is transcendental in order to avoid falling into the trap of thinking that this is an absolute technology or that it is "divorced" from the socio-cultural dimensions and is adequate for all contexts (Foster, 1973). Therefore, a co-design process supported by ethnographic methods was implemented, involving the Central Cooperativa Indígena del Cauca (CENCOIC) and the coffee businesswomen of southern Huila, as initial validators of this research. This multidisciplinary symbiosis strategy was developed from a socio-cultural and technical vision, because it was important to unveil the existing cultural roots associated with the use of traditional plastic bags.

3.2.1 Co-creating a solution

Technology transfer practices for rural environments, understood from the farmers' processes, are few, since these initiatives have traditionally been imposed by the State, its agencies, universities and industries, leaving aside the real participation of farmers with their traditions (Reina-Rozoy and Ortiz, 2019). Consequently, it was established that it was appropriate for coffee growers to participate in the process of designing and/or adjusting the packaging, before conducting scaling and validation tests of the material in the field, thereby seeking to ensure that the final product is useful and usable for Colombian coffee growing. The following is the case study "Bioalmácigos" which compiles the process.

3.2.1.1 Validation of environmental issues

An approach and knowledge workshop was held with the objective of listening, talking and observing the participants, approaching from an integral and participatory perspective, the environmental problems associated with the use of traditional plastics, as a problematic social situation that requires some change or development (Sandoval, 2002). Here it was possible to confirm, that the use of biodegradable packaging was a potential solution that they had identified for this problem, but that they did not have the tools to develop them.

At the same time, it was concluded that producers tend to use the bag suggested by technicians in coffee production, with a dimension of 17 × 23 cm (Farfan et al., 2016), but with the desire or preference to use bags of greater length and smaller diameter, due to the difficulty that the suggested ones imply at the time of transport; in addition, to requirements of greater labor, land and care.

3.2.1.2 Preliminary production of biodegradable packaging

Tests were carried out at laboratory level for the production of the biodegradable tubular, where a binary mixture was made with native starch, by means of extrusion in a twin-screw equipment, mixing starch and plasticizer, in a ratio of 70/30, stearic acid, with polylactic acid, maleic anhydride and other couplants. The process conditions for the extrusion of the biodegradable tubular were: average extrusion temperature of 161°C and screw speed between 20 rpm and 40 rpm.

According to the improvements made, a package with the following technical characteristics was developed: 0.04 mm (thickness), side gusset, 12 cm wide by 23 cm long and between 8 to 12 perforations. The biodegradable bags produced at this stage were the first step in bringing the technology closer to the coffee-growing community.

3.2.1.3 First approach to biodegradable packaging

Different coffee growers from the associations CENCOIC and Empresarias Cafeteras del Sur del Huila were given the biodegradable bags developed at the previous stage, so that they could carry out a first validation, in the coffee seedling stage, in accordance with their traditional practices.

The main conclusions associated with the technical validation of the packaging were as follows:

- The bags before sowing suffer more manipulations, therefore, they must be resistant to support them.
- The bag should be kept without degrading for at least 4 months in the seedling stage.
- In the process, the seedbed should include shade, since according to their experiences the bag degrades faster in the open air.
- The biodegradable bag requires more care and time for bagging and must be more manageable.
- At the time of socializing the use of biodegradable bags with more coffee growers, it should be supported with training on their added value and conditions of use.

3.2.1.4 Prototyping and scale-up of biodegradable packaging according to local requirements

The prototyping of the packaging was started based on the knowledge generated with the organizations in the previous stage, and the bag was re-elaborated adjusted to the new requirements, with the objective of prolonging its degradability and resistance.

For the scale-up, the starch ratio in the final matrix was 50% and the following temperature profile was defined for obtaining the pellets: feeding zone between 90°C and 100°C, compression zone between 110°C and 130°C, dosing zone between 110°C and 120°C and exit zone (die) between 100°C and 110°C, controlling the screw speed between 35 rpm and 40 rpm. At the same time, the following temperature profile was determined for the production of the biodegradable tubular: feeding zone between 100°C to 120°C, compression zone between 110°C to 140°C, dosing zone between 110°C to 140°C and exit zone (die) between 100°C to 120°C.

Finally, biodegradable bags were made according to the findings and recommendations identified in the ethnographic work, with the technical characteristics specified in the following sheet:

3.2.2 Implementing a solution from social innovation

After the initial validation of the biodegradable packaging and the scaling conditions for it were defined, it was tested with coffee growers of the allied associations and from an ethnographic exercise, the methodology for the design and assembly of the demonstrative and participatory plots was built, from the seed to the seedling stage, where from the debate between the expertise of coffee growers, the articulation with sociocultural factors and what is reported in literature, a consensus was reached that would not affect their traditional practices.

3.2.2.1 Participatory and demonstration plots

A randomized complete block design was implemented for the seedling stage, with six treatments and three replications for a total of 18 experimental units, where three types of bags were evaluated:

- *B1*: Biodegradable bag validated in the preliminary stage by coffee growers.
- *B2*: Biodegradable bag resulting from scaling.
- *BT*: Traditional plastic bag.

The variables to be evaluated were:

- Mechanical properties, according to ASTM D882-10, applied to gaskets.
- Root volume (mL), Stem length (cm), Tap root length (cm) and number of leaves (Gil and Diaz, 2016), were taken into account in the seedlings.

The germinator stage was carried out in order to guarantee the same quality and type of seed during the study. Subsequently, the construction of the seedbed was advanced according to the indicated design. Initially, the seedlings were removed and selected from the germinator for transplanting and subsequent construction of the seedbed.

The plots were monitored for 2 months, with the following results:

Regarding the mechanical properties, it was found that for the departments of Huila and Cauca, there were significant differences between the B2 and BT treatments, finding that the values of B2 are superior to those of BT for the variables of modulus of elasticity and maximum tensile strength in the longitudinal and transverse directions. Regarding the maximum elongation at the breaking point, there were also differences in the longitudinal direction, but the values for the traditional bag were higher than those found for the B2 bag. Only in the case of maximum elongation in the transverse direction, differences were found in the department of Cauca, with those of the traditional bag being higher than those of B2, but no significant differences were found in the department of Huila.

Likewise, monitoring variables associated with the growth of coffee seedlings associated with root volume (mL), stem length (cm), tap root length (cm) and number of leaves were also monitored. For the three study bags, it was realized that for none of the study sites, the dependent variables of root volume (mL), stem length (cm), tap root length (cm) and number of leaves, the significance was less than 0.025, therefore the Ho is not rejected, indicating that "There are no significant differences between bags B1, B2 and traditional bag on the development of the seedlings", a result similar to that indicated by the farmers in the ethnographic study, given that they mentioned that they found no differences in the development of the seedlings in the three study bags and believed that there was no influence of the packaging on their development.

3.2.3 Biodegradable packaging validation and feedback

Christensen and Knezeh (2001), highlight six possible stages in an appropriation or adoption process: first stage of awareness, but its applications and functions are unknown; second stage of knowledge/understanding, where one begins to know how the new technology works; third stage of basic use/appropriation; fourth stage of familiarity and confidence in the use of the new technology; fifth stage of application where use of the technology is given and its use is encouraged; sixth stage of integration of the technology in everyday practices (Jamaica, 2016).

In agreement with the authors, in this case, a level of appropriation was observed up to the fourth stage, where the first level was reached when it was corroborated that the environmental problem was identified by the coffee growers; the second stage of appropriation associated with knowledge and understanding was achieved when the understanding and reconciliation of key concepts for appropriation took place; the third stage was achieved by managing to articulate diverse knowledge, putting it into dialogues to make decisions based on the observations that the producers made in each follow-up, in order to implement a participatory appropriation process; the fourth stage of an appropriation process allowed achieving familiarity and confidence for the use of this new technology among the coffee growers, because they could see improvements in the packaging associated with the adoption of their opinions, which led them to interact about the project with other coffee growers and to have the conviction that the technology will continue to be improved.

Conclusions

The applications of biodegradable polymers in agriculture have as many expectations and possibilities as the current use of synthetic plastics in agriculture. Although there is a trend towards the search for alternatives to the mulch technology used in different parts of the world and with different crops, there are important possibilities for the development of biodegradable plastics as an alternative in applications such as greenhouses and plastic irrigation devices.

One application of biodegradable plastics in agriculture that still requires further exploration has to do with their use in seedbeds of different crops, such as coffee in the case of Colombia, where it is possible to identify an important demand associated with the consumption of traditional polyethylene plastic bags used for this purpose.

The development of new applications of biodegradable plastics in agriculture also depends on the massification of analyses adjusted to the particular conditions of the territories, since the degradation of these materials is subject not only to the chemical structure of the materials, but also to the conditions of the environment and soils, in addition to the particularities of cultural and technical practices.

Finally, it was recognized that the process suffered several difficulties in reconciling the practices of coffee growers, because they are linked to the level of expertise of each one. It is also recognized that the doubts expressed by the coffee growers, associated with the availability of biodegradable packaging and its final cost, is an opportunity to complete the appropriation process, according to the resulting suggestions that seek to improve the packaging and obtain a material with definitive technical specifications and use for its implementation in coffee growing.

References

Andrade, C., Da Graça Palha, M. and Duarte, E. (2014). Biodegradable mulch films performance for autumn-winter strawberry production. *Journal of Berry Research,* 4: 193–202.

Arboleda, G., Montilla, C., Villada, H. and Varona, G. (2015). Obtaining a flexible film elaborated from cassava thermoplastic starch and polylactic acid. *International Journal of Polymer Science,* 2015: 1–9.

Arizabaleta, M. and Pire, R. (2008). Respuesta de plántulas de cafeto al tamaño de la bolsa y fertilización con nitrógeno y fósforo en vivero. *Agrociencia,* 42(1): 47–55.

Barragán, D., Pelacho, A. and Martin, L. (2016). Degradation of agricultural biodegradable plastics in the soil under laboratory conditions. *Soil Research,* 54: 216–224.

Bilck, A., Olivato, J., Yamashita, F. and Pinto de Souza, J. (2014). Biodegradable bags for the production of plant seedlings. *Polímeros,* 24(5): 547–553.

Briassoulis, D. and Giannoulis, A. (2018). Evaluation of the functionality of bio-based plastic mulching films. *Polymer Testing,* 67: 99–109.

Brodhagen, M., Peyron, M., Miles, C. and Inglis, D. (2014). Biodegradable plastic agricultural mulches and key features of microbial degradation. *Applied Microbiology and Biotechnology,* 99(3): 1039–1056.

Chaparro, F. (2001). Conocimiento, aprendizaje y capital social como motor de desarrollo. *Ciência da Informação,* 30(1): 19–31.

Christensen, R. and Knezek, G. (2001). Las etapas de adopción como medida de integración de la tecnología. En: Morales, C., Ávila, P., Knezek, G. y Christensen, R. (eds.). *El punto de vista de los usuarios de las nuevas tecnologías en educación: estudio de diversos países.* México: ILCE.

Combrzyński, M. and Mościcki, L. (2013). The physical properties of starch biocomposite containing PLA. *Teka. Commission of Motorization and Energetics in Agriculture,* 13(2): 3–6.

Costa, R., Saraiva, A., Carvalho, L. and Duarte, E. (2014). The use of biodegradable mulch films on strawberry crop in Portugal. *Scientia Horticulturae,* 173: 65–70.

Cowan, J., Inglis, D. and Miles, C. (2013). Deterioration of three potentially biodegradable plastic mulches before and after soil incorporation in a broccoli field production system in Northwestern Washington. *HortTechnology,* 23(6): 849–858.

De polanía, i. and peña, f. (2013). Plásticos en la agricultura: beneficio y costo ambiental: una revisión. Revista u.d.c.a actualidad y divulgación científica, 16(1): 139–150.

Fahrngruber, B., Siakkou, E., Wimmer, R., Kozich, M. and Mundigler, N. (2017). Malic acid: A novel processing aid for thermoplastic starch/poly(butylene adipate-co-terephthalate) compounding and blown film extrusion. *Journal of Applied Polymer Science,* 134(48): 1–12.

Farfan, F. and Sánchez, P. M. (2016). Densidad de siembra del café variedad Castillo en sistemas agroforestales en el departamento de Santander Colombia. *Cenicafé,* 67(1): 55–62.

Farfán, F., Serna, C. and Sánchez, P. (2015). Almácigos para caficultura orgánica. Alternativas y costos. *Gerencia Técnica/Programa de Investigación Científica. Fondo Nacional del Café,* 452: 1–8.

Federación Nacional de Cafeteros. (2017). *85 Congreso Nacional de Cafeteros. Rentabilidad de la caficultura es de 25%.* https://www.federaciondecafeteros.org/static/files/Periodico_CNC2017.pdf.

Foster, G. M. (1973). *Las culturas tradicionales y los cambios técnicos.* México: Fondo de Cultura Económica.

Gaitán, A., Villegas, C., Rivillas, C., Hincapié, E. and Arcila, J. (2011). Almácigos de Café: Calidad fitosanitaria, manejo y siembra en el campo. *Gerencia Técnica/Programa de Investigación Científica. Fondo Nacional del Café,* 404: 1–8.

Gil, A. and Díaz, L. (2016). Evaluación de tipos de contenedores sobre el crecimiento radical de café (Coffea arabica L. cv. Castillo) en etapa de vivero. *Revista Colombiana de Ciencias Hortícolas,* 10(1): 125–136.

González, D. (2001). Comparación entre la bolsa y el "cono macetero" o "tubete" en la producción de plantas de café. Universidad Zamorano, Zamorano, Honduras.

Hayes, D., Wadsworth, L., Sintim, H., Flury, M., English, M., Schaeffer, S. and Saxton, A. (2017). Effect of diverse weathering conditions on the physicochemical properties of biodegradable plastic mulches. *Polymer Testing,* 62: 454–467.

Iapichino, G., Mustazza, G., Sabatino, L. and D'Anna, F. (2014). Polyethylene and biodegradable starch-based mulching films positively affect winter melon production in sicily. *Acta Horticulturae,* 1015(1015): 225–232

Innocenti, F. (2005). Biodegradation behaviour of polymers in the soil. pp. 57–102. En: Bastioli, C. (ed.). *Handbook of Biodegradable Polymers.* Shawbury, Reino Unido: Rapra Technology Limited.

Jamaica, F. (2016). *Los beneficios de la capacitación y el desarrollo del personal de las pequeñas empresas.* Thesis. Universidad Militar Nueva Granada, Bogotá, Colombia.

Li, C., Moore-Kucera, J., Miles, C., Leonas, K., Lee, J., Corbin, A. and Inglis, D. (2014). Degradation of potentially biodegradable plastic mulch films at three diverse U.S. locations. *Agroecology and Sustainable Food Systems,* 38: 861–889.

Liu, D. and Yao, F. (2018). Multi-functional characteristics of novel biodegradable mulching films from citric acid fermentation wastes. *Waste Biomass Valorization,* 9(8): 1379–1387.

Lleras, S. and Moreno, A. (2001). Desarrollo y evaluación de bolsas biodegradables para almácigos de café. *Cenicafé (Colombia),* 52(1): 20–28.

Morra, L., Bilotto, M., Cerrato, D., Coppola, R., Leone, V., Mignoli, E., Pasquariello, M., Petriccione, M. and Cozzolino, E. (2016). The Mater-Bi biodegradable film for strawberry (Fragaria x ananassa Duch.) mulching: effects on fruit yield and quality. *Italian Journal of Agronomy,* 11(731): 203–206.

Moreno, M., Gonzalez-Mora, S., Villena, J., Campos, J. and Moreno, C. (2017). Deterioration pattern of six biodegradable, potentially low-environmental impact mulches in field conditions. *Journal of Environmental Management,* 200: 490–501.

Muñoz, G. (2014). Editorial. Caficultura sostenible, moderna y competitiva. *Ensayos sobre Economía cafetera,* 30: 5–9.

Palacios, L., Niño, D., Villada, H. and Arboleda, G. (2016). Estudio exploratorio sobre la aceptación de una bolsa biodegradable para almácigos de cafés especiales. *Agronomía Colombiana,* 34: S217–S219.

Palechor, J., Cerón, A., Villada, H. and Salazar, M. (2016). Deterioro de una bolsa biodegradable de almidón de yuca con ácido poliláctico en un vivero. *Vitae,* 23(1Supl): S585–S589.

Pardo, R. and Natalia, L. (2012). Propuesta de un Sistema de Gestión Ambiental para el sistema de producción cafetera de la finca Las Palmas, La Vega-Cundinamarca, bajo los requisitos de la norma ISO 14001: 2004. Thesis. Pontificia Universidad Javeriana, Bogotá, Colombia.

Reina-Rozo, J. and Ortiz, J. (2019). Ecosistemas de Innovación local para fortalecer la agroecología en Colombia: El caso preliminar del Lab Campesino de Tierra Libre. *Social Innovations Journal,* 53.

Saglam, M., Sintim, H., Bary, A., Miles, C., Ghimire, S., Inglis, D. and Flury, M. (2017). Modeling the effect of biodegradable paper and plastic mulch on soil moisture dynamics. *Agricultural Water Management,* 193: 240–250.

Salazar, J. (1996). Efecto del tamaño de la bolsa del almácigo sobre la producción de café. Cenicafé, 47(3): 115–120.

Sandoval, C. (2002). *Investigación cualitativa.* Bogotá, Colombia: Arfo Editores e Impresores.

Sartore, L., Vox, G. and Schettini, E. (2013). Preparation and performance of novel biodegradable polymeric materials based on hydrolyzed proteins for agricultural application. *Journal of Polymers and the Environment,* 21(3): 718–725.

Sinha, A. and Bousmina. (2005). Ageing of starch based systems as observed with FT-IR and solid state NMR spectroscopy. *Starch/Stärke,* 50: 478–483.

Solano, J. (2015). Evaluación de los factores de riesgo producidos por la degradación de los suelos por cultivos de café en la vereda El Cascajo, Municipio de Concordia (Antioquia). *Cuaderno Activa,* 7(7): 85–98.

Tan, Z., Yi, Y., Wang, H., Zhou, W., Yang, Y. and Wang, C. (2016). Physical and degradable properties of mulching films prepared from natural fibers and biodegradable polymers. *Applied Sciences,* 6(5): 147.

Touchaleaume, F., Martin, L., Angellier, H., Chevillard, A., Cesar, G., Gontard, N. and Gastaldi, E. (2016). Performance and environmental impact of biodegradable polymers as agricultural mulching films. *Chemosphere,* 144: 433–439.

Transparency market research. (2019). Agricultural films market - global industry analysis, size, share, growth, trends, and forecast 2018–2026. Disponible en: https://www.transparencymarketresearch.com/agricultural-film.html.

Virtanen, S., Reddy, R., Irmak, S., Honkapää, C. and Isom, L. (2017). Food Industry co-streams: potential raw materials for biodegradable mulch film applications. *Journal of Polymers and the Environment,* 25(4): 1110–1130.

Wortman, S., Kadoma, I. and Crandall, M. (2015). Assessing the potential for spunbond, nonwoven biodegradable fabric as mulches for tomato and bell pepper crops. *Scientia Horticulturae,* 193: 209–217.

Yang, N., Sun, Z., Feng, L., Zheng, M., Chi, D., Meng, W., Hou, Z., Bai. and Li, K. (2015). Plastic film mulching for water-efficient agricultural applications and degradable films materials development research. *Materials and Manufacturing Processes,* 30(2): 143–154.

Zuo, Y., Wu, Y., Gu, J. and Zhang, Y. (2017). The UV aging properties of maleic anhydride esterified starch/polylactic acid composites. *Journal Wuhan University of Technology, Materials Science Edition,* 32(4): 971–977.

Chapter 21

Agriculture Applications of Biodegradable Polymers

Rocio Yaneli Aguirre-Loredo,[1,5] *Lluvia de Abril Alexandra Soriano Melgar,*[2,5] *Luis Valencia,*[3] *Gonzalo Ramírez García*[4] and *Alma Berenice Jasso-Salcedo*[2,5,*]

1. Introduction

For a plastic to be considered biodegradable, it must meet the condition of decomposing with the help of microorganisms in aerobic conditions to give rise to carbon dioxide, water and new biomass; or if the process occurs in anaerobic conditions, its decomposition also produces methane. According to the material of manufacture, two types of plastics are considered biodegradable: (1) bioplastics and (2) plastics with biodegradable additives.

Plastics with biodegradable additives are manufactured based on petrochemical products but contain a portion of renewable raw materials or additives, which improve their ability to biodegrade. Bioplastics are obtained using almost 100% renewable raw materials as a base. They can serve as a matrix of active substances that improve the quality of the stored products.

Bioplastics can be prepared from biopolymers alone or in mixtures of two or more, which are known as composite polymers, as well as with some additives such as plasticizers and preservatives. Biopolymers used to make these bioplastics materials can be of synthetic origin, natural origin and those synthesized using microorganisms. Among synthetic sources, one finds polylactic acid (PLA), polyvinyl alcohol (PVOH), and polycaprolactone (PCL); those of natural origin are starch, pectin, cellulose, gelatin, alginate, zein and chitosan, etc.; and those obtained by microorganisms are polyhydroxybutyrate (PHBs), xanthan gum and bacterial cellulose. The recent applications of these biopolymers in the pre-harvest and post-harvest phases of agricultural products are reviewed in this chapter and illustrated in Fig. 21.1.

[1] Departamento de Procesos de Polimerización, Centro de Investigación en Química Aplicada (CIQA). Blvd. Enrique Reyna Hermosillo 140, San José de los Cerritos, 25294, Saltillo, Coahuila de Zaragoza, México.

[2] Departamento de Biociencias y Agrotecnología, Centro de Investigación en Química Aplicada (CIQA). Blvd. Enrique Reyna Hermosillo 140, San José de los Cerritos, 25294, Saltillo, Coahuila de Zaragoza, México.

[3] Biofiber Tech Sweden AB, Norrsken Hourse, Birger Jarlsgatan 57 C, SE -113 56 Stockholm.

[4] Centro de Física Aplicada y Tecnología Avanzada, Universidad Nacional Autónoma de México, 3001, Boulevard Juriquilla, 76230, Querétaro, México.

[5] Research-fellow Cátedras-CONACyT.

[*] Corresponding author: alma.jasso@ciqa.edu.mx

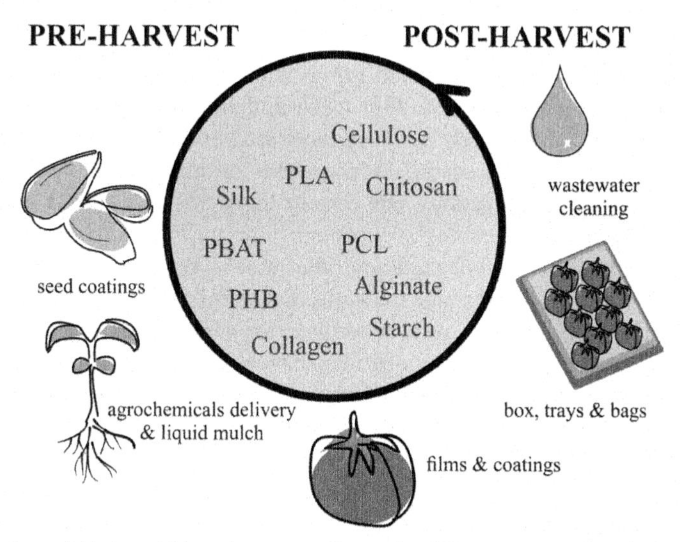

Figure 21.1. Illustration of biodegradable polymers trends on the different stages of agricultural production cycle. Biopolymers like cellulose and starch are booming in post-harvest, while bioplastics like PBAT and PHB thrive on pre-harvest applications.

2. Preharvest applications

2.1 Seeds protection

A common problem with seed preservation is that the shelf life is shortened by microorganisms and mold activity. A coating strategy can keep them stable longer during storage, distribution and application. Biodegradable seed coatings can also provide significant germination or seeding growth rates, which should be compatible with other agrochemicals such as fertilizers and pesticides or biocontrol agents, and it should not be persistent in the soil after planting or it should degrade in a manner that provides a nutrient source to plants or other involved biocontrol agents (Accinelli et al., 2016).

A sprayable liquid bioplastic formulation was also developed for multifunctional film-coating applications and tested on corn and canola seeds. The coating consisted of starch-based bioplastic dispersed in water and incorporating spores of *Trichoderma harzianum*, a plant-growth-promoting fungus. The obtained results demonstrated that germination was unaffected, while the coatings containing spores significantly stimulated the growth of both species. Furthermore, more intense colonization of the soil rhizosphere by *T. harzianum* was detected, which prevented negative effects on germination due to a mixture of the insecticide imidacloprid and the fungicide metalaxyl-M incorporated along with the spores in the biodegradable film (Accinelli et al., 2016). Similarly, Zvinavashe et al. (2019) reported a silk and trehalose matrix applied to bean seeds containing plant growth-promoting rhizobacteria able to release their content within 10 minutes into saline environments. The resulting plant roots were effectively inoculated with *Rhizobium tropici* that increases assimilation of nutrients and establishment even after 4 weeks stored seeds. These engineered microenvironments help both to the effective dosage of plant growth-promoting microorganisms and to enlarge the shelf-life of seeds.

Seed germination requires a sterile environment, controlled temperature and humidity conditions but the options for *in vitro* supporting materials are limited to germination paper. A biodegradable and inert foam and aerogel may innovate the field. For instance, the foaming process using supercritical liquids produces biopolymers with cavities useful for seed germination and root establishment. The most experienced are cellulose-based foams but pectin, alginate, starch, chitosan and vegetal proteins foams have also been reported and exhibit interesting characteristics for vegetable propagation (Nita et al., 2020).

2.2 Mulching for intensive agriculture

Biodegradable agricultural mulches in form of film or non-woven fabric are mainly used for weed and disease control and retain humidity. With these mulches, the optimal crop development is improved, and for this reason, research has been motivated in the development of new materials. For instance, the number of biodegradable mulch patents of one or multiple layers has increased in the last decade. Among the innovations, one can see that a tree layer PBAT and PLA mulch has distributed functionalities so that the outer layer contains a resin for pest control, while the inner layer has a stabilizer to play with their biodegradation timeframe as required from 130 to 90 days (Guo et al., 2021). Another strategy is to manage the molecular weight and molecular weight distribution for better control of the degradation cycle as proven for PHB mulch (Ze et al., 2020). This is crucial for long-term crops in which the biodegradable mulch should last till the harvest period. For example, few commercially available PBAT mulch were used for tomato production reaching up to 65 days before the degradation signals started, compared with 22 days of a PLA mulch (Abduwaiti et al., 2021). The results indicate that the yield and profit were similar to the PE mulch, but showed a 70% of disintegration by the end of the cycle, thus avoiding plastics removal using heavy machinery which could bring an additional economic advantage. Even so, few considerations can be taken like using biodegradable mulch under covers to control the degradation as well as a selection of suitable crops growing far from residual bioplastics. Esthetical issues may arise, for instance, if black pieces of biodegradable mulch end up in the out layer of pumpkins or peanuts.

Liquid film mulching is an alternative administration allowing for the widest range of formulations using temperature-sensitive biopolymers. Traditional mulch films processed by extrusion, melt and blow requires a high temperature that is no longer essential in liquid film mulching. For instance, Dang et al. (2016) optimized a starch and gelatin liquid formulation that was sprayed on the soil surface and tested for rapeseed seed germination. The results indicate an increased plant survival rate and yield compared with simple watered pots. The liquid mulch film shows water absorption, moisture retention and temperature control needed for crop growth. In another study, a leather industry residual like collagen was mixed with polyethylene glycol (PEG) and soybean oil to prepare a sprayable mulch. A short cultivation test was done in lettuce under greenhouse conditions. The soil showed no pH changes but an electrical conductivity slightly high that helps with nitrogen availability during mulch decomposition (Ze et al., 2020). A previous study by the same authors resulted in weed suppression and moisture control over 12 months for a collagen formulation added with carbon black and cellulose microfibers (Sartore et al., 2013). Recently a sprayable and biodegradable PCL based mulch successfully retained humidity and weed control over 5 months on two of three different soils (Borrowman et al., 2020). The biodegradation rate was soil type-dependent being faster on acidic soil with higher organic matter content.

Current efforts in the field are devoted to designing multifunctional mulch. For instance, the fertilizer-infused mulch paper modified with biodegradable poly(hexamethylene succinate) (PHS) fabricated through a solution blade coating process showed dual functions of mulching and controlled release of fertilizers. This optimized configuration demonstrated improved water absorption, better resistance to enzymatic degradation and slow-release of fertilizers. Plant growth trials demonstrated acceleration of the blossom and root growth of marigold plants (Bi et al., 2021).

2.3 Controlled administration of agricultural inputs

Significant portions of agrochemicals are lost during the field application, resulting not only in serious environmental concerns but in large economic losses. Biodegradable polymeric compounds and their composites have shown interesting approaches as delivery systems for the controlled and/or sustained release of agrochemicals for different purposes such as improving plant nutrition and plant development (fertilizers), pest management (pesticides, antibiotics and pheromones) or soil conditioning (water super-absorbents). The aim is to improve the productivity of crops, warrant innocuity for plants and consumers, reduction of the required dose and cost of agrochemicals and water and depletion of the carbon footprint impact.

The inclusion of additives into those biodegradable polymer composites could enhance the chemical and mechanical properties required for their expanded use and applications. Some additives can be, for instance, plant derivatives like seeds (Das Lala et al., 2018; Hongsriphan and Pinpueng, 2019) and fibers (Appusamy et al., 2021; Dada et al., 2019; Hwang, 2021; Maraveas, 2020), animal fibers (Das Lala et al., 2018; Sonjan et al., 2021), micro and nanoparticles at varied configurations (graphene oxide (Lee and Ke, 2020), cobalt (Krishnamoorthy and Rajiv, 2017), ZnO/SiO_2 (Dispat et al., 2020) or minerals like clays (Abd El-Aziz et al., 2018; Prudnikova et al., 2021; Yakubu et al., 2018) and montmorillonite (Azarian et al., 2018; Kenawy et al., 2018; Malik et al., 2018). However, there usually exists a reinforcing limit, above which the further incorporation of additives would compromise the achieved mechanical reinforcements instead. Thereby, works found in literature are commonly devoted to optimizing the ratio of the precursors in the biodegradable composite materials. Representative studies were reviewed to exemplify the progress in this field and are briefly mentioned below.

2.3.1 Fertilizer

Formulation of biodegradable polymers could be applied for micro- and macronutrient supply. In this perspective, these nutrients could be available to the plants in two different ways: (i) after conventional load/release processes, in which polymers are "impregnated" with the fertilizer, and then released through slow or controlled diffusion mechanisms, or (ii) as an inherent component of the composite, whose decomposition can release the nutrients as a subproduct of the degradation. Considering the first case, biodegradable polymers are prepared in progressively complex structures to improve control over releasing of impregnated agrochemicals. For instance, a novel biodegradable composite film was prepared by hot pressing poly(butylene succinate) (PBS) with the natural sorbent lemon basil seeds (LB) along with a fixed content of ammonium sulfate as a fertilizer. This configuration enhanced moisture sorption and ammonium sulfate release for composites with the higher LB concentrations evaluated. Although the addition of LB improved crystallinity, the composite films were more brittle than pure PBS films. Thereby, the soil burial test resulted in accelerated biodegradation in the presence of LB and ammonium sulfate (Hongsriphan and Pinpueng, 2019).

In another work, a single-layered hollow PLA, and double-layered hollow nanofiber yarns composed of PHB and PLA nanofiber as the outer and inner layers, respectively, were developed to encapsulate and release urea, the most common form of nitrogenous fertilizers. Results indicated that nitrogen release from double-layered hollow nanofiber yarns was significantly slower, resulting in a prolonged urea release rate compared to single-layered nanofibers (Javazmi et al., 2020). In another work, Sobkowicz research group proposed a reliable method to extrude PHS along with Poly(Butylene Succinate-co-Hexamethylene Succinate) P(BS-co-HS) copolyesters at different ratios and investigated the influence of the copolymer structure on the urea phosphate fertilizer (UP) release from composite tablets (Bi et al., 2021). Accordingly, the presence of smaller UP crystals dispersed in the polymer matrix allows an enhanced diffusion rate of phosphate, which is usually limited in other slow-release fertilizer coatings. Similarly, an interpenetrated biodegradable system (IPN) for the controlled release of agrochemicals was prepared via the impregnation of methyl methacrylic acid onto gum tragacanth and a polyacrylic-based hydrogel with glutaraldehyde as a crosslinking agent. The slow release of calcium nitrate was demonstrated, as well as the degradation of up to 91.62% of the synthesized IPN within 11 weeks under composting methods and 78.83% under the soil burial method. Furthermore, the degraded IPN showed a positive impact on the fertility of the soil, enhancing the contents of organic carbon, phosphorus and potassium, and thereby, the growth of *Phaseolus vulgaris* plants (beans) with respect to control studies (Saruchi et al., 2019). On the other hand, the release of nutrients as a part of the degradation mechanism of the biodegradable polymer was exemplified by the integration of the multi-purpose phosphate-containing monomers within acrylate/organophosphorus-based hydrogels. Furthermore, these crosslinkers are an additional source of phosphorus, since the hydrolysis of phosphate units trigger their release overtime at the exact location next to the root system of plants, facilitating nutrient assimilation (Glowinska et al., 2019). By this strategy, some unwanted effects of common fertilization can be avoided, e.g., eutrophication.

The state of the art in the field also makes clear that the activity of the novel materials should be evaluated under real conditions, as well as their biodegradability, rather than only in simulated conditions. A relevant example is the effects of heavy metals in the soil on the degradation of biodegradable polymers and the release of their composites have been only scarcely studied. In the work of Hu et al. (2019), the relevance of the soil conditions (normal vs contaminated) on the fertilization effects of biodegradable polymers was evaluated. Notably, poly(aspartic acid) (PASP) increased nutrient efficiency, which caused greater biomass productivity under normal soil conditions. Furthermore, PASP demonstrated regulating the dynamics of minerals in the substrate-plant system, restricting the translocation ability of Cu and Cd from the roots to shoots of tomato plants, but promoting the absorption and translocation of some essential elements like K, Ca and Mg. Therefore, it is important to consider that biodegradable polymer fertilizers could also impact the absorption of non-desirable elements. Xiang et al. (2021) studied the influence of heavy metals (Cu, Zn, Cd and Pb) on the inhibition of degradation and nutrient release characteristics of biodegradable and superabsorbent polymers. They demonstrated that these effects could be effectively reduced by combining the fertilizing source with super-absorbent polymers, especially in a semi-interpenetrating polymer network. As the higher concentration and activities of functional groups contained in the superabsorbent polymer component, the lesser was the inhibition effect on biodegradation and nutrient release caused by heavy metals. This general observation is essential for developing biodegradable polymer composites devoted to water holding and agrochemicals supplying in polluted soils.

2.3.2 Pesticides, antibiotics and pheromones

The purpose of biodegradable polymers as carriers is to effectively protect a load of agrochemicals, however many of these advanced materials exert an additional function, which consists of improving interactions between the plant-carrier and/or carrier-pathogen systems. In this sense, poly (lactic-co-glycolic acid) (PLGA) nanoparticles have been used to encapsulate cyazofamid, a pesticide applied to control the disease caused by *Phytophthora infestans*. While *in vitro* assays demonstrated that these nanocarriers were taken into sporangia, encysted zoospores and germinated cysts of *P. infestans*, the *in vivo* studies resulted in enhanced adhesion of the cyazofamid-loaded PLGA-NPs to tomato leaves (Fukamachi et al., 2019). To avoid plant damage a method was developed for the effective incorporation of 20% of the sex pheromones of *Bactrocera oleae* and *Prays oleae* as pesticide agents into PCL, PHB and CA nanofibers (Kikionis et al., 2017). The employed electrospinning process demonstrated the capability to produce field traps for pheromones release at an effective rate during at least two pest generations which represent about 16 weeks. The pheromone showed chemical stability over that time avoiding the use of antioxidants that may hinder their effectiveness.

2.3.3 Water super-absorbents

Considering the physiological basis of hydration, water is the primary nutrient. Nevertheless, clean water scarcity is a major issue that should be dealt with promptly. Materials containing water-superabsorbent biodegradable polymers represent an important trend for the restoration of degraded soils from drought. Despite its high water absorption capacity and mechanical resistance, attempts are being made to replace synthetic acrylamides and acrylate derivatives currently used by biodegradable polymers such as cellulose, starch, alginate, agar and chitosan, to achieve more sustainable materials (Skrzypczak et al., 2020). Some biodegradable absorbent materials are already commercially available, e.g., cellulose-based polymers. However, they commonly present some disadvantages in comparison to the petroleum-derivative superabsorbents, namely, inferior liquid absorption capacity, lack of complete biodegradability and higher cost (Alam and Christopher, 2018). As an example of recently developed materials to overcome these issues, a novel polysaccharide-based superabsorbent hydrogel was prepared from wood-derived cellulosic fibers cross-linked with carboxymethylated chitosan. Due to the high porous architecture and specific surface area, the hydrogel facilitates rapid mass penetration, allowing a water retention value of 610 (g/g gel), which is several times higher than any already reported neat cellulose-based superabsorbent material (Alam and Christopher, 2018).

In addition to the desired properties, researchers should be careful about potential side-effects derived from the application of novel materials. For instance, a lignin-based hydrogel was used to alleviate drought stress in maize, and their effects compared with a synthetic sodium polyacrylate hydrogel. On application of hydrogels to soils at the same rates (0.3 or 0.6%), both of them increased water availability, reducing the drought stress in maize leaves. However, lignin hydrogel enhanced the P uptake in maize shoots, while Na ions were released by the polyacrylate hydrogel, increasing soil sodicity and pH, both of them detrimental for the crop yields (Mazloom et al., 2020). In another work, the cassava starch granules were modified by ZnO/tetraethyl orthosilicate and polyacrylate (MS-PA) to provide an antibacterial capacity and longer service life. A balance between activity and biodegradability should be ensured during the development of novel materials. First, the water absorption of the MS-PA composite was demonstrated and gradually decreased down to 85% of the original level after 10 absorption-desorption cycles, considering an acceptable reusability potential for agricultural applications.

3. Post-harvest applications

3.1 Biodegradable packaging in the postharvest handling and transport

The package consists of any external cover that contains the product, as well as the packaging materials whose function is to support, protect, manipulate and conserve the product (Ait-Oubahou et al., 2019). Therefore, sustainable and biodegradable packaging is the key in the food supply that allows extending the shelf life, protecting food from damage and degradation, as well as ensuring food safety, thus reducing food waste. The packaging production will continue to grow due to the high demand, up to about 6% by 2023 (Mali, 2018). A few of the wide range of biodegradable packaging in the form of bags, containers and boxes are highlighted here.

3.1.1 Biodegradable bags

The bags should be strong, resistant and flexible, as well as support breakage, temperature and mechanical damage. Biodegradable bags, in addition, to offering safe food storage, are an ecological and environmentally friendly option. Biodegradable bags generated from natural sources show improved physical properties (Ivanković et al., 2017). For instance, bags from PBAT/PLA (commercially known as Mater-Bi) reinforced with 5% banana plant residue show softness, as well as biodegradability and compostability. These bags were also able to accelerate the ripening of banana fruits compared to conventional bags. They can be applied for earlier harvest when used to protect the fruits during plant development (Bordón et al., 2021). Bof et al. (2021) obtained bags formulated with corn starch and chitosan (75:25) added with grapefruit seed extract (3%). These bags presented good gas barrier properties and a decrease in the incidence of blueberry fruit rot due to the presence of grapefruit seed extract (Bof et al., 2021). Biodegradable bags of chitosan and graphene oxide (0.25%) decreased water vapor permeability and microbiological growth as well as increased the melon fruits shelf-life storage at 7°C (de Paiva et al., 2020). Therefore, biopolymeric materials and additives used for the generation of biodegradable bags are a viable option, with important advantages over conventional materials.

3.1.2 Biodegradable boxes

Boxes are materials for the transportation of a large number of items and products, including food and post-harvest fruit and vegetable products (Kale and Nath, 2020). Boxes can be made of different materials, including cardboard, however, they can release contaminants. Biodegradable boxes (bio-boxes) maintain food quality and protect food during transport and storage (Ait-Oubahou et al., 2019). In a recent study, PLA bio-boxes demonstrated mechanical stability inside the cold chamber (6°C), while sugarcane bagasse-based bio-boxes may need wax coatings to preserve their shape and stability (Kumar et al., 2021). Tapioca starch was used as the vehicle of essential oils (peppermint and lime) to control microbial activity when applied as wood box coating. The treated boxes were used for the storage and transport of mangosteen (*Garcinia mangostana*) at 25°C for 1 week. This packaging decreased the moldiness on fruits and maintained fruit quality (Owolabi et al., 2021). Packaging such as boxes can be modified by altering the surface of the materials and thereby improving their functionality, among these techniques ultraviolet radiation and cold plasma have been reported. Even biopolymers can coat other materials to generate a package with improved characteristics (Hanani, 2018).

Active packaging uses a variety of active compounds such as antimicrobial and antioxidants agents added during the formulation of biodegradable polymers. Other active components focus on the removal of gases (Ait-Oubahou et al., 2019) and humidity control (moisture absorbers) (Gaikwad et al., 2019). On the other hand, smart packaging interacts with the product it contains as well as its environment. For instance, it can contain a wide variety of active compounds and sensors both in and outside to advise the consumer about the quality and safety of the contents (Nešic et al., 2020).

3.1.3 Biodegradable foams

Biodegradable foams are used to replace synthetic polymer foams, mainly expanded polystyrene or styrofoam. Their main characteristic is lightness due to the generation of air or gas bubbles, which is why some of the properties of these containers are given as a function of the size and density of these cells or the mass and volume acquired during processing, as well as the type of polymer and its formulations (Mali, 2018). The most commonly used biodegradable polymers for the production of foams are starch, PVOH, PLA mixed with other polymers and additives, modifying their manufacturing process (Mali, 2018). However, some of the polymers are unstable in conditions of high relative humidity, so research has been carried out on foams composed of other materials such as clays, cellulose and even nanomaterials, which give them better mechanical and thermal stability (Araque et al., 2018). In addition, foams based on different agricultural wastes have recently been reported, such as cassava starch and rice husk as a macro filler to store cherry tomatoes (Corralo et al., 2020), starch and paper waste (Yudanto and Diponegoro, 2020), banana pseudostems as a transport cushioning material for fruit transportation (Yoga Milani et al., 2020), and coconut husk, cassava bagasse, cassava starch, sugarcane bagasse, among others (Regubalan et al., 2018). These new materials are more environmentally friendly and seek to generate added value to the agricultural wastes.

3.2 Biodegradable films and coatings

Biodegradable packaging is designed so that living organisms use them as a carbon source; that is, they feed on it, decomposing it and therefore consuming the plastic material. In this way, it reduces the plastics' presence in the environment and can be quickly and safely degraded by microorganisms (Pirsa, 2020; Solís-Contreras et al., 2021; Youssef et al., 2019). The package acts as a moisture barrier, gas barrier (O_2, CO_2, ethylene and water vapor) and avoids aroma loss (Falcó et al., 2019; Guitián et al., 2019; Xu et al., 2020). It has also been shown to maintain the organoleptic quality of agricultural products, physical integrity and in many cases, control the growth of microorganisms and fungi, leading to significantly increase the life of the produce, thus minimizing waste of post-harvest produce. These packaging materials can be used in the form of a coating or pre-formed film with the uniqueness that they can be edible depending on their composition.

A coating is obtained when a thin layer of the biopolymer is applied in liquid form on the food, which can be obtained by immersing the product in the polymer solution or by a spray method, making a coating over the entire surface of the product. A film is a pre-formed solid polymer sheet used to cover food, creating a barrier between the food and the environment or a wall between various food product components. Films can be obtained by methods such as casting or extrusion. These biopolymers are also used in the pre-harvest phase, where it is sought to reduce the problems or damages that could arise in agricultural products in the post-harvest phase. For instance, chitosan has been used before and after harvest in fruits such as grapes (Nia et al., 2021).

The biodegradable packaging materials also serve as carriers of additives or beneficial substances that can increase the shelf life of food, maintain its safety and enrich the products' quality. Among these additives, there are antimicrobial and antioxidant agents, vitamins, fatty acids, minerals, prebiotics, among others. Several biodegradable films and coatings were demonstrated to extend the shelf-life of fruits and vegetables (Falcó et al., 2019; Pellá et al., 2020). Preservative agents are incorporated, which can be chemical (Jeong et al., 2020; Youssef et al., 2019), synthesized by microorganisms (Guitián et al., 2019; Pérez-Arauz et al., 2021; Xu et al., 2020) or from natural sources. Many active compounds have been obtained from plants, which have active molecules that help these polymeric materials increase the product's shelf life (Falcó et al., 2019; Friedrich et al., 2020; Gómez-Aldapa et al., 2021; Solís-Contreras et al., 2021). Due to this, several investigations have been developing new materials friendly to nature, from renewable and biodegradable sources and many particularly since they can be edible.

Films based on natural polymers can store vegetable and fruit products, both in thin-film and more conveniently as an edible coating. Coatings are applied to fresh fruits and vegetables to increase their shelf life by promoting regulation of water and gas (oxygen, ethylene and carbon dioxide) exchange. Coatings reduce the weight loss of the product during the storage period and reduce post-harvest physiological damage, such as those caused mechanically or cold. Coatings can also provide gloss to enhance the external appearance of the fruit.

Table 21.1 summarizes the composition and effect of some films and coatings based on natural polymers used during the conservation of various agricultural products during the post-harvest phase. While Fig. (21.2) shows two final films and coatings products and enumerates positive functionalities in agricultural products, among which can be highlighted: (1) reduction of gas

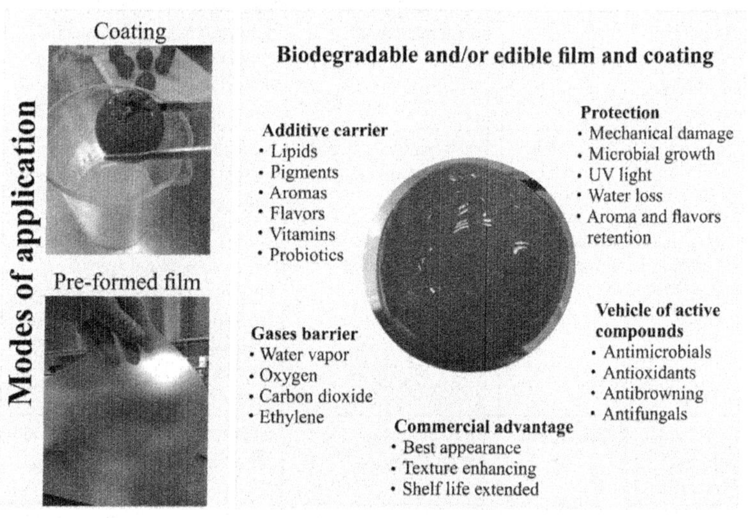

Figure 21.2. Application modes and main functions of biodegradable films and coatings used to extend the shelf life and packaging of food products.

Table 21.1. Formulation and physical characteristics of coatings and films made from biodegradable polymers used for the preservation of post-harvest products.

Biopolymers used	**Elaboration method**	**Post-harvest product**	**Effects**	**Reference**
Coating				
Agar (2%) and pomegranate seed oil (0.4 mL L^{-1}).	Immersion	Papaya Formosa fresh fruits stored at 17.5°C.	Conservation of the quality papaya fruits.	(Morais et al., 2020)
Cassava starch (2.5%), kefir and chitosan.	Immersion	Blueberry fruits refrigerated at 3°C and relative humidity 85–95%.	Edible coatings maintain the physicochemical characteristics of blueberry.	(Oliveira et al., 2018)
Nanocellulose (0.5%).	Immersion	Fresh-cut spinach leaves stored at 25°C.	The coating maintains the color, appearance, moisture and chlorophyll content, reduces the respiration rate of spinach.	(Pacaphol et al., 2019)
Chitosan (1%), glycerol and cinnamon essential oil (0.06%).	Immersion	Minimally processed apple slices.	Coating reduces weight loss, maintains firmness and phenolic compounds and delays brown index, the respiration rate of the fruit, and the growth of microorganisms.	(Solís-Contreras et al., 2021)
Chitosan (2%) – *Aloe vera* gel (33%) and glycerol (1%).	Immersion	Table grapes.	The coating extended the storage life of fruits maintaining their biochemical characteristics for 15 days.	(Nia et al., 2021)
Cordia myxa gum, glycerol and calcium dichloride ($CaCl_2$) (1%).	Immersion	Fresh artichoke bottoms refrigerated at 2°C.	The coating can reduce weight loss, browning index and control the mesophilic and psychrotrophic bacteria and mold growth.	(El-Mogy et al., 2020)
Mucilage from cactus (*Opuntia ficus indica*), and glycerol.	Immersion	Fresh breba figs refrigerated at 4°C	The coating extends the figs shelf-life, maintaining brightness, firmness and good visual appearance during 10 days.	(Allegra et al., 2017)
Pectin (high methyl ester), sunflower lecithin, sucrose monopalmitate and Citrus extract (Biosecur F440D) or mixture of essential oils.	Immersion	Pre-cut carrots refrigerated at 4°C	Coating with antimicrobial activity. Coating with pectin gels improves the stability of bioactive compounds with a controlled release.	(Ben-Fadhel et al., 2020)
Shellac gum, glycerol and tannic acid.	Immersion	Fresh mangoes stored at room temperature.	The edible coating delays weight loss, maintains firmness and aromas, reduces browning and lipid peroxidation.	(Ma et al., 2021)
Soy protein isolate, glycerol and cinnamaldehyde (CIN), zinc oxide nanoparticle (ZnONP).	Immersion	Bananas stored at 25°C.	Material with antifungal capacity. The coating delays ripening and weight loss, maintaining the firmness, soluble sugar, acidity and sensory quality of bananas.	(Pacaphol et al., 2019)
Starch (cassava) – gelatin, sorbitol (30%) and *Tetradenia riparia* extract.	Immersion	Strawberries in refrigeration.	Material with antimicrobial and antioxidant capacity.	(Friedrich et al., 2020)

Table 21.1 contd. ...

...Table 21.1 contd.

Biopolymers used	Elaboration method	Post-harvest product	Effects	Reference
Zein (10%), glycerol, oleic acid and tannic acid (4%)	Immersion	Guavas stored at 23°C.	Coating reduces product weight and moisture loss.	(Santos et al., 2018)
Films				
Agar (2.5%), glycerol (30%) and zinc oxide nanoparticles (ZnONPs) (4%).	Casting	Green grapes stored at 37°C.	Films with ZnONPs presented improved thermal and mechanical behavior and increased the grapes' lifetime.	(Kumar et al., 2019)
Carboxymethyl cellulose (CMC), cinnamaldehyde (CIN) and zinc oxide nanoparticles (ZnONPs).	Casting	Fresh cherry tomatoes.	CMC/CIN/ZnONPs film exhibited good mechanical behavior, excellent gas barrier capacity and antifungal activity, reducing weight loss and maintaining the firmness of the product.	(Guo et al., 2020)
Casein (rennet, acid caseins and sodium caseinate) and glycerol (13.2 or 24.2%).	Extrusion	N.A.	Rennet and acid casein sheets are less sensitive to water at low plasticizer concentration.	(Chevalier et al., 2018)
Cellulose nanocrystals, chitosan nanoparticle and polyvinyl alcohol.	Casting	Mango fruits stored at 26°C.	Composite increased tensile strength, thermostability, biodegradability and antifungal properties.[a]	(Dey et al., 2021)
Chitosan (2%) and gelatin (4%) nanocomposite hybrid with ZnONPs (2 and 4%).	Casting	N.A.	The presence of ZnO nanoparticles improved the mechanical behavior and thermal stability of the films.	(Kumar et al., 2020)
Chitosan-rice starch with Ag and ZnONPs.	Casting	Peach fruits refrigerated at 4°C.	Nanocomposite inhibiting the spoilage microflora.	(Kaur et al., 2017)
Nanocellulose with polyvinyl alcohol.	Casting	Black grapes fruits stored at 4 and 27°C.	Nanocellulose with polyvinyl alcohol 5% is effective to packaged grapes.[b]	(Ghosh and Ghosh, 2020)
Pectin, nanoclay, Methylene Blue (MB) and glycerol.	Casting	Sensing film for the detection of vitamin C.	The addition of clay improved the gas barrier and the addition of MB increased the antioxidant activity. Pectin/Clay/MB films change color in the presence of ascorbic acid.	(Pirsa, 2020)
Persimmon (*Diospyros kaki* L.) with glycerol and pectin.	Casting	Cucumber, carrot and beetroot minimally processed and stored at 4°C.	Glycerol and pectin modified the properties of films.	(Rabelo et al., 2021)
PLA (v) and PLA-corn starch (90:10, 80:20) with benzyl peroxide and glycidyl methacrylate.	Blown extrusion	Capsicum and fresh fruits stored at 25 and 8°C.	Corn starch increases flexibility.	(Mangaraj et al., 2019)
Polycaprolactone (PCL) and Sodium Metabisulfite (SM) (0, 10, 20 and 40%).	Melt-extrusion	Freshly-cut apple.	SM/PCL films exhibited oxygen scavenging, antioxidant, antifungal, anti-browning and antimicrobial capabilities.	(Jeong et al., 2020)

Table 21.1 contd. ...

...Table 21.1 contd.

Biopolymers used	Elaboration method	Post-harvest product	Effects	Reference
Soy protein isolate, rapeseed oil (0 to 3%) and glycerol (40%).	Casting	N.A.	The presence of rapeseed oil improves the gas barrier, the mechanical performance and hydrophobicity of the films.	(Galus, 2018)
Starch (cassava), gelatin and plasticizers.	Casting	Strawberries fresh fruits stored at environment temperature.	Sorbitol increased the transparency and flexibility properties.	(Franco et al., 2017)
Starch (cassava) and hydroxyethyl cellulose.	Casting	Guava fruits at environment temperature.	Hydroxyethyl cellulose increased the transparency and hygroscopicity.	(Francisco et al., 2020)
Starch (cassava), sorbitol and microcrystalline cellulose.	Casting	N.A.	Microcrystalline cellulose increased physical and biodegradability properties.[c]	(Abel et al., 2021)
Starch (corn), glycerol (30%), and *Hibiscus sabdariffa* extract (HSE).	Casting	N.A.	The addition of HSE produces films with antimicrobial activity, improves the mechanical behavior and the water vapor barrier capacity.	(Gómez-Aldapa et al., 2021)
Starch (corn)-gelatin with by-products (mango and pineapple pomace) (5–15%).	Casting	N.A.	Agricultural by-products can be used to improve the physicochemical properties of materials. Less susceptibility to hydration and better optic properties.[d]	(Susmitha et al., 2021)
Whey protein isolate/ Psyllium Seed Gum (PSG), and glycerol (40%).	Casting	N.A.	PSG improved mechanical behavior and hydrophobicity. Composite films have lower oxygen permeability and higher degree of crystallization.	(Zhang et al., 2020)
Films and coatings				
Alginate (sodium) and cyclolipopeptides (CL) from *Bacillus subtilis* (0 to 3%).	Casting for films. Coating (immersion) and soaked in $CaCl_2$ for crosslinking (for application).	Fresh blueberries stored (between −1°C and 0°C).	Antifungal edible film and coating with the addition of CL. Easily washable coating.	(Xu et al., 2020)
Carrageenan (1%) and green tea extract (0.7%).	Casting for films and coating (immersion) for application.	Blueberries and raspberries stored at 10°C and 25°C.	Enhanced antiviral activity against Murine norovirus (MNV-1) and Hepatitis A Virus (HAV).	(Falcó et al., 2019)
Chitosan (2%), glycerol (30%), Litchi Peel Extract (LPE) and nano-titanium dioxide (nano-TiO_2).	Casting for films and coating (immersion) for application.	Whole Fuji apples refrigerated at 0–1°C during 180 days.	The chitosan-LPE-TiO_2 film presented higher gas barrier capacity and mechanical resistance. The active coating inhibited the respiratory rate and the weight loss of the apples.	(Liu et al., 2021)

Table 21.1 contd. ...

...Table 21.1 contd.

Biopolymers used	Elaboration method	Post-harvest product	Effects	Reference
Chitosan – whey protein isolate and glycerol.	Casting for films and double coating (immersion) for application.	Strawberries refrigerated at 5°C.	The films showed good barrier performance and antioxidant capacity. The coating maintained the biochemical characteristics of the fruits and an increase in their post-harvest shelf life.	(Muley and Singhal, 2020)
Starch (cassava) – gelatin-casein and sorbitol (30%).	Casting for films and coating (immersion) for application.	Fresh guavas	More starch and casein and low concentration of gelatin improve properties, lowers opacity and water vapor transmission rate. Fruit shelf life increasing with the coating.	(Pellá et al., 2020)
Starch (purple yam) – chitosan and glycerol (2%).	Casting for films and coating (immersion) for application.	Fresh apples	Film WVP increases with chitosan content. Coating improves shelf life of apples.	(Martins da Costa et al., 2020)
Multicomposite vegetable matrix or fruit and vegetable residues.	Casting for films and coating (immersion) for application.	Carrots minimally processed stored at 5°C in a dark cold chamber.	Films increased the tensile strength and elongation at break.	(Fai et al., 2016)

N.A.: Not applicable; WVP: Water Vapor Permeability; WVTR: Water Vapor Transmission Rate; PLA: polylactic acid.

[a] Weight loss in two types of soil (dry and wet soil) moisturized daily, pieces into 5 cm × 5 cm placed in a plastic box for 30 days.

[b] Microbiological degradation test on agar plates.

[c] Weight loss in soil under environmental conditions, pieces into 2 cm × 2 cm. The film without MCC presented a degradation percentage higher than 55% after 2 weeks.

[d] Weight loss placed on an iron wired gauze and buried, pieces into 1 cm × 2 cm for 15 days.

permeability, (2) change atmosphere for fresh coated products, (3) delayed weight loss and drying of the surface, (4) maintaining the integrity of the original structure, and (5) reduction of rotten products.

3.3 Water purification from the agricultural sector

Water pollution is a global problem and the agriculture sector accounts for 70% of the world's water abstraction, playing a major role in the depletion of water. Most water pollutants are organic matter and agrochemicals runoffs. As it degrades, organic matter absorbs dissolved oxygen in water, which largely leads to hypoxia in water bodies, in addition to increasing eutrophication and algal blooms in lakes (Turunen et al., 2019). Excessive fertilization and irrigation cause salinity and soil conductivity owing to the excessive presence of sodium, potassium, chloride, calcium, sulfate or bicarbonate as well as an excess of nitrogen, phosphates resulting in a negative impact on different ecosystems and their biodiversity (EPA, 2019; Fao and Iwmi, 2017; Wato and Amare, 2020). Over the years, numerous techniques have been made to incorporate different wastewater treatment technologies to remove these pollutants, some of them are chemical precipitation, adsorption, reverse osmosis, electrocoagulation, sedimentation, filtration and coagulation-flocculation. As a material selection for water purification, it is important to use reliable, robust, lower cost to decontaminate and disinfect water from sources without further environmental stress or endangering human health (Dongre et al., 2019; Figoli et al., 2017; Obotey and Rathilal, 2020). A compilation of interesting biopolymers like cellulose derivatives, PLA, chitosan and starch as building blocks to form membranes, as well as acting as flocculants and coagulants is presented here.

3.3.1 Flocculants

Polysaccharide-based flocculants have proved that accelerating the agglomeration helps to remove suspended solids from agricultural wastewater. The effectivity is dependent on the type of biopolymer coagulant, the initial pollution category, pH and the optimal dose applied. For instance, tannin and chitosan performed similarly with reductions up to 95 and 98% of total phosphorus and turbidity, in contrast to starch, for which the best reduction was 80 and 82%, respectively (Turunen et al., 2019). Total organic carbon reductions were found around 20% for the three biopolymers. The mucilage removes over 92% turbidity using small amounts of iron coagulant (Lim et al., 2018). Nanocellulose alone or combined with ferric sulfate showed a turbidity reduction of 40–80% and a Chemical Oxygen Demand (COD) reduction of 40–60% (Suopajärvi et al., 2013).

3.3.2 Membranes

Biopolymer membranes to selectively remove water contaminants are of some interest because of their biodegradability and/or low carbon footprint and also because of their wide range of functionalities and properties that could open up the possibilities of new applications (Tomietto et al., 2021). Cellulose and its derivatives, form strong porous networks on water evaporation, and the pore size can be tuned by varying the fiber diameter. Cellulose membranes have been extensively studied and already commercialized, to remove organic and inorganic materials in water. However, further incorporation of organic or inorganic substances is required for more improvement of functionality (Mahdavi and Shahalizade, 2015; Su et al., 2010). A cellulose acetate (CA) based ultrafiltration membrane was prepared for the removal of bovine serum albumin (BSA). Since CA membranes are susceptible to fouling by chemicals and microorganisms present in the feed contaminant, the presence of Hydrous Manganese Dioxide (HMO) improved their hydrophilic nature and anti-fouling capacity. CA/HMO membranes had a pure water flux of 143 L m^{-2} h^{-1}, 95.9% BSA rejection, and improved anti-fouling behavior by means of lower irreversible fouling (Vetrivel et al., 2019). The phase inversion process was used to incorporate PBS into biodegradable CA/PBS membranes. The presence of PBS altered the membrane mesopore structure and CA improved tensile strength, according to the research. The membrane efficiency tests indicate that a higher PBS content improved turbidity and COD rejection (Ghaffarian et al., 2015). A casting method was used to obtain a lanthanum interconnected carboxymethylcellulose-bentonite membrane. This resulted in phosphate and nitrate anions adsorption of approximately 80 and 68 mg/g, respectively. The electrostatic forces and ion exchange play a major role in removal efficiency (Karthikeyan et al., 2020).

Other biopolymers like PHB and its copolymerization with 3-hydroxyvalerate (PHBHV) have been used to prepare membranes incorporating polyethylene glycols (PEGs) as additives. The results showed that PEGs increased water permeabilities by over 200 L m^{-2} bar^{-1} h^{-1} and bacteria (*E. coli*) rejection by over 99%, which could be explained by a smaller pores size distribution (Tomietto et al., 2020). PLA, derived from corn or sugarcane, is also promising for membrane fabrication. With the aid of surfactant, nonsolvent mediated phase separation membranes of PLA have been prepared to improve mechanical properties. Keawsupsak et al. (2014) prepared flat-sheet membranes with PLA and other biodegradable polymers such as PBS, PBAT and PHBHV. According to the results the PLA-PHBHV blend membrane had pure water flux of 65 L m^{-2} h^{-1} and 79% BSA rejection as high as PLA-PBS and PLA-PBAT blend membranes, as well as higher tensile strength and elongation at break (Keawsupsak et al., 2014).

Alginate membranes have furthermore been tested for the removal of diquat (DQ) and difenzoquat (DF) herbicides. The authors prepared membranes with and without chitosan, and found that the herbicides were adsorbed on the alginate (reaching 95 and 62% for DQ and DF, respectively) and chitosan/alginate membranes (reaching 95 and 12% for DQ and DF, respectively), suggesting that the adsorption occurs in the alginate layer through coulombic interactions (Agostini de Moraes et al., 2013).

3.4 Perspectives

Even if the development of biodegradable polymers for agricultural applications is an effective way to partly solve the problem of white pollution in this field, some important shortcomings have to been addressed. A key point for the generation of degradable polymers is the amount of biomass available to meet the requirements for the generation of biomaterials. Biodegradable and non-biodegradable materials should be compared face-to-face in terms of effectiveness and ecotoxicological impacts. The release of micro-, nano-plastics and lixiviates are also of critical concern due to the inclusion of toxic components on the local ecosystems and the food chain. Therefore, most fundamental studies should include the analysis of the production, behavior, fate and transport of these residuals. Furthermore, competitive production costs of biodegradable polymers should be considered to envisage commercial or practical applications rather than just demonstrations at the laboratory level. The reusability and stability of a biodegradable polymer in typical agricultural environments are crucial for continuous applications. Applicable regulations should be also considered during the design of novel biodegradable materials, integrating functionality with a systematic assessment of environmental compatibility under various exposure conditions. From the other perspective, social awareness is not enough to motivate consumers to pay more for biodegradable products. Thereby, ideal biodegradable polymers for agricultural applications should be economically, environmentally and technically sustainable.

Conclusions

Biodegradable polymers have demonstrated numerous advantages for agricultural applications. Since they do not accumulate for extended periods and can be obtained from agro-industrial waste, biodegradable polymers provide an excellent opportunity to substantially decrease the plastic fraction of solid wastes. Accelerated rising demand for high-quality food, climate change, soil erosion, droughts, enhanced resistance of pathogens against pesticides and some global economic factors have to be faced by farmers, making necessary the development of more effective production strategies.

Nowadays, the efforts in the field are devoted to multifunctional biodegradable materials, for instance, those used in the pre-harvest period that significantly improves the seed conservation and germination process, which impacts the crop's yields and quality. Biopolymers as carriers and dispensers of pesticides, antibiotics, fertilizers and super-absorbents of humidity facilitate their safe handling and dosage control. In the post-harvest period, biodegradable polymers can be used in various ways for food preservation, storage and transport of agricultural products with bags, boxes, foamed products, films and coatings. It is of interest the way that active and smart biodegradable packaging will impact its interaction with consumers with positive economic, social and environmental advantages.

Furthermore, water reclamation in agroindustry is possible using biopolymers as flocculants and membranes to remove pathogens and contaminants present in agro-wastewater, mitigating environmental pollution. Overall, the opportunities for biodegradable materials and biopolymers polymers in agricultural applications at the pre- and post-harvest stages are vast. The increasingly multidisciplinary research approach is important to have a more complete and timely set of information to solve the current challenges.

Acknowledgements

The doctors Aguirre-Loredo, Soriano-Melgar and Jasso-Salcedo thank CONACYT for their appointment as research fellows assigned to CIQA. LIASM acknowledges CONACYT funding No. 314907_LABMYN 2020 and No. 316010.

References

Abd El-Aziz, M. E., Kamal, K. H., Ali, K. A., Abdel-Aziz, M. S. and Kamel, S. (2018). Biodegradable grafting cellulose/clay composites for metal ions removal. *International Journal of Biological Macromolecules*, 118(Part_B): 2256–2264.

Abduwaiti, A., Liu, X., Yan, C., Xue, Y., Jin, T., Wu, H., … Liu, Q. (2021). Testing biodegradable films as alternatives to plastic-film mulching for enhancing the yield and economic benefits of processed tomato in xinjiang region. *Sustainability*, 13(6): 1–13.

Abel, O. M., Chinelo, A. S. and Chidioka, N. R. (2021). Enhancing cassava peels starch as feedstock for biodegradable plastic. *Journal of Materials and Environmental Science*, 12(02): 169–182.

Accinelli, C., Abbas, H. K., Little, N. S., Kotowicz, J. K., Mencarelli, M. and Shier, W. T. (2016). A liquid bioplastic formulation for film coating of agronomic seeds. *Crop Protection*, 89: 123–128.

Agostini de Moraes, M., Cocenza, D. S., da Cruz Vasconcellos, F., Fraceto, L. F. and Beppu, M. M. (2013). Chitosan and alginate biopolymer membranes for remediation of contaminated water with herbicides. *Journal of Environmental Management*, 131: 222–227.

Ait-Oubahou, A., Hanani, Z. A. N. and Jamilah, B. (2019). Packaging. pp. 375–399. *In*: Yahia, E. M. (ed.). *Postharvest Technology of Perishable Horticultural Commodities*. Kidlington, UK: Woodhead Publishing.

Alam, M. N. and Christopher, L. P. (2018). Natural cellulose-chitosan cross-linked superabsorbent hydrogels with superior swelling properties. *ACS Sustainable Chemistry & Engineering*, 6(7): 8736–8742.

Allegra, A., Sortino, G., Inglese, P., Settanni, L., Todaro, A. and Gallotta, A. (2017). The effectiveness of Opuntia ficus-indica mucilage edible coating on post-harvest maintenance of 'Dottato' fig (*Ficus carica* L.) fruit. *Food Packaging and Shelf Life*, 12: 135–141.

Appusamy, A. M., Eswaran, P., Subramanian, M., Rajamanickam, A., Selvakumar, V. K. and Chandrasekar, V. P. (2021). Invasive Parthenium weed as reinforcement in polymer matrix composite to reduce its environmental impact. *Materials Today: Proceedings*, 45: 1112–1118.

Araque, L. M., Alvarez, V. A. and Gutiérrez, T. J. (2018). Composite foams made from biodegradable polymers for food packaging applications. pp. 347–355. *In*: Gutiérrez, T. J. (ed.). *Polymers for Food Applications*. Cham, Switzerland: Springer Nature Switzerland AG.

Azarian, M. H., Kamil Mahmood, W. A., Kwok, E., Bt Wan Fathilah, W. F. and Binti Ibrahim, N. F. (2018). Nanoencapsulation of intercalated montmorillonite-urea within PVA nanofibers: Hydrogel fertilizer nanocomposite. *Journal of Applied Polymer Science*, 135(10): 45957.

Ben-Fadhel, Y., Maherani, B., Manus, J., Salmieri, S. and Lacroix, M. (2020). Physicochemical and microbiological characterization of pectin-based gelled emulsions coating applied on pre-cut carrots. *Food Hydrocolloids*, 101: 105573.

Bi, S., Pan, H., Barinelli, V., Eriksen, B., Ruiz, S. and Sobkowicz, M. J. (2021). Biodegradable polyester coated mulch paper for controlled release of fertilizer. *Journal of Cleaner Production*, 294: 126348.

Bof, M. J., Laurent, F. E., Massolo, F., Locaso, D. E., Versino, F. and García, M. A. (2021). Bio-packaging material impact on blueberries quality attributes under transport and marketing conditions. *Polymers*, 13(4): 1–20.

Bordón, P., Paz, R., Peñalva, C., Vega, G., Monzón, M. and García, L. (2021). Biodegradable polymer compounds reinforced with banana fiber for the production of protective bags for banana fruits in the context of circular economy. *Agronomy*, 11(2): 242.

Borrowman, C. K., Johnston, P., Adhikari, R., Saito, K. and Patti, A. F. (2020). Environmental degradation and efficacy of a sprayable, biodegradable polymeric mulch. *Polymer Degradation and Stability*, 175: 109126.

Chevalier, E., Assezat, G., Prochazka, F. and Oulahal, N. (2018). Development and characterization of a novel edible extruded sheet based on different casein sources and influence of the glycerol concentration. *Food Hydrocolloids*, 75: 182–191.

Corralo, J., Amanda, S., Isabel, J. and Tessaro, C. (2020). Biodegradable cassava starch based foams using rice husk waste as macro filler. *Waste and Biomass Valorization*, 11(8): 4315–4325.

Dada, O. R., Abdulrahman, K. O. and Akinlabi, E. T. (2019). Production of biodegradable composites from agricultural waste. pp. 39–48. *In*: Kumar, K. and Davim, J. P. (eds.). *Biodegradable Composites: Materials, Manufacturing and Engineering*. Berlin/Boston: Walter de Gruyter GmbH & Co KG.

Dang, X., Shan, Z. and Chen, H. (2016). The preparation and applications of one biodegradable liquid film mulching by oxidized corn starch-gelatin composite. *Applied Biochemistry and Biotechnology*, 180(5): 917–929.

Das Lala, S., Deoghare, A. B. and Chatterjee, S. (2018). Effect of reinforcements on polymer matrix bio-composites—an overview. *Science and Engineering of Composite Materials*, 25(6): 1039–1058.

de Paiva, C. A., Vilvert, J. C., de Menezes, F. L. G., Leite, R. H. de L., dos Santos, F. K. G., de Medeiros, J. F. and Aroucha, E. M. M. (2020). Extended shelf life of melons using chitosan and graphene oxide-based biodegradable bags. *Journal of Food Processing and Preservation*, 44(11): 1–12.

Dey, D., Dharini, V., Periyar Selvam, S., Rotimi Sadiku, E., Mahesh Kumar, M., Jayaramudu, J. and Nath Gupta, U. (2021). Physical, antifungal, and biodegradable properties of cellulose nanocrystals and chitosan nanoparticles for food packaging application. *Materials Today: Proceedings*, 38: 860–869.

Dispat, N., Poompradub, S. and Kiatkamjornwong, S. (2020). Synthesis of ZnO/SiO_2-modified starch-graft-polyacrylate superabsorbent polymer for agricultural application. *Carbohydrate Polymers*, 249: 116862.

Dongre, R. S., Sadasivuni, K. K., Deshmukh, K., Mehta, A., Basu, S., Meshram, J. S., … Karim, A. (2019). Natural polymer based composite membranes for water purification: a review. *Polymer-Plastics Technology and Materials*, 58(12): 1295–1310.

El-Mogy, M. M., Parmar, A., Ali, M. R., Abdel-Aziz, M. E. and Abdeldaym, E. A. (2020). Improving postharvest storage of fresh artichoke bottoms by an edible coating of Cordia myxa gum. *Postharvest Biology and Technology*, 163: 111143.

EPA. (2019). The Sources and Solutions: Agriculture. *United States Environmental Protection Agency.*

Fai, A. E. C., Alves de Souza, M. R., de Barros, S. T., Bruno, N. V., Ferreira, M. S. L., Gonçalves, T. C. B. D. A. and Branco de Andrade, É. C. (2016). Development and evaluation of biodegradable films and coatings obtained from fruit and vegetable residues applied to fresh-cut carrot (Daucus carota L.). *Postharvest Biology and Technology*, 112: 194–204.

Falcó, I., Randazzo, W., Sánchez, G., López-Rubio, A. and Fabra, M. J. (2019). On the use of carrageenan matrices for the development of antiviral edible coatings of interest in berries. *Food Hydrocolloids*, 92: 74–85.

Fao and Iwmi. (2017). Water pollution from agriculture: a global review Executive summary. *FAO and IWMI.*

Figoli, A., Marino, T., Galiano, F., Dorraji, S. S., Di Nicolò, E. and He, T. (2017). Sustainable route in preparation of polymeric membranes. pp. 97–120. *In*: Alberto Figoli and Criscuoli, A. (eds.). *Sustainable Membrane Technology for Water and Wastewater Treatment*. Singapore: Springer.

Francisco, C. B., Pellá, M. G., Silva, O. A., Raimundo, K. F., Caetano, J., Linde, G. A., … Dragunski, D. C. (2020). Shelf-life of guavas coated with biodegradable starch and cellulose-based films. *International Journal of Biological Macromolecules*, 152: 272–279.

Franco, M. J., Martin, A. A., Bonfim, L. F., Caetano, J., Linde, G. A. and Dragunski, D. C. (2017). Effect of plasticizer and modified starch on biodegradable films for strawberry protection. *Journal of Food Processing and Preservation*, 41(4): 1–9.

Friedrich, J. C. C., Silva, O. A., Faria, M. G. I., Colauto, N. B., Gazzin, Z. C., Colauto, G. A. L., … Dragunski, D. C. (2020). Improved antioxidant activity of a starch and gelatin-based biodegradable coating containing Tetradenia riparia extract. *International Journal of Biological Macromolecules*, 165: 1038–1046.

Fukamachi, K., Konishi, Y. and Nomura, T. (2019). Disease control of Phytophthora infestans using cyazofamid encapsulated in poly lactic-co-glycolic acid (PLGA) nanoparticles. *Colloids and Surfaces, A: Physicochemical and Engineering Aspects*, 577: 315–322.

Gaikwad, K. K., Singh, S. and Ajji, A. (2019). Moisture absorbers for food packaging applications. *Environmental Chemistry Letters*, 17(2): 609–628.

Galus, S. (2018). Functional properties of soy protein isolate edible films as affected by rapeseed oil concentration. *Food Hydrocolloids*, 85: 233–241.

Ghaffarian, V., Mousavi, S. M. and Bahreini, M. (2015). Effect of blend ratio and coagulation bath temperature on the morphology, tensile strength and performance of cellulose acetate/poly(butylene succinate) membranes. *Desalination and Water Treatment*, 54(2): 473–480.

Ghosh, M. and Ghosh, P. (2020). Storage study of grapes (Vitis vinifera) using the nanocomposite biodegradable film from banana pseudostem. *Journal of Food Processing and Preservation*, 44(12): 1–12.

Glowinska, A., Trochimczuk, A. W. and Jakubiak-Marcinkowska, A. (2019). Novel acrylate/organophosphorus-based hydrogels for agricultural applications. New outlook and innovative concept for the use of 2-(methacryloyloxy) ethyl phosphate as a multi-purpose monomer. *European Polymer Journal*, 110: 202–210.

Gómez-Aldapa, C. A., Díaz-Cruz, C. A., Castro-Rosas, J., Jiménez-Regalado, E. J., Velazquez, G., Gutierrez, M. C. and Aguirre-Loredo, R. Y. (2021). Development of antimicrobial biodegradable films based on corn starch with aqueous extract of Hibiscus sabdariffa L. *Starch-Stärke*, 73(1–2): 2000096.

Guitián, M. V., Ibarguren, C., Soria, M. C., Hovanyecz, P., Banchio, C. and Audisio, M. C. (2019). Anti-Listeria monocytogenes effect of bacteriocin-incorporated agar edible coatings applied on cheese. *International Dairy Journal*, 97: 92–98.

Guo, X., Wang, X. and Yin, J. (2021). Three-layer composite high-barrier controllable full-biodegradable mulching film and preparation method thereof. Chinese Patent CN202011150609A.

Guo, X., Chen, B., Wu, X., Li, J. and Sun, Q. (2020). Utilization of cinnamaldehyde and zinc oxide nanoparticles in a carboxymethylcellulose-based composite coating to improve the postharvest quality of cherry tomatoes. *International Journal of Biological Macromolecules*, 160: 175–182.

Hanani, Z. A. N. (2018). Surface properties of biodegradable polymers for food packaging. pp. 131–147. *In*: Gutiérrez, T. J. (ed.). *Polymers for Food Applications*. Cham, Switzerland: Springer.

Hongsriphan, N. and Pinpueng, A. (2019). Properties of agricultural films prepared from biodegradable poly(butylene succinate) adding natural sorbent and fertilizer. *Journal of Polymers and the Environment*, 27(2): 434–443.

Hu, M., Dou, Q., Cui, X., Lou, Y. and Zhuge, Y. (2019). Polyaspartic acid mediates the absorption and translocation of mineral elements in tomato seedlings under combined copper and cadmium stress. *Journal of Integrative Agriculture*, 18(5): 1130–1137.

Hwang, S. H. (2021). Method for producing modified cellulose fibers with epoxidized soybean oil and biodegradable polymer composites containing modified cellulose fibers. Patent WO2021015334A1.

Ivanković, A., Zeljko, K., Talić, S., Bevanda, A. M. and Lasić, M. (2017). Biodegradable packaging in the food industry. *Journal of Food Safety and Food Quality*, 68(2): 26–38.

Javazmi, L., Low, T., Ash, G. and Young, A. (2020). Investigation of slow release of urea from biodegradable single- and double-layered hollow nanofibre yarns. *Scientific Reports*, 10(1): 19619.

Jeong, S., Lee, H. G., Cho, C. H. and Yoo, S. R. (2020). Characterization of multi-functional, biodegradable sodium metabisulfite-incorporated films based on polycarprolactone for active food packaging applications. *Food Packaging and Shelf Life*, 25: 100512.

Kale, S. J. and Nath, P. (2020). Application of plastics in postharvest management of crops. pp. 1–16. *In*: Kumar, R., Singh, V. P., Jhajharia, D. and Mirabbasi, R. (eds.). *Applied Agricultural Practices for Mitigating Climate Change* (Volume 2). Florida: CRC Press.

Karthikeyan, P., Banu, H. A. T., Preethi, J. and Meenakshi, S. (2020). Performance evaluation of biopolymeric hybrid membrane and their mechanistic approach for the remediation of phosphate and nitrate ions from water. *Cellulose*, 27(8): 4539–4554.

Kaur, M., Kalia, A. and Thakur, A. (2017). Effect of biodegradable chitosan–rice-starch nanocomposite films on post-harvest quality of stored peach fruit. *Starch/Staerke*, 69(1–2): 1600208.

Keawsupsak, K., Jaiyu, A., Pannoi, J., Somwongsa, P., Wanthausk, N., Sueprasita, P. and Eamchotchawalit, C. (2014). Poly(lactic acid)/biodegradable polymer blend for the preparation of flat-sheet membrane. *Jurnal Teknologi (Sciences and Engineering)*, 69(9): 99–102.

Kenawy, E.-R., Azaam, M. M. and El-nshar, E. M. (2018). Preparation of carboxymethyl cellulose-g-poly (acrylamide)/montmorillonite superabsorbent composite as a slow-release urea fertilizer. *Polymers for Advanced Technologies*, 29(7): 2072–2079.

Kikionis, S., Ioannou, E., Konstantopoulou, M. and Roussis, V. (2017). Electrospun micro/nanofibers as controlled release systems for pheromones of Bactrocera oleae and Prays oleae. *Journal of Chemical Ecology*, 43(3): 254–262.

Krishnamoorthy, V. and Rajiv, S. (2017). Potential seed coatings fabricated from electrospinning hexaaminocyclotriphosphazene and cobalt nanoparticles incorporated polyvinylpyrrolidone for sustainable agriculture. *ACS Sustainable Chemistry & Engineering*, 5(1): 146–152.

Kumar, S., Boro, J. C., Ray, D., Mukherjee, A. and Dutta, J. (2019). Bionanocomposite films of agar incorporated with ZnO nanoparticles as an active packaging material for shelf life extension of green grape. *Heliyon*, 5(6): e01867.

Kumar, S., Mudai, A., Roy, B., Basumatary, I. B., Mukherjee, A. and Dutta, J. (2020). Biodegradable hybrid nanocomposite of chitosan/gelatin and green synthesized zinc oxide nanoparticles for food packaging. *Foods*, 9(9): 1143.

Kumar, Sasi, Leitão, F., Gaspar, P. D. and Silva, P. D. (2021). Experimental tests of the thermal behaviour of new sustainable bio-packaging food boxes. *Procedia Environmental Science, Engineering and Management*, 8(1): 215–223.

Lee, L.-T. and Ke, Y.-L. (2020). Superior crystallization kinetics caused by the remarkable nucleation effect of graphene oxide in novel ternary biodegradable polymer composites. *ACS Omega*, 5(47): 30643–30656.

Lim, B.-C., Lim, J.-W. and Ho, Y.-C. (2018). Garden cress mucilage as a potential emerging biopolymer for improving turbidity removal in water treatment. *Process Safety and Environmental Protection*, 119: 233–241.

Liu, Z., Du, M., Liu, H., Zhang, K., Xu, X., Liu, K., ... Liu, Q. (2021). Chitosan films incorporating litchi peel extract and titanium dioxide nanoparticles and their application as coatings on watercored apples. *Progress in Organic Coatings*, 151: 106103.

Ma, J., Zhou, Z., Li, K., Li, K., Liu, L., Zhang, W., ... Zhang, H. (2021). Novel edible coating based on shellac and tannic acid for prolonging postharvest shelf life and improving overall quality of mango. *Food Chemistry*, 354: 129510.

Mahdavi, H. and Shahalizade, T. (2015). Preparation, characterization and performance study of cellulose acetate membranes modified by aliphatic hyperbranched polyester. *Journal of Membrane Science*, 473: 256–266.

Mali, S. (2018). Biodegradable foams in the development of food packaging. pp. 329–345. *In*: Gutiérrez, T. J. (ed.). *Polymers for Food Applications*. Cham, Switzerland: Springer.

Malik, N., Shrivastava, S. and Ghosh, S. B. (2018). Moisture absorption behaviour of biopolymer polycapralactone (PCL)/organo modified montmorillonite clay (OMMT) biocomposite films. *IOP Conference Series: Materials Science and Engineering*, 346: 012027.

Mangaraj, S., Mohanty, S., Swain, S. and Yadav, A. (2019). Development and characterization of commercial biodegradable film from PLA and corn starch for fresh produce packaging. *Journal of Packaging Technology and Research*, 3(2): 127–140.

Maraveas, C. (2020). Production of sustainable and biodegradable polymers from agricultural waste. *Polymers*, 12(5): 1127.

Martins da Costa, J. C., Lima Miki, K. S., da Silva Ramos, A. and Teixeira-Costa, B. E. (2020). Development of biodegradable films based on purple yam starch/chitosan for food application. *Heliyon*, 6(4): e03718.

Mazloom, N., Khorassani, R., Zohury, G. H., Emami, H. and Whalen, J. (2020). Lignin-based hydrogel alleviates drought stress in maize. *Environmental and Experimental Botany*, 175: 104055.

Morais, F. A. de, Araújo, R. H. C. R., Oliveira, A. M. F. de, Alves, K. de A., Vitor, R. C. L., Morais, S. K. Q., ... Oliveira, A. G. de. (2020). Agar and pomegranate seed oil used in a biodegradable coating composition for formosa papaya. *Food Science and Technology*, 40: 280–286.

Muley, A. B. and Singhal, R. S. (2020). Extension of postharvest shelf life of strawberries (Fragaria ananassa) using a coating of chitosan-whey protein isolate conjugate. *Food Chemistry*, 329: 127213.

Nešic, A., Cabrera-Barjas, G., Dimitrijevic-Brankovic, Suzana Davidovic, S., Radovanovic, N. and Delattre, C. (2020). Prospect of polysaccharide-based materials as advanced food packaging. *Molecules*, 25(135): 1–35.

Nia, A. E., Taghipour, S. and Siahmansour, S. (2021). Pre-harvest application of chitosan and postharvest Aloe vera gel coating enhances quality of table grape (Vitis vinifera L. cv. 'Yaghouti') during postharvest period. *Food Chemistry*, 347: 129012.

Nita, L. E., Ghilan, A., Rusu, A. G., Neamtu, I. and Chiriac, A. P. (2020). New trends in bio-based aerogels. *Pharmaceutics*, 12(5): 499.

Obotey Ezugbe, E. and Rathilal, S. (2020). Membrane technologies in wastewater treatment: A review. *Membranes*, 10(5): 89.

Oliveira, Í., Ribeiro, A., Mello-Farias, P., Malgarim, M., Machado, M. and Lamela, C. (2018). Biodegradable coatings on blueberries postharvest conservation refrigerated in a modified atmosphere. *Journal of Experimental Agriculture International*, 20(4): 1–11.

Owolabi, I. O., Songsamoe, S., Khunjan, K. and Matan, N. (2021). Effect of tapioca starch coated-rubberwood box incorporated with essential oils on the postharvest ripening and quality control of mangosteen during transportation. *Food Control*, 126: 108007.

Pacaphol, K., Seraypheap, K. and Aht-Ong, D. (2019). Development and application of nanofibrillated cellulose coating for shelf life extension of fresh-cut vegetable during postharvest storage. *Carbohydrate Polymers*, 224: 115167.

Pellá, M. C. G., Silva, O. A., Pellá, M. G., Beneton, A. G., Caetano, J., Simões, M. R. and Dragunski, D. C. (2020). Effect of gelatin and casein additions on starch edible biodegradable films for fruit surface coating. *Food Chemistry*, 309: 125764.

Pérez-Arauz, Á. O., Rodríguez-Hernández, A. I., del Rocío López-Cuellar, M., Martínez-Juárez, V. M. and Chavarría-Hernández, N. (2021). Films based on Pectin, Gellan, EDTA, and bacteriocin-like compounds produced by Streptococcus infantarius for the bacterial control in fish packaging. *Journal of Food Processing and Preservation*, 45(1): e15006.

Pirsa, S. (2020). Biodegradable film based on pectin/Nano-clay/methylene blue: Structural and physical properties and sensing ability for measurement of vitamin C. *International Journal of Biological Macromolecules*, 163: 666–675.

Prudnikova, S., Streltsova, N. and Volova, T. (2021). The effect of the pesticide delivery method on the microbial community of field soil. *Environmental Science and Pollution Research*, 28(7): 8681–8697.

Rabelo, J., Matheus, V., Assis, R. M. De, Correia, T. R., Regina, M., Christina, M., ... Fai, C. (2021). Biodegradable and edible film based on persimmon (Diospyros kaki L.) used as a lid for minimally processed vegetables packaging. *Food and Bioprocess Technology*, 14: 765–779.

Regubalan, B., Pandit, P., Maiti, S. and Nadathur, G. T. (2018). Potential bio-based edible films, foams, and hydrogels for food packaging. pp. 105–123. *In*: Ahmed, S. (ed.). *Bio-based Materials for Food Packaging: Green and Sustainable Advanced Packaging Materials*. Singapore: Springer.

Santos, T. M., Souza Filho, M. de S. M., Silva, E. de O., Silveira, M. R. S. d., Miranda, M. R. A. d., Lopes, M. M. A. and Azeredo, H. M. C. (2018). Enhancing storage stability of guava with tannic acid-crosslinked zein coatings. *Food Chemistry*, 257: 252–258.

Sartore, L., Vox, G. and Schettini, E. (2013). Preparation and performance of novel biodegradable polymeric materials based on hydrolyzed proteins for agricultural application. *Journal of Polymers and the Environment*, 21(3): 718–725.

Saruchi, Kumar, V., Pathak, D., Mittal, H. and Alhassan, S. M. (2019). Experimental assessment of the utilization of a novel interpenetrating polymer network in different processes in the agricultural sector. *Journal of Applied Polymer Science*, 136(28): 47739.

Skrzypczak, D., Mikula, K., Kossinska, N., Widera, B., Warchol, J., Moustakas, K., ... Witek-Krowiak, A. (2020). Biodegradable hydrogel materials for water storage in agriculture—review of recent research. *Desalination and Water Treatment*, 194: 324–332.

Solís-Contreras, G. A., Rodríguez-Guillermo, M. C., Reyes-Vega, M. de la L., Aguilar, C. N., Rebolloso-Padilla, O. N., Corona-Flores, J., ... Ruelas-Chacon, X. (2021). Extending shelf-life and quality of minimally processed golden delicious apples with three bioactive coatings combined with cinnamon essential oil. *Foods*, 10(3).

Sonjan, S., Ross, G. M., Mahasaranon, S., Sinkangam, B., Intanon, S. and Ross, S. (2021). Biodegradable hydrophilic film of crosslinked PVA/silk sericin for seed coating: the effect of crosslinker loading and polymer concentration. *Journal of Polymers and the Environment*, 29(1): 323–334.

Su, J., Zhang, S., Chen, H., Chen, H., Jean, Y. C. and Chung, T. S. (2010). Effects of annealing on the microstructure and performance of cellulose acetate membranes for pressure-retarded osmosis processes. *Journal of Membrane Science*, 364(1–2): 344–353.

Suopajärvi, T., Liimatainen, H., Hormi, O. and Niinimäki, J. (2013). Coagulation-flocculation treatment of municipal wastewater based on anionized nanocelluloses. *Chemical Engineering Journal*, 231: 59–67.

Susmitha, A., Sasikumar, K., Rajan, D., Padmakumar M, A. and Nampoothiri, K. M. (2021). Development and characterization of corn starch-gelatin based edible films incorporated with mango and pineapple for active packaging. *Food Bioscience*, 41: 100977.

Tomietto, P., Loulergue, P., Paugam, L. and Audic, J. L. (2020). Biobased polyhydroxyalkanoate (PHA) membranes: Structure/performances relationship. *Separation and Purification Technology*, 252: 117419.

Tomietto, P., Loulergue, P., Paugam, L. and Audic, J. L. (2021). Polyhydroxyalkanoates (PHAs) for the fabrication of filtration membranes. pp. 177–195. *In*: Zhang, Z., Zhang, W. and Chehimi, M. M. (eds.). *Membrane Technology Enhancement for Environmental Protection and Sustainable Industrial Growth*. Switzerland: Springer Cham.

Turunen, J., Karppinen, A. and Ihme, R. (2019). Effectiveness of biopolymer coagulants in agricultural wastewater treatment at two contrasting levels of pollution. *SN Applied Sciences*, 1(3): 210.

Vetrivel, S., Rana, D., Sri Abirami Saraswathi, M. S., Divya, K., Kaleekkal, N. J. and Nagendran, A. (2019). Cellulose acetate nanocomposite ultrafiltration membranes tailored with hydrous manganese dioxide nanoparticles for water treatment applications. *Polymers for Advanced Technologies*, 30(8): 1943–1950.

Wato, T. and Amare, M. (2020). The agricultural water pollution and its minimization strategies—a review. *Journal of Resources Development and Management*, 64: 10–22.

Xiang, Y., Li, C., Hao, H., Tong, Y., Chen, W., Zhao, G. and Liu, Y. (2021). Performances of biodegradable polymer composites with functions of nutrient slow-release and water retention in simulating heavy metal contaminated soil: Biodegradability and nutrient release characteristics. *Journal of Cleaner Production*, 294: 126278.

Xu, L., Zhang, B., Qin, Y., Li, F., Yang, S., Lu, P., ... Fan, J. (2020). Preparation and characterization of antifungal coating films composed of sodium alginate and cyclolipopeptides produced by *Bacillus subtilis*. *International Journal of Biological Macromolecules*, 143: 602–609.

Yakubu, A., Ucheoma, E. U., Muhammad, M. A., Paiko, Y. B., Adenike, F. O., Damilare, S. O., ... Makunsidi, Y. B. (2018). Composites based on poly(vinyl chloride) and organically modified clay. *Asian Journal of Chemistry*, 30(8): 1902–1908.

Yoga Milani, M. D., Samarawickrama, D. S., Perera, P. S. D., Wilson Wijeratnam, R. S. and Hewajulige, I. G. N. (2020). Eco friendly packaging material from banana pseudo stem for transportation of fruits and vegetables. *Acta Horticulturae*, 1278: 59–64.

Youssef, A. M., Assem, F. M., Abdel-Aziz, M. E., Elaaser, M., Ibrahim, O. A., Mahmoud, M. and Abd El-Salam, M. H. (2019). Development of bionanocomposite materials and its use in coating of Ras cheese. *Food Chemistry*, 270: 467–475.

Yudanto, Y. A. and Diponegoro, U. (2020). Characterization of physical and mechanical properties of Biodegradable foam from maizena flour and paper waste for Sustainable packaging material. *International Journal of Engineering Applied Sciences and Technology*, 5(8): 1–8.

Ze, K., Zhibo, L. and Yon, S. (2020). Biodegradable poly (4-hydroxybutyrate) agricultural mulching film and preparation method thereof. Chinese Patent CN202010846226A.

Zhang, X., Zhao, Y., Li, Y., Zhu, L., Fang, Z. and Shi, Q. (2020). Physicochemical, mechanical and structural properties of composite edible films based on whey protein isolate/psyllium seed gum. *International Journal of Biological Macromolecules*, 153: 892–901.

Zvinavashe, A. T., Lim, E., Sun, H. and Marelli, B. (2019). A bioinspired approach to engineer seed microenvironment to boost germination and mitigate soil salinity. *Proceedings of the National Academy of Sciences*, 116(51): 25555–25561.

Chapter 22

Insight on Polymeric Hydrogel Networks

A Sustainable Tool for the Isolation of Enzymes and Bioremediation

Bárbara Bosio, Paola Camiscia, Guillermo Picó and *Nadia Woitovich Valetti**

1. Introduction

Flexible Chain Polymers (FCP) of biological origin constitute a universe of molecules found in different types of biomass: from microorganisms (bacteria, fungi and microalgae) to macro algae, plants and higher animals. FCPs have a series of common characteristics, which gives them interesting properties that can be applied in different biotechnological areas:

i) With the sole exception of gelatin, which is a polypeptide of animal origin, all natural FCPs are polysaccharides, some of them with covalently linked electrical charged chemical groups (SO^{3-}, NH_2, COOH) which confers them with the property of polyelectrolytes. Others, on the other hand, are electrically neutral since they are only composed by polysaccharide chains (Haug and Draget, 2011; Seymour, 1971; Woitovich Valetti et al., 2016).
ii) They all come from natural sources, but in some cases they must be subjected to earlier chemical and physical treatments for their extraction, partially degrading the polysaccharide chains or introducing acidic or basic groups to transform them into soluble forms (Bakshi et al., 2020). The most popular is cellulose, which is insoluble in water and must be subjected to chemical and/or physical procedures to obtain soluble forms (Wernersson et al., 2015).
iii) As a consequence of a high number of OH groups in their structure, FCP are soluble and have an extraordinary ability to interact with water dipoles by forming hydrogen bonds, obtaining highly viscous solutions, which is the basis of many of its applications (Dobrynin and Rubinstein, 2005; José García de la et al., 2008).
iv) FCPs with electrically charged chemical groups tend to act as weak acids or bases. Therefore, according to the pH of the medium, and their pKa value, more or less rigid structures are formed (Dobrynin and Rubinstein, 2005).

Instituto de Procesos Biotecnológicos y Químicos. Consejo Nacional de Investigaciones Científicas y Técnicas de Argentina. Facultad de Ciencias Bioquímicas y Farmacéuticas, Universidad Nacional de Rosario. Rosario, Argentina.

* Corresponding author: woitovichvaletti@iprobyq-conicet.gob.ar

2. The most used natural flexible chain polymers

Despite the large number of polysaccharide derivatives that exist, several of them stand out for their physical and chemical properties in an aqueous solution, which allows them to be used in different biotechnological processes.

2.1 Alginate

Alginate (Alg) is a non-toxic, biodegradable, naturally occurring polysaccharide obtained from marine brown algae and certain species of bacteria (Pawar and Edgar, 2012). It is formed by two monomeric units: β-(1-4) linked D-mannuronic acid (M) residues and α-(1-4)-linked L-guluronic acid (G) residues, being its basic structure linear unbranched units of monomers arranged in blocks of M and G residues interspersed with regions containing alternating MG sequence (Hernández-González et al., 2020). Alg are available in the form of sodium, potassium salt or acid free, with molecular weights from 30 to 400 kDa depending on the extraction method (Lee and Mooney, 2012). Although its solubility is affected by the ionic strength of the medium, which also has an important effect on the extension of the polymer chain and its viscosity, the factor that mainly affects its solubility is the pH of the solvent, which will determine the presence or absence of charges in its lateral chains (Pawar and Edgar, 2012). The dissociation constants of the mannuronic and guluronic acid monomers are 3.38 and 3.65, respectively. Therefore, Alg tends to be negatively charged over a wide pH range. Most of its applications are related to their ability to form gels in the presence of divalent cations such as alkaline earth metals (calcium ions are the most common) (Hernández-González et al., 2020; Lee and Mooney, 2012; Pawar and Edgar, 2012; Tønnesen and Karlsen, 2002). Gelation has been demonstrated to result from strong interactions between Ca^{2+} ions and G residues, which leads to the chain–chain associations and to the distinguishing, so-called, junction zone. These associations are also responsible for gel formation.

Numerous experimental and theoretical studies have been performed to obtain information about the structural features of the junction zone. The most popular "egg-box" model is usually used to describe the formation of Alg gels in the presence of alkaline earth metals (Cao et al., 2020; Grant et al., 1973). This model indicates that the G residues along the alginate chain adopt a 2_1 helical conformation, and a pair of such chains are packed with calcium ions located between them (Plazinski, 2011). Figure 22.1 shows the mechanism by which Alg gelling occurs.

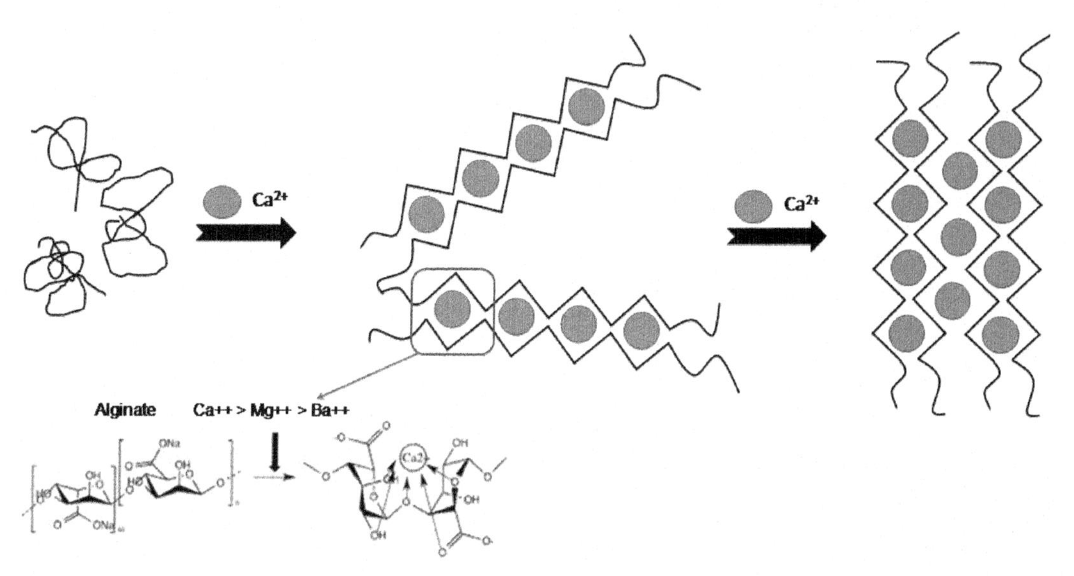

Figure 22.1. The molecular three-dimensional mechanism by which Alg gelling occurs.

2.2 Carrageenan

The carrageenans (Car) are commercially important hydrophilic colloids (water-soluble gums) that are found as matrix material in numerous red seaweed species (*Rhodophyta*) wherein they serve a structural function analogous to that of cellulose in land plants (Campo et al., 2009). Car are sulfated polygalactan with 15 to 40% of ester-sulfate content and an average relative molecular mass well above 100 kDa (Ghanbarzadeh et al., 2018). They are formed by alternated units of d-galactose and 3,6-anhydro-galactose joined by α-1,3 and β-1,4-glycosidic linkage (Chakraborty, 2017). The sulfate groups are strongly anionic (pKa = 1), being comparable to inorganic sulfate. Car is classified into various types such as: λ, κ, ι, ε, μ. The different types of Car differ from each other by their 3,6-anhydrogalactose and sulfate content, which both can be interconverted with each other using an alkali treatment (Campo et al., 2009).

There are three most important types of commercially available Car: monosulfated κ-Car, bisulfated ι-Car and trisulfated λ-Car (Ghanbarzadeh et al., 2018). Kappa- and iota-Car form hydrogels in the presence of different ions, whereas λ-Car does not. Structure differences between i and k-Car result in different gelling properties (Günter et al., 2020). The gel formation of k-Car involves a transition from a random coil to an ordered helical conformation. After the conformational transition, the helices may further associate to form a gel, a process generally stimulated by the presence of salts combined with a reduction in temperature. It is generally accepted that K^+, Rb^+, Cs^+, NH_4^+ stabilizes the helix and promotes the gelation (Bercea and Wolf, 2019). The presence of divalent cations also favors gelation in a more complicated process (Liu and Li, 2016).

i-Car will gel in the presence of K^+, Rb^+, Cs^+ under suitable conditions of polymer concentration, ionic strength and temperature. The thermo reversible gelation involves a coil to helix conformational transition which may be followed by the aggregation of the ordered molecules to form an infinite network (Michel et al., 1997).

2.3 Pectin

Pectins are structural polysaccharides, which are present in all higher plants cell walls (Yapo, 2009). Pectin consists primarily of linear chains of partly methyl esterified α-d-galacturonic acid, interrupted and bent in places by rhamnose, galactose and arabinose units (Moreira et al., 2014). Depending on the degree of esterification of the carboxyl group of the galacturonic acid with methanol, pectin can be categorized into two types. The species of pectin with degrees of methylation (molar ratio of methanol to galacturonic acid) higher than 50% is called high methoxyl pectin, while pectin samples with degrees of methylation less than 50% are called low methoxyl pectin (Chan et al., 2017).

The gelation of low methoxyl pectin depends upon the pH of the system and the presence of Ca^{2+} or other multivalent ions. In the presence of divalent cations like Ca^{2+}, pectin gels can form via Ca^{2+} ion bridges between two carboxyl groups of two different chains forming well organized structures known as the "egg-box" model (Basak and Bandyopadhyay, 2014; Cao et al., 2020; Moreira et al., 2014).

High methoxyl pectin gelation requires high amounts of sugar and is very sensitive to acidity. Forms gel at low pH values and at high concentrations of soluble solids due to the presence of hydrogen bonding and hydrophibic interactions between the pectin chains (Noreen et al., 2017).

2.4 Chitosan

More of the natural occurring polysaccharides are neutral or acidic in nature. Chitin and chitosan (Chi) are the only examples of highly basic polysaccharides (Arvanitoyannis, 2008). Chitin is one of the world´s most abundant renewable organic sources, it's composed by β(1 → 4) linked 2-acetamido-2-deoxy-β-D-glucose units (or N-acetyl-D-glucosamine), forming a long linear polymer chain (Zargar et al., 2015). It is insoluble in most solvents. Chi, the principal derivative of chitin, is obtained by N-deacetylation to a varying extent which is later characterized by its degree of deacetylation, and is consequently a copolymer of N-acetyl-D-glucosamine and D-glucosamine (Krajewska, 2004). Chi can be defined as a sufficiently deacetylated chitin, which forms soluble amine salts. The degree of deacetylation necessary to obtain a soluble product is 80–85% or higher. The amino-groups of the D-glucosamine units present in Chi are mostly protonated in an acid aqueous medium while they are mostly non-protonated in basic conditions, favoring macromolecular association and resulting in a physicochemical gelation when pH is higher than the pKa of its amino-groups (pKa = (6.2–7.1)) (Pillai et al., 2009).

3. Hydrogels based on natural polymers

Hydrogels are very swollen, hydrophilic polymer networks that can absorb large amounts of water and drastically increase in volume (Chai et al., 2017). Hydrogels can be prepared from synthetic or natural polymers, involving a wide range of chemical compositions and with different mechanical, physical and chemical properties (Hamedi et al., 2018). It is well known that the physicochemical properties of a hydrogel depends not only on the molecular structure, gel structure and the degree of crosslinking, but also on the content and state of the water in the hydrogel (Augst et al., 2006; Bruck, 1973). Polymers that form hydrogels may have hydrophilic or hydrophobic functional groups. Hydrophilic functional groups enable the hydrogel to absorb water leading to a hydrogel expansion, which is known as swelling (Peppas et al., 2000). Hydrogels with hydrophobic chains have lower swelling capacity than hydrophilic lattices.

Although polymers from different sources can be used to obtain hydrogels, hydrogels composed of natural polymers have several advantages:

- Since they come from biomass, they are biodegradable and do not have a negative effect on the environment (Varaprasad et al., 2017). In addition, this type of source is available in large quantity, inexhaustible and many can be obtained from waste, solving an environmental and waste management problem.
- They are biocompatible and non-toxic (Berger et al., 2004; Chai et al., 2017).
- Its obtaining process is simple and inexpensive.

Among the disadvantages they present, one can highlight:

- Due to the fact that they are polysaccharides, they are a source of carbon for organisms, which is why they tend to become quickly contaminated, leading to their degradation (Kulkarni Vishakha et al., 2012). This problem can be overcome by using additives and bacteriostats.
- Like most products derived from natural sources, its composition tends to vary according to how the biomass was produced (time of collection, climate, place of origin, etc.) (Kulkarni Vishakha et al., 2012).
- Polysaccharides in solution are usually unstable, their chains tend to fragment varying their molecular weight, influencing on the viscosity of their solutions.

3.1 Basic principles for obtaining Hydrogels

Obtaining polymeric hydrogels is based mainly on the characteristics of the polymer or polymers used to obtain them. Depending on the composition of the gels, it will be the basic principle by which they are obtained. The main methodologies are based on ionotropic or isoionic gelation.

Ionotropic gelation is based on the ability of polyelectrolytes to cross-link in the presence of counter ions to form hydrogels as are the cases of Alg, pectin, Chi and Car (Patil et al., 2010). It is one of the most used methods for the production of hydrogels, but its application is limited to polymers that have the ability to coordinate with counter ions to form gels.

In the case of Alg and pectin, as indicated above, their ionotropic gelation occur in the presence of different divalent cations, their affinity increases in the order of Mn < Zn, Ni, Co < Fe < Ca

< Sr < Ba < Cd < Cu < Pb (Ching et al., 2017). From this list, the most used cation is Ca^{2+} since it is non-toxic, which increases the range of applications of the gels obtained (Hu et al., 2021). Within the Car, the iota and kappa varieties have shown their ability to gel in the presence of monovalent cations depending on the working conditions (Günter et al., 2020). Chi, on the other hand, gels in the presence of tripolyphosphate (TPP) forming stable gels (Sacco et al., 2014).

Isoionic precipitation, on the other hand, is based on the ability of certain polymers to gel when varying the pH of the medium. Such ability appears due to the presence of some functional groups which can be protonated/deprotonated at a critical transition pH (Vakurov et al., 2012). This characteristic is widely used to obtain Chi gels, despite also being observed in other polymers such as Alg.

In all the cases mentioned above, gelation is carried out by contacting the polymer solution with the gelling solution. Depending on the way in which the solutions are contacted, different shapes and sizes of gels can be obtained. Among the simplest methodologies for the preparation of hydrogels using ionotropic or isoionic precipitation, the following stand out:

i) *Simple dripping (extrusion)*: It is one of the most used mechanisms to obtain macrogels. The basis of this method involves the drop-wise extrusion of polymers solution from a syringe into a gelling solution bath (Patil et al., 2010). When the polymer flows from the opening of the syringe, a droplet is formed and falls toward the gelling bath. When it encounters the solution, the polymer gels shaping up to spherical gels with diameters between 1–2 mm depending on the working conditions, such as: polymer concentration, dripping rate, temperature, stirring speed, among others. A diagram of the technique is shown in Fig. 22.2.

ii) *Modified extrusion method*: Relatively smaller droplets can be formed using a vibration system or air atomization method to extrude the polymer solution. Pressurized air is fed to the polymer solution mix, forcing tiny liquid droplets throughout the orifice of the nozzle (Patil et al., 2010). The micro droplets thus obtained, the gel when they come into contact with the gelling agent solution obtaining microspheres within the size range of 5–200 μm.

iii) *Emulsification*: In the emulsification method, microspheres are formed in a non-aqueous continuous phase. The polymer solution is dispersed into oil to form a water-in-oil emulsion. The gelling solution is introduced slowly and with continuous stirring in order to maintain the emulsion. Contact between the dispersed polymer droplets and previously mentioned solution leads to gelation (Sahil et al., 2011). The size of the spheres obtained by this method is between 100–1000 μm.

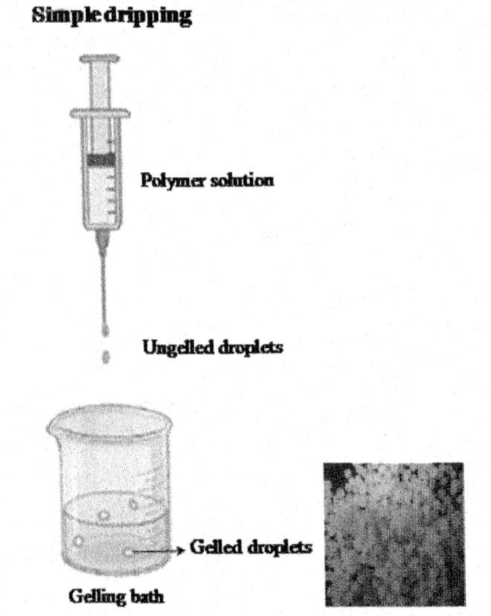

Figure 22.2. Simple dripping method for obtaining hydrogels.

3.2 Cross-linking of the hydrogel

The gels obtained by the earlier methods, allow to obtain macro, micro and nano spheres. These gels are unstable to variations of its surrounding conditions, for example Chi gels obtained by isoionic precipitation, are stable at alkaline pH, but below pH 6.4 (pKa of the amino groups of Chi) protonation of the amino groups occur which induces its dissolution. Alg gel obtained in the presence of Ca^{2+}, are dissolved in the presence of a quelator of Ca^{2+}. Therefore, one way to stabilize the hydrogel, at any pH or in the presence of other solutes is by introducing covalent bonds between the hydrocarbonate chains of the polysaccharides (Varaprasad et al., 2017).

The most popular covalent cross-linking agents are epichlorohydrin and glutaraldehyde (Dumitriu et al., 1996; Hennink and van Nostrum, 2012). The properties of the cross-linked polysaccharide depend largely on the chemical nature of the crosslinking agent, since there is a chemical reaction between the FCP and the cross-linker. The reaction conditions such as pH of the medium, solvent, temperature, reaction time and polymer/cross-linking agent ratio have a marked influence on the final properties of the gel obtained and must be determined for each particular system.

3.3 Physicochemical characterization of the gels obtained

Regardless of the composition and the methodology used to obtain them, the main methods used to characterize the gels obtained are the following:

3.3.1 Swelling

One of the main characteristics of hydrogels is their high water adsorption capacity. The determination of the amount of water embedded in the gel is one of the first characteristics to be determined due to its importance and simplicity (Bajpai and Giri, 2002). The amount of water present inside the hydrogels will be an important point to define their possible applications and will depend on various variables such as: the polymer concentration, the method employed to obtain the gel, the concentration of gelling agent, temperature, among others (Catoira et al., 2019).

The determination of the swelling is generally carried out using the weight difference method, according to the following equation (Benhalima et al., 2017):

$$\%S = (Ws–Wd)/Wd \times 100$$

where: W_s is the weight of the swollen gel and W_d is the weight of the dry gel.

Higher swelling percentage of the gel implies that it has greater amount of water which favors, among other things, the diffusion of solutes within them, a characteristic that is important for compound delivery or molecule adsorption applications.

3.3.2 Electrical surface charge

Point of zero charge ($pH_{z=0}$) has been defined traditionally as a pH value where the net electrical surface charge is equal to zero (Bakatula et al., 2018). It is a useful parameter when studying the adsorption of small ions onto adsorbents since it gives an idea about ionization and the interactions between the adsorbent surface and the adsorbate (Oussalah et al., 2019). The groups that are present on the hydrogel surface can either accept or donate an additional proton from the solution depending on its pH and the type and number of acid and basic groups. The surface becomes positively charged by accepting protons from the solution; on the contrary, it becomes negatively charged due to the loss of protons. The determination of this parameter is of interest because it provides valuable information on the state of charges of the material that can be widely used when applied as an ion exchanger, for example (Brassesco et al., 2018).

3.3.3 Porosity

Hydrogels have a porous nature that allows solutes to diffuse into and out of them. The porosity of gels is the key for many of their applications, mainly in the immobilization of cells and enzymes field. The porosity of the material obtained and the size of the pores depend on variables such as: polymer concentration, method of obtaining the gel, concentration of the gelling agent. For the determination of the porosity of hydrogels, various techniques can be used, each of which has its advantages and disadvantages. Various authors have determined the porosity of polymeric gels using scanning electron microscopy (Antonietti et al., 1999; Chung et al., 2002; de Moura et al., 2005). This methodology is not optimal because the gels must be previously lyophilized, which can alter the pores of the material and lead to erroneous conclusions (Kaberova et al., 2020). Another technique used to determine porosity is the adsorption and desorption isotherm of liquid N_2 at 77 K and its subsequent adjustment to different types of isotherms (Oussalah et al., 2019).

3.3.4 Infrared spectroscopy

Since the presence of different functional groups play an important role in the possible applications of the hydrogels, it becomes necessary to analyze their presence in newly synthesized hydrogels. In this sense, one of the most used techniques for this analysis is infrared spectroscopy (IR) (Abreu et al., 2020). IR spectra are performed with previously lyophilized or dried gels to eliminate any remaining water in their composition. The number, type and position of the bands of the spectrum obtained allows to analyze the composition of the gel, the interactions that may occur in it and, in the case that covalent crosslinking has been made between polymeric chains, the new bonds formed can be observed. The amount of information provided by the realization of these spectras is reflected by the large number of publications that uses this technique for the characterization of hydrogels (Bajpai and Giri, 2002; Bárbara et al., 2020; Benhalima et al., 2017; Oussalah et al., 2019).

3.3.5 Microscopies

Hydrogels can be morphologically characterized using various types of microscopies that, depending on their resolution, will allow conclusions to be drawn on different aspects of the materials (Wang et al., 2010). Among the most used for the analysis of hydrogels formed by natural polymers, one can highlight optical microscopy, which allows a superficial analysis of microspheres in which their shape and size distribution can be determined (Wang et al., 2010). Scanning electron microscopy allows the analysis of macro, micro and nano spheres and gives a much more detailed information about the surface morphology: roughness of the gel, variations produced by altering its composition or by chemical treatments. As indicated above, it also determines porosity (Bárbara et al., 2020; Benhalima et al., 2017; Brassesco et al., 2018; Chung et al., 2002). Figure 22.3 shows characterization results of hydrogels using some of the techniques detailed above.

3.4 Application of natural polymer hydrogels as an economical, ecological and simple tool for the recovery of molecules present in aqueous solutions by adsorption

3.4.1 The adsorption unit operation

Adsorption is one of the most widely used unit operations in the early stages of concentration and purification of macromolecules from complex mixtures and for the remediation of polluted waters (Bárbara et al., 2020; Nakanishi et al., 2001). This is mainly due to its simplicity and easily scalable operation, which allows continuous work. Adsorbents are a key point in the adsorption process. Different types of adsorbents exist and its selection is based on the process under study. The choice of which kind of adsorbent to use will depend on: adsorbate under study, maximum cost of the process, availability, etc. The main limitation for the application of this type of technique is the high cost of commercial adsorption beds such as activated carbon, DEAE, Sephadex®, Streamline®, among others (Gupta et al., 2018; Roy and Gupta, 2002).

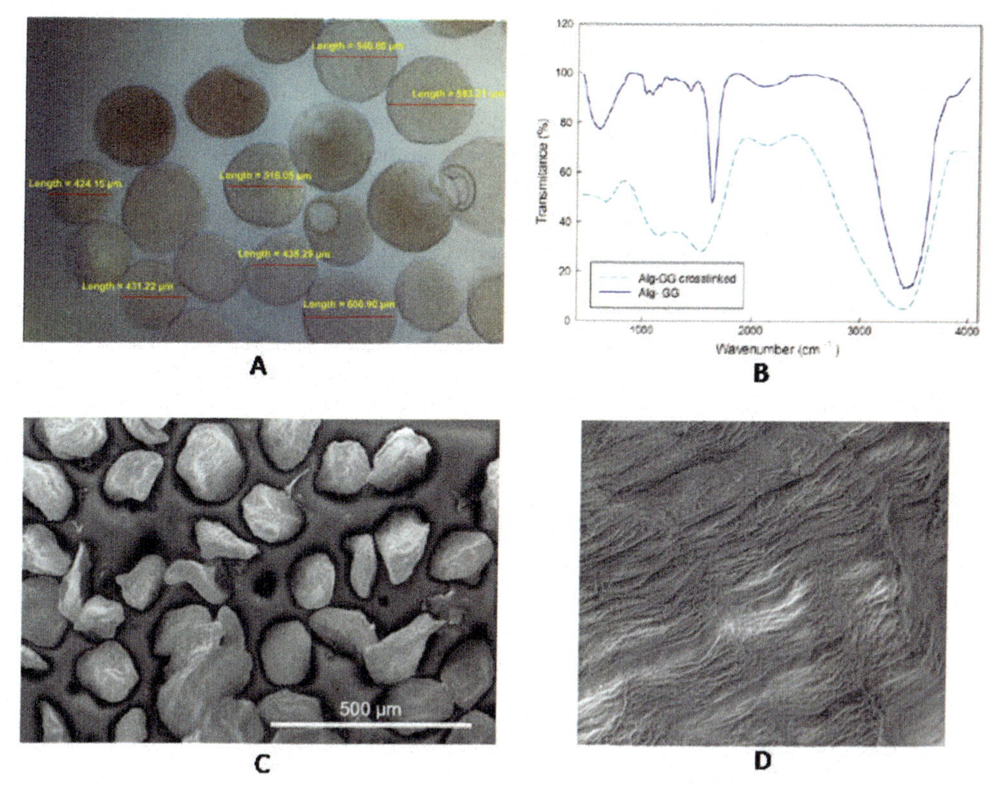

Figure 22.3. Alg-guar gum hydrogels characterization. (A) Optical microscopy; (B) FT-IR; (C) and (D) Scanning electron microscopy.

An alternative, emerging in recent years, is the use of natural polymer gels as adsorbents for proteins and other molecules in solution. Electrically charged natural polymers such as Chi, Alg, carboxymethyl cellulose, Car, etc., have shown the ability, alone or combined, to form in appropriate mediums: solid precipitates, films, fibers and spheres of various sizes and porosities with bioadsorbents properties. This is feasible since the acidic and/or basic groups remain unchanged (Bárbara et al., 2020; Brassesco et al., 2017; Chiou et al., 2004; Li et al., 2018; Woitovich Valetti and Picó, 2016; Zengin Kurt et al., 2017). These natural polymers present qualities such as high capacity to bind selectively to proteins, regenerability and the possibility of being prepared with different geometric configurations, which allows to obtain materials with the desired permeability and specific surface. In addition, they are natural, non-toxic, biodegradable, abundant and very economical substances (Lee and Mooney, 2012; Rinaudo, 2006). The groups exposed on their surface show reactivity which makes it possible to bind ligands on the surface of the matrix that can further increase their adsorption capacity and affinity (Azlan et al., 2009; Chung et al., 2002; Eser et al., 2012; Zhuang and Wang, 2018).

3.4.2 Use of polymeric hydrogels for the purification and concentration of proteins

The design of new bioseparative strategies is one of the greatest challenges of modern biotechnology. If it is taken into account that the process of purification and concentration of a molecule can represent up to 80% of its final cost, the reduction of cost and time are fundamental factors when developing new bioseparation methodologies (Gavara et al., 2015; Somasundaram et al., 2018). Developing efficient enzyme purification strategies on an industrial scale means obtaining protocols with minimum unit operations. An effective alternative, that started emerging in recent years, is the expanded bed chromatography technique. However, this methodology has a critical point: the matrixes used are commercial and have high costs. Hydrogel beds are inexpensive and easy-to-prepare systems, which is one of the main advantages over commercial beds used to isolate enzymes both in the laboratory and at a scaled-up level (Woitovich and Picó, 2016). When obtaining hydrogels to be applied as adsorption beds, polyelectrolytes such as Alg, Car and Chi are generally used, since they have the ability to gel under mild conditions and they function as anionic or cationic exchangers due to their acidic or basic groups present on their surface (Abdella et al., 2020; Roy and Gupta, 2002; Spelzini et al., 2011). Many times, different combinations of polymers are used in order to change the properties of the bed to improve its adsorption capacity or its porosity (Brassesco et al., 2018; Rodrigues et al., 2013; Woitovich and Picó, 2016).

Various authors have demonstrated the effectiveness of polymeric hydrogels for protein purification, where they demonstrate the feasibility of using Alg hydrogels for the purification of enzymes (Somers et al., 1989). By using gelled Alg spheres in the presence of calcium, they purified three times an endo-polygalacturonase present in an enzyme extract using a fluid bed process. They also found that it is possible to reuse the adsorbent 100 times without considerable losses in its adsorption capacity, thus demonstrating that polymeric hydrogels are a good possibility for the development of new inexpensive and reusable adsorbents. Roy et al., 2002 demonstrated the feasibility of using spherical Alg or Chi hydrogels for the purification of α-amilase and cellulase, respectively. In this work, the efficiency of the beds for the adsorption of different types of enzymes is confirmed by the high yields and purification factors shown. In 2002, the same research group studied the application of Alg hydrogels for the purification of pullulanase using a packed bed and a fluidized bed (Roy and Gupta, 2002). The results obtained showed that the capacity of the beds for this enzyme compares favorably with the binding capacities reported with other enzymes and their affinity matrixes in the expanded bed format. They also managed to purify the enzyme 59 times working in a fluid bed, demonstrating the effectiveness of Alg hydrogels for the purification of enzymes.

In addition to the beds formed by a single polymer, different works are found in literature that uses a combination of polymers to improve the adsorption capacity, the kinetics or the affinity of the beds. In this sense, Roy et al., 2005 developed Alg and guar gum beds that have been successfully applied for the purification of Jacalin present in a plant extract. In this work, the authors combined the efficiency of Alg to form hydrogels in the presence of calcium with the affinity of guar gum for the enzyme to be purified. In this way, they managed to purify the Jacalin 50 times with a yield of 88% in a single step using a fluidized bed of Alg and guar gum cross-linked with epichlorohydrin. Santos Leite da Silva et al., 2016 used Alg-guar gum bed to study the adsorption in the equilibrium of horseradish peroxidase. The characterization carried out with the commercial enzyme determined that the beds have an adsorption efficiency of 80% and that thc process is highly influenced by the pH and ionic strength of the working medium, which allows one to assume that the interaction is influenced by an electrostatic component. This last observation makes it possible to further extend the applications of polymeric hydrogels since they function both as anion/cation exchangers and as affinity beds, depending on their composition and working conditions. The same bed was shown to be efficient to adsorb lysozyme working in a packed bed column (Brassesco et al., 2019; Brassesco et al., 2017). In this work, the working conditions were optimized using commercial protein and then the enzyme was purified from egg white. The lysozyme recovery from egg white achieved was of 75% with a purification factor around 15, under the following working conditions: 9 cm of bed height, 0.4 mL/min flow rate and total protein inlet of 10 mg/mL (Brassesco et al., 2019).

Regarding mixed polymeric hydrogels, Alg-Chi beds have shown their efficiency for the adsorption of enzymes in literature. Gondim et al., 2012 immobilized Cibacron Blue F3GA onto an Alg-Chi hydrogel to achieve an affinity adsorbent for IgG purification. The polymers are used in this case as support for the affinity ligand in order to reduce the costs of the process and due to the facility with which gels are obtained from them. Rodrigues et al., 2013 used alginate-chitosan hydrogels chemically cross-linked with epichlorohydrin for the purification of isolated cellulase from *Aspergillus niger* cultures. Although a good performance was not obtained when working in a fixed bed, they managed to demonstrate the feasibility of reusing the beds which is the key in reducing costs and environmental impact.

Car-Alg hydrogels have also been designed and used to isolate aspartic proteases (Bosio et al., 2020). This type of combination takes advantage of the capabilities that each polymer presents separately: the excellent ability of Alg to form hydrogels is combined with the high negative electrical charge density present in Car allowing the hydrogel to act as an ion exchanger in a wide pH range. Bosio et al., 2020 used these beds for the purification of chymotrypsin from bovine pancreas, optimizing the adsorption conditions using statistical tools and studying the behavior of the system both in equilibrium and in packed bed. This work shows that Alg-Car hydrogels have a great adsorption capacity and that they can be used in different operation modes with purification factors for chymotrypsin of 3.8 and 6.9 working in a stirred tank and packed bed, respectively. Another highlight of this work is that the beds allowed the use of a pancreatic homogenate without previous treatment, for which the adsorption with this hydrogel functioned as both a clarification and purification step.

Table 22.1 summarizes different systems in which protein adsorption has been studied using polymer hydrogels.

Table 22.1. Adsorption capacity of polymeric hydrogels for different proteins.

Adsorbent	Protein	Maximum adsorption capacity in equilibrium	Purification factor	Yield	Times reused	Reference
Calcium Alg beads	Chymotrypsin	2.7 UA/g dry beds	9 (batch System)	62% (batch System)	NA	(Spelzini et al., 2011)
Calcium Alg beads	Pullulanase	1476 UA/ml	46–59 (fluidized bed)	85% (fluidized bed)	NA	(Roy and Gupta, 2002)
Alg-guar gum beads	Jacalin	35×10^4 UA/ml bed	50 (fluidized bed)	88% (fluidized bed)	NA	(Roy et al., 2005)
Alg-Guar gum beads	Chymotrypsin	7 mg/g hydrated bed	NA	NA	4	(Woitovich and Picó, 2016)
Alg-Guar gum beads	Peroxidase	4.5×10^{-4} UA/mg hydrated bed	NA	NA	NA	(Santos Leite da Silva et al., 2016)
Alg-Guar gum beads	Lysozyme	2.4 mg/g hydrated bed	NA	NA	6 (batch) 10 (packed bed)	(Brassesco et al., 2019; Brassesco et al., 2017)
Alg-Car beads	Chymotrypsin	11 mg/g hydrated bed	3.8 (batch system) 6.9 (packed bed)	49% (batch system) 67.9% (packed bed)	NA	(Bosio et al., 2020)
Alg-Chi beads	Cellulase	11 mg/g hydrated bed	NA	30% (packed bed)	NA	(Rodrigues et al., 2013)
Chi-Carboxymetilchitosan membranes	Ovoalbumine Lyzozyme	1021 mg/g 98 mg/g	NA	NA	NA	(Feng et al., 2009)
Glutaraldehyde cross-linked Chi	Bovine serum albumine	9.24 mg/g	NA	NA	NA	(Torres et al., 2007)
Epiclorohydrin cross-linked chi	Lysozyme	2.84 mg/g	NA	NA	NA	(Torres et al., 2007)

3.4.3 Use of polymeric hydrogels to remove contaminants from water

Large amounts of toxic metals and non-metals (Cr (III), Pb (II), Hg (II), As (V), etc.) and diazo textile dyes are discharged into the ecosystem daily, as wastewaters from different types of industries (Oussalah et al., 2019; Sharma and Bhattacharya, 2017). Great efforts have been made to eliminate these toxic substances using different methodologies, one of the most widespread and efficient is adsorption in packed bed and expanded bed column (Oussalah et al., 2019; Wan Ngah et al., 2011). Additionally, adsorption has been found to be superior to other techniques for water re-use in terms of initial cost, design simplicity and easiness of operation.

The adsorbent historically used has been activated carbon since it has a high adsorption capacity and efficiency. However, it is a costly adsorbent, which creates an impediment to its massive use and additionally there is a 10–15% mass loss in each cycle (Papageorgiou et al., 2009). That is why various research groups and industries have focused their efforts on the search for new adsorbents with lower cost and comparable efficiency. In this field, hydrogel beds have proven to be an emerging alternative since they are easily obtained, biodegradable and economically beneficial. Chi hydrogels have shown over the years to be highly efficient for wastewater treatment. In this sense, various groups have shown the usefulness of Chi and functionalized Chi beds for the adsorption of different metals. Ngah and Fatinathan, 2008 studied the adsorption of Cu^{2+} on Chi hydrogels without treatment and with different types of cross-linking. This work has demonstrated the capacity of the different hydrogels for the adsorption of Cu^{2+} in aqueous solutions under different experimental conditions, obtaining a maximum adsorption of 80.71 mg of Cu^{2+}/g chitosan under the best-determined conditions. The results obtained showed efficiencies comparable to those found when using activated carbon with the advantage of easily regenerating the adsorbent when using Chi spheres. In 2008, the same research group studied the use of Chi hydrogels, Chi cross-linked with glutaraldehyde and Chi-Alg for the adsorption of Cu^{2+} (Ngah and Fatinathan, 2008). In this study, they developed and characterized the different hydrogels and compared their efficiency in terms of adsorption.

Iron is the fourth most abundant metal on earth. It is present in various industrial effluents in its Fe^{2+} and Fe^{3+} forms in concentrations considered toxic, which is why various groups have focused on the study of the adsorption of these cations. Ali et al., 2018 demonstrated the usefulness of chitosan nanoparticles for the adsorption of Fe^{2+} and Mn^{2+}. The removal efficiency and maximum adsorption capacity of Fe^{2+} and Mn^{2+} were 99.8%, 116.2 mg/g and 95.3%, 74.1 mg/g, respectively.

Environmental applications of Alg hydrogels hinge partly on the fact that the surface is rich in functional groups (e.g., carboxyl and hydroxyl) and could capture metallic or cationic ions via ion exchange between the cross-linking cations and target pollutants such as heavy metals or dyes. Park and Chae, 2004 studied the adsorption of Pb^{2+} on Alg spheres and capsules. Alg capsules showed the greatest Pb^{2+} uptake capacity of 1560 mg/g of dry sodium Alg. This work also demonstrated the ease of regeneration of the adsorbent and the possibility of using it in repeated adsorption cycles. Papageorgiou et al., 2009 demonstrated the feasibility of using calcium Alg hydrogels for the adsorption of Cu^{2+}, Cd^{2+} and Pb^{2+} in binary mixtures. The prepared beds exhibit high adsorption capacities for the corresponding heavy metal ions, following the order $Pb^{2+} > Cu^{2+} > Cd^{2+}$ (in mmol/g). They demonstrated that the prepared Alg beds are a promising material to be used *per se* as a heavy metal sorbent exhibiting high adsorption capacity for divalent metals with predictable behavior in complex wastewater feeds.

Lastly, polymeric hydrogels have shown efficacy for the adsorption of textile dyes present in industrial effluents. Dyes represent a serious environmental problem both due to their toxicity and to the visual pollution generated by their presence in water. Various natural polymers have been used to treat effluents contaminated with this type of compounds. Chiou Li, 2003 demonstrated the applicability of Chi spheres gelled in the presence of tripolyphosphate and cross-linked with different reagents for the adsorption of Reactive Red 189 in stirred tank systems. The best adsorption capacity was obtained when using Chi spheres treated with epichlorohydrin reaching 1800 mg of dye/g of bed. Aravindhan et al., 2007 studied the potential use of pure calcium Alg beds for the

Table 22.2. Use of polymeric hydrogels for the adsorption of different water contaminants.

Adsorbent	Adsorbate	Maximum adsorption capacity in equilibrium	Reference
Alg beads	Pb^{2+} Cd^{2+} Cu^{2+}	2.85 mmol/g 2.44 mmol/g 3.60 mmol/g	(Benettayeb et al., 2017)
Alg-urea beads	Pb^{2+} Cd^{2+} Cu^{2+}	4.91 mmol/g 4.12 mmol/g 4.94 mmol/g	(Benettayeb et al., 2017)
Chi beads	As^{+3} As^{+5}	1.83 mg/g 1.94 mg/g	(Chen and Chung, 2006)
Magnetic Glutaraldehyde cross-linking Chi beads	Cs^{+1}	3.86 mg/g	(Chen and Chung, 2006)
Epiclorohydrin cross-linking Chi beads	Reactive black 5	0.43 mmol/g	(Kim et al., 2012)
Chi beads	Reactive black 5	0.46 mmol/g	(Kim et al., 2012)
Ba-Alg beads	Victoria Blue	1.65 mg/g	(Kumar et al., 2013)

removal of black dyes in a dynamic batch mode. The adsorption isotherm suggested a Langmuir adsorption capacity of 57.70 mg/g.

Table 22.2 summarizes the polymer hydrogels' performance for the adsorption of different types of contaminants from water.

4. Conclusions

Natural polymers have a wide range of applications in different industrial, biotechnological and research fields due to the properties they present. Various polymers have the ability to form hydrogels under mild conditions, which have characteristics that make them of great interest in processes such as drug delivery, wastewater treatment and protein purification. The development of new adsorbents is based on economic issues. Commercial beds for adsorption are formed by polysaccharides chemically cross-linked to make it non-soluble. Several commercial trademarks are available, such as Streamline (™) and Sepharose (™), but they cost approximately USD 2000 per L, which significantly affects the final cost of the production process when scaling up. Polymeric hydrogels have properties similar to those of commercial adsorbents, but are cheaper and easier to prepare. These advantages allow their use at a escalated level for enzyme purification and effluent treatment.

Abbreviations

Alg	Alginate
Car	carrageenan
Chi	chitosan
FCP	Flexible chain polymer

References

Abdella, A. A., Ulber, R. and Zayed, A. (2020). Chitosan-toluidine blue beads for purification of fucoidans. *Carbohydrate Polymers,* 231: 115686.

Abreu, F. O. M. S., Bianchini, C., Forte, M. M. C. and Kist, T. B. L. (2008). Influence of the composition and preparation method on the morphology and swelling behavior of alginate–chitosan hydrogels. *Carbohydrate Polymers,* 74(2): 283–289.

Ali, M. E. A., Aboelfadl, M. M. S., Selim, A. M., Khalil, H. F. and Elkady, G. M. (2018). Chitosan nanoparticles extracted from shrimp shells, application for removal of Fe (II) and Mn (II) from aqueous phases. *Separation Science and Technology,* 53(18): 2870–2881.

Antonietti, M., Caruso, R. A., Göltner, C. G. and Weissenberger, M. C. (1999). Morphology variation of porous polymer gels by polymerization in lyotropic surfactant phases. *Macromolecules,* 32(5): 1383–1389.

Aravindhan, R., Fathima, N. N., Rao, J. R. and Nair, B. U. (2007). Equilibrium and thermodynamic studies on the removal of basic black dye using calcium alginate beads. *Colloids and Surfaces A: Physicochemical and Engineering Aspects,* 299(1-3): 232–238.

Arvanitoyannis, I. S. (2008). 6 - The use of chitin and chitosan for food packaging applications. pp. 137–158. *In*: Chiellini, E. (ed.). *Environmentally Compatible Food Packaging.* Woodhead Publishing.

Augst, A. D., Kong, H. J. and Mooney, D. J. (2006). Alginate hydrogels as biomaterials. *Macromolecular Bioscience,* 6(8): 623–633.

Azlan, K., Wan Saime, W. N. and Lai Ken, L. (2009). Chitosan and chemically modified chitosan beads for acid dyes sorption. *Journal of Environmental Sciences,* 21(3): 296–302.

Bajpai, A. K. and Giri, A. (2002). Swelling dynamics of a macromolecular hydrophilic network and evaluation of its potential for controlled release of agrochemicals. *Reactive and Functional Polymers,* 53(2): 125–141.

Bakatula, E. N., Richard, D., Neculita, C. M. and Zagury, G. J. (2018). Determination of point of zero charge of natural organic materials. *Environmental Science and Pollution Research,* 25(8): 7823–7833.

Bakshi, P. S., Selvakumar, D., Kadirvelu, K. and Kumar, N. S. (2020). Chitosan as an environment friendly biomaterial—a review on recent modifications and applications. *International Journal of Biological Macromolecules,* 150: 1072–1083.

Bárbara, B., Emilia, B. M., Gastón, K., Guillermo, P. and Nadia, W. V. (2020). Design and optimization of an alternative chymotrypsin purification method by adsorption onto non-soluble alginate–carrageenan bed. *Polymer Bulletin,* 1–20.

Basak, R. and Bandyopadhyay, R. (2014). Formation and rupture of Ca^{2+} induced pectin biopolymer gels. *Soft Matter,* 10(37): 7225–7233.

Benettayeb, A., Guibal, E., Morsli, A. and Kessas, R. (2017). Chemical modification of alginate for enhanced sorption of Cd (II), Cu (II) and Pb (II). *Chemical Engineering Journal,* 316: 704–714.

Benhalima, T., Ferfera-Harrar, H. and Lerari, D. (2017). Optimization of carboxymethyl cellulose hydrogels beads generated by an anionic surfactant micelle templating for cationic dye uptake: Swelling, sorption and reusability studies. *International Journal of Biological Macromolecules,* 105: 1025–1042.

Bercea, M. and Wolf, B. A. (2019). Associative behaviour of κ-carrageenan in aqueous solutions and its modification by different monovalent salts as reflected by viscometric parameters. *International Journal of Biological Macromolecules,* 140: 661–667.

Berger, J., Reist, M., Mayer, J. M., Felt, O., Peppas, N. A. and Gurny, R. (2004). Structure and interactions in covalently and ionically crosslinked chitosan hydrogels for biomedical applications. *European Journal of Pharmaceutics and Biopharmaceutics,* 57(1): 19–34.

Brassesco, M. E., Valetti, N. W. and Picó, G. A. (2018). Control of the adsorption properties of alginate-guar gum matrix functionalized with epichlorohydrin through the addition of different flexible chain polymers as toll for the chymotrypsinogen isolation. *International Journal of Biological Macromolecules,* 115: 494–500.

Brassesco, M. E., Valetti, N. W. and Picó, G. (2019). Prediction of breakthrough curves in packed-bed column as tool for lysozyme isolation using a green bed. *Polymer Bulletin,* 76(11): 5831–5847.

Brassesco, M. E., Woitovich Valetti, N. and Picó, G. (2017). Molecular mechanism of lysozyme adsorption onto chemically modified alginate guar gum matrix. *International Journal of Biological Macromolecules,* 96: 111–117.

Bruck, S. D. (1973). Aspects of three types of hydrogels for biomedical applications. *Journal of Biomedical Materials Research,* 7(5): 387–404.

Campo, V. L., Kawano, D. F., Silva Jr, D. B. d. and Carvalho, I. (2009). Carrageenans: Biological properties, chemical modifications and structural analysis—A review. [doi: 10.1016/j.carbpol.2009.01.020]. *Carbohydrate Polymers,* 77(2): 167–180.

Cao, L., Lu, W., Mata, A., Nishinari, K. and Fang, Y. (2020). Egg-box model-based gelation of alginate and pectin: A review. *Carbohydrate Polymers,* 116389.

Catoira, M. C., Fusaro, L., Di Francesco, D., Ramella, M. and Boccafoschi, F. (2019). Overview of natural hydrogels for regenerative medicine applications. *Journal of Materials Science: Materials in Medicine,* 30(10): 115.

Chai, Q., Jiao, Y. and Yu, X. (2017). Hydrogels for biomedical applications: their characteristics and the mechanisms behind them. *Gels,* 3(1): 6.

Chakraborty, S. (2017). Carrageenan for encapsulation and immobilization of flavor, fragrance, probiotics, and enzymes: A review. *Journal of Carbohydrate Chemistry,* 36(1): 1–19.

Chan, S. Y., Choo, W. S., Young, D. J. and Loh, X. J. (2017). Pectin as a rheology modifier: Origin, structure, commercial production and rheology. *Carbohydrate Polymers,* 161: 118–139.

Chen, C.-C. and Chung, Y.-C. (2006). Arsenic removal using a biopolymer chitosan sorbent. *Journal of Environmental Science and Health, Part A,* 41(4): 645–658.

Ching, S. H., Bansal, N. and Bhandari, B. (2017). Alginate gel particles–A review of production techniques and physical properties. [doi: 10.1080/10408398.2014.965773]. *Critical Reviews in Food Science and Nutrition,* 57(6): 1133–1152.

Chiou, M. S. and Li, H. Y. (2003). Adsorption behavior of reactive dye in aqueous solution on chemical cross-linked chitosan beads. *Chemosphere,* 50(8): 1095–1105.

Chiou, M.-S., Ho, P.-Y. and Li, H.-Y. (2004). Adsorption of anionic dyes in acid solutions using chemically cross-linked chitosan beads. *Dyes and Pigments,* 60(1): 69–84.

Chung, T. W., Yang, J., Akaike, T., Cho, K. Y., Nah, J. W., Kim, S. I. et al. (2002). Preparation of alginate/galactosylated chitosan scaffold for hepatocyte attachment. *Biomaterials,* 23(14): 2827–2834.

de Moura, M. R., Guilherme, M. R., Campese, G. M., Radovanovic, E., Rubira, A. F. and Muniz, E. C. (2005). Porous alginate-Ca^{2+} hydrogels interpenetrated with PNIPAAm networks: Interrelationship between compressive stress and pore morphology. *European Polymer Journal,* 41(12): 2845–2852.

Dobrynin, A. V. and Rubinstein, M. (2005). Theory of polyelectrolytes in solutions and at surfaces. [doi: 10.1016/j.progpolymsci.2005.07.006]. *Progress in Polymer Science,* 30(11): 1049–1118.

Dumitriu, S., Vidal, P. F. and Chornet, E. (1996). Hydrogels based on polysaccharides. *Polysaccharides in Medicinal Applications,* 125–262.

Eser, A., Tirtom, V. N., Aydemir, T., Becerik, S. and Dinçer, A. (2012). Removal of nickel (II) ions by histidine modified chitosan beads. *Chemical Engineering Journal,* 210: 590–596.

Feng, Z., Shao, Z., Yao, J., Huang, Y. and Chen, X. (2009). Protein adsorption and separation with chitosan-based amphoteric membranes. *Polymer,* 50(5): 1257–1263.

Gavara, P. R., Bibi, N. S., Sanchez, M. L., Grasselli, M. and Fernandez-Lahore, M. (2015). Chromatographic characterization and process performance of column-packed anion exchange fibrous adsorbents for high throughput and high capacity bioseparations. *Processes,* 3(1): 204–221.

Ghanbarzadeh, M., Golmoradizadeh, A. and Homaei, A. (2018). Carrageenans and carrageenases: Versatile polysaccharides and promising marine enzymes. *Phytochemistry Reviews,* 17(3): 535–571.

Gondim, D. R., Lima, L. P., de Souza, M. C., Bresolin, I. T., Adriano, W. S., Azevedo, D. C. et al. (2012). Dye ligand epoxide chitosan/alginate: a potential new stationary phase for human IgG purification. *Adsorption Science & Technology,* 30(8-9): 701–711.

Grant, G. T., Morris, E. R., Rees, D. A., Smith, P. J. C. and Thom, D. (1973). Biological interactions between polysaccharides and divalent cations: The egg-box model. [doi: 10.1016/0014-5793(73)80770-7]. *FEBS Letters,* 32(1): 195–198.

Günter, E. A., Martynov, V. V., Belozerov, V. S., Martinson, E. A. and Litvinets, S. G. (2020). Characterization and swelling properties of composite gel microparticles based on the pectin and κ-carrageenan. *International Journal of Biological Macromolecules,* 164: 2232–2239.

Gupta, M., Gupta, H. and Kharat, D. (2018). Adsorption of Cu (II) by low cost adsorbents and the cost analysis. *Environmental Technology & Innovation,* 10: 91–101.

Hamedi, H., Moradi, S., Hudson, S. M. and Tonelli, A. E. (2018). Chitosan based hydrogels and their applications for drug delivery in wound dressings: A review. *Carbohydrate Polymers,* 199: 445–460.

Haug, I. and Draget, K. (2011). *Gelatin Handbook of Food Proteins*. pp. 92–115. Elsevier.

Hennink, W. E. and van Nostrum, C. F. (2012). Novel crosslinking methods to design hydrogels. *Advanced Drug Delivery Reviews,* 64: 223–236.

Hernández-González, A. C., Téllez-Jurado, L. and Rodríguez-Lorenzo, L. M. (2020). Alginate hydrogels for bone tissue engineering, from injectables to bioprinting: A review. *Carbohydrate Polymers,* 229: 115514.

Hu, C., Lu, W., Mata, A., Nishinari, K. and Fang, Y. (2021). Ions-induced gelation of alginate: Mechanisms and applications. *International Journal of Biological Macromolecules,* 177: 578–588.

José García de la, T., José, G. H. C. and Martínez, M. C. L. (2008). Prediction of solution properties of flexible-chain polymers: a computer simulation undergraduate experiment. *European Journal of Physics,* 29(5): 945.

Kaberova, Z., Karpushkin, E., Nevoralová, M., Vetrík, M., Šlouf, M. and Dušková-Smrčková, M. (2020). Microscopic structure of swollen hydrogels by scanning electron and light microscopies: artifacts and reality. *Polymers,* 12(3): 578.

Kim, T.-Y., Park, S.-S. and Cho, S.-Y. (2012). Adsorption characteristics of Reactive Black 5 onto chitosan beads cross-linked with epichlorohydrin. *Journal of Industrial and Engineering Chemistry,* 18(4): 1458–1464.

Krajewska, B. (2004). Application of chitin- and chitosan-based materials for enzyme immobilizations: a review. *Enzyme and Microbial Technology,* 35(2): 126–139.

Kulkarni Vishakha, S., Butte Kishor, D. and Rathod Sudha, S. (2012). Natural polymers—A comprehensive review. *International Journal of Research in Pharmaceutical and Biomedical Sciences,* 3(4): 1597–1613.

Kumar, M., Tamilarasan, R. and Sivakumar, V. (2013). Adsorption of Victoria blue by carbon/Ba/alginate beads: kinetics, thermodynamics and isotherm studies. *Carbohydrate Polymers,* 98(1): 505–513.

Lee, K. Y. and Mooney, D. J. (2012). Alginate: Properties and biomedical applications. *Progress in Polymer Science,* 37(1): 106–126.

Li, J., Ma, J., Chen, S., Huang, Y. and He, J. (2018). Adsorption of lysozyme by alginate/graphene oxide composite beads with enhanced stability and mechanical property. *Materials Science and Engineering: C,* 89: 25–32.

Liu, S. and Li, L. (2016). Thermoreversible gelation and scaling behavior of Ca^{2+}-induced κ-carrageenan hydrogels. *Food Hydrocolloids,* 61: 793–800.

Michel, A., Mestdagh, M. and Axelos, M. (1997). Physico-chemical properties of carrageenan gels in presence of various cations. *International Journal of Biological Macromolecules,* 21(1-2): 195–200.

Moreira, H. R., Munarin, F., Gentilini, R., Visai, L., Granja, P. L., Tanzi, M. C. et al. (2014). Injectable pectin hydrogels produced by internal gelation: pH dependence of gelling and rheological properties. *Carbohydrate Polymers,* 103: 339–347.

Nakanishi, K., Sakiyama, T. and Imamura, K. (2001). On the adsorption of proteins on solid surfaces, a common but very complicated phenomenon. *Journal of Bioscience and Bioengineering,* 91(3): 233–244.

Ngah, W. S. W. and Fatinathan, S. (2008). Adsorption of Cu(II) ions in aqueous solution using chitosan beads, chitosan–GLA beads and chitosan–alginate beads. *Chemical Engineering Journal,* 143(1): 62–72.

Noreen, A., Akram, J., Rasul, I., Mansha, A., Yaqoob, N., Iqbal, R. et al. (2017). Pectins functionalized biomaterials; a new viable approach for biomedical applications: A review. *International Journal of Biological Macromolecules,* 101: 254–272.

Oussalah, A., Boukerroui, A., Aichour, A. and Djellouli, B. (2019). Cationic and anionic dyes removal by low-cost hybrid alginate/natural bentonite composite beads: Adsorption and reusability studies. *International Journal of Biological Macromolecules,* 124: 854–862.

Papageorgiou, S. K., Katsaros, F., Kouvelos, E. and Kanellopoulos, N. (2009). Prediction of binary adsorption isotherms of Cu^{2+}, Cd^{2+} and Pb^{2+} on calcium alginate beads from single adsorption data. *Journal of Hazardous Materials,* 162(2-3): 1347–1354.

Papageorgiou, S. K., Katsaros, F. K., Kouvelos, E. P. and Kanellopoulos, N. K. (2009). Prediction of binary adsorption isotherms of Cu^{2+}, Cd^{2+} and Pb^{2+} on calcium alginate beads from single adsorption data. *Journal of Hazardous Materials,* 162(2): 1347–1354.

Park, H. G. and Chae, M. Y. (2004). Novel type of alginate gel-based adsorbents for heavy metal removal. *Journal of Chemical Technology & Biotechnology: International Research in Process, Environmental & Clean Technology,* 79(10): 1080–1083.

Patil, J., Kamalapur, M., Marapur, S. and Kadam, D. (2010). Ionotropic gelation and polyelectrolyte complexation: the novel techniques to design hydrogel particulate sustained, modulated drug delivery system: a review. *Digest Journal of Nanomaterials and Biostructures,* 5(1): 241–248.

Pawar, S. N. and Edgar, K. J. (2012). Alginate derivatization: A review of chemistry, properties and applications. *Biomaterials,* 33(11): 3279–3305.

Peppas, N. A., Bures, P., Leobandung, W. and Ichikawa, H. (2000). Hydrogels in pharmaceutical formulations. *European Journal of Pharmaceutics and Biopharmaceutics,* 50(1): 27–46.

Pillai, C. K., Paul, W. and Sharma, C. P. (2009). Chitin and chitosan polymers: Chemistry, solubility and fiber formation. *Progress in Polymer Science,* 34(7): 641–678.

Plazinski, W. (2011). Molecular basis of calcium binding by polyguluronate chains. Revising the egg-box model. *Journal of Computational Chemistry,* 32(14): 2988–2995.

Rinaudo, M. (2006). Chitin and chitosan: Properties and applications. *Progress in Polymer Science,* 31(7): 603–632.

Rodrigues, E., Bezerra, B., Farias, B., Adriano, W., Vieira, R., Azevedo, D. et al. (2013). Adsorption of cellulase isolated from *Aspergillus niger* on chitosan/alginate particles functionalized with epichlorohydrin. *Adsorption Science & Technology,* 31(1): 17–34.

Roy, I., Sardar, M. and Gupta, M. N. (2000). Exploiting unusual affinity of usual polysaccharides for separation of enzymes on fluidized beds. *Enzyme and Microbial Technology,* 27(1): 53–65.

Roy, I. and Gupta, M. N. (2002). Purification of a bacterial pullulanase on a fluidized bed of calcium alginate beads. *J. Chromatogr A,* 950(1-2): 131–137.

Roy, I., Sardar, M. and Gupta, M. N. (2005). Cross-linked alginate–guar gum beads as fluidized bed affinity media for purification of jacalin. *Biochemical Engineering Journal,* 23(3): 193–198.

Sacco, P., Borgogna, M., Travan, A., Marsich, E., Paoletti, S., Asaro, F. et al. (2014). Polysaccharide-based networks from homogeneous chitosan-tripolyphosphate hydrogels: Synthesis and characterization. *Biomacromolecules,* 15(9): 3396–3405.

Sahil, K., Akanksha, M., Premjeet, S., Bilandi, A. and Kapoor, B. (2011). Microsphere: A review. *Int. J. Res. Pharm. Chem,* 1(4): 1184–1198.

Santos Leite da Silva, A. C., Woitovich Valetti, N., Brassesco, M. E., Teixeira, J. A. and Picó, G. (2016). Adsorption of peroxidase from Raphanus sativus L. onto alginate–guar gum matrix: Kinetic, equilibrium and thermodynamic analysis. *Adsorption Science & Technology,* 34(6): 388–402.

Seymour, R. B. (1971). *Introduction to Polymer Chemistry*. New York: Mcgraw-Hill.

Sharma, S. and Bhattacharya, A. (2017). Drinking water contamination and treatment techniques. *Applied Water Science,* 7(3): 1043–1067.

Somasundaram, B., Pleitt, K., Shave, E., Baker, K. and Lua, L. H. (2018). Progression of continuous downstream processing of monoclonal antibodies: Current trends and challenges. *Biotechnology and Bioengineering,* 115(12): 2893–2907.

Somers, W., Van't Reit, K., Rozie, H., Rombouts, F. and Visser, J. (1989). Isolation and purification of endo-polygalacturonase by affinity chromatography in a fluidized bed reactor. *The Chemical Engineering Journal,* 40(1): B7–B19.

Spelzini, D., Farruggia, B. and Picó, G. (2011). Purification of chymotrypsin from pancreas homogenate by adsorption onto non-soluble alginate beads. [doi: 10.1016/j.procbio.2010.11.011]. *Process Biochemistry,* 46(3): 801–805.

Tønnesen, H. H. and Karlsen, J. (2002). Alginate in drug delivery systems. *Drug Development and Industrial Pharmacy,* 28(6): 621–630.

Torres, M., Beppu, M. and Santana, C. (2007). Characterization of chemically modified chitosan microspheres as adsorbents using standard proteins (bovine serum albumin and lysozyme). *Brazilian Journal of Chemical Engineering,* 24(3): 325–336.

Vakurov, A., Pchelintsev, N. A., Gibson, T. and Millner, P. (2012). Development of polymeric nanoparticles showing tuneable pH-responsive precipitation. *Journal of Nanoparticle Research,* 14(12): 1–9.

Varaprasad, K., Raghavendra, G. M., Jayaramudu, T., Yallapu, M. M. and Sadiku, R. (2017). A mini review on hydrogels classification and recent developments in miscellaneous applications. *Materials Science and Engineering: C,* 79: 958–971.

Wan Ngah, W. S., Endud, C. S. and Mayanar, R. (2002). Removal of copper(II) ions from aqueous solution onto chitosan and cross-linked chitosan beads. *Reactive and Functional Polymers,* 50(2): 181–190.

Wan Ngah, W. S., Teong, L. C. and Hanafiah, M. A. K. M. (2011). Adsorption of dyes and heavy metal ions by chitosan composites: A review. *Carbohydrate Polymers,* 83(4): 1446–1456.

Wang, X., Li, X., Stride, E., Huang, J., Edirisinghe, M., Schroeder, C. et al. (2010). Novel preparation and characterization of porous alginate films. *Carbohydrate Polymers,* 79(4): 989–997.

Wernersson, E., Stenqvist, B. and Lund, M. (2015). The mechanism of cellulose solubilization by urea studied by molecular simulation. *Cellulose,* 22(2): 991–1001.

Woitovich Valetti, N., Brassesco, M. E. and Picó, G. A. (2016). Polyelectrolytes–protein complexes: A viable platform in the downstream processes of industrial enzymes at scaling up level. *Journal of Chemical Technology & Biotechnology,* 91(12): 2921–2928.

Woitovich Valetti, N. and Picó, G. (2016). Adsorption isotherms, kinetics and thermodynamic studies towards understanding the interaction between cross-linked alginate-guar gum matrix and chymotrypsin. *Journal of Chromatography B,* 1012–1013: 204–210.

Yapo, B. M. (2009). Pectin quantity, composition and physicochemical behaviour as influenced by the purification process. *Food Research International,* 42(8): 1197–1202.

Zargar, V., Asghari, M. and Dashti, A. (2015). A review on chitin and chitosan polymers: structure, chemistry, solubility, derivatives, and applications. *ChemBioEng Reviews,* 2(3): 204–226.

Zengin Kurt, B., Uckaya, F. and Durmus, Z. (2017). Chitosan and carboxymethyl cellulose based magnetic nanocomposites for application of peroxidase purification. *International Journal of Biological Macromolecules,* 96: 149–160.

Zhuang, S. and Wang, J. (2018). Modified alginate beads as biosensor and biosorbent for simultaneous detection and removal of cobalt ions from aqueous solution. *Environmental Progress & Sustainable Energy,* 37(1): 260–266.

Index

R

S

T

U

Editors Biography

Margarita del Rosario Salazar Sánchez is BSc. in Biology in 2008 from the University of Cauca, Colombia, and MSc. in Continental Hydrobiological Resources in 2013 and PhD in Agrarian and Agroindustrial Sciences at the University of Cauca, Colombia in 2019. Her topics of interest as a professor and researcher include ecotoxicity, polymer biodegradation, ecology, food biotechnology, bioprocesses and food preservation and processing; topics in which several articles have been published in peer-reviewed journals. She is currently a researcher and professor at the Universidad Popular del Cesar, Colombia.

José Fernando Solanilla Duque is BSc. Eng. in Agroindustrial Engineering in 1996 from the Universidad Gran Colombia, Colombia, and his PhD in Colloids and Interface Science and Technology from the Universidad Pablo de Olavide (Spain), in 2009. He has been a professor since February 1995 at the Universidad del Quindío, Universidad del Tolima and currently at the Universidad del Cauca. His topics of interest as a teacher and researcher include food biotechnology, interfacial properties of biopolymers, bioprocesses and food preservation, where he has published several articles in peer-reviewed journals.

Aidé Sáenz Galindo: Research Professor at the Autonomous University of Coahuila, Saltillo, Coahuila Mexico. Degree in Chemical Sciences with a PhD in Polymers from the Centro de Investigación en Química. D. in Polymers from the Centro de Investigación en Química and a post doctorate in the Department of Advanced Materials at the Centro de Investigación en Química. The research line of work is related to developing synthetic and natural polymeric composites and nano-materials, specializing in the surface modification of carbon-based nanostructures. Lecturer at the undergraduate level of subjects related to Organic Chemistry, in the educational programs of Chemist, QFB and Eng. Chemistry, a member of the Basic Academic Nucleus of the Postgraduate Master and PhD in Materials Science and Technology and Master in Chemical Science and Technology, currently Coordinator of the Master in Materials Science and Technology, and has SNI recognition from 2008 to date, In 2013 she was awarded first place at national level in the Green Chemistry competition, awarded by the Chemical Society of Mexico-BASF-ULAP. She has supervised theses at Bachelor, Master and PhD level, as well as author of books, book chapters and scientific articles.

Raúl Rodríguez Herrera is a professor at the Universidad Autónoma de Coahuila, México. He has published 310 referred papers, 6 books, 25 patents and 120 book chapters. He has been distinguished with: twice PCCMCA Congress prize (1988 and 1990), Researcher-level 3 Mexico (2018), Guest of Honor-Kannur University, prize in Food Science and Technology, (2003), Agro-Bio Prize (2005). Dr. Rodriguez has held conferences in different countries and served as a reviewer for more than 25 JCR Journals. His research focuses on: 1. plant genome, 2. isolation and cloning of genes codifying for enzymes with industrial applications from microbial and plant genomes, 3. transgenic detections in processed food and 4. functional foods. E-mail: raul.rodriguez@uadec.edu.mx. ORCID: 0000-0002-6428-4925.